Fuzzy Logic
Intelligence, Control, and Information

John Yen and Reza Langari

Center for Fuzzy Logic, Robotics, and Intelligent Systems
Texas A&M University

PRENTICE HALL
UPPER SADDLE RIVER, NEW JERSEY 07458

Library of Congress Cataloging-in-Publication Data

Yen, John.
    Fuzzy logic: intelligence, control, and information
John Yen and Reza Langari
        p.   cm.
    Includes bibliographical references and index.
    ISBN: 0-13-525817-0
    1. Fuzzy logic. I. Langari, Reza. II. Title.
    QA9.64 .Y46 1998
    511.3--dc21   98-40882
                                                        CIP

Publisher: *Tom Robbins*
Production editor: *Edward DeFelippis*
Editor-in-chief: *Marcia Horton*
Managing editor: *Eileen Clark*
Assistant vice president of production and manufacturing: *David W. Riccardi*
Art director: *Jayne Conte*
Cover designer: *Bruce Kenselaar*
Manufacturing buyer: *Pat Brown*
Editorial assistant: *Dan DePasquale*

©1999 by Prentice-Hall, Inc.
Simon & Schuster / A Viacom Company
Upper Saddle River, New Jersey 07458

Printed in the United States of America

10 9 8 7 6 5 4 3 2

ISBN 0-13-525817-0

Prentice-Hall International (UK) Limited, *London*
Prentice-Hall of Australia Pty. Limited, *Sydney*
Prentice-Hall Canada Inc., *Toronto*
Prentice-Hall Hispanoamericana, S.A., *Mexico*
Prentice-Hall of India Private Limited, *New Delhi*
Prentice-Hall of Japan, Inc., *Tokyo*
Simon & Schuster Asia Pte. Ltd., *Singapore*
Editora Prentice-Hall do Brasil, Ltda., *Rio de Janeiro*

For Michelle (Taiyu) — J. Y.

For my parents — R. L.

CONTENTS

10   Analytical Issues in Fuzzy Logic Control   247

IMPORTANT

FOREWORD

Superb. This one-word assessment summarizes my view of *Fuzzy Logic: Intelligence, Control, and Information*, (or *FL:ICI* for short) — a long-awaited text by Professors John Yen and Reza Langari. Written with authority and high expository skill, *FL:ICI* reflects the extensive experience of its authors in both teaching and research in fuzzy logic and related fields.

Like some recent texts, *FL:ICI* also reflects the emergence of soft computing (SC) as the prime platform for the conception, design, and analysis of information/intelligent/ control systems. In essence, soft computing is an association of fuzzy logic, neurocomputing, evolutionary computing, and probabilistic computing.

The basic premise of SC is that, in the main, its constituent methodologies are complementary and synergistic rather than competitive. It is this premise that motivates the inclusion in *FL:ICI* of succinct expositions of neurocomputing, evolutionary computing, and their combinations with fuzzy logic.

This inclusion serves an important pedagogical function and enhances the reader's ability to decide which methodology or combination of methodologies in SC is likely to be most effective in solving a problem. In my view, it is almost certain that in coming years most information/intelligent/control systems will be of hybrid type, i.e., will employ a combination of methodologies — rather than a single methodology — drawn from SC or other fields.

To view the organization of *FL:ICI* in a proper perspective, it is important to have a clear understanding of the role of fuzzy logic in SC and the substance of its principal contributions. Underlying this issue is the basic question: What is fuzzy logic? The question does not have a simple answer because fuzzy logic has a multiplicity of facets — many of which overlap and have unsharp boundaries. In what follows, I shall take the liberty of sketching my current perception of fuzzy logic with the aim of relating it to *FL:ICI* and clarifying some of the common misconceptions.

Basically, fuzzy logic has four principal facets: (i) the logical facet, *L*; (ii) the set-theoretic facet, S; (iii) the relational facet, R; and (iv) the epistemic facet, *E*.

The logical facet, L, is a logical system or, more generally, a collection of logical systems which includes as a special case both two-valued and multiple-valued systems. A distinguishing feature of L is that the rules of inference in L serve as rules which govern propagation of fuzzy constraints. The logical facet of fuzzy logic plays a pivotal role in applications to knowledge representation and inference from information which is imprecise, incomplete, uncertain, or partially true.

The set-theoretic facet of fuzzy logic is concerned with classes or sets whose boundaries are not sharply defined. The initial development of fuzzy logic was focused on this facet. Most of the applications of fuzzy logic in mathematics have been and continue to be related to the set-theoretic facet.

The relational facet of fuzzy logic is concerned in the main with representation and manipulation of imprecisely defined functions and relations. It is this facet of fuzzy logic that is central to its applications to information and control, and is the facet that receives the most attention in *FL:ICI*.

The epistemic facet of fuzzy logic is linked to its logical facet and is focused on the applications of fuzzy logic to knowledge representation, information systems, fuzzy databases, and the theories of possibility and probability. In essence, the epistemic facet is concerned with meaning, knowledge, and decision.

At the core of fuzzy logic, which is shared by all of its facets, lie two basic concepts: fuzziness and fuzzy granularity, with fuzzy granularity and, more particularly, fuzzy information granularity, lying at the epicenter. In this context, a fuzzy granule is a clump of points (objects) drawn together by indistinguishability, similarity, proximity, and functionality with words in a natural language playing the role of labels of fuzzy granules.

Taken together, the four principal facets of fuzzy logic and the core concepts of fuzziness and fuzzy information granularity provide a foundation for the methodology of computing with words (CW). In my view, in coming years the methodology of computing with words will emerge as the principal and unique contribution of fuzzy logic, especially in the realm of applications.

In essence, computing with words is inspired by the remarkable human ability to preform a wide variety of physical and mental tasks without any measurements and any computations. Underlying this ability is another remarkable ability—the ability of the human mind to manipulate perceptions—perceptions of distance, weight, color, time, taste, and likelihood, among others. A crucial difference between measurements and perceptions is that, in general, measurements are crisp whereas perceptions are fuzzy. In a natural language, words serve preponderantly as labels of perceptions.

Viewed in this perspective, the importance of computing with words derives in the main from its provision of a machinery for manipulation of both measurements and perceptions. The ability to manipulate perceptions plays a crucial role in our quest for machines with high MIQ (Machine Intelligence Quotient)—machines which can mimic the unique human ability to make rational decisions in an environment of imprecision, uncertainty, and partial truth.

As was alluded to already, the organization of *FL:ICI* is consistent with the view of fuzzy logic sketched here. The stress on linguistic variables and the calculus of fuzzy rules serves to lay the groundwork for the machinery of computing with words, though CW as a methodology is not treated as a distinct entity. The historical notes, the bibliographies and the exercises contribute substantially to the reader-friendliness of *FL:ICI* and enhance its value as a text.

In sum, *FL:ICI* is in every respect an outstanding contribution to the literature of fuzzy logic and soft computing, and is a must for anyone who is interested in the conception, design, and utilization of information/intelligent/control systems. Professors Yen and Langari deserve our thanks and congratulations.

Lotfi A. Zadeh
Berkeley, California

PREFACE

Success of fuzzy logic in a wide range of applications has inspired much interest in fuzzy logic among undergraduate and graduate students, academic professionals, and practicing engineers. This presents an opportunity as well as a challenge for introducing a textbook on fuzzy logic that meets the needs of these audience with *diverse background and interests*. The goal of this book is thus to introduce fuzzy logic and its applications to upper-level undergraduate students, beginning graduate students, and practicing engineers so that they can develop a reasonably in-depth understanding of both the principle and the practice of the technology, as well as a working knowledge of how to use the technology themselves.

To achieve the first objective, the book focuses on aspects of the theory that serve as the basis for fuzzy logic-based control, pattern classification, and, in general, information processing strategies in relation to complex systems. Important theoretical issues (e.g., stability of fuzzy logic control) is also treated at a level that is suitable for a classroom discussion.

To achieve the second objective, the book not only focuses on various important application areas of fuzzy logic but also crosses over the boundaries of individual application domains and discusses *design methodology* in a rather general setting. We relate design issues to concrete applications where possible so as to help the reader develop a more thorough understanding of the impact of these issues on the design and implementation process. In particular, the book uses MATLAB, SIMULINK and the Fuzzy Logic Toolbox for MATLAB developed by The Mathworks extensively to accomplish these objectives. More specifically, the Fuzzy Logic Toolbox is used in several ways:

1. To demonstrate applications of concepts and techniques when they are introduced.

2. To illustrate design issues, design guidelines, and strategies for tuning membership functions.

3. To provide simulation codes (i.e., Simulink blocks) and preliminary designs for homework problems.

Book Features

This text has the following features that distinguish it from other texts on fuzzy logic and fuzzy control.

1. *Provides a modern perspective on the fuzzy logic technology.*
   The text aims to present a balanced coverage on three important perspectives of fuzzy logic: (1) machine intelligence, (2) control engineering, and (3) information technology. By presenting materials related to these three areas in a single text, we hope to not only help students better understand and appreciate the synergistic interactions between these complementary viewpoints, but also to foster innovation for shaping information technology of the 21st century. The treatment of two types of fuzzy rules (i.e., fuzzy mapping rules and fuzzy implication rules) also reflects the modern perspective, rather than the traditional perspective, regarding fuzzy rules.

2. *Offers a comprehensive and up-to-date coverage.*
   The text covers areas that are sometimes missing or underemphasized in other texts, including hierarchical intelligent control, analytical issues of fuzzy logic control, fuzzy database and information systems, and fuzzy pattern recognition. The book also covers many of the more recent topics in the field, including intelligent agents, fuzzy model identifications, neuro-fuzzy systems, and GA-fuzzy systems.

3. *Provides examples and exercises that are related to real world problems.*
   Use good examples to motivate learning and demonstrate concepts and techniques. Whenever possible, the concepts are initially introduced in a more or less concrete setting and elaborated on progressively. For instance, Chapter 2 uses a flow mixing control problem and a washing machine cycle selection problem to demonstrate fuzzy rule-based inference.

4. *Introduces core concepts and techniques gently in two steps.*
   Chapter 2 introduces four core concepts in fuzzy logic (fuzzy sets, linguistic variables, possibility distributions, and fuzzy if-then rules) without elaborating on their theoretical issues, which are discussed in detail in later chapters. Introducing these core concepts together using simple applications helps students to see how these concepts work together. It also avoids the potential problem of "seeing the trees, but not the forest" when these concepts (especially the theoretical aspects) are treated thoroughly.

5. *Uses MATLAB/SIMULINK and the Fuzzy Logic Toolbox for MATLAB to demonstrate exemplary applications and to develop hands-on exercises.*
The use of Fuzzy Logic Toolbox for MATLAB allows the instructors and students to reuse relevant codes (e.g., Simulink, fuzzy inference systems), which are available from the book's Web page, for reimplementing exemplary applications and for working on hands-on homework assignments.

6. *Provides design guidelines and design methods for developing fuzzy logic applications.*
Design guidelines are recommended whenever appropriate. The icon below has been used to highlight design guidelines in the book. In addition, several chapters (e.g., the chapters on fuzzy model identification, neuro-fuzzy systems, and GA-fuzzy systems) introduce techniques for automating various decisions in the design process using training data (i.e., examples) and learning algorithms.

7. *Includes relevant background material so that students from a wide range of disciplines can easily understand the text.*
The textbook introduces important background material at a suitable level so that students lacking the background can quickly develop a basic understanding of the major concepts and techniques without being distracted and burdened by irrelevant details. For instance, Chapter 3 introduces classical set theory before introducing fuzzy sets; Chapter 8 introduces fundamental issues in control engineering before discussing fuzzy logic control.

8. *Discusses open research issues and their implications.*
Even though fuzzy logic has been successfully applied to a wide range of applications, there are open issues that remain to be addressed by future research. Therefore, we hope this book helps the reader to understand not only what fuzzy logic can do, but also what it may not be able to accomplish using the current technology. In addition to pointing out open issues and limitations of fuzzy logic technology, we also identify major trends of research activities and refer to other related resources whenever appropriate. We use the light bulb icon to highlight notes about research works.

9. *Introduces the connection between fuzzy logic and related ideas, methods, and theories developed in other disciplines.*
The book refers to related theories and methodologies and discusses their similarities and differences compared to the corresponding theory or methodology in fuzzy logic. For example, the relationship between possibility theory and

probability theory is treated in Chapter 7. The connection between nonlinear control theory and fuzzy logic control is discussed in Chapter 8. The relationships between fuzzy logic control, supervisory control and gain scheduling are discussed in Chapter 9. We believe introducing these related technologies allows the students to put fuzzy logic into the correct global perspective, and encourage them to explore novel combinations of these techniques.

This book is primarily intended for use in an undergraduate course in fuzzy logic or fuzzy control. It is also suitable for a graduate-level course in fuzzy logic and/or fuzzy control. Because the book emphasizes both theory and practice, it can also serve as a self-contained self-study text for engineers and technical professionals. In addition, the book can be used as supplementary text or reference material for courses on intelligent control, soft computing, and artificial intelligence.

Overview of the Book

The book consists of 17 chapters, organized in four parts. Part I "Fuzzy Logic" introduces all the core concepts and techniques. Chapter 2 introduces several key concepts within the context of concrete applications. Chapters 5 and 6 elaborate on the theoretical foundation and the distinction between two types of fuzzy rules: fuzzy mapping rules and fuzzy implication rules. Part II "Fuzzy Logic Control" discusses the basics of fuzzy logic controllers, the relevant analytical issues, and their roles in advanced hierarchical control systems. Part III "Fuzzy Logic in Intelligent Information Systems" discusses the impact of fuzzy logic in three areas in information technology: artificial intelligence, database and information systems, and pattern recognition. Part IV "Fuzzy Model Identification and Soft Computing" focuses on techniques for automated design of fuzzy rule-based systems. Even though this part is positioned last, we encourage the instructor to cover Chapter 14 earlier in the course. Other chapters in this part can also be introduced earlier. In fact, Part II, III, and IV are relatively independent; therefore, the instructor can freely arrange them based on the emphasis of the course.

Using this Book

This book can be used for different emphases by selecting suitable chapters. While the ultimate decision will be made by the instructor, we list below several recommended sequences of chapters and the corresponding emphases:

- *One-semester general introductory course on fuzzy logic:*
  Chapters 1, 2, 3, 4, 5, 6, 7, 8, 14

- *One-semester general introductory course on fuzzy control*:
  Chapters 1, 2, 3, 4, 5, 7, 8, 14, 9, 10

- *One-semester course on fuzzy logic and intelligent information systems*:
  Chapters 1, 2, 3, 4, 5, 6, 7, 11, 12, 13, 14

- *One-semester course on fuzzy logic and fuzzy modeling:*
  Chapters 1, 2, 3, 4, 5, 8, 14, 7, 15, 16, 17

- *One-semester course on fuzzy logic and soft computing:*
  Chapters 1, 2, 3, 4, 5, 6, 13, 14, 7, 16, 17,

- *One-semester course on fuzzy logic and intelligent control:*
  Chapters 1, 2, 3, 4, 5, [6,] 8, 14, 9, [11,] 10, 7, 16, 17

- *One-semester graduate course on fuzzy logic, intelligent control, and information systems:*
  Chapters 1, 2, 3, 4, 5, 6, 7, 8, 14, 9, 11, 12, 13, 16, 17

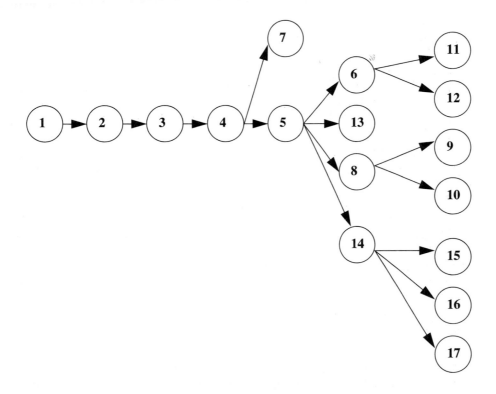

The dependency relationship between chapters is shown in the above figure. We think that Chapter 14 should be covered early in the course, even though it is in the last part of the book. This way, the basic idea of identifying a fuzzy rule-based model from data can be applied to all rule-based applications introduced later in the course, whether they are in control, intelligent agents, information systems, or pattern recognition.

Additional Information on the Web

Additional information regarding fuzzy logic and the textbook can be found on the World Wide Web at the following URL:
    http://www.prenhall.com/yen

Acknowledgments

This book would not have been born without the help of many mentors, colleagues, friends, and students. First of all, we would like to thank Professor Lotfi A. Zadeh, who has been our mentor since we were graduate students at University of California, Berkeley. More importantly, he was the first to encourage us to write this textbook. It was the Fall of 1991. We did not realize at that time it will take us seven years to make this dream come true. Before we acknowledge specific individuals who have contributed to the writing of this text, we would like to first acknowledge fuzzy logic researchers and engineers worldwide, especially those who have passed away and those in Japan, for their persistent effort in planting seeds in a land considered desert by many others. Even though some of them are mentioned in Chapter 1, many may not be known by us. Yet, their contributions to the theory and the application of fuzzy logic, directly or indirectly, are as valuable as those of the familiar names.

Tom Robbins, our Editor from Prentice Hall, has been both supportive and patient about this book project. Prof. James Yao provided valuable information regarding educational issues and civil engineering applications discussed in Chapter 1. Liang Wang wrote most of Chapters 14, 15, and 16. Magy El-Self helped the writing of Chapter 12 and the emotional agent part of Chapter 11. She also typed and formatted most of the manuscript. We have benefited from details comments from Hao Ying, Valerie Cross, and other reviewers. An earlier version of the manuscript was used as the main reading material for a class on fuzzy logic and intelligent systems at Texas A&M University in the Spring of 1997. Many students in that class pointed out mistakes and suggested improvements that we have adopted. They include Saad Biaz, Cary Butler, Dan Ragsdale, Buck Surdu, Timothy McKinney, Manuel Jimenez-Monteon, Sung Kyung Hong, Dongyoon Hyun, and Anna Zacchi. The page layout of the text is inspired by that of the AI textbook by Stuart Russell and Peter Norvig (*Artificial Intelligence: A Modern Approach*).

John would also like to thank David Reed, S. J., a long-time close friend of my family, for proof reading the entire manuscript. He is also indebted to Bernard Chu, S. J., who has taught and inspired him so much in self-growth, which helped him to overcome major "writer's blocks" he encountered in writing this book. John wishes to thank his mother (Bernadette Fong) for giving him abundant love. Finally, but not least, he wishes to thank his wife (Michelle Taiyu Lu) for giving him warmest support and insightful suggestions. To her, he dedicates this book.

Reza would like to thank all his friends and colleagues in the "fuzzy community" for encouraging him to co-write this book. Their support and encouragement has been invaluable.

John Yen and Reza Langari
College Station, Texas, USA

PART I

FUZZY LOGIC

1   INTRODUCTION

After being mostly viewed as a controversial technology for two decades, fuzzy logic has finally been accepted as an emerging technology since the late 1980s. This is largely due to a wide array of successful applications ranging from consumer products, to industrial process control, to automotive applications. Before we engage in an in-depth discussion of technical issues concerning fuzzy logic, however, we must first place this paradigm in perspective. For this, we first clarify two meanings of the term "fuzzy logic" and present a brief history of the development of fuzzy logic technology and its applications. We will then discuss the insights that motivated the birth of this technology. This is followed by a clarification of some of the common misunderstandings about fuzzy logic.

1.1   What Is Fuzzy Logic?

The term fuzzy logic has been used in two different senses. It is thus important to clarify the distinctions between these two different usages of the term. In a narrow sense, fuzzy logic refers to a logical system that generalizes classical two-valued logic for reasoning under uncertainty. In a broad sense, fuzzy logic refers to all of the theories and technologies that employ fuzzy sets, which are classes with unsharp boundaries.

For instance, the concept of "warm room temperature" may be expressed as an interval (e.g., [70° F, 78° F]) in classical set theory. However, the concept does not have a well-defined natural boundary. A representation of the concept closer to human interpretation is to allow a gradual transition from "not warm" to "warm." In order to achieve this, the notion of membership in a set needs to become a matter of degree. This is the essence of fuzzy sets. An example of a classical set and a fuzzy set is shown in Fig 1.1, where the vertical axis represents the degree of membership in a set.

FIGURE 1.1   A Classical Set and a Fuzzy Set Representation of "Warm Room Temperature"

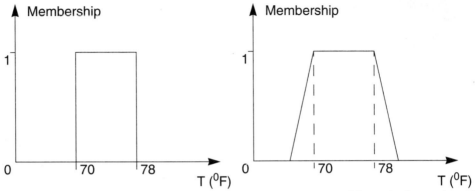

The broad sense of fuzzy logic includes the narrow sense of fuzzy logic as a branch. Other areas include fuzzy control, fuzzy pattern recognition, fuzzy arithmetic, fuzzy mathematical programming, fuzzy probability theory, fuzzy decision analysis, fuzzy neural networks theory, and fuzzy topology, etc. In all these areas, a conventional black-and-white concept is generalized to a matter of degree. By doing this, one accomplishes two things: (1) ease of describing human knowledge involving vague concepts, and (2) enhanced ability to develop a cost-effective solution to real-world problems.

The term fuzzy logic in this book is most frequently used in the broad sense. Whenever it is used in the narrow sense, we will explicitly state so.

1.2 The History of Fuzzy Logic

1.2.1 The Birth of Fuzzy Set Theory

The idea of fuzzy sets was born in July 1964. Lofti A. Zadeh is a well-respected professor in the department of electrical engineering and computer science at University of California, Berkeley. In the fifties, Professor Zadeh believed that all real-world problems could be solved with efficient, analytical methods and/or fast (and big) electronic computers. In this direction, he has made significant contributions in the development of system theory (e.g., the state variable approach to the solution of simultaneous differential equations) and computer science. In early 1960s, however, he began to feel that traditional system analysis techniques were too precise for many complex real-world problems. In a paper written in 1961, he mentioned that a different kind of mathematics was needed:

> *We need a radically different kind of mathematics, the mathematics of fuzzy or cloudy quantities which are not described in terms of probability distributions. Indeed, the need for such mathematics is becoming increasingly apparent..., for in most practical cases the a priori data as well as the criteria by which the performance of a man-made system is judged are far from being precisely specified or having accurately known probability distributions.*

In July 1964, Zadeh was in New York City visiting his parents and had plans to leave soon for Southern California, where he had been invited by Richard Bellman to spend part of the summer at Rand Corp. to work on problems in pattern classification and system analysis. With this upcoming work on his mind, his thoughts often turned to the use of imprecise categories for classification. One night in New York, Zadeh had a dinner engagement with some friends. It was canceled, and he spent the evening by himself in his parents' apartment. The idea of grade of membership, which is the concept that became the backbone of fuzzy set theory, occurred to him then. This important event led to the publication of his seminal paper on fuzzy sets in 1965 and the birth of fuzzy logic technology.

The concept of fuzzy sets encountered sharp criticism from the academic community. Some rejected it because of the name, without knowing the content in detail. Others rejected it because of the theory's emphasis on imprecision — a major departure from the Western scientific discipline's focus on precision. In the late 1960s, it even was suggested to Congress as an example of the waste of government funds (much of Zadeh's research was being funded by the National Science Foundation). A flurry of correspondence from Zadeh and emerged in defense of the work. However, the controversy concerning the fuzzy logic remains.

1.2.2 A Decade of Theory Development (1965 – 1975)

Even though there was strong resistance to fuzzy logic, many researchers around the world became Zadeh's followers. While Zadeh continued to broaden the foundation of fuzzy set theory, scholars and scientists in a wide variety of fields — ranging from psychology, sociology, philosophy, and economics to natural sciences and engineering — were exploring this new paradigm during the first decade after the birth of fuzzy set theory. Important concepts introduced by Zadeh during this period include fuzzy multistage decision-making, fuzzy similarity relations, fuzzy restrictions, and linguistic hedges. Other contributions include R.E. Bellman's work (with Zadeh) on fuzzy multistage decision making [33], G. Lakoff's work from a linguistic view [346], J. A. Goguen's work on the category-theoretic approach to fuzzify mathematical structure [213,212], L. J. Kohout and B. R. Gains on the foundation of fuzzy logic [200, 325], the work on fuzzy measures by R. E. Smith and M. Sugeno [543, 455, 557, 558], G. J. Klir, Sols and Meseguer's work on fuzzified algebraic and topological systems [545, 411], C. L. Chang's work on fuzzy topology [108], Dunn and J. C. Bezdek's work on fuzzy clustering [181], C. V. Negoita's work on fuzzy information retrieval [443-445], the work by M. Mizumoto and K. Tanaka on fuzzy automata and fuzzy grammars [421-424], A. Kandel's work on the fuzzy switching function [294-295], and H. J. Zimmermann's work on fuzzy optimization [718].

During the first decade, many mathematical structures were *fuzzified* by generalizing the underlying sets to be fuzzy. These structures include logics, relations, functions, graphs, groups, automata, grammars, languages, algorithms, and programs. One of the two early fuzzy logic journals in the world is actually a Chinese journal on fuzzy mathematics. It is rather unfortunate that fuzzy logic research in China, like all other academic research, suffered from the Cultural Revolution.

In the late 1970s, a few small university research groups on fuzzy logic were established in Japan. Professor T. Terano and Professor H. Shibata from Tokyo University led one such group in Tokyo. A second research group in the Kanasai area was led by Professor K. Tanaka from Osaka University and Professor K. Asai from the University of Osaka Prefecture. These researchers encountered an "anti-fuzzy" atmosphere in Japan during those early days. However, their persistence and hard work would prove to be worthwhile a decade later. These Japanese researchers, their students, and the students of their students would make many important contributions to the theory as well as to the applications of fuzzy logic.

An important milestone in the history of fuzzy logic control was established by Assilian and E. Mamdani in the United Kingdom in 1974. They developed the first fuzzy logic controller, which was for controlling a steam generator. They were initially comparing learning algorithms for adaptive control of a nonlinear, multidimensional plant for a physical steam engine but found that many learning schemes failed to even begin to converge on a reasonable time scale (running out of steam!). A fuzzy linguistic method was developed to prime the learning controller with an initial policy to speed the adaptation — the verbal statements of engineers were transcribed as fuzzy rules and used under fuzzy logic to form a control policy. The performance of these fuzzy linguistic controllers was so good in their own right, however, that they became central to a range of studies that subsequently took place. As early as 1975, E. Mamdani and Baaklini already showed that fuzzy control rules may be tuned automatically by fuzzy linguistic adaptive strategies. Automatic learning and tuning of fuzzy rules would prove to be a very important area in the next two decades.

Pioneering efforts to use fuzzy logic applications in civil engineering were made by C. B. Brown, D. Blockley, and D. Dubois. In April 1971, C. Brown and R. Leonard [90] introduced and discussed civil engineering applications of fuzzy sets during the ASCE Structural Engineering Meeting in Baltimore, Maryland. In 1975, D. Blockley [60] published a paper on the likelihood of structural accidents, which was followed by a continous flow of stimulating papers [59, 58] and a thought-provoking book [57]. In 1979, C. Brown [87] presented a fuzzy safety measure, with which more realistic failure rates were obtained by utilizing both subjective information and objective calculations. Later, Brown treated entropy constructed probabilities [88].

In 1977, Dubois applied fuzzy sets in a comprehensive study of traffic conditions [163, 178]. The general problem of uncertainty and fuzziness in engineering decisionmaking was discussed in a comprehensive manner by J. Munro [431]. J. Yao summarized the development of civil engineering applications of fuzzy sets during the seventies in a 1985 NSF workshop on civil engineering applications of fuzzy sets held at Purdue University in memory of Professor King-Sun Fu, who died in April 1985.

One way to get an overall picture of the growth of fuzzy logic during the first decade is to study the number of papers published on the subject. Based on a survey conducted by B. Gaines, the number of papers increased by about 40 percent annually during the mid 1970s. Works accomplished during this period established the foundation of fuzzy logic technology and led to the development of application of this technology in the following years.

1.2.3 Pioneers of Industrial Applications (1976 - 1987)

In 1976, the first industrial application of fuzzy logic was developed by Blue Circle Cement and SIRA in Denmark. The system is a cement kiln controller that incorporates the "know-how" of experienced operators to enhance the efficiency of a clinker through smoother grinding. The system went to operation in 1982.

After eight years of persistent research, development, and deployment efforts, Seiji Yasunobu and his colleagues at Hitachi put a fuzzy logic-based automatic train operation control system into operation in Sendai city's subway system in 1987. Another early successful industrial application of fuzzy logic is a water-treatment system developed by Fuji Electric. We should make a few important points about these applications. First, after a successful demonstration of these approaches, it took years for both projects to be deployed in real-world operation due to various concerns about this new technology from government officials. The roads traveled by these engineers were not easy at all. Second, these two applications became the major "success stories" of fuzzy logic technology in Japan. Consequently, many more Japanese engineers and companies started to investigate fuzzy logic applications. Third, both applications made significant contributions to the technology of fuzzy logic. The development of the Sundai subway system introduced an interesting architecture for using fuzzy logic for *predictive control*. It also used fuzzy logic together with mathematical modeling, the former for recommending control options and for evaluating control options, and the latter for simulating control options to predict their effects. The development of water treatment systems enabled Fuji Electric to introduce the first Japanese general-purpose fuzzy logic controller (named FRUITAX) into the market in 1985.

1.2.4 The Fuzzy Boom (1987 - present)

The fuzzy boom in Japan was a result of the close collaboration and technology transfer between universities and industries. In 1988, the Japanese government launched a careful feasibility study about establishing national research projects on fuzzy logic involving both universities and industry. Two large-scale national research projects were established by two agencies — the Ministry of International Trade and Industry (MITI) and the Science and Technology Agency (STA). The project established by MITI was a consortium called the Laboratory for International Fuzzy Engineering Research (LIFE), which involved 50 companies with a six-year total budget of $5,000,000,000.

Matsushita Electric Industrial Co. (also known as Panasonic outside Japan) was the first to apply fuzzy logic to a consumer product, a shower head that controlled water temperature, in 1987. In late January 1990, Matsushita Electric Industrial Co. named their newly developed fuzzy controlled automatic washing machine "Asai-go (beloved wife) Day Fuzzy" and launched a major commercial campaign for the "fuzzy" product. This campaign turned out to be a successful marketing effort not only for the product, but also for the fuzzy logic technology. A foreign word pronounced "fuzzy" was thus introduced to Japan with a new meaning — intelligence. Many other home electronic companies followed Panasonic's approach and introduced fuzzy vacuum cleaners, fuzzy rice cookers, fuzzy refrigerators, and others. This resulted in a fuzzy vogue in Japan. As a result, consumers (including children) in Japan all recognized the Japanese word "fuzzy," which won the gold prize for a new word in 1990.

This fuzzy boom in Japan triggered a broad and serious interest in this technology in Europe, and, to a lesser extent, in the United States, where fuzzy logic was invented. Several major European companies formed fuzzy logic task forces within their corporate Research and Development (R&D) divisions. They include SGS-Thomson of Italy, Siemens, Daimler-Benz, and Klockner-Moeller in Germany. In the United States, the General Electric Corporate Research Division and Rockwell International Science Center have both developed advanced fuzzy logic technology as well as their industrial applications.

Fuzzy logic has found applications in other areas. The first financial trading system using fuzzy logic was Yamaichi Fuzzy Fund. It handles 65 industries and a majority of the stocks listed on Nikkei Dow and consists of approximately 800 fuzzy rules. Rules are determined monthly by a group of experts and modified by senior business analysts as necessary. The system was tested for two years, and its performance in terms of the return and growth exceeds the Nikkei Average by over 20 percent. While in testing, the system recommended "sell" 18 days before the Black Monday in 1987. The system went into commercial operation in 1988.

1.2.5 Tools for Implementing Fuzzy Logic Applications (1986 - present)

Another important milestone in the history of fuzzy logic is the first VLSI chip for performing fuzzy logic inferences developed by M. Togai and H. Watanabe in 1986 [590]. These special-purpose VLSI chips can enhance the performance of fuzzy rule-based systems for real-time applications. Togai later formed a company (Togai Infralogic) that sold hardware and software packages for developing fuzzy logic applications. Several other companies (e.g., APTRONIX, INFORM) were formed in the late 1980s and early 1990s. Even though these companies had some initial success, several did not survive through the mid- 1990s. This is partially due to the fact that vendors of conventional control design software such as MathWorks started introducing add-on toolboxes for designing fuzzy systems. *The Fuzzy Logic Toolbox* for MATLAB was introduced as an add-on component to MATLAB in 1994.

1.2.6 Fuzzy Logic in Education (1980 - present)

The important issue about introducing fuzzy sets to undergraduate engineering and science curricula was discussed during a panel discussion of the first conference of NAFIPS (North American Fuzzy Information Processing Society) [660]. Participants include C. B. Brown, K. S. Fu, L. A. Zadeh, R. L. Yager, P. Smets, J. Bezdek, and T. Whalen. Colin Brown concluded the panel discussion with the following summary remarks:

> *Engineering consists largely of recommending decisions based on insufficient information and even ignorance on the basis of subjective acceptance criteria. It is essential that these students be exposed to ways of treating uncertainty and vagueness. This also requires that existing faculty utilize these methods. The problem is not just the provision of material on fuzzy sets and probability; the material has to be developed in professional courses.*

Indeed, the need to develop courses and textbooks for undergraduates and graduates, as well as training courses for practitioners in the industry has been voiced many times during conferences and workshops since then. We hope this book responds to such a need.

1.2.7 Toward A More Principled Design (1984 - present)

The development of fuzzy systems in the early days required the manual tuning of the system parameters based on observing system performance. This drawback has become one of the major criticisms of fuzzy logic. Even though Mamdani and Baaklini introduced self-adaptive fuzzy logic control as early as 1975, the most common citation of the first work in this area is a paper by T. J. Procyk and E. H. Mamdani published in 1979. This was followed by Japanese researchers in the 1980s. T. Takagi and his advisor M. Sugeno together took an important step by developing the first approach for constructing (not tuning) fuzzy rules using training data. Their approach developed fuzzy rules for controlling a toy vehicle by observing how a human operator controlled the vehicle. Even though this important work did not gain as much immediate attention as it did later, it laid the foundation for a popular subarea in fuzzy logic, which is now referred to as *fuzzy model identification* in the 1990s.

Another trend that contributed to research in fuzzy model identification is the increasing visibility of neural network research in the late 1980s. Because of certain similarities between neural networks and fuzzy logic, researchers began to investigate ways to combine the two technologies. The most important outcome of this trend is the development of various techniques for identifying the parameters in a fuzzy system using neural network learning techniques. A system built this way is called a *neuro-fuzzy system*. Bart Kosko has been known for his contribution to neuro-fuzzy systems. His books on neural networks and fuzzy logic (his was the first book on this topic) also introduced fuzzy logic to many readers who were not aware of the existence of the technology previously [332].

The 1990s is an era of new computational paradigms. In addition to fuzzy logic and neural networks, a third nonconventional computational paradigm has also become popular—*evolutionary computing*, which includes *genetic algorithms, evolutionary strategies,* and *evolutionary programming*. Genetic Algorithms(GAs) and evolutionary strategies are optimization techniques that attempt to avoid being easily trapped in local minima by simultaneously exploring multiple points in the search space and by generating new points based on the Darwinian theory of evolution—survival of the fittest. The popularity of GA in the 1990s inspired the use of GA for optimizing parameters in fuzzy systems.

The various combinations of neural networks, genetic algorithms, and fuzzy logic help people to view them as complementary. To distinguish them from the conventional methodologies based on precise formulations, Zadeh introduced the term *soft computing* in the early 1990s.

The history of fuzzy logic research and application development continues as you read this book. We hope that you can contribute to this history in the near future as well.

1.3 Motivations

Fuzzy logic was motivated by two objectives. First, it aims to alleviate difficulties in developing and analyzing complex systems encountered by conventional mathematical tools. Second, it is motivated by observing that human reasoning can utilize concepts and knowledge that do not have well-defined, sharp boundaries (i.e., vague concepts). The first motivation is directly related to solving real-world problems, while the second motivation is related to *artificial intelligence*—the discipline in computer science involved with developing computer systems that exhibit intelligent behaviors similar to those of human beings. The former motivation requires fuzzy logic to work in quantitative and numeric domains, while the latter motivation enables fuzzy logic to have a descriptive and qualitative form because vague concepts are often described qualitatively by words. These two motivations together not only make fuzzy logic unique and different from other technologies that focus on only one of these goals but also enable fuzzy logic to be a natural bridge between the quantitative world and the qualitative world. As will become clear at a later point, this unique characteristic of fuzzy logic allows this technology to offer an important benefit—it not only provides a cost-effective way to model complex systems involving numeric variables but also offers a qualitative description of the system that is easy to comprehend.

1.3.1 The Underpinning: The Principle of Incompatibility

The underpinning of a fuzzy logic approach to achieve the two objectives described above is based on an observation about a fundamental trade-off between precision and cost — which is referred to by L. A. Zadeh as *the principle of incompatibility*. In one of his seminal papers, Zadeh explained this observation:

> *Stated informally, the essence of this principle is that as the complexity of a system increases, our ability to make precise yet significant descriptions about its behavior diminishes until a threshold is reached beyond which precision and significance (or relevance) become almost mutually exclusive characteristics.*

In other words, one has to pay a cost for high precision. Therefore, the cost for precise modeling and analysis of a complex system can be too high to be practical. An example often used by Zadeh to illustrate this trade-off is the problem of parking a car. Usually, it takes a driver less than half a minute to parallel park. However, if we were asked to park a car in a parking space such that the outside wheels are precisely within 0.01 mm from the side lines of a parking space, and the wheels are within 0.01 degree from a specified angle, how long do you think it would take you to park the car? It would take me a very long time. In fact, I would probably give up after trying ten minutes or so. The point is that the cost (i.e., the time required) to park a car increases as the precision of the car parking task increases. This trade-off between precision and cost exists not only in car parking but also in control, modeling, decision making, and almost any kind of problem.

Conceptually, we can use Fig 1.2 to depict this trade-off for many systems. The horizontal axis represents the degree of precision, while the vertical axis serves the dualpurpose of representing both the cost and the degree of utility. As the precision of a system increases, the cost for developing the system also increases, typically in an exponential manner. On the other hand, the utility (i.e., usefulness) of the system does not increase proportionally as its precision increases — it usually saturates after a certain point. This insight about the trade-off between precision, cost, and utility inspired Zadeh and his followers to exploit the gray area in Fig 1.2, which resulted in a revolutionary way of thinking for developing approximate solutions that are both cost-effective and highly useful. In other words, the fundamental principle of fuzzy logic is to develop cost-effective approximate solutions to complex problems by exploiting the tolerance for imprecision.

Traditionally, fuzzy logic has been viewed as a theory for dealing with uncertainty about complex systems. From the discussion above, however, it seems even more suitable to view fuzzy logic as an approach for approximation theories. This perspective on fuzzy logic brings to the surface one of its ultimate concerns — low cost. Indeed, providing a low cost solution to a wide range of real-world problems is the primary reason that fuzzy logic has found so many successful applications in industry to date. Understanding this driving force of the success of fuzzy logic will prevent us from falling into the trap of debating with critics "whether fuzzy logic can accomplish what X cannot accomplish" where X is an alternative technology such as probability theory, control theory, etc. Such a debate is usually not fruitful because it ignores one important issue — cost. A better question to ask is, "What is the difference between the cost of a fuzzy logic approach and the cost of an approach based on X to accomplish a certain task?" In a panel discussion on fuzzy logic and neural networks held at the NASA Johnson Space Center in 1991, K. Hirota, an internationally known fuzzy logic researcher from Japan, made a short, yet sharp, comment by saying, "There is a fundamental difference between the theoretically possible and the practically feasible." People with this insight can more easily recognize and appreciate the benefits offered by fuzzy logic.

FIGURE 1.2   The Cost-Precision Trade-off

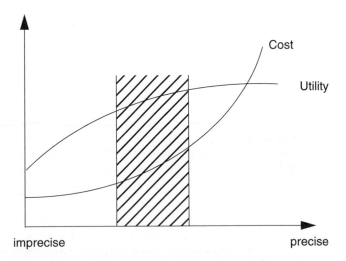

1.3.2 A Quest for Precision Forever?

Even though the precision-cost trade-off principle is intuitive, the idea of not pursuing precision all the time is actually not easy. Precision has often been viewed as "desirable," while imprecision has been considered "undesirable." This is especially true in the field of science and engineering. It is true that precise problem formulation enables scientists and engineers to use a wide range of mathematical tools to develop and analyze solutions in a rigorous way. It is also true that this quest for precision has contributed to the rapid development of basic science and applied technology in the twentieth. century. However, as technology advances, so does the complexity of the problems facing us. While the quest for precision will continue and will remain beneficial to human society to a certain extent, fuzzy logic brings a complementary viewpoint into the world — a view in which cost-effectiveness rather than precision is the ultimate concern.

1.4 Why Use Fuzzy Logic for Control?

One common answer to this question is that fuzzy logic can be used for controlling a process (i.e., a *plant* in control engineering terminology) that is too nonlinear or too ill-understood to use conventional control designs. Another answer is that fuzzy logic enables control engineers to easily implement control strategies used by human operators. The first answer echoes the first motivation for fuzzy logic — to deal with complex systems — while the second answer is related to the second motivation — the ease of describing human knowledge. We will illustrate these points using a real-world application below. Before we do that, though, we would like to point out that fuzzy logic control broadens the range of applications of control engineering tools and methods. It is not meant to replace but rather to augment existing methodologies for control design. We will return to this important point in a later chapter.

1.4.1 Modus Operandi of Fuzzy Logic

The distinguishing mark of fuzzy logic in rule-based systems is its ability to deal with situations in which making a sharp distinction between the boundaries of application in the use of rules or constraints is very difficult. An example that clearly illustrates this fact is one that is taken from an actual application of fuzzy logic control and one of its seminal ones, namely fuzzy logic-based image stabilization in camcorders. Here there are no known algorithms that would in a structured manner determine the compensation strategy. For this reason, one would rely on rules or *heuristics* as the basis for implementation of a control strategy. In particular, in order to determine whether the present image is a function of the movement of the camera or of the movement of objects in the field of view, the designers[1] have chosen to mark certain objects in the field of view as *references*. Further, the position of reference objects in two subsequent frames is related

---

[1.] We describe the following based on descriptions of several such designs made public. The actual implementations by any specific manufacturer may differ from this description.

in terms of four *motion vectors,* one for each subarea in the image. An example of heurstic rules is given below:

If all motion vectors are almost parallel and their time differential is small, then the hand jittering is detected and the direction of the hand movement is the direction of the moving vectors.

Heuristic rules of this kind are used to deduce whether the differences in two consecutive frames are due primarily to the motion of the camera, or represent true motion of objects in the field of view. Thus as shown in Fig 1.3, the image on the left would be interpreted as being based on the motion of the camera since all motion vectors are in the same direction, while the image on the right is interpreted as being caused by the motion of reference objects.

In general, however, it is not evident whether an image should absolutely be accepted or rejected. This is where fuzzy logic enables the designers to incorporate both the uncertainty regarding the applicability of the rules as well as inherent imprecision in the rule set itself in a sensible manner and thus to improve the performance of the product. The actual performance improvement can be traced back to both the rule based descriptions where cleverly designed rules help deal with a potentially intractable situation, as well as the use of fuzzy logic to help implement the rules in a sensible way. We should point out that the purpose of fuzzy logic and its contribution is not merely to soften logical transition in a syntactic sense as it may be supposed. It is the implications of this softening at the level of rule interpretation, or semantics, that has a critical role in performance.

FIGURE 1.3   Image Stabilization via Fuzzy Logic

To summarize, we can identify two important points from this example.

1. Absence of a readily available algorithm for control leads to the use of rules or heuristics as the basis for implementation of control strategy.

2. Fuzzy logic deals with uncertainty regarding the applicability of heuristic rules and accounts for imprecision in measurements.

Note that as suggested earlier the term control here is interpreted in a broad sense and includes classification and not merely feedback control in the classical sense. We shall deal with this issue in Chapter 8 in relation to conventional notions in control engineering. Further, the control algorithm incorporates linguistic rules or heuristics that are interpreted on the basis of fuzzy logic. In other words, the effectiveness of the algorithm is not merely due to the application of fuzzy logic. Rather fuzzy logic enhances the prospect of a linguistic algorithm forming an effective control strategy. In other words, fuzzy logic in and of itself does not solve a problem, much as a hammer in and of itself does not build a house. In the hands of an expert carpenter it builds a beautiful house. Misused, it fails to achieve anything.

There are other applications of fuzzy logic that are more closely related to conventional approaches in control engineering. We shall discuss these later, however. Table 1.1 summarizes applications of fuzzy logic control in four areas.

TABLE 1.1   Applications of Fuzzy Logic Control[a]

Consumer Products	Automotive and Power Generation	Industrial Process Control	Robotics and Manufacturing
Cameras and Camcorders (Canon, Minolta, Ricoh, Sanyo)	Power Train and Transmission Control (GM-Saturn, Honda, Mazda)	Cement Kiln, Incineration Plant	Electrical Discharge Machine (Mitsubishi)
Washing Machines (AEG, Sharp, Goldstar, Siemens, General Electric), Refrigerators (Whirlpool), Vacuum Cleaners (Phillips, Siemens)	Engine Control (Nissan)	Refining, Distillation, and other Chemical Processes	

a. The information here is based on many sources including advertisements in *Manufacturing Engineering*, feature articles in trade magazines such as *Electronic Design News, Computer Design, Control* among others. Where no manufacturer is specifically named, there is no particular name associated with the given area.

1.5 Myths about Fuzzy Logic

There have been several myths about fuzzy logic. We will try to clarify some of these myths below and point out related chapters that provide a more detailed discussion on these topics.

- *Fuzzy logic is a clever disguise of the probability theory.*
  Definitely not. In short, probability measures the likelihood of an event before the actual outcome is known, while fuzzy logic measures the degree to which an outcome

belongs to an event that does not have a well-defined sharp boundary. This topic will be further discussed in detail in Chapter 7.

- *Fuzzy logic and probability are competing technologies. Only one of them can be used to solve a given problem.*
  No, they can be complementary, as we shall see in Chapter 7.

- *The behavior of a fuzzy logic system is fuzzy nondeterministic.*
  This confusion is usually caused by the word "fuzzy." As we shall see in the following chapters, the behavior of a fuzzy system is completely deterministic. There is nothing fuzzy, ambiguous, or mysterious about a fuzzy system, despite its name.

- *Fuzzy logic in the narrow sense is basically a kind of multivalued logic.*
  Even though fuzzy logic in the narrow sense is related to multivalued logic, it differs from multivalued logic by introducing concepts such as linguistic variables and hedges that are important for fuzzy logic's agenda to capture human linguistic reasoning.

- *Fuzzy logic is a model-less approach and therefore does not require a good understanding of the problem.*
  This inappropriate claim is sometimes made by fuzzy logic enthusiasts. The term modelless refers to the fact that the design of a fuzzy logic control system does not require a mathematical model of the plant. Even though this is true, the design of a fuzzy system can often benefit from (and may sometimes require) a good understanding of the problem. Some fuzzy logic systems, such as the Sendai subway control system, actually use fuzzy logic together with a model of the plant. We will elaborate on these points in Chapter 10 when we discuss the design of fuzzy logic control systems.

- *Fuzzy logic is no more than a table lookup technique.*
  This false conclusion is sometimes drawn by people who read a few articles about simple applications of fuzzy logic control. It is true that a simple fuzzy rule-based controller can be compiled into a form similar to table lookup. However, fuzzy logic offers a high-level language that is much easier to comprehend and modify than the underlying table. The relationship between fuzzy rules and their compiled (table-like) form is analogous to that between a high-level programming language (such as C or Java) and their compiled machine code. You would not want to modify the table underlying a fuzzy rule-based system just as you would not want to touch the machine code directly. In addition, there are rather specific concerns (myths) regarding fuzzy logic control. We shall refer to and address them at some length in Chapter 8. However, we will briefly recount some of the more critical of these concerns as follows:

- *Fuzzy logic control is ill-founded, not rigorous, and may lead to potential disasters*
  This myth is based on two false conceptions. First, it is often assumed that the apparently rigorous and even meticulous manner in which control theory is presented belies the fact that formulations and developments that thus emerge do not represent or apply to reality. The gap is often filled with ad-hoc adjustments (in the field tuning) and ultimately a leap of faith is necessary in the development of any control system. Second, fuzzy logic control systems thus developed in industry have even been subjected to rigorous testing and validation and this trend will continue.

- *Fuzzy logic controllers cannot be shown to be stable*
  This myth is a more specific version of the one above and can be addressed in two ways. First, there are indeed proofs of stability for at least a class of fuzzy logic control systems that take advantage of the very same techniques as in conventional control theory. Second, from a practical standpoint the ultimate proof of the stability of a control system is its empirical validation. This holds true for fuzzy logic control as well as conventional control techniques. In this sense one cannot hold fuzzy logic control to a more rigorous (and unfair) standard than conventional techniques.

- *Fuzzy logic control is just nonlinear Propotional Integral PI/ Proportional Derivative PD control.*
  This myth is indeed based on partial truth. It is indeed true that a class of fuzzy logic control algorithms do resolve into nonlincar PI/PD controllers. This fact does not, however, in itself imply that fuzzy logic control has nothing new to offer as long as no effective design methodology exists that can efficiently be used to design nonlinear PI/ PD controllers (and they indeed do not exist). Fuzzy logic control offers a usable means of developing nonlinear PI/PD controllers. The fact that the outcome of the design process is equivalent to a class of conventional control algorithms, does not undermine fuzzy logic control's role in the design process.

1.6 Intelligence, Control, and Information

Fuzzy logic is a technology for developing intelligent control and information systems. We summarize below three facets of fuzzy logic from the viewpoint of machine intelligence, control systems, and information technology.

Fuzzy logic achieves *machine intelligence* by offering a way for representing and reasoning about human knowledge that is imprecise by nature. Even though fuzzy logic is not the only technique for developing AI systems, it is unique in its approach for explicit representation of the impreciseness in human knowledge and problem solving techniques. This facet of fuzzy logic will be revealed throughout the book. A focal point of this facet is presented in Chapter 11. In addition to achieving artificial intelligence, fuzzy logic also benefits from machine learning techniques because they can be used to construct fuzzy systems automatically (Chapter 14). Neural networks and genetic algorithms are two machine learning techniques that are particularly useful for this purpose. These are the topics of Chapter 16 and 17.

Fuzzy logic offers a practical way for designing nonlinear *control systems*. It achieves nonlinearity through piece-wise linear approximation. The basic building block of a fuzzy logic control system is set of fuzzy if-then rules (i.e., fuzzy rule-based models) that approximate a functional mapping. In fact, fuzzy rule-based models are useful not only for control systems, but also for decision making and pattern recognition. The basics of fuzzy rule-based models will be introduced in Chapter 5. Fuzzy logic control will be discussed in Chapter 8. There are at least two important related topics: (1) their role in higher levels of hierarchical intelligent control systems, and (2) the analytical issues of fuzzy logic controllers. We will discuss these two subjects in Chapters 9 and 10 respectively.

Fuzzy logic is becoming increasingly important in advanced <$nopage> information *systems*. In addition to its potential for developing intelligent software agents (Chapter 11), it has been used for storing and retrieving imprecise information from databases and information sources (Chapter 12), as well as for extracting interesting patterns from a large amount of data (Chapter 13). The latter can be applied to medical image analysis, computer vision, and data mining.

Like facets of a diamond, the three facets of fuzzy logic technology shed light on each other and create brilliant and vivid colors. We hope that the remaining Chapters of this book help you to visualize such an exciting image.

1.7 Summary

We summarize below major points covered in this chapter:

- We explained two difference usages of the term "fuzzy logic."
- We described the history and major milestones of the development of fuzzy logic technology.
- We introduced two motivations for developing fuzzy logic approaches and their under-pinning — the principle of incompatibility.
- The rationale of using fuzzy logic for control is discussed using an industrial application (i.e., image stabilization for a video camcorder).
- We attempted to clarify some myths about fuzzy logic and fuzzy logic control.
- Finally, we gave an overview of fuzzy logic technology from the perspectives of machine intelligence, control systems, and information technology.

Bibliographical and Historical Notes

Honoring Zadeh's receipt of the IEEE Medal of Honor Award in 1995, the most prestigious IEEE Award, *IEEE Spectrum* published an article that describes Zadeh's personal journey as well as how the concept of fuzzy sets first came to him [471].

A good summary of fuzzy logic research progress during the first decade can be found in a collection edited by Gupta, Saridis, and Gaines [227]. In this volume, M.M. Gupta describe some of the events that took place during the first decade of fuzzy logic [224]; E.H. Mamdani gives a survey of fuzzy logic control and points out several important issues regarding the stability and the design of fuzzy logic controllers [398]; and B.R. Gaines and L.J. Kohout give a detailed bibliography of the first decade of fuzzy logic research [201].

K. Hirota presents a history of the development of fuzzy logic technology in Japan in [249]. The fuzzy washing machine that triggered the fuzzy boom is discussed by N. Wakami et al. [617] and by S. Kondo et. al. [327]. S. Yasunobu and his colleague at Hitachi describe their design of the Sendai subway controller in [662, 661]. D. G. Schwartz and G. Klir discuss several key milestones of fuzzy logic technology development and applications in the 1980's [527]. Constantin von Altrock summarizes the historic development and the industrial applications of fuzzy logic in Europe [11]. H. Takagi surveys the applications of fuzzy logic and neuro-fuzzy systems in consumer products in [571, 570]. Industrial fuzzy control applications have been published in collected volumes edited by M. Sugeno [554] and by J. Yen, R. Langari, and L. Zadeh [673]. Fuzzy logic applications in other areas have been collected by T. Terano, K. Asai, and M. Sugeno [587], and by K. Hirota and M. Sugeno [251]. An update on fuzzy logic applications in civil engineering has been compiled by F. Wong, K. Chou, and J. Yao [643].

Exercises

1.1   Illustrate the trade-off between precision and cost using a problem that you are familiar with.

1.2   What is the principle of incompatibility that motivated the development of fuzzy logic technology?

1.3   What are the two different usages of the term "fuzzy logic"? Identify the appropriate usage for each sentence below.

- Fuzzy logic has found many real-world applications.
- Fuzzy logic generalizes classical two-valued logic.
- An important rationale underlying the development of fuzzy logic is the principle of incompatibility.

1.4   What are the two major motivations for developing fuzzy logic? Describe these motivations using one or two problems that you encounter in your daily life.

1.5   What is the motivation for using fuzzy logic in control engineering?

1.6   Fuzzy logic was motivated by difficulty in analyzing complex systems. However, it is been used to solve many not-so-complex problems in consumer products. Is there a conflict between the two?

1.7   Do a literature survey to identify the first paper on each of the following topics:

- Stability of fuzzy logic controllers
- Fuzzy relations
- Fuzzy functions
- Fuzzy graphs
- Fuzzy algorithms
- Fuzzy clustering

1.8   Do a literature survey to identify a significant fuzzy logic application in the following areas:

- Control systems
- Diagnostic system
- Consumer electronics
- Manufacturing systems
- Mobile robots
- Automobile applications
- Image processing systems

1.9   Find five Web sites that contain valuable information about fuzzy logic. Describe briefly the information on these sites.

2   BASIC CONCEPTS OF FUZZY LOGIC

2.1 Introduction

Even though the broad sense of fuzzy logic introduced in the previous chapter covers a wide range of theories and techniques, its core technique is based on four basic concepts: (1) *fuzzy sets*: sets with smooth boundaries; (2) *linguistic variables*: variables whose values are both qualitatively and quantitatively described by a fuzzy set; (3) *possibility distributions*: constraints on the value of a linguistic variable imposed by assigning it a fuzzy set; and (4) *fuzzy if-then rules*: a knowledge representation scheme for describing a functional mapping or a logic formula that generalizes an implication in two-valued logic. The first three concepts are fundamental for all subareas in fuzzy logic. The fourth concept is also important because it is the basis for most industrial applications of fuzzy logic developed to date, including many fuzzy logic control systems.

Even though the theories behind these concepts deserve separate detailed treatments, the very basic ideas underlying these concepts can be understood by focusing first on their motivations and usages, rather than on their theoretical foundations. Furthermore, presenting these four basic concepts together helps us to understand their close relationship as well as to develop a coherent picture about them in our mind. We hope that such a two-pass introduction to these concepts also helps you to relate them quickly to applications, and to facilitate your understanding about their theoretical foundations when they are introduced in later chapters.

2.2 Two Exemplary Problems

To illustrate various basic concepts and methods in fuzzy logic, we will use two example problems throughout this chapter.

Problem A.   Rudimentary Air Flows Mixing Control

The first problem is the control of air flow mixing for room temperature control. It is a "rudimentary" control problem because no feedback about the current room temperature is provided. We chose it for its simplicity and its connection to fuzzy logic control. An extended version of the problem (i.e., closed-loop feedback control) will be revisited when we discuss fuzzy logic controllers in Chapter 8. The given task of this problem is to control the amount of hot air flow and cold air flow based on a set target temperature. In this rudimentary version of the problem, we assume that there is no feedback signal about the current room temperature. The amount of hot air flow and cold air flow is controlled by adjusting the voltage to the pump in the mixing stage. The lowest and highest voltage settings are denoted by $\underline{V}$ and $\overline{V}$. If the voltage is set at $\underline{V}$, maximal cold air flow will be allowed. If the voltage is set at $\overline{V}$, maximal hot air flow will be generated. A voltage between $\underline{V}$ and $\overline{V}$ mixes the hot flow and the cold flow proportionally. This problem is depicted in Fig 2.1.

FIGURE 2.1   A Rudimentary Air Flow Mixing Control System

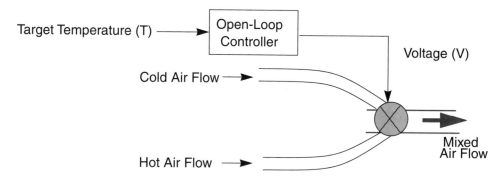

Problem B.   Automated Washing Machine Control

The second problem is about the automatic selection of cycles and washing time for a washing machine. This problem allows us to illustrate certain fuzzy logic techniques that cannot be demonstrated using the first problem. We chose to use this problem among many candidate applications for two additional reasons. First, the nature of human decision making involved in this problem is easy to understand due to our familiarity with the tasks in our daily lives. Second, a fuzzy logic approach to the problem developed by Matsushita Electric Industrial Co. [the automatic washing machine is called "Aisai-go (beloved wife) Day Fuzzy"], as mentioned in Chapter 1, generated a fuzzy vogue in 1990 which triggered the introduction of many other successful consumer products using fuzzy logic [327, 617]. The word "fuzzy" even won the 1990 gold prize for a new word in Japan. This application therefore has an additional historical significance.

The goal of this problem is to simplify the operation of the washing machine while ensuring high-quality results. More specifically, we wish to automate the selection of the wash cycle and wash time based on the quantity and the softness of the laundry. To simplify our discussion, we assume that the quantity and the softness of the laundry are provided directly by two sensing techniques. In reality, however, these measures are not measurable directly. They are often inferred from other sensor inputs. Fig 2.2 depicts the problem graphically.

FIGURE 2.2   Fully Automatic Washing Machine

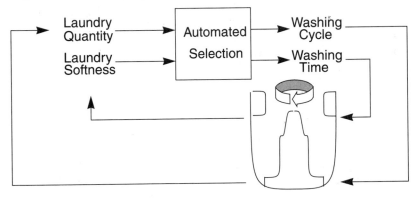

2.3 Fuzzy Sets

A fuzzy set is a set with a smooth boundary. Fuzzy set theory generalizes classical set theory to allow partial membership. The best way to introduce fuzzy sets is to start with a limitation of classical sets. A set in classical set theory always has a sharp boundary because membership in a set is a black-and-white concept — an object either completely belongs to the set or does not belong to the set at all. For instance, suppose we wish to represent the set of high-income families within the framework of classical set theory. This would require us to choose a particular value as a threshold (e.g., yearly income) such that all families with a yearly income greater than the threshold (say $80,000) are considered to be a member of the set of high-income families, and all families with a yearly income lower than the threshold are not members of the set. However, the fact that a family with a yearly income of $79,999 is not considered high income whereas a family with $1 more yearly income is considered high income is counterintuitive, to say the least. Even though unnatural and sometimes disturbing, such artificially created sharp boundaries have often been used in defining classical sets, primarily due to a lack of a better methodology.

Even though some sets do have sharp boundaries (e.g., the set of married people), many others do not have sharp boundaries (e.g., the set of happily married couples, the set of good graduate schools). Fuzzy set theory directly addresses this limitation by allowing membership in a set to be a matter of degree. The degree of membership in a set is expressed by a number between 0 and 1; 0 means entirely not in the set, 1 means completely in the set, and a number in between means partially in the set. This way, a smooth and gradual transition from the regions outside the set to those in the set can be described. A fuzzy set is thus defined by a function that maps objects in a domain of concern to their membership value in the set. Such a function is called the *membership function* and is usually denoted by the Greek symbol μ for ease of recognition and consistency. For example, a more realistic representation of the high-income family can now be expressed by the membership function in Fig 2.3. The membership function of a fuzzy set A is denoted as μ_A, and the membership value of x in A is denoted as $\mu_A(x)$. The domain of the membership function, which is the domain of concern from which elements of the set are drawn, is called the *universe of discourse*. For instance, the universe of discourse of the fuzzy set High Income can be the positive real line $[0, \infty)$.

In addition to membership functions, a fuzzy set is also associated with a linguistically meaningful term. For instance, the fuzzy set in Fig 2.3 is associated with the linguistic term "High." In this book, we capitalize such linguistic terms when used as labels of fuzzy sets to distinguish this usage. Associating a fuzzy set to a linguistic term offers two important benefits. First, the association makes it easier for human experts to express their knowledge using the linguistic terms. This point will soon become clear when we introduce fuzzy if-then rules in Section 2.6. Second, the knowledge expressed using linguistic terms is easily comprehensible. This benefit often results in significant savings in the cost of designing, modifying, and maintaining a fuzzy logic system. An important concept in fuzzy logic that enables these two benefits is that of the *linguistic variable*. This concept will be introduced in Section 2.4.

One concern often raised regarding the association of fuzzy sets and linguistic terms is that the meaning of a linguistic term often changes when the term is used in a different context. For instance, the meaning of a short person in the United States is different from that in Taiwan (i.e., Republic of China). Even within the United States, the meaning of a short person may change in the context of a specific subgroup whose height distribution differs from that of the general public. For instance, "a short person in the NBA" differs in meaning from "short American." Because the meaning of a term depends on the context, it is important to remember that *a fuzzy set is always defined in a context*, even though this context may not be explicit. The context of defining a linguistic term is usually implicitly specified within the application in which it is used and therefore rarely misunderstood.

2.3.1 Designing Membership Functions

One of the questions you may have by now is, "How do we determine the exact shape of the membership function for a fuzzy set?" Before answering this question we will point out a few important points regarding membership functions in order to put this question in a larger perspective. What is important about membership function is that it provides a *gradual transition* from regions completely outside a set to regions completely in the set.

A membership function can be designed in three ways: (1)Interview those who are familiar with the underlying concept and later adjust it based on a tuning strategy. (2) Construct it automatically from data; (3) learn it based on feedback from the system performance. The first approach was the main approach used by fuzzy logic researchers and practitioners until the late 80s. Due to a lack of systematic tuning strategies in those days,

FIGURE 2.3   Membership Function of High Annual Income

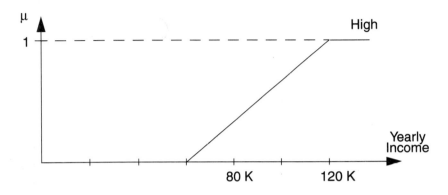

most fuzzy systems were tuned through a trial-and-error process. This has become one of the main criticisms of fuzzy logic technology. Fortunately, many techniques in the second two categories have been developed since the late 80s using statistical techniques, neural networks, and genetic algorithms. More systematic tuning strategies have also been developed for adjusting membership functions. We will postpone the discussion of these topics to later chapters. Chapter 14 will discuss general issues regarding learning membership functions; Chapters 16 and 17 will introduce specific approaches based on neural network learning and genetic algorithms, respectively.

We have a word of caution about the design of membership functions. Even though one may attempt to define a membership function of arbitrary shape, **we strongly recommend the use of parameterizable functions that can be defined by a small number of parameters**. Using parameterized membership functions cannot only reduce the system design time, it can also facilitate the automated tuning of the system because desired changes to the membership function (e.g., widening vs. narrowing a membership function) can be directly related to corresponding changes in the related parameters. Even though the specific strategy for tuning membership functions will not be introduced until Chapter 8, the impact of membership function design on its tuning is too important to be ignored here.

The parameterizable membership functions most commonly used in practice are the *triangular* membership function and the *trapezoid* membership function. The former has three parameters and the latter has four parameters.[1] An example of these two kinds of membership function and their parameters are shown in Figs 2.4 and 2.5.

[1] Here we assume that the peak of the membership function is 1. Otherwise, an additional parameter would be needed.

Simplicity is the main advantage of triangular and trapezoidal membership functions. To understand why this simple membership function is sufficient for most practical
applications, we refer you to an important principle underlying fuzzy logic discussed in
Chapter 1 — *exploring cost-effective approximate solutions*. Therefore, a membership
function is intended to *approximate* a smooth transition between two regions (the region

FIGURE 2.4   A Triangular Membership Function and Its Parameters

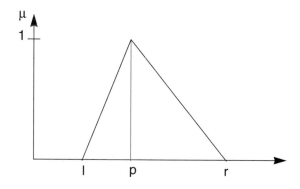

FIGURE 2.5   A Trapezoid Membership Function and Its Parameters

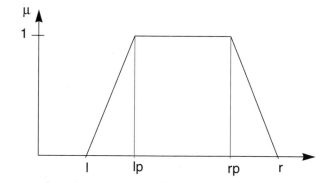

outside the set and that inside the set). An attempt to accurately model such transitions (if
at all possible) would undermine the exact rationale of using fuzzy logic. Even though
there are situations in which nonlinear membership functions are more suitable, most
practitioners have found that triangular and trapezoidal membership functions are sufficient for developing good approximate solutions for the problems they wish to solve.
Membership functions that are differentiable have certain advantages in their application
to *neuro-fuzzy systems*—systems that learn membership functions using neural network
learning techniques. Gaussian membership functions have been a popular choice for these
systems, as we will see in Chapter 16. Other forms of membership functions will be introduced in the next chapter.

To summarize, we present the following guidelines for membership function design:

1. Always use parameterizable membership functions. Do not define a membership function point by point
2. Use a triangular or trapezoidal membership function, unless there is a good reason to do otherwise.
3. If you want to learn the membership function using neural network learning techniques, choose a differentiable (or even continuous differentiable) membership function (e.g., Gaussian).

2.3.2   Basic Operations in Fuzzy Sets

The three fundamental operations in classical sets are union, intersection, and complement. The union of two sets A and B (denoted as $A \cup B$) is the collection of those objects that belong to either A or B. The intersection of A and B (denoted as $A \cap B$) is the collection of those objects that belong to both A and B. The complement of a set A (denoted as A^C or $\bar{A}$) is the collection of objects not belonging to A. For instance, let A and B be two sets of annual personal income defined as below:

$$A = \{x \mid 100K \leq x \leq 200K, \, x \in U\}$$

$$B = \{x \mid 50K \leq x \leq 120K, \, x \in U\}$$

where U is the universe of discourse [0, 1000K]. We get

$$A \cap B = \{x \mid 100K \leq x \leq 120K, \, x \in U\}$$

$$A \cup B = \{x \mid 50K \leq x \leq 200K, \, x \in U\}$$

$$A^C = \{x \mid 0 \leq x < 100K \text{ or } 200K < x \leq 1000K\}$$

These operations are also illustrated in Fig 2.6.

Since membership in a fuzzy set is a matter of degree, set operations should be generalized accordingly. We introduce here only the most commonly used fuzzy set operations that are essential for the discussion in this chapter. A more detailed discussion about the theoretical issues involved and additional fuzzy set operations will be given in the next chapter.

The fuzzy intersection operation turns out to be mathematically equivalent to the fuzzy conjunction (AND) operation, because they have identical desired properties. While we will give a detailed discussion about these operations in the next chapter, we will explain the connection between set operations and logic operations here. To do this, let us first review basic operations in classical logic. A statement in classical logic has only two possible truth values: T (true) or F (false). Two logic statements can be combined using logic connectives such as AND (conjunction, denoted by $\wedge$), OR (disjunction, denoted by $\vee$), NOT (negation, denoted by $\neg$) and IMPLY (implication, denoted by $\rightarrow$). The truth values of these compound statements are listed in Table 2.1 for all possible cases. We use p and q in the table to denote two arbitrary logic statements (called propositions).

FIGURE 2.6   Basic Set Operations

a) Intersection of Two Sets

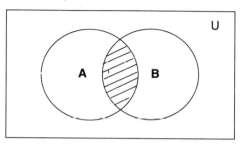

b) Union of Two Sets

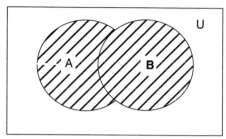

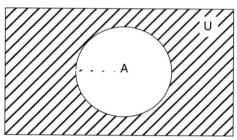

c) Complement of a Set

TABLE 2.1   Truth Table of Classical Logic Connectives

p	q	$\neg p$	$p \wedge q$	$p \vee q$	$p \rightarrow q$
F	F	T	F	F	T
F	T	T	F	T	T
T	F	F	F	T	F
T	T	F	T	T	T

A conjunctive composite statement $p \wedge q$ is true if and only if both p and q are true. A disjunctive composite statement $p \vee q$ is true if and only if either statement is true. A negated statement is true if and only if the original statement is false. We will discuss the implication statement in Section 2.5 because it is related to fuzzy rules.

Suppose we use the proposition p to represent the sentence "x is in the set A," i.e.,

$$p \text{ is true iff } x \, \varepsilon \, A$$

and another proposition to represent the sentence "x is in the set B," i.e.,

$$q \text{ is true iff } x \, \varepsilon \, B$$

where iff stands for "if and only if". It is easy to see that p and q are both true when x is in the intension of A and B, i.e.,

$$(p \wedge q) \text{ is true iff } x \in A \cap B$$

and that p or q is true when x is in the union of A and B, i.e.,

$$(p \vee q) \text{ is true iff } x \in A \cup B$$

Finally, p is false when x is in A's complement, i.e., is true iff

$$\neg p \text{ is true iff } x \in A^c$$

Hence, intersection, union, and complement operations in set theory are similar to conjunction, disjunction, and negation in logic. A common fuzzy disjunction operator is the maximum operator. Hence, fuzzy union is often defined as

$$\mu_{A \cup B}(x) = max\{\mu_A(x), \mu_B(x)\} \tag{EQ 2.1}$$

and is illustrated in Fig 2.7.

FIGURE 2.7   Fuzzy Set Union

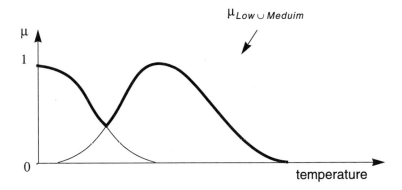

A common fuzzy conjunction (AND) operation is the minimum operator. Fig 2.8 shows the result of using min to form the intersection of two fuzzy sets, i.e.,

$$\mu_{A \cap B}(x) = min\{\mu_A(x), \mu_B(x)\} \tag{EQ 2.2}$$

The complement of a fuzzy set A is defined by the difference between one and the membership degree in A. This is illustrated in Fig 2.9 using the fuzzy set Medium.

$$\mu_{A^c}(x) = 1 - \mu_A(x) \tag{EQ 2.3}$$

2.4 Linguistic Variables

Having introduced the fundamental concept of a fuzzy set, it is natural to see how it can be used. Like a conventional set, a fuzzy set can be used to describe the value of a variable. For example, the sentence "The amount of trading is heavy" uses a fuzzy set "Heavy" to describe the quantity of the stock market trading in one day. More formally, this is expressed as

TradingQuantity   is   Heavy

FIGURE 2.8   Fuzzy Set Intersection

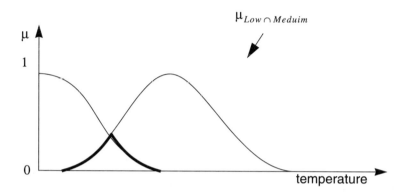

FIGURE 2.9   The Complement of a Fuzzy Set *t*

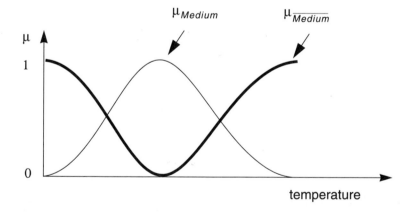

The variable *TradingQuantity* in this example demonstrates an important concept in fuzzy logic: the **linguistic variable**. A linguistic variable enables its value to be described both qualitatively by a linguistic term (i.e., a symbol serving as the name of a fuzzy set) and quantitatively by a corresponding membership function (which expresses the meaning of the fuzzy set). The linguistic term is used to express concepts and knowledge in human communication, whereas membership function is useful for processing numeric input data, as we will see in Section 2.6.

A linguistic variable is like a composition of a symbolic variable (a variable whose value is a symbol) and a numeric variable (a variable whose value is a number). An example of a symbolic variable is

Shape = Cylinder

where Shape is a variable indicating the shape of an object. An example of numeric variable is

Height = 4'

Numeric variables are frequently used in science, engineering, mathematics, medicine, and many other disciplines, while symbolic variables play an important role in artificial intelligence and decision sciences. Using the notion of the linguistic variable to combine these two kinds of variables into a uniform framework is, in fact, one of the main reasons that fuzzy logic has been successful in offering intelligent approaches in engineering and many other areas that deal with continuous problem domains.

In our example about stock market trading activities, there are certainly many other linguistic descriptions about the trading quantity such as "light," "moderate," "somewhat heavy," "very heavy," and so on. One of the important concepts in fuzzy logic is that instead of enumerating all these different descriptions, they can be generated from a core set of linguistic terms (called a **term set**) using modifiers (e.g., "very," "more or less") and connectives (e.g., "and," "or"). In fuzzy logic, we call these modifiers **hedges**. We will return to these concepts for a more detailed discussion in the next chapter.

2.5 Possibility Distributions

Assigning a fuzzy set to a linguistic variable constrains the value of the variable, just as a crisp set does. The difference between the two, however, is that the notion of possible versus impossible values becomes a matter of degree. We will illustrate this using the following example. Suppose the police report that the suspect of a bombing is a terrorist, between 20 and 30 years old. This can be expressed by assigning the interval [20, 30] to a variable representing the suspect's age:

Age(suspect) = [20, 30].

This interval-valued assignment constrains the age of the suspect. More specifically, it states that the suspect's age can be 20, 21,..., 29, or 30. It is, on the other hand, impossible for the suspect to be 19, 31, or any other age not included in the set [20, 30]. Thus, the assignment above introduces a sharp boundary between possible values and impossible values. For situations where such a sharp boundary is undesirable, fuzzy logic offers an appealing alternative — it generalizes the binary distinction between possible vs. impossible to a matter of degree called the **possibility**. For instance, if we assign the fuzzy set *YOUNG*, whose membership function is shown in Fig 2.10, to the age of the suspect, we obtain a distribution about the possibility degree of the suspect's age (e.g., the possibility that the suspect is 19 years old is 0.7, while the possibility of 21 through 28 is 1), i.e.,

$$\Pi_{Age(Suspect)}(x) = \mu_{Young}(x) \qquad \textbf{(EQ 2.4)}$$

where Π denotes a possibility distribution of the suspect's age, and x is a variable representing a person's age. In general, when we assign a fuzzy set A to a variable X, the assignment results in a possibility distribution of X, which is defined by A's membership function:

$$\Pi_X(x) = \mu_A(x)$$

(EQ 2.5)

One of the questions commonly raised about possibility distribution is its relationship with probability distribution. While the subject is too complex for a comprehensive discussion here, it is important enough to descrve some clarification. The best way to understand the relationship between possibility distribution and probability distribution is to compare intervalvalues with probability. As we have mentioned earlier, an interval-valued assignment constrains the possible value of a variable without indicating the likelihood that the variable takes a specific value in the interval. Similarly, a possibility distribution states the degree of ease (i.e., possibility) for the variable to take a certain value without indicating the likelihood that the variable has such a value. Even though possibility distribution and probability distribution are different, they are also related -- if a value is impossible, it is obviously improbable. We will discuss the relationship between probability and possibility in depth in Chapter 7.

FIGURE 2.10   A Possibility Distribution of the Fuzzy Set *Young*

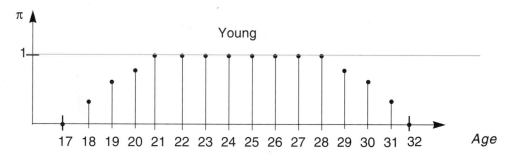

2.6 Fuzzy Rules

Among all the techniques developed using fuzzy sets, the fuzzy if-then rule (or, in short, the fuzzy rule) is by far the most visible one due to its wide range of successful applications. Fuzzy if-then rules have been applied to many disciplines such as control systems, decision making, pattern recognition, and system modeling. Fuzzy if-then rules also play a critical role in industrial applications ranging from consumer products, robotics, manufacturing, process control, medical imaging, to financial trading.

Fuzzy rule-based inference can be understood from several viewpoints. Conceptually, it can be understood using the metaphor of drawing a conclusion using a panel of experts. Mathematically, it can be viewed as an interpolation scheme. Formally, it is a generalization of a logic inference called *modus ponens*. We elaborate on each of these viewpoints below.

Conceptually, fuzzy rule-based inference can be understood using a multi-expert decision making metaphor. Let us consider an ancient kingdom in which only seven masters have knowledge about calculating the square root of a real number. Unfortunately, none of them knows a general procedure for calculating the square root of an arbitrary real number. Each one of them, nevertheless, is specialized in solving the problem for a fuzzy region of a real number. For instance, one is an expect for calculating the square root of numbers around 100. One day, the king is challenged to solve a square root square problem by his enemy. He immediately assembles the seven masters in his palace. Each master will provide a solution to the king, which is attached with a confidence measure between 0 and 1 (1 means most confident, 0 means no confidence at all). The king will combine all their solutions, each weighted by their confidence measure, into a final solution, which he will present to his enemy. Even though the king and the masters do not know about fuzzy logic, this process turns out to be very close to fuzzy rule-based inference. Each master is like a fuzzy rule; the confidence measure of a master's solution corresponds to the degree a fuzzy rule's condition is satisfied; and the way the masters' recommendations are combined by the king is similar to the way conclusions of multiple fuzzy rules are combined in fuzzy logic.

FIGURE 2.11   The King and the Seven Masters

Mathematically, fuzzy rule-based inference can be viewed as an interpolation scheme because it enables the fusion of multiple fuzzy rules when their conditions are all satisfied to a degree. The degree to which each rule is satisfied determines the weight of the rule's conclusion. Using these weights, fuzzy rule-based inference combines the conclusions of multiple fuzzy rules in a manner similar to linear interpolation.

From a logical viewpoint, fuzzy rule-based inference is a generalization of a logical reasoning scheme called *modus ponens*. While this important formal foundation of fuzzy inference will be fully explored in Chapter 5, we wish to introduce the basics of this connection to classical logic here. In classical logic, if we know a rule is true and we also know the antecedent of the rule is true, then it can be inferred that the consequent of the rule is true. This is referred to as *modus ponens*. For example, suppose we know that rule R1 below is true:

R1:IF the annual income of a person is greater than 120K

   THEN the person is rich

we also know that the following statement is true:

Jack's annual income is 121K

Based on *modus ponens*, classical logic can deduce that the following statement is also true:

Jack is rich.

One limitation of *modus ponens* is that it cannot deal with partial matching. To illustrate this, let us consider rule R1 and a different case:

Bob's annual income is $119,999.

People would usually say that Bob is somewhat rich. However *modus ponens* cannot infer whether Bob is rich or not using rule R1, because Bob's annual income does not satisfy the antecedent (i.e., the "IF" part) of R1, even though it is only one dollar short! The problem has two causes:

(1) The antecedent of R1 does not represent a smooth transition into the rich category that is often exhibited in human reasoning.

(2) *Modus ponens* cannot deal with a situation where the antecedent of a rule is partially satisfied.

Viewing such a limitation, fuzzy rule-based inference generalizes *modus ponens* to allow its inferred conclusion to be modified by the degree to which the antecedent is satisfied. This is the essence of fuzzy rule-based inference.

In this section, we will first introduce the structure of fuzzy rules. A description of the mechanism of fuzzy rule-based inference then follows. Finally, we suggest several guidelines regarding the design of membership functions used in fuzzy rules.

2.6.1 Structure of Fuzzy Rules

A fuzzy rule is the basic unit for capturing knowledge in many fuzzy systems. A fuzzy rule has two components: an if-part (also referred to as the **antecedent**) and a then-part (also referred to as the **consequent**):

IF < antecedent > THEN < consequent >

The antecedent describes a condition, and the consequent describes a conclusion that can be drawn when the condition holds. We introduce the basics of these two components below

2.6.1.1 The Antecedent of Fuzzy Rules

The structure of a fuzzy rule is identical to that of a conventional rule in artificial intelligence. The main difference lies in the content of the rule antecedent — the antecedent of a fuzzy rule describes an elastic condition (a condition that can be satisfied to a degree) while the antecedent of a conventional rule describes a rigid condition (a condition that is either satisfied or dissatisfied). For instance, consider the two rules below:

R1: IF the annual income of a person is greater than 120K,

THEN the person is rich.

R2: IF the annual income of a person is *High*,

THEN the person is *Rich*.

where *High* is a fuzzy set defined by the membership function shown in Fig 2.3.

The rule R1 is a conventional one, because its condition is rigid. In contrast, R2 is a fuzzy rule because its condition can be satisfied to a degree for those people whose income lies in the boundary of the fuzzy set *High* representing high annual income. Like conventional rules, the antecedent of a fuzzy rule may combine multiple simple conditions into a complex one using three logic connectives: AND (conjunction), OR (disjunction), and NOT (negation). For instance, a loan approval system may contain the following fuzzy rule:

> IF the annual income of a person is *High* AND
> > (the credit report of the person is *Fair* OR
> > the person has a *Valuable* real estate asset) AND
> > the amount of the loan requested is NOT *Jumbo*
>
> THEN recommend approving the loan

2.6.1.2   The Consequent of Fuzzy Rules

The consequent of fuzzy rules can be classified into three categories:

1. *Crisp Consequent*: IF... THEN $y = a$
where a is nonfuzzy numeric value or symbolic value.

2. *Fuzzy Consequent*: IF... THEN y is A
where A is a fuzzy set.

3. *Functional Consequent*:
IF x_1 is A_1 AND x_2 is A_2 AND ... x_n is A_n THEN $y = a_0 + \sum_{i=1}^{n} a_i \times x_i$

where a_0, a_1,..., a_n are constants.

Each type of rule consequent has its merit. Generally speaking, fuzzy rules with a crisp consequent can be processed more efficiently. A rule with a fuzzy consequent is easier to understand and more suitable for capturing imprecise human expertise. Finally, rules with a functional consequent can be used to approximate complex nonlinear models using only a small number of rules. We will elaborate on these points in Chapter 5.

As we shall see in the next section, the fuzzy rules for flow mixing control introduced in the previous section have crisp consequents. The rules for an automatic washing machine, on the other hand, use fuzzy consequents.

It is worthwhile to point out an important difference between the fuzzy sets used in a rule's antecedent and those in a a rule's consequent. The fuzzy sets in a rule's antecedent define a fuzzy region of the input space covered by the rule (i.e., the input situations that fit the rule's condition completely or partially), whereas the fuzzy sets in a rule's consequent describe the vagueness of the rule's conclusion. This difference has an important impact on the design of membership functions. Generally speaking, the membership functions of an input variable should cover the entire input space, whereas the membership functions of an output variable are not subject to such a constraint. We will discuss some general guidelines for designing antecedent membership functions (i.e., membership functions used in rules antecedents) in Section 2.6.4.

2.6.2   Rules for the Two Exemplary Problems

We will now turn to the two exemplary problems described in Section 2.2. We will use these two problems to illustrate fuzzy rule-based inference in the rest of the chapter.

One possible approach to problem A in Section 2.2 is to program a rudimentary controlled pump controller to enact the following fuzzy rules:

R3: IF the target temperature T is *Low*,
    THEN set the voltage to $\underline{V}$ (i.e., turn on the cold flow).

R4: IF the target temperature T is *High*,
    THEN set the voltage to $\overline{V}$ (i.e., turn on the hot flow).

where the membership functions of *High* and *Low* are defined in Fig 2.12.

FIGURE 2.12   Membership Functions of the Target Temperature.

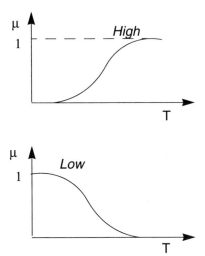

For the automatic washing machine problem, rules should combine conditions on laundry quantity and laundry softness using conjunction (AND). Two rules for selecting the wash cycle are listed below:

R5: IF Laundry Quantity is *Large* AND
Laundry Softness is *Hard*
THEN Washing Cycle is *Strong*

R6: IF Laundry Quantity is *Normal* AND
Laundry Softness is *Normal Hard*
THEN Washing Cycle is *Normal*

The membership functions for fuzzy conditions about laundry softness and laundry quantity are depicted in Figs 2.13 and 2.14, respectively. The membership functions for fuzzy decisions regarding the wash cycle are shown in Fig 2.15. A complete set of fuzzy rules for the selection of a wash cycle is shown in Table 2.2.

FIGURE 2.13   Membership Functions of Laundry Softness

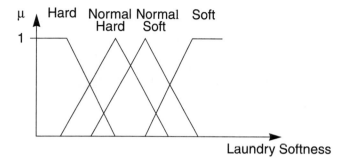

FIGURE 2.14   Membership Functions of Laundry Quantity

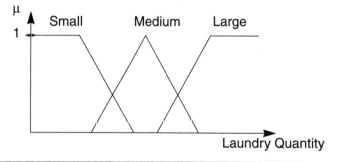

FIGURE 2.15   Membership Functions of Washing Cycles

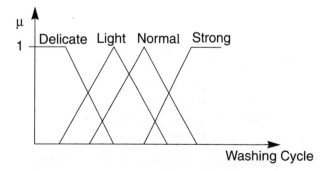

TABLE 2.2   Fuzzy Rules for Washing Cycle

Laundry Softness \ Laundry Quantity	Small	Medium	Large
Soft	Delicate	Light	Normal
Normal Soft	Light	Normal	Normal
Normal Hard	Light	Normal	Strong
Hard	Light	Normal	Strong

2.6.3 Fuzzy Rule-based Inference

The algorithm of fuzzy rule-based inference consists of three basic steps and an additional optional step.

1. **Fuzzy Matching**: Calculate the degree to which the input data match the condition of the fuzzy rules.
2. **Inference**: Calculate the rule's conclusion based on its matching degree.
3. **Combination**: Combine the conclusion inferred by all fuzzy rules into a final conclusion.
4. (Optional) **Defuzzification**: For applications that need a crisp output (e.g., in control systems), an additional step is used to convert a fuzzy conclusion into a crisp one.

We will illustrate the basics of these four steps using the flow mixing problem and the automatic washing machine problem.

2.6.3.1 Fuzzy Matching

Let us first consider fuzzy matching for the flow mixing control rules. The rudimentary flow mixing control is described in Section 2.6.1.1. The degree to which the input target temperature satisfies the condition of rule R3 "target temperature is Low" is the same as the degree to which the input target temperature T belongs to the fuzzy set Low. For the convenience of our discussion, we denote the degree of matching between input data d and rule r as Matching Degree(d, r). Thus, we summarize the discussion above as

$$MatchingDegree(T, R3) = \mu_{Low}(T)$$

where "=" represents assignment (not equality test). Similarly, we have

$$MatchingDegree(T, R4) = \mu_{High}(T)$$

Fig 2.16 illustrates this step for the case of input target temperature $T = 80°F$.

When a rule has multiple conditions combined using AND (conjunction), we simply use a fuzzy conjunction operator to combine the matching degree of each condition. One of the most commonly used fuzzy conjunction operators is the min operator. Another fuzzy conjunction operator used for this purpose is the product (i.e., multiplication) operator.

Fig 2.17 illustrates the use of the min operator for combining the degree of matching of conjunction conditions for washing cycle selection rules[2]. For the first rule in the figure the condition "Laundry Quantity is Heavy" matches the sensor input 0.5 degree; and the condition "Laundry Softness" matches the sensor input 0.2 degree. Combining the two matching degrees using min, we obtain 0.2 as the degree to which the input data match the antecedent of the following rule:

IF Laundry Quantity is *Heavy* AND
   Laundry Softness is *Hard*
   THEN....

Similarly, we can find that the degree the input data match the second rule in Fig. 2.17 is 0.5.

FIGURE 2.16   Fuzzy Matching for Simple Conditions

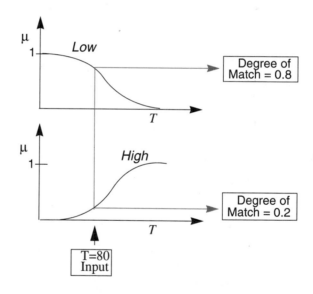

In general, the degree to which a rule in the form of

$$\text{IF } X_1 \text{ is } A_{i_1}^1 \text{ AND } X_2 \text{ is } A_{i_2}^2 \text{ AND } ... \ X_n \text{ is } A_{i_n}^n \qquad \text{THEN...}$$

matches the input data

$$X_1 = x_1^0, \quad X_2 = x_2^0, ... \ X_n = x_n^0$$

---

[2] Fig 2.17 shows only two of the four rules that will be invoked for the given input. We omit the other two rules to simplify the discussion.

is computed by the following formula:

$$\text{Matching Degree} = \min\left\{\mu_{A_{i_1}^1}(x_1^0),\ \mu_{A_{i_2}^2}(x_2^0),\ \ldots\ \mu_{A_{i_n}^n}(x_n^0)\right\} \qquad \textbf{(EQ 2.6)}$$

If the antecedent of a rule includes conditions connected by OR (disjunction), we use a fuzzy disjunction operator (e.g., max) to combine matching degrees accordingly.

2.6.3.2 Inference

After the fuzzy matching step, a fuzzy inference step is invoked for each of the relevant rules to produce a conclusion based on their matching degree. How should the conclusion be produced? There are two methods: (1) the clipping method and (2) the scaling method. Both methods generate an inferred conclusion by suppressing the *membership function* of the consequent. The extent to which they suppress the membership function depends on the degree to which the rule is matched. The lower the matching degree, the more severe the suppression of membership functions.

FIGURE 2.17   Fuzzy Matching for Conjunctive Conditions

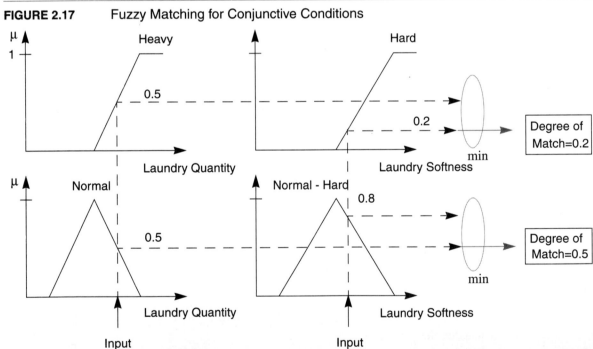

The clipping and scaling methods produce their inferred conclusion by suppressing the membership function of the consequent differently. The clipping method cuts off the top of the membership function whose value is higher than the matching degree. The scaling method scales down the membership function in proportion to the matching degree. These two methods are illustrated in Figs 2.18 and 2.19 using a hypothetical fuzzy consequent "*Y* is *A*." It is easy to see that both methods are consistent with *modus ponens* in classical logic:

- When the matching degree is 1 (i.e., the antecedent is completely satisfied), the inferred conclusion is identical to the rule's consequent.
- When the matching degree is 0 (i.e., the antecedent is not satisfied at all), no conclusion can be inferred from the rule.

FIGURE 2.18   The Clipping Method for Fuzzy Inference

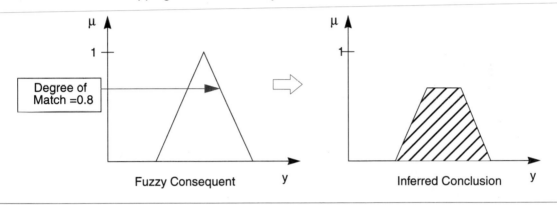

FIGURE 2.19   The Scaling Method for Fuzzy Inference

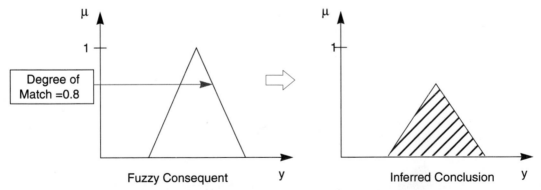

In the case of a crisp consequent, the two methods degenerate into an identical one. We will use the air flow mixing control problem to illustrate this. Recall that the rules for this problem are

$$\text{R3: If } T \text{ is Low THEN } V = \underline{V}.$$
$$\text{R4: If } T \text{ is High THEN } V = \overline{V}.$$

To apply fuzzy inference to these rules with crisp consequents, we need to first convert them into an equivalent fuzzy set representation. For instance, the crisp value $\underline{V}$ is equivalent to a membership function that assigns value 1 to $\underline{V}$, and 0 to all other values as shown in Fig 2.20(a). Similarly, we can construct the membership function of crisp value $\overline{V}$ as shown in Fig 2.20(b).

FIGURE 2.20   Membership Functions of Crisp Consequents

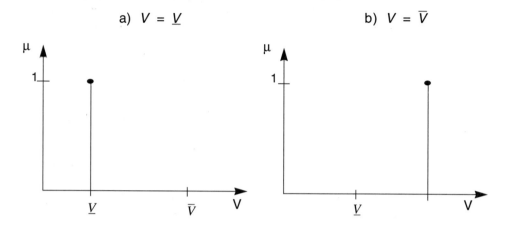

a) $V = \underline{V}$ b) $V = \overline{V}$

Suppose the input target temperature is T We have shown in the previous section that the matching degrees between T and the two rules are

$$MatchingDegree(T, R3) = \mu_{Low}(T)$$

$$MatchingDegree(T, R4) = \mu_{High}(T)$$

It is then straightforward to show that the conclusion $V = V'$ inferred by rule R3 and the conclusion $V = V''$ inferred by R4 have the following membership function for both the clipping method and the scaling method:

$$\mu_{v'}(V) = \begin{cases} \mu_{LOW}(T) & V = \underline{V} \\ 0 & V = \overline{V} \end{cases} \qquad \text{(EQ 2.7)}$$

$$\mu_{v''}(V) = \begin{cases} 0 & V = \underline{V} \\ \mu_{HIGH}(T) & V = \overline{V} \end{cases} \qquad \text{(EQ 2.8)}$$

Fig 2.21 illustrates this for R3 and a target temperature that is 0.8 degree in the fuzzy set Low. Hence, the conclusion inferred by R3 in this case is

$$\mu_{v'}(v) = \begin{cases} 0.8 & V = \underline{V} \\ 0 & V = \overline{V} \end{cases} \qquad \text{(EQ 2.9)}$$

FIGURE 2.21   A Fuzzy Inference for a Crisp Consequent

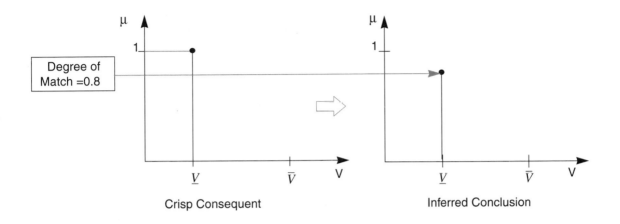

Crisp Consequent   Inferred Conclusion

2.6.3.3 Combining Fuzzy Conclusions

The two steps in fuzzy inference described so far enable each fuzzy rule to infer a fuzzy statement about the value of the consequent variable. Because a fuzzy rule-based system consists of a set of fuzzy rules with partially overlapping conditions, a particular input to the system often "triggers" multiple fuzzy rules (i.e., more than one rule will match the input to a nonzero degree). Therefore, a third step is needed to combine the inference results of these rules. This is accomplished typically by superimposing all fuzzy conclusions about a variable. This is illustrated in Fig 2.22 for combining two fuzzy conclusions obtained using the scaling method, and in Fig 2.23 for combining fuzzy conclusions inferred using the clipping method.

Combining fuzzy conclusions through superimposition is based on applying the max fuzzy disjunction operator to multiple possibility distributions of the output variable. We will discuss the theoretical foundation of the three fuzzy inference steps in Chapter 5.

FIGURE 2.22   Combining Fuzzy Conclusions Inferred by the Scaling Method

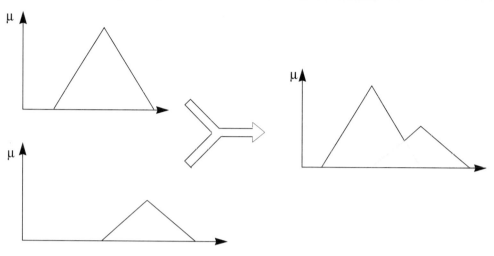

2.6.3.4   Defuzzification

For a fuzzy system whose final output needs to be in a crisp (nonfuzzy) form, a fourth step is needed to convert the final combined fuzzy conclusion into a crisp one. This step is called the *defuzzification*. A fuzzy rule-based controller, for instance, uses such a step to generate a crisp control command.

There are two major defuzzification techniques: (1) the Mean of Maximum (MOM) method and (2) the Center of Area (COA) or the centroid method. The mean of maximum defuzzification calculates the average of all variable values with maximum membership degrees. Fig 2.24 shows the result of applying MOM defuzzification to a combined fuzzy conclusion.We will introduce other defuzzification techniques in Chapter 5.

FIGURE 2.23   Combining Fuzzy Conclusions Inferred by the Clipping Method

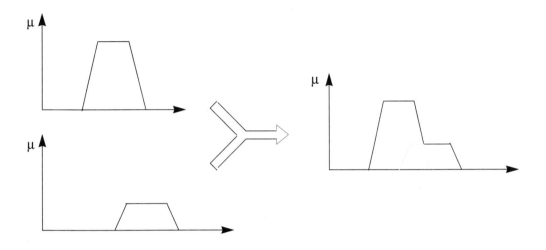

FIGURE 2.24   An Example of MOM Defuzzification

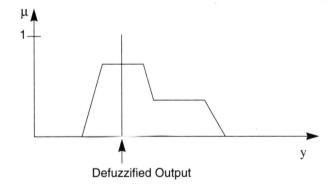

Defuzzified Output

Suppose "y is A" is a fuzzy conclusion to be defuzzified. We can express the MOM defuzzification method using the following formula:

$$MOM(A) = \frac{\sum\limits_{y^* \in P} y^*}{|P|} \qquad \text{(EQ 2.10)}$$

where $P(A)$ is the set of output values y with the highest membership degree in A, i.e.,

$$P = \left\{ y^* \mid \mu_A(y^*) = \sup_y \mu_A(y) \right\} \qquad \text{(EQ 2.11)}$$

where *sup* is an operator that returns the maximum value of a continuous function (i.e., the continous version of the *max* operator).

FIGURE 2.25   An Example of COA Defuzzification

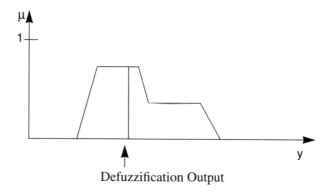

Defuzzification Output

The centroid (or COA) defuzzification method calculates the weighted average of a fuzzy set. The result of applying COA defuzzification to a fuzzy conclusion "*Y* is *A*" can be expressed by the formula

$$y = \frac{\sum_i \mu_A(y_i) \times y_i}{\sum_i \mu_A(y_i)}$$ **(EQ 2.12)**

if *y* is discrete, and by the formula

$$y = \frac{\int \mu_A(y_i) \times y_i \ dy}{\int \mu_A(y_i) \ dy}$$ **(EQ 2.13)**

if *y* is continous.

2.6.3.5   Putting All Four Steps Together

Figs 2.26 and 2.27 show the result of combining all four steps for the flow mixing controller and the automatic washing machine, respectively. The clipping inference method and the centroid defuzzification technique are used in these two examples.

FIGURE 2.26   Fuzzy Rule-based Inference for the Rudimentary Flow Mixing Controller

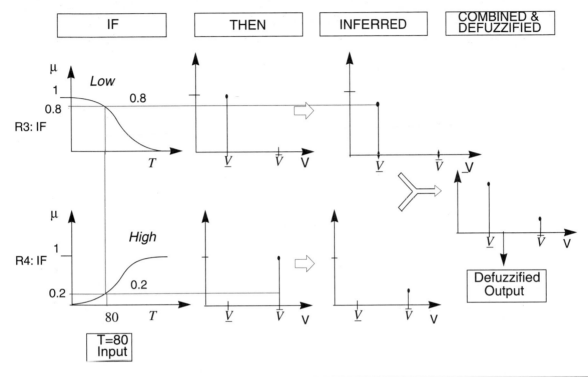

FIGURE 2.27   Fuzzy Inference of Fuzzy Logic Washing Machine

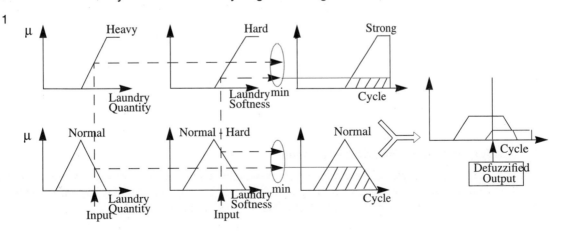

The defuzzified output for the flow mixing controller can be expressed as a function of the input target temperature using the formula below:

$$V = \frac{\mu_{LOW}(T) \times \underline{V} + \mu_{HIGH}(T) \times \overline{V}}{\mu_{LOW}(T) + \mu_{HIGH}(T)}$$

(EQ 2.14)

This formula clearly shows that fuzzy rule-based inference indeed performs a kind of interpolation, as we mentioned earlier in the section.

2.6.4 Designing Antecedent Membership Functions

Since the membership functions of an antecedent variable determine the coverage of rules, we recommend that you adopt the following design principles used by many fuzzy system developers.

 The membership functions of an input variable's fuzzy sets should usually be designed in a way such that the following two conditions are satisfied: (1) Each membership function overlaps only with the closest neighboring membership functions; (2) for any possible input data, its membership values in all relevant fuzzy sets should sum to 1 (or nearly so).

Let us use A_i to denote fuzzy sets of an input variable x. The two guidelines above may be expressed formally as the two equations below:

$$1. \quad A_i \cap A_j = \phi \quad \forall \ j \neq i, i+1, i-1$$

(EQ 2.15)

$$2. \quad \sum_i \mu_{A_i}(x) \cong 1$$

(EQ 2.16)

Two examples of membership functions that do not follow these design principles are shown in Figs 2.28 and 2.29. Fig 2.28 obviously violates the second principle because the membership values of 10 in three fuzzy sets do not sum to 1, i.e.,

$$\mu_{A_1}(10) + \mu_{A_2}(10) + \mu_{A_3}(10) = 0.5 < 1$$

FIGURE 2.28   A Membership Function Design That Violates the Second Principle

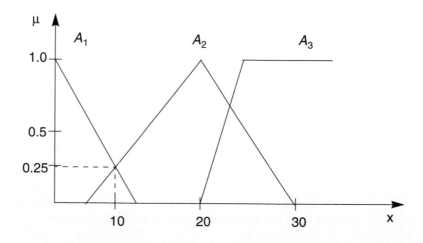

FIGURE 2.29 A Membership Function Design That Violates Both Principles

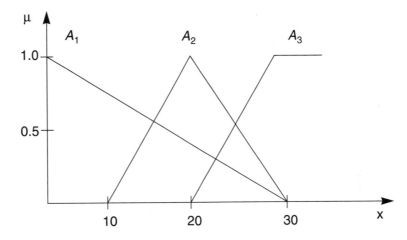

Fig 2.29 violates both design principles because $A_1 \cap A_3 \neq \phi$. Two examples of membership functions that follow these guidelines are shown in Figs 2.30 and 2.31. The former uses five symmetric membership functions, whereas the latter uses five asymmetric membership functions. Most industrial applications use symmetric functions similar to those in Fig 2.30. Applications that may benefit from the use of asymmetric membership functions will be introduced later in Chapter 6.

FIGURE 2.30 A Symmetric Membership Function Design Following the Guidelines

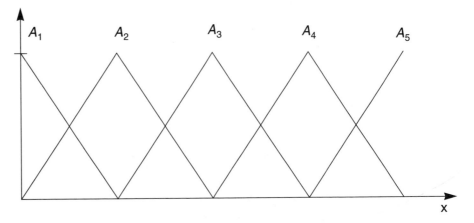

FIGURE 2.31   An Asymmetric Membership Function Design Following the Guidelines

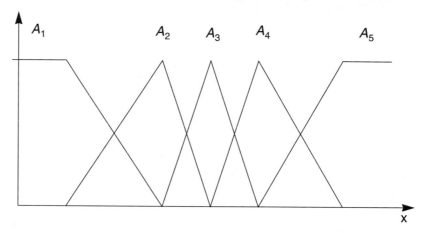

Based on the discussion above, we recommend the following design guideline:

Design Guideline:
Unless there is a good reason, use symmetric membership functions.

It has been shown that a fuzzy system that follows this design guideline has an additional benefit from the viewpoint of stability analysis.

2.6.5 Designing Fuzzy Rule-based Systems Using the Fuzzy Logic Toolbox

The Fuzzy Logic Toolbox for MATLAB provides several built-in membership functions (including triangular, trapezoid, Gaussian, and other membership functions that we will introduce in the next chapter). See the *Fuzzy Logic Toolbox User's Guide* for a detailed description. The washing machine cycle selection system has been implemented using MATLAB. Fig 2.32 shows the control surface of the system displayed in the Surface Viewer of the Fuzzy Logic Toolbox. These figures show that, like most other fuzzy systems, the fuzzy system captures complex nonlinear input-output relationships using a relatively simple rule set. The fuzzy system is implemented using the following FIS (Fuzzy Inference System) properties:

> And method: min
> Or method: max
> Implication: min
> Aggregation: max
> Defuzzification: centroid

FIGURE 2.32   The Control Surface of the Washing Machine Cycle Selection System Using Clipping Inference Method (MATLAB)

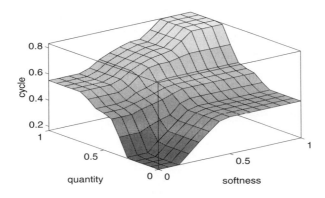

The inference process of the system is shown in the **Rule View** window of the **Fuzzy Logic Toolbox**. Fig 2.33 shows the **Rule View** window for an exemplary input (i.e., quantity = 0.6, softness = 0.6). As the figure shows clearly, the fuzzy system employs the clipping inference method. In fact, the clipping inference method corresponds to choosing "min" for the **Implication** operation in the **FIS Editor**.

We have also implemented an alternative fuzzy system for selecting the washing cycle. The second system is identical to the first one except that the following FIS properties were chosen:

> And method: prod
> Or method: max
> Implication: prod
> Aggregation: sum
> Defuzzification: centroid

Fig 2.34 shows the control surface of this fuzzy system, and Fig 2.35 shows the rule view of the system for an input identical to that of Fig 2.33. The system employs the scaling inference method, because it uses "product" for the implication operation.

Design Guideline:
We recommend that the clipping inference method (i.e., the min implication) be used together with the max aggregation operator and the min And method, while the scaling inference method (i.e., the product implication) is used together with the sum aggregation operator, and the product And method. The rationale underlying these choices will be discussed in more detail in Chapter 5. However, it suffices to mention at this point that these combinations of fuzzy inference choices have well-defined formal foundations.

FIGURE 2.33   The Rule View of the Washing Machine Cycle Selection System (MATLAB)

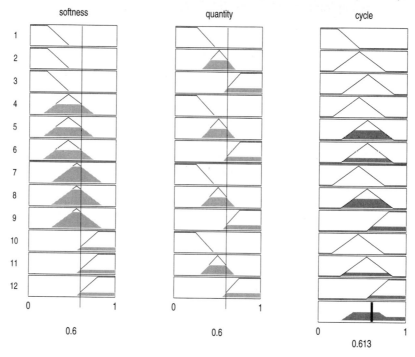

FIGURE 2.34   The Control Surface View of the Washing Machine Cycle Selection System Using
Scaling Inference Method (MATLAB)

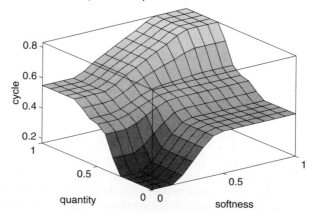

We would also like to point out the default membership functions generated by MATLAB for a variable (by specifying the type of the functions and the number of fuzzy sets) follow all three guidelines discussed in the previous section.

FIGURE 2.35   The Rule View of the Washing Machine Cycle Selection System Using Scaling Inference Method (MATLAB)

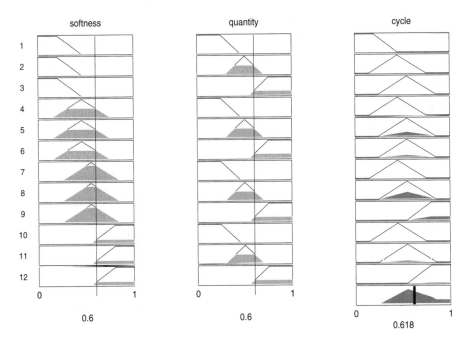

2.7 Summary

We introduced four core concepts in fuzzy logic: fuzzy sets, linguistic variables, possibility distribution, and fuzzy if-then rules. We summarize major points regarding these concepts below.

- A *fuzzy set* is a set with a smooth boundary such that membership in the set becomes a matter of degree.
- A fuzzy set has a dual representation: a qualitative description using a *linguistic term* and a quantitative description through a *membership function*, which maps elements in a universe of discourse (i.e., a domain of interest) to their membership degree in the set.
- A *linguistic variable* is a variable whose values are an expression involving fuzzy sets.

- When a fuzzy set is assigned to a variable whose precise value is unknown, the fuzzy set serves as a constraint on the degree of ease for the variable to take a certain value. This degree of ease is called the *possibility degree.*

- A *possibility distribution* of a variable is a function that maps elements in the variable's universe of discourse to their possibility degree.

- A *fuzzy if-then rule* is a scheme for representing knowledge and association that is inexact and imprecise in nature.

- The if-part of a fuzzy rule is called the rule's *antecedent,* and the then-part of a rule is called its *consequent.*

- Reasoning using fuzzy if-then rules has three major features. First, it enables a rule that partially matches the input data to make an inference. Second, it typically infers the possibility distribution of an output variable from the possibility distribution of an input variable. Third, the system combines the inferred conclusions (e.g., possibility distributions of an output variable) from all rules to form an overall conclusion.

- Many fuzzy rule-based systems that need to produce a precise output (e.g., fuzzy logic control systems) use a *defuzzification* process to convert the inferred possibility distribution of an output variable to a representative precise value.

- Most applications of fuzzy logic (including fuzzy logic control) use fuzzy if-then rules.

- We have also introduced some design guidelines regarding the design of membership functions.

    We will further discuss fuzzy sets, linguistic variables, and possibility distribution in the next chapter. The relationship between possibility distribution and probability distribution, as well as other topics related to fuzzy logic and probability, will be discussed in Chapter 7. Fuzzy if-then rules will be discussed in the two following chapters. Chapter 5 introduces two types of fuzzy rules and emphasizes the type most often used in fuzzy logic control (called fuzzy mapping rules). Chapter 6 discusses the other type of fuzzy rules (called fuzzy implication rules) that are closely related to implications in two-valued logic.

Exercises

2.1   Find two examples of fuzzy sets in newspaper articles or magazines. Cut and paste the paragraphs and highlight the words that mean fuzzy sets. Identify linguistic variables in the paragraphs.

2.2   Find two examples of overly precise description of an intrinsically imprecise concept from newspapers or magazines.

2.3   Use the Fuzzy Logic Toolbox for MATLAB to implement a fuzzy system for controlling the cycle of a washing machine. You don't need to implement a simulator for this assignment. Print the rule viewer and control surface of the system for the following two fuzzy inference units:

   a. A Mamdani model using min for And Method and Implication, max for Or Method and Aggregation, and centroid Defuzzification.

b. A Mamdani model using product for And Method and Implication, max for Or Method, sum for Aggregation, and centroid Defuzzification. (This is equivalent to Kosko's Fuzzy Associative Memory.)

Compare the results of the two systems.

2.4   Define the following linguistic terms as fuzzy subsets of an appropriate universe of discourse. State your answer in the following format. Implement them using a tool such as MATLAB. Adjust some of the parameters and describe how different changes affect the membership function differently.

Universe of discourse: Set of real numbers, $\Re$.

Fuzzy Subset: Numbers close to 5, $\mu(x) \equiv \dfrac{1}{1 + (x - 5)^2}, x \in \Re.$

Interpretation: Numbers as large as 1000, or as small as -1000 are in the *support set*; in other words, have a nonzero membership value in the given fuzzy subset, rendering a nonrealistic definition for the given term.
Alternative Definition:?

Linguistic terms: Healthy Family, Good Graduate Courses.

3   FUZZY SETS

We have introduced the basic concept of fuzzy sets in the previous chapter. This chapter provides a more thorough discussion on fuzzy sets.

3.1   Classical Sets

A classical set is a collection of objects in a given domain. An object either belongs to the set or does not belong to the set. Therefore, there is a sharp boundary between members of the set and those not in the set. As we mentioned in Section 2.2, the difficulty in expressing sets with smooth boundaries motivated the development of fuzzy sets.

A set may be defined in two ways: (1) by enumerating its elements, or (2) by describing the common properties of its elements. The former is called an *extensional* definition and the latter, an *intensional* definition. An intensional definition is usually preferred to an extensional definition for three major reasons. First, an intensional definition of a set is usually more concise than the set's extensional definition. In fact, enumerating all elements of a set is sometimes impossible (e. g., the set of real numbers between 1 and 10 has an infinite number of elements). Second, an intensional definition explicitly expresses the meaning of a set. Third, an intensional definition can be used to recognize new elements of a set when the properties of the element change or when the defining properties change.

Suppose we wish to represent the set of universities in Texas whose total student population is greater than 40,000. Let the universe of discourse U be the set of universities in the United States. The extensional definition would list two such universities: University of Texas at Austin (UT), and Texas A&M University (TAMU).

$$A = \{ UT, TAMU \}$$

An intensional definition of the set is given below for comparison.

A = {u | total-student-population(u) > 40,000 AND state(u) = Texas}

Membership of an element in a set is denoted by the symbol $\in$. For instance, the fact that Texas A&M University is a member of set A can be expressed as

$TAMU \in A$

In contrast, the nonmembership relation is denoted by $\notin$. For instance, $Rice \notin A$ because the student population of Rice University is less than 40,000. However, the result is a set of precise but artificially constrained definitions which may be appealing for this very reason— at least to those who favor arbitrary precision. Moreover, this apparent flaw may not appear problematic until one has to make a decision that relies on such definitions.

A set S_1 is a subset of set S_2, denoted as $S_1 \subset S_2$, if every element of S_1 is also an element of S_2. For instance, let us define a set B to be the set of public universities in Texas. Since both UT and TAMU are public universities in Texas, set A is a subset of B, i.e., $A \subset B$. The inverse of the subset relation is a superset, also referred to as inclusion. The set B defined earlier includes (is a superset of) the set A, denoted as $B \supset A$. A set without any elements is an empty set, denoted by the symbol ϕ. The empty set is a subset of any set.

The cardinality of a set, denoted as $|S|$, is the number of elements in the set. For instance, the set of Texas universities having at least 40,000 students has exactly two elements. Therefore, its cardinality is 2, i.e., $|A| = 2.$.

3.1.1   Set Operations

We have introduced the three basic set operations in the previous chapter. We will thus summarize them below:

$$A \cap B = \{x | x \in A \text{ and } x \in B\} \qquad \textbf{(EQ 3.1)}$$

$$A \cup B = \{x | x \in A \text{ or } x \in B\} \qquad \textbf{(EQ 3.2)}$$

$$A^C = \{x | x \in U \text{ and } x \notin A\} \qquad \textbf{(EQ 3.3)}$$

where U is a universe of discourse and A and B are sets in U.

The set complement is in fact a special case of an operation called *difference*, denoted by \, which is defined as

$$A \backslash B = \{x | x \in A \text{ and } x \notin B\} \qquad \textbf{(EQ 3.4)}$$

Therefore the complement of A is in fact the difference between the universe U and set A, i.e.,

$$A^c = U \backslash A \qquad \textbf{(EQ 3.5)}$$

Intersection, union, and complement operations have several fundamental properties. They are summarized below.

Commutative Laws	$A \cup B = B \cup A$
	$A \cap B = B \cap A$
Associative Laws	$A \cap (B \cap C) = (A \cap B) \cap C$
	$A \cup (B \cup C) = (A \cup B) \cup C$
Distributive Laws	$A \cup (B \cap C) = (A \cup B) \cap (A \cup C)$
	$A \cap (B \cup C) = (A \cap B) \cup (A \cap C)$
Law of Double Complementation	$\overline{(\overline{A})} = A$
DeMorgan's Laws	$\overline{(A \cup B)} = \overline{A} \cap \overline{B}$
	$\overline{(A \cap B)} = \overline{A} \cup \overline{B}$
Law of the Excluded Middle	$A \cup \overline{A} = U$
Law of Contradiction	$A \cap \overline{A} = \phi$
Laws of Tautology	$A \cup A = A$
	$A \cap A = A$
Laws of Absorption	$A \cap (A \cup B) = A$
	$A \cup (A \cap B) = A$

We have informally discussed the relationship between set operations and logic operations in the previous chapter. This relationship is not coincidental. If A, B, and C represent the truth values of a statement, $\cap$, $\cup$, and $\overline{}$ refer to AND, OR, and NOT respectively, ϕ refers to FALSE, and U refers to TRUE, all the properties below become the properties of a branch of logic called Boolean logic. In fact, both set theory and Boolean logic belong to a branch of mathematics known as Boolean algebra, which is characterized by the fundamental laws above (i.e., axioms).

An interesting and useful property of Boolean algebra is the *principle of duality*. For any law listed above, if we replace each union with intersection, each intersection with union, ϕ with U, and U with ϕ, the resulting equation is another valid law in Boolean algebra. For instance, the law of excluded middle and the law of contradiction are dual laws and vice versa.

3.2 Fuzzy Sets

In the evening of July 1964, the idea of fuzzy sets occurred to Professor Lotfi A. Zadeh while he was visiting his parents in New York City. He later published the seminal paper on fuzzy sets in 1965. Even though he encountered many criticisms, a group of researchers around the world have followed in his footsteps and developed a whole new discipline based on this revolutionary idea.

A *fuzzy set* is a set with a smooth (unsharp) boundary. It generalizes the notion of membership from a black-and-white binary categorization in classical set theory into one that allows partial membership. The notion of membership in fuzzy sets thus becomes a matter of degree, which is a number between 0 and 1. A membership degree of 0 represents complete nonmembership, while a membership degree of 1 represents a complete membership. Mathematically speaking, a *fuzzy set* is characterized by a mapping from its universe of discourse into the interval, $[0, 1]$. This mapping is the *membership function* of the set. Usually, the membership or characteristic function is denoted by the Greek lowercase letter μ. (For instance, we may define the fuzzy set High Temperature using the membership function shown in Figure 3.1.)

FIGURE 3.1   Membership Function of High Temperature.

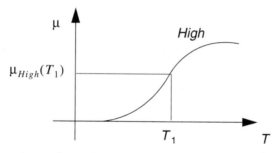

A set of core fuzzy sets about a variable forms a *term set*, which is the basis for generating composite fuzzy sets using hedges. For instance, Low, Medium, and High, whose membership functions are shown in Figure 3.2, form a complete term set for the temperature T of a process. We say that a term set is *complete* if, for any element in the given universe of discourse, there is at least one term in the term set that covers the given element.

3.2.1   Representation of Fuzzy Sets

A fuzzy set can be defined in two ways: (1) by enumerating membership values of those elements in the set (completely or partially), or (2) by defining the membership function mathematically.

Obviously, the first approach is possible only if the set is discrete, because a continuous fuzzy set has an infinite number of elements. Generally speaking, a fuzzy set A can be defined through enumeration using the expression

$$A = \sum_i \mu_A(x_i)/x_i \qquad \text{(EQ 3.6)}$$

FIGURE 3.2 A Complete Term Set for Temperature.

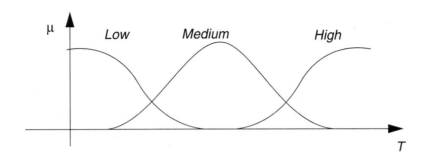

where the summation and addition operators refer to the union (disjunction) operation and the notation $\mu_A(x_i)/x_i$ refers to a fuzzy set containing exactly one (partial) element x with a membership degree $\mu_A(x_i)$. For brevity, we do not list those elements x_i whose membership degree in set A is zero.

For instance, let us consider a subset of natural numbers, say from 1 to 20, as the universe of discourse, U. We may define the terms *Small* and *Medium* by enumeration as follows.

EXAMPLE 1 Given the discrete universe of discourse, $U = \{1, 2, ..., 20\}$, *Small* and *Medium* are fuzzy subsets of U characterized by the following membership functions:

$Small \equiv 1/1 + 1/2 + 0.9/3 + 0.6/4 + 0.3/5 + (0.1/6)$,

$Medium \equiv 0.1/2 + 0.3/3 + 0.7/4 + 1/5 + 1/6 + 0.7/7 + 0.5/8 + 0.2/9$

The two fuzzy sets are also depicted graphically in Figure 3.3.

Let us return to the example about Texas universities. Suppose we wish to represent the set of Texas universities with a large student population. This can be defined using the two approaches described above.

(1)By enumeration: We can define the set by listing those Texas universities considered to have, to an extent, large student populations as well as their membership degrees.

$\quad LargeTexasUniv = 1/UT + 1/TAMU + 0.5/UH + 0.3/Rice$

(2)Alternatively, we can define the set by explicitly describing its characteristics: Texas Universities with *Large* total student population:

$$\mu_{LargeTexasUniv}(x) = \mu_{Large}(TotalStudentPopulation(x))$$

$$= \begin{cases} 0 & P < 20000 \\ \dfrac{(P - 20000)}{20000} & 20000 \leq P \leq 40000 \\ 1 & P > 40000 \end{cases}$$

FIGURE 3.3   Definitions of the Terms *Small* and *Medium*.

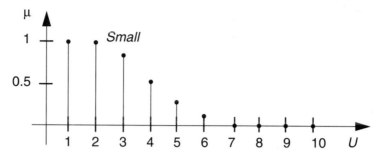

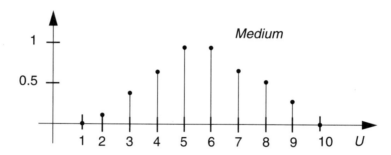

In the example above, the first approach is an extensional definition of the set, while the second approach is an intensional definition of the set.

3.2.2 Types of Membership Functions

Whereas there exist numerous types of membership functions, the most commonly used in practice are triangles, trapezoids, bell curves, Gaussian, and signoidal functions. In the following we will introduce these four types of membership functions.

Triangular Membership Function

A triangular membership function is specified by three parameters $\{a, b, c\}$ as follows:

$$triangle(x: a, b, c) = \begin{cases} 0 & x < a \\ (x-a)/(b-a) & a \le x \le b \\ (c-x)/(c-b) & b \le x \le c \\ 0 & x > c \end{cases} \qquad \textbf{(EQ 3.7)}$$

The precise appearance of the function is determined by the choice of parameters a, b, and c. Fig. 3.4(a) illustrates an example of a triangular membership function defined by *triangle*$(x; 20, 60, 80)$.

FIGURE 3.4   Examples of Four types of Membership Functions: (a) *Triangle(x;* 20, 60, 80); (b) *Trapezoid(x;* 10, 20, 60, 95); (c) *Bell(x;* 20, 4, 50); (d) *Gaussian(x;* 50, 20)

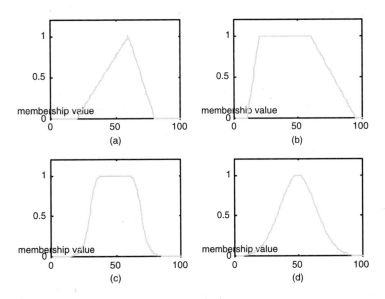

Trapezoidal Membership Function

A trapezoidal membership function is specified by four parameters $\{a, b, c, d\}$ as follows:

$$trapezoid(x: a, b, c, d) = \begin{cases} 0 & x < a \\ (x-a)/(b-a) & a \le x < b \\ 1 & b \le x < c \\ (d-x)/(d-c) & c \le x < d \\ 0 & x \ge d \end{cases} \quad \textbf{(EQ 3.8)}$$

Fig. 3.4(b) illustrates an example of a trapezoidal membership function defined by *trapezoid(x;* 10, 20, 60, 95). Obviously, the triangular membership function is a special case of the trapezoidal membership function.

Due to their simple formulas and computational efficiency, both triangular and trapezoidal membership functions have proven popular with fuzzy logic practitioners and been used extensively, particularly in control.

Gaussian Membership Functions

A Gaussian membership function is specified by two parameters $\{m, \sigma\}$ as follows:

$$gaussian(x: m, \sigma) = \exp\left(-\frac{(x-m)^2}{\sigma^2}\right) \quad \textbf{(EQ 3.9)}$$

where m and σ denote the center and width of the function, respectively. We can control the shape of the function by adjusting the parameter σ. A small σ will generate a "thin" membership function, while a big σ will lead to a "flat" membership function. This feature is used in Song et al. (1993) to fuse and delete rules from a given rule base. Fig. 3.4(d) illustrates an example of a Gaussian membership function defined by $gaussian(x;\ 50, 20)$.

Bell-shaped Membership Function

A bell-shaped membership function is specified by three parameters $\{a, b, c\}$ as follows:

$$bell(x: a, b, c) = \frac{1}{1 + \left|\dfrac{x - c}{a}\right|^{2b}} \qquad \textbf{(EQ 3.10)}$$

where the parameter b is usually positive. Note that this membership function is a direct generalization of the Cauchy distribution used in probability theory. A desired bell-shaped membership function can be obtained by a proper selection of the parameters a, b, and c. Specifically, we can adjust c and a to vary the center and width of the function and then use b to control the slopes at the crossover points. Fig.3.4(c) illustrates an example of a bell-shaped membership function defined by $bell(x;\ 20, 4, 50)$.

Sigmoidal Membership Functions

Terms on the boundaries of the given domain of definition[1] have a profile that is most readily described in terms of a *sigmoidal* function as shown in the next example.

FIGURE 3.5   Sigmoidal Membership Function.

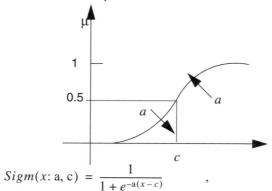

$$Sigm(x: a, c) = \frac{1}{1 + e^{-a(x - c)}} \qquad ,$$

As parameter increases, the transition from 0 to 1 becomes sharper. The function thus approximates the step function as approaches positive infinity. Finally, note that $\mu(c) = 0.5$ for all sigmoidal membership functions, indicating the break-even point of the definition where transition from membership to nonmembership occurs.

---

[1] We use the terms *domain of definition* and *universe of discourse* interchangeably.

All the membership functions introduced so far are available in the **Membership Function Editor** of the **Fuzzy Logic Toolbox for MATLAB.** We list below the names of these built-in membership functions. When the parameters in these functions differ from those in our formula, we explain the differences.

> trimf: Triangular Membership Function
>
> trapmf: Trapezoid Membership Function
>
> gbellmf: Bell-shaped Membership Function
>
> gaussmf(x, [sig c]): Gaussian Membership Function ($2 \times \text{sig}^2 = \sigma^2$ and $c = m$ where c and m are parameters in the formula of our Gaussian membership function.
>
> gauss2mf: Two Gaussian Membership Functions, one for each side
>
> sigmf: Sigmoidal Membership Function
>
> dsigmf: Difference of Two Sigmoidal Membership Functions
>
> psigmf: Product of Two Sigmoidal Membership Functions

We introduce below several additional types of membership functions.

S Membership Functions

The S membership function is a smooth membership function with two parameters: a and b. The shape of the function is shown in Figure 3.6. The membership value is 0 for points below a, 1 for points above b, and 0.5 for the midpoint between a and b. The name of this type of membership function, as you may have guessed, comes from the "S" shape of the function.

$$S(x: a, b) = \begin{cases} 0 & x < a \\ 2\left(\dfrac{x-a}{b-a}\right)^2 & a \leq x \leq \dfrac{a+b}{2} \\ 1 - 2\left(\dfrac{x-b}{b-a}\right)^2 & \dfrac{a+b}{2} \leq x < b \\ 1 & x \geq b \end{cases}$$

(EQ 3.11)

FIGURE 3.6   S Membership Function

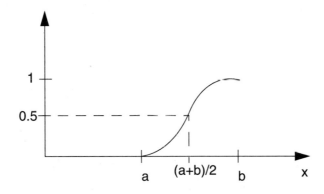

Π Membership Function

There are two Π membership functions. The first Π membership function is defined with two parameters: a and b. The function has a membership value 1 at point a, membership value 0.5 at $a - b$ and $a + b$, respectively. Unlike the S function, the Π function decreases toward zero asymptotically as we move away from point a, as shown in Figure 3.7

$$\Pi_1(x: a, b) = \frac{1}{1 + \left(\dfrac{x - a}{b}\right)^2}$$ **(EQ 3.12)**

FIGURE 3.7 Π_1 Membership Function

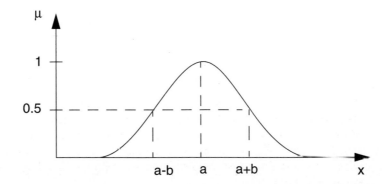

The other membership function that has the "Π" shape is given in Eq. 3.13 and shown in Figure 3.8. This membership function has four parameters.

$$\Pi_2(x: lw, lp, rp, rw) = \begin{cases} \dfrac{lw}{lp + lw - x} & x < lp \\ 1 & lp \le x \le rp \\ \dfrac{rw}{x - rp + rw} & x > rp \end{cases} \qquad \textbf{(EQ 3.13)}$$

FIGURE 3.8 Π_2 Membership Function

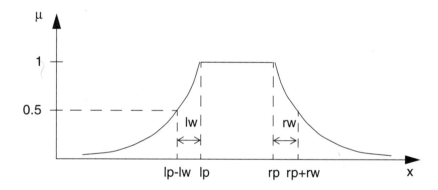

This membership function was used to define the fuzzy set "Good Stopping Accuracy" in the Sendai Subway Train Control System:

$$\mu_{GoodStoppingAccuracy}(x) = \Pi_2(x;20, 20, 30, 30)$$

where x is the distance (in cm) between actual stopping position and target stopping position. Based on this definition, a train stopped within 20 cm of the target position is considered 100 percent good stopping accuracy, while a train stopped at 50 cm away is considered to satisfy good stopping accuracy up to 50 percent.

3.2.3 Failing of the Law of Excluded Middle and the Law of Contradiction

Because fuzzy set theory generalizes the black-and-white situations of set membership to allow various "grey areas," it has to violate two fundamental laws of set theory — the law of excluded middle, $A \cup \bar{A} = U$, and the law of contradiction, $A \cap \bar{A} = \phi$. In other words, it is possible for an element to partially belong to both a fuzzy set and the set's complement. For instance, suppose John is a person who belongs to the set of bald people to degree 0.2, i.e.,

$$\mu_{Bald}(John) = 0.2$$

Based on the definition of the complement operator, we know that John belongs to the complement of bald people to degree 0.8, i.e.,

$$\mu_{\overline{Bald}}(John) = 0.8$$

Therefore, John partially belongs to the set of bald people as well as the set of people who are not bald.

Generally speaking, any element partially belonging to a fuzzy set is also a partial member of its complement. Knowing this difference between classical sets and fuzzy sets is important because it helps us to formally compare and analyze the two theories. Due to the fact that the law of excluded middle and the law of contradiction are not axioms of fuzzy set theory, formula equivalents in classical set theory are not necessarily equivalent in fuzzy set theory. A potential danger of ignoring such a difference is to reject fuzzy sets (and fuzzy logic in general) based on an inappropriate set of axioms.

A paper presented in the Eleventh National Conference on Artificial Intelligence (AAAI 93) by Charles Elkan entitled — *The Paradoxical Success of Fuzzy Logic* — is an example of such a danger [184]. Assuming that logically equivalent formulas are also equivalent in fuzzy logic, Elkan claimed that the only possible truth values in fuzzy logic are 0 and 1. The fundamental flaw of such a claim is that it adopts all the axioms of two-valued logic (including the law of excluded middle and the law of contradiction) as the axioms of his "fuzzy logic," which are different from the axioms of real fuzzy logic. A collection of responses to Elkan's paper written by a set of prominent researchers were published in a special issue of *IEEE Expert* entitled "A Fuzzy Logic Symposium" in 1994 [37].

3.2.4 Hedges

A *hedge* is a modifier to a fuzzy set. It modifies the meaning of the original set to create a *compound* fuzzy set. "Very" and "More or Less" are two commonly used hedges. Their definitions are listed below.

Very:

$$\mu_{Very\ A}(x) = [\mu_A(x)]^2 \qquad\qquad \textbf{(EQ 3.14)}$$

More or Less:

$$\mu_{MoreOrLess\ A}(x) = \sqrt{\mu_A(x)} \qquad\qquad \textbf{(EQ 3.15)}$$

We illustrate these hedges in Figure 3.9 through their application to the concept High temperature. As shown by the figure, "Very" has the effect of narrowing the membership function, while "More or Less" widens the membership function. This is intuitively appealing because the criteria for "Very High" should be more stringent than those for "High", while the criteria for "More or Less high" should be relaxed. In general, we should have

$$\mu_{Very\ A}(x) \le \mu_A(x) \le \mu_{MoreOrLess\ A}(x)$$

In principle, a hedge can be applied to any fuzzy sets. In practice, however, it is used only when the compound term is meaningful. For instance, "very medium temperature" does not seem to make much sense. Consequently, the hedge "very" is primarily used with fuzzy sets on the two ends of a universe (e.g., low vs. high, small vs. large, short vs. long).

FIGURE 3.9   Definition of *Very High*

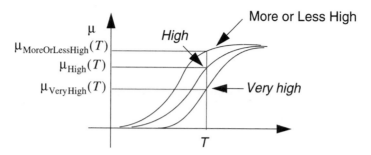

3.3 Operations of Fuzzy Sets

As we have seen in Section 3.1, three major operations on sets are union, intersection, and complement. Since the notion of set membership in classical set theory has been generalized into a matter of degree, these operations should be extended accordingly.

3.3.1 Intersection and Union of Fuzzy Sets

The set operations *intersection* and *union* correspond to logic operations, *conjunction (and)* and *disjunction (or),* respectively. We have mentioned in the previous chapter that there are multiple choices for the fuzzy conjunction and the fuzzy disjunction operators. In addition to using min for fuzzy conjunction, max for fuzzy disjunction, another common pair is the algebraic product (for fuzzy conjunction) and algebraic sum (for fuzzy disjunction):

$$\mu_{(A \cap B)}(x) = \mu_A(x) \times \mu_B(x)$$ **(EQ 3.16)**

$$\mu_{A \cup B}(x) = \mu_A(x) + \mu_B(x) - \mu_A(x) \times \mu_B(x)$$ **(EQ 3.17)**

There are an infinite number of other choices.

The choice of a fuzzy conjunction operator determines the choice of the fuzzy disjunction operator, and vice versa. This is due to the principle of duality between the two operators. More specifically, a fuzzy conjunction operator, denoted as $t(x, y)$ and a fuzzy disjunction operator, denoted as $s(x, y),$ form a dual pair if they satisfy the following condition:

$$1 - t(x, y) = s(1 - x, 1 - y)$$ **(EQ 3.18)**

In fact, this duality condition ensures that

$$\overline{A \cap B} = \overline{A} \cup \overline{B} \qquad \text{(EQ 3.19)}$$

still holds in fuzzy set theory. Can you remember what fundamental law this is? (Refer to Section 3.1.1 in case you have forgotten.)

The set of candidate fuzzy conjunction operators, called *triangular norms* or *t-norms,* is defined by a set of axioms. Similarly, the set of candidate fuzzy disjunction operators called *triangular conorms, t-conorms,* or *s-norms* is defined by a set of dual axioms. We formally define t-norms and t-conorms using their axioms below.

DEFINITION 1 A t-norm operator, denoted as $t(x, y)$ is a function mapping from $[0, 1] \times [0, 1]$ to $[0, 1]$ that satisfies the following conditions for any $w, x, y, z \in [0, 1]$:
1. $(0, 0) = 0$, $t(x, 1) = t(1, x) = x$
2. $t(x,y) \le t(z,w)$ if $x \le z$ *and* $y \le w$ (monotonicity)
3. $t(x, y) = t(y, x)$ (commutativity)
4. $t(x, t(y, z)) = t(t(x, y), z)$ (associativity)

DEFINITION 2 A t-conorm operator, denoted as $s(x,y)$, is a function mapping from $[0, 1] \times [0, 1]$ to $[0, 1]$ the following conditions for any $w, x, y, z \in [0, 1]$.
1. $(1, 1) = 1$, $s(x, 0) = s(0, x) = x$
2. $s(x,y) \le s(z,w)$ if $x \le z$ *and* $y \le w$ (monotonicity)
3. $s(x, y) = s(y, x)$ (commutativity)
4. $s(x, s(y, z)) = s(s((x, y), z))$ (associativity)

Typical pairs of t-norms and t-conorms are listed below:

1) Drastic Product:

$$t_1(x, y) = \begin{cases} min\{x, y\} & \text{if } max\{x, y\} = 1 \\ 0 & x, y < 1 \end{cases} \qquad \text{(EQ 3.20)}$$

Drastic Sum:

$$s_1(x, y) = \begin{cases} max\{x, y\} & \text{if } min\{x, y\} = 0 \\ 1 & x, y > 0 \end{cases} \qquad \text{(EQ 3.21)}$$

2) Bounded Difference:

$$t_2(x, y) = max\{0, x + y - 1\} \qquad \text{(EQ 3.22)}$$

Bounded Sum:

$$s_2(x, y) = min\{1, x + y\} \qquad \text{(EQ 3.23)}$$

3) Einstein Product:

$$t_3(x, y) = \frac{x \cdot y}{2 - [x + y - (x \cdot y)]} \tag{EQ 3.24}$$

Einstein Sum:

$$s_3(x, y) = \frac{x + y}{[1 + x \cdot y]} \tag{EQ 3.25}$$

4) Algebraic Product

$$t_4(x, y) = x \cdot y \tag{EQ 3.26}$$

Algebraic Sum:

$$s_4(x, y) = x + y - x \cdot y \tag{EQ 3.27}$$

5) Hamacher Product:

$$t_5(x, y) = \frac{x \cdot y}{x + y - (x \cdot y)} \tag{EQ 3.28}$$

Hamacher Sum:

$$s_5(x, y) = \frac{x + y - 2xy}{1 - (x \cdot y)} \tag{EQ 3.29}$$

6) Minimum:

$$t_6(x, y) = min\{x, y\} \tag{EQ 3.30}$$

Maximum:

$$s_6(x, y) = max\{x, y\} \tag{EQ 3.31}$$

An important property about t-norms is that all t-norms are bounded above by min and bounded below by drastic product. Similarly, all t-conorms are bounded above by drastic sum and bounded below by max. We state these formally below.

THEOREM 1   All t-norm operators, denoted t, are bounded below by drastic product t_1 and bounded above by *min*:

$$t_1(x, y) \le t(x, y) \le min(x, y) \tag{EQ 3.32}$$

THEOREM 2   All t-conorm operators, denoted s, are bounded below by *max* and bounded above by drastic sum s_1.

$$max(x, y) \le s(x, y) \le s_1(x, y) \tag{EQ 3.33}$$

We will leave the proof of these theorems as exercises.

3.3.2 Complement of a Fuzzy Set

The idea of complement reflects negation. We may define the complement of a fuzzy set in terms of the algebraic complement of its membership function, μ.

DEFINITION 3   Let A be a fuzzy set defined over U. Then its complement, $\neg A$, is defined, in terms of $\mu_A(u)$, as

$$\mu_{\neg A}(u) = 1 - \mu_A(u). \tag{EQ 3.34}$$

Figure 3.10 shows the complement of the membership function *High*. Note that the above definition overlaps with the definition of the term *High*. In fact, there exists some T_0 such that $\mu_{NotHigh}(T_0) = \mu_{High}(T_0)$, which we may call the *break-even* point of the membership function. It is easy to see that the break-even point of a fuzzy set A always has 0.5 membership degree in A.

FIGURE 3.10   Definition of *Not High*

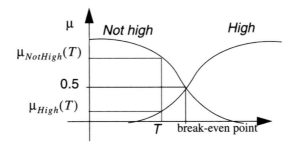

3.3.3 Subsethood

A fuzzy set A in the universe U is a subset of another fuzzy set B if, for every element x in U, its membership degree in A is less or equal to its membership degree in B. This can be formally stated as

$$A \subseteq B \Leftrightarrow \forall x \in U \;\; \mu_A(x) \le \mu_B(x) \tag{EQ 3.35}$$

The notion of subset itself can also be extended to a matter of degree, which is called fuzzy subsethood.

There are two types of approaches to defining a fuzzy subsethood (i.e., a measure of subset between two fuzzy sets). One is based on logic; the other has a probabilistic foundation. The logic-based approach to fuzzy subsethood is motivated by the following observation in classical set theory: If A is a subset of B, then membership in A implies membership in B, i.e.,

$$A \subseteq B \Leftrightarrow (\forall x \in U \qquad x \in A \to x \in B)$$

If A and B are fuzzy sets, the fuzzy implication x is $A \rightarrow x$ is B forms a fuzzy implication relation, as we will see in Chapter 5. If A is a strict subset of B, all entries in the implication relation are 1. Hence, subsethood can be defined as the minimum value in the implication relation:[2]

$$s(A, B) = \inf_{x}(A \rightarrow B) \qquad \textbf{(EQ 3.36)}$$

Due to the fact that the material implication $A \rightarrow B$ is defined as $\neg A \vee B$ in classical logic, we can rewrite the equation above as

$$s(A, B) = \inf_{x}[(1 - \mu_A(x)) \oplus \mu_B(x)] \qquad \textbf{(EQ 3.37)}$$

where $\oplus$ is a fuzzy disjunction operator. For instance, if we use the max operator for disjunction, we have

$$s_1(A, B) = \inf_{x}[max((1 - \mu_A(x)), \mu_B(x))] \qquad \textbf{(EQ 3.38)}$$

Other fuzzy disjunction operators such as bounded sum have also been used to define alternative fuzzy subsethood measures.

The second class of fuzzy subsethood measures are analogous to conditional probability in probability theory. The rationale of these approaches is that the probability of B given A (i.e., $P(B|A)$ is 1 if A is a subset of B.[3] Recall that conditional probability is defined as

$$P(A|B) = \frac{P(A \cap B)}{P(A)} = \frac{|A \cap B|}{|A|} \qquad \textbf{(EQ 3.39)}$$

We can hence define the degree that A is a subset of B as

$$S(A, B) = \frac{|A \cap B|}{|A|} \qquad \textbf{(EQ 3.40)}$$

If we use the min operator for intersection and the definition of cardinality to be introduced in next section, we get the following subsethood measure:

$$S(A, B) = \frac{\sum_{x \in U} min(\mu_A(x), \mu_B(x))}{\sum_{x \in U} \mu_A(x)} \qquad \textbf{(EQ 3.41)}$$

[2.] We should point out that fuzzy implications typically involve fuzzy sets from two different universes of discourse. Hence, the fuzzy implication used here for defining fuzzy subsethood is nontypical in that it relates two fuzzy sets from one universe of discourse.

[3.] The probability of A given B (i.e., $P(A|B)$) is not necessarily one, however.

3.4 Properties of Fuzzy Sets

3.4.1 The Cardinality of Fuzzy Sets

The cardinality of a set is the total number of elements in the set. Since an element can partially belong to a fuzzy set, a natural generalization of the classical notion of cardinality is to weigh each element by its membership degree, which gives us the following formula for calculating the cardinality of a fuzzy set:

$$Card(A) = \sum_{x_i} \mu_A(x_i) \qquad \textbf{(EQ 3.42)}$$

For instance, the cardinality of the fuzzy set LargeTexasUniversity is
$$Card(\text{Large Texas University}) = 1 + 1 + 0.5 + 0.3 = 2.8$$
The cardinality of fuzzy sets is useful for answering questions such as "How many large universities are there in Texas?" Therefore, it plays an important role in fuzzy databases and information systems. The cardinality of a set is also used in defining other properties, serving as a "normalization factor." In fact, the denominator (normalization factor) of the centroid defuzzification formula discussed in the previous chapter is the cardinality of the fuzzy set being defuzzified.

3.4.2 Height: Normal Versus Subnormal

The height of a fuzzy set is the highest membership value of its membership function:

$$Height(A) = \max_{x_i} \mu_A(x_i) \qquad \textbf{(EQ 3.43)}$$

For example, the height of the fuzzy set in Figure 3.11 is 0.5. A fuzzy set with height 1 is called a *normal* fuzzy set. In contrast, a fuzzy set whose height is less than 1 is called a *subnormal* fuzzy set. The fuzzy set in Figure 3.11, for example, is subnormal.

In classical set theory, a set is either nonempty or empty. The notion of subnormal fuzzy sets introduces the grey area between these two extremes — a set with full members (i.e., nonempty sets) and a set with no members (i.e., empty sets). A subnormal fuzzy set is a fuzzy set containing only partial members, but no full members. Hence, it is somewhere between empty sets and nonempty sets (in the classical sense).

Most of the concepts used in human reasoning correspond to normal fuzzy sets. An example of meaningful subnormal fuzzy sets is "the set of perfect people," because we know there is no perfect person.

Subnormal fuzzy sets are usually generated during the fuzzy rule-based reasoning process. As we have seen in Chapter 2, both the clipping method and the scaling method in fuzzy inference often produce subnormal fuzzy sets. These subnormal fuzzy sets are actually possibility distributions of the consequent variable (i.e., variables that occur in a fuzzy rule's then-part).

FIGURE 3.11   A Subnormal Fuzzy Set

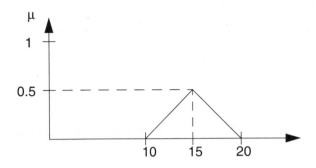

3.4.3 Support and Alpha-level Cuts

The *support* of a fuzzy set A is the set of elements whose degree of membership in A is greater than 0. For instance, the support of the fuzzy set in Figure 3.11 is the open interval (10, 20). Let A be a fuzzy set in the universe of discourse U. We can formally define support as follows:

$$Spt(A) = \{x \in U \mid \mu_A(x) > 0\} \qquad \textbf{(EQ 3.44)}$$

It should be pointed out that the support of a fuzzy set is a crisp (classical) set.

FIGURE 3.12   The Membership Function of Moderately Approved Presidential Candidate

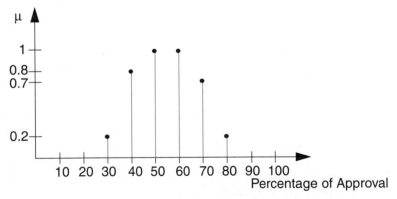

Percentage of Approval

The notion of α-cut (also called α-*level*) is more general than that of support. Let α_0 be a number between 0 and 1. The α-*cut* of a fuzzy set A at α_0, denoted as A_{α_0}, is the set of elements whose degree of membership in A is no less than α_0. Mathematically, the α-cut of a fuzzy set A in U is defined as

$$A_{\alpha_0} = \{x \in U \mid \mu_A(x) \geq \alpha_0\} \qquad \textbf{(EQ 3.45)}$$

For example, let us consider the concept "moderately approved" regarding the public's opinion of a presidential candidate. The universe of discourse is the percentage of those people supporting the candidate in a national poll. To simplify our discussion, we use a discrete universe of discourse: $U = \{0\%, 10\%, 20\%,..., 100\%\}$ The membership function of the fuzzy set Moderately Approved is shown in Figure 3.12. We can construct the α-cut of the fuzzy set at 0.7 as

$$\text{ModeratelyApproved}_{0.7} = \{40\%, 50\%, 60\%, 70\%\}$$

Similarly, we can construct the following α-cuts:

$$\text{ModeratelyApproved}_{0.2} = \{30\%, 40\%, 50\%, 60\%, 70\%, 80\%\}$$
$$\text{ModeratelyApproved}_{0.8} = \{40\%, 50\%, 60\%\}$$

Obviously, as the alpha value increases, the set generated by α-cutting a fuzzy set becomes smaller. It can be shown that if $\alpha_1 < \alpha_2$, we have $A_{\alpha_1} \supseteq A_{\alpha_2}$. Where $\supseteq$ denotes a crisp superset (i.e., suppositions) relation. We will leave the proof of this as an exercise.

3.4.4   Resolution Identity

Based on the notion of α-cuts, a fuzzy set can be decomposed into multiple crisp sets (i.e., α-level sets) using different α values. Intuitively, each α-level specifies a slice of the membership function. Therefore, it is conceivable that the original membership function can be reconstructed by "piling up" these slices in order.

FIGURE 3.13   Resolution Identify of a Fuzzy Set

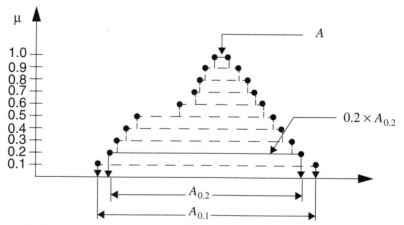

The foundation for reconstructing a membership function from its α-cuts is the *resolution identity* in fuzzy set theory. Let A be a discrete fuzzy set. We can order the nonzero membership values of the elements in the support set into a list in increasing order (without duplicates): $(\alpha_0, \alpha_1, ..., \alpha_n)$. The *resolution identity principle* in fuzzy set theory states that

$$A = \alpha_0 \times A_{\alpha_0} + \alpha_1 \times A_{\alpha_1} + ... + \alpha_n \times A_{\alpha_n} \qquad \textbf{(EQ 3.46)}$$

where $\alpha_i \times A_{\alpha_i}$ represents a fuzzy set such as the one below:

$$\mu_{\alpha_i \times A_{\alpha_i}} = \begin{cases} \alpha_i & \text{if } \mu_A(x) \geq \alpha_i \\ 0 & \text{otherwise} \end{cases} \qquad \textbf{(EQ 3.47)}$$

and + represents the disjunction operator. In other words, the original fuzzy set A can be viewed as the union of all those fuzzy sets constructed from an α_i in the list $(\alpha_0, \alpha_1, ..., \alpha_n)$ such that (1) their support is the α_i level cut of A, and (2) their membership function is flat at the alpha value α_i. The shape of the membership function $\alpha_i \times A_{\alpha_i}$ is hence a rectangle with a height of α_i as illustrated in Figure 3.13.

A typical choice of the fuzzy disjunction operator in resolution identity is the max operator, which in effect superimposes one membership function on top of another one. Figure 3.14 shows the result of superimposing $0.1 \times A_{0.1}$ and $0.2 \times A_{0.2}$. If we continue superimposing other scaled α-cuts such as $0.3 \times A_{0.3} ... 1 \times A_1$, we can reconstruct the membership function of the discrete fuzzy set A in Figure 3.13

FIGURE 3.14   Combining Two Scaled α-cuts

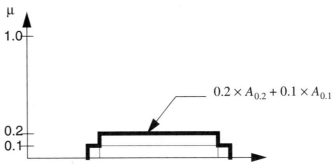

It should be noted that the α-values in the resolution identity of a discrete fuzzy set may not be equally separated. They are determined by the membership function of the set. For instance, the resolution identity of the fuzzy set in Figure 3.12 consists of four α-cuts with α-values 0.2, 0.7, 0.8, and 1, respectively, as shown in Figure 3.15.

FIGURE 3.15   Resolution Identity of the Fuzzy Set Moderately Approved

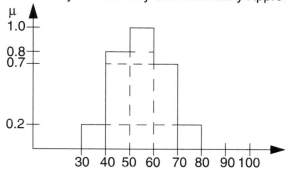

The resolution identity is important in fuzzy set theory because it establishes a bridge between fuzzy sets and crisp sets. This bridge provides a convenient way of generalizing various concepts associated with nonfuzzy sets with regard to fuzzy sets.

The resolution identity of a continuous fuzzy set can be expressed as

$$A = \bigcup_{\alpha} \alpha \times A_{\alpha}$$ **(EQ 3.48)**

where α is a continuous variable ranging from 0 to 1 and $\cup$ denotes the "sup" operator (i.e., the maximum of an infinite number of arguments).

3.4.5 Convex Fuzzy Sets

Another useful concept in fuzzy set theory is the notion of convexity. Intuitively, a fuzzy set is convex if its membership function does not have a "valley." The fuzzy set in Figure 3.16, for example, is not convex because it has a valley around a. Formally, a *convex fuzzy set* can be described as follows.

DEFINITION 4   Let U be a universe of discourse of a variable x. Let A be a fuzzy subset of U. The set A is *convex* if and only if

$$\mu_A(\lambda a + (1 - \lambda)b) \geq min\{\mu_A(a), \mu_A(b)\}$$ **(EQ 3.49)**

for all $a, b \in U$ and $0 \leq \lambda \leq 1$.

This condition states that the membership value of any given element in the interval $[a, b]$ should be no less than the membership value of either end points.

FIGURE 3.16   A Nonconvex Fuzzy Set

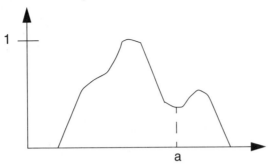

3.5 A Geometric Interpretation of Fuzzy Sets

Bart Kosko developed a complete geometric interpretation of a fuzzy set in which a fuzzy set is a point in a space.[4] This is best understood using an example. Let us consider a universe of discourse U containing two elements u_1 and u_2, i.e., $U = \{u_1, u_2\}$. Let A be a fuzzy subset of U with the following membership functions:

$$\mu_A(u_1) = a$$
$$\mu_A(u_2) = b$$

(EQ 3.50)

where a and b are in [0, 1]. The membership function is shown in Figure 3.17(a). An alternative way to view the membership function is to consider a two-dimensional unit square in which the horizontal and vertical axes correspond to the possible membership values of u_1 and u_2, respectively. A fuzzy subset of U corresponds to a point in the square. For instance, the fuzzy set A above is represented as the point (a, b) in Figure 3.17(b).

Several interesting observations can be made about this geometric interpretation.

- A classical set corresponds to a vertex in the square. For instance, the vertices in Figure 3.17(b) (i.e., (0, 0), (0, 1), (1, 0), and (1, 1)) corresponds to ϕ, $\{u_2\}$, $\{u_1\}$, and $\{u_1, u_2\}$, respectively.

- A normal fuzzy set corresponds to a point on the boundary of the square.

- A subnormal fuzzy set corresponds to an interior point of the square.

- The cardinality of a fuzzy set A is the Hamming distance between the point A and the origin of the cube. The Hamming distance between two vectors $\hat{y} = (x_1, ..., x_n)$ and $\hat{y} = (y_1, ..., y_n)$ is

---

[4.] L. A. Zadeh was the first to introduce the concept that a fuzzy set can be viewed as a point in a hypercube (p. 486 in [704])

$$d_H(\hat{x}, \hat{y}) = \sum_i |x_i - y_i| \qquad\qquad \textbf{(EQ 3.51)}$$

It is easy to show that

$$|A| = d_H(A, \phi) \qquad\qquad \textbf{(EQ 3.52)}$$

We will leave this as an exercise.

- This geometric interpretation can be extended to any finite universe of discourse U. A fuzzy set in such a universe corresponds to a point in an N-dimensional hypercube, where N is the number of elements in U.

FIGURE 3.17   Two Views of a Fuzzy Set; (a) a Normal View of its Membership Function, and (b) an Alternative View Based on a Geometric Interpretation of the Fuzzy Set.

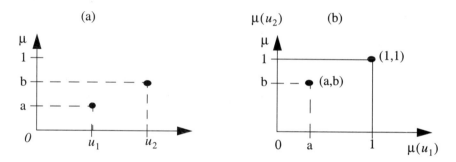

3.6 Possibility Theory

We have introduced the basic idea of possibility distribution in the previous chapter. Even though the idea of possibility distribution parallels that of probability distribution in conventional mathematics, their differences cannot be overlooked. In particular, a possibility distribution need not satisfy the additivity property of axiomatic probability. Whereas a probability distribution states the probability that the given variable takes a certain value, a possibility distribution states the *possible* value of the variable or the possibility that the variable takes a certain value.

A *possibility distribution*, π, maps a given domain of definition into the interval $[0, 1]$. We can view a possibility distribution as a mechanism for interpreting factual statements involving fuzzy sets. The statement, "Temperature is *High*," where *High* is defined as $\mu_{High}: T \to [0, 1]$, translates into a possibility distribution, $\pi(T) = \mu_{High}(T)$. Complex statements involving more than one fuzzy set translate into possibility distributions as well. In fact this is precisely how we interpret linguistic statements, given a priori or derived through an inference process.

EXAMPLE 2   The statement, "Temperature is *High but not too High*," translates into

FIGURE 3.18   A possibility distribution of High AND NOT VERY High

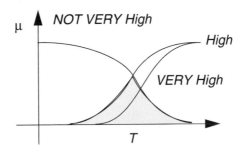

a possibility distribution in terms of conjunction of the terms *High* and *NOT VERY High*:

$$\pi(T) \ = \ \min(\mu_{High}(T), \mu_{NOT\ VERY\ High}(T))\,.$$

$$= \ \min[\mu_{High}(T), 1 - (\mu_{High}(T))^{2}]$$

3.6.1 Possibility Measures and Necessity Measures

Once we describe the value of a variable using a possibility distribution, we often need to determine to what degree it matches a given condition. Such needs arise in fuzzy logic control as well as in fuzzy databases. In fuzzy logic control (and other fuzzy rule-based models), an imprecise sensor/model input (described as a possibility distribution) needs to match conditions of fuzzy rules, as we shall see in Chapter 5. In fuzzy database systems, an imprecise information may be represented using possibility distributions. Retrieving information from a database involves posting a question (called a query) to the database management system. The question typically involves conditions about the kind of information we are interested in retrieving. Hence, processing these database queries for retrieving imprecise information requires the capability to match conditions in a query with possibility distributions in a fuzzy database, which we shall discuss in detail in Chapter 12.

In general, matching a possibility distribution of a variable x with a condition "*x is A*" involves asking the following two related questions:

(Q1) Given the possibility distribution of x (denoted $\prod(x)$), is it **possible** for x to be A?

(Q2) Given the possibility distribution $\prod(x)$, is it **necessary** for x to be A?

We will first illustrate these two questions using a nonfuzzy example. Suppose we know the following:

Jack's age is between 20 and 25 years old.

Consider the following condition:

Whether a person's age exceeds 22 years old.

Does the possibility distribution of Jack's age match the condition? Due to the imprecise information about Jack's age, we cannot find a black-and-white answer to the question. More specifically, it is possible for Jack's age to exceed 22; however, this is not necessarily the case. Hence, the answer to the first question listed above (Q1) is "yes," but the answer to the second question (Q2) is "no." Obviously, if we are interested in the condition that "whether a person's age is between 21 and 23 years old," our answers to both questions are positive for Jack (i.e., it is possible as well as necessary for Jack's age to be in the interval [21, 23]).

In the example above, the answers to Q1 and Q2 are either "yes" or "no" because neither Jack's age nor the conditions about a person's age is fuzzy. In general, however, the answer to each question is a matter of degree. We call the answer to Q1 a "possibility", and the answer to Q2 a "necessity." For the convenience of our discussion, we will use $Pos(A|\Pi_X)$ to denote the *possibility* of the condition "X is A" given the possibility distribution Π_X, and $Nec(A|\Pi_X)$ to denote the *necessity* of the condition "X is A" given the possibility distribution Π_X.

FIGURE 3.19   An Illustration of the Relationship between Necessity and Possibility

NOT A

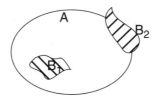

Possibility and necessity are two related measures (i.e., they are not totally independent). There are the relationships between the extreme values (i.e., 0 or 1) of these two measures. First of all, **total necessity implies total possibility**. If a variable is necessarily A, then it is possibly A. The reverse is not true for obvious reasons. Second, **no possibility implies no necessity**. We will leave the proof as an exercise. Third, **a variable is not possible to be *NOT A* if and only if it is necessarily A**. To show this informally, let us consider the Venn diagram in Figure 3.19. If a variable is not possible to be "*NOT A*," its possibility distribution must not have any intersection with "*NOT A*." Hence, it must be entirely contained in A. B_1 is such an example, but B_2 is not. It is easy to show that the reverse is also true. Hence, we get the third relationship. The last relationship is a dual to the third relationship — a variable is possible to be *NOT A* if and only if it is not necessarily A. We can formally state the four relationships as follows:

(1a) $Nec(A|\Pi_X) = 1 \rightarrow Pos(A|\Pi_X) = 1$

(1b) $Pos(A|\Pi_X) = 0 \rightarrow Nec(A|\Pi_X) = 0$

(2a) $1 - Pos(\neg A|\Pi_X) = 1 \leftrightarrow Nec(A|\Pi_X) = 1$

(2b) $Pos(\neg A | \Pi_X) = 1 \leftrightarrow 1 - Nec(A | \Pi_X) = 1$

Both (2a) and (2b) interpret "NOT" as the fuzzy negation operator. We can rewrite (2b) as

(2b') $1 - Pos(\neg A | \Pi_X) = 0 \leftrightarrow Nec(A | \Pi_X) = 0$

So far, we have limited our discussion on the relationship between extreme values of possibility and necessity. Nevertheless, these observations can provide insights on the general relationships between the two measures. The relationships (1a) and (1b) can be generalized to

$$Nec(A | \Pi_X) \leq Pos(A | \Pi_X) \qquad \textbf{(EQ 3.53)}$$

The relationships (2a) and (2b') can be generalized to

$$1 - Pos(\neg A | \Pi_X) = Nec(A | \Pi_X) \qquad \textbf{(EQ 3.54)}$$

In fact, Equation 3.54 can be used to define a necessity measure using a possibility measure. Hence, we only need to define one of the two measures; the other is automatically derived.

We are now ready to define the possibility measure $Pos(\neg A | \Pi_X)$. To do so, we first make the following observations. (1) If Π_X and μ_A have no intersection at all, the possibility for X to be A is 0. (2) If Π_X and μ_A has "full intersection" (i.e., having at least a common element completely belonging to both), the possibility for X to be A is 1. Based on these observations, the possibility measure of A given Π_X is defined to be the height of their intersection.

DEFINITION 5 The *possibility measure* for a variable X to satisfy the condition "X is A" given a possibility distribution Π_X is defined to be

$$Pos(A | \Pi_X) = \sup_{x_i \in U} (\mu_A \otimes \Pi_X) $$

$$\textbf{(EQ 3.55)}$$

where $\otimes$ denotes a fuzzy intersection (i.e., a fuzzy conjunction) operator, and U denotes the universe of discourse of the variable X.

A common choice of the fuzzy intersection operator for calculating the possibility measure is the min operator. Equation 3.55 thus becomes

$$Pos(A | \Pi_X) = \sup_{x_i \in U} [(\mu_A(x_i), \Pi_X(x_i))] \qquad \textbf{(EQ 3.56)}$$

An example of possibility measure is shown in Fig 3.20.

Based on Equation 3.54, it is easy to derive the corresponding formula for the necessity measure:

$$Nec(A | \Pi_X) = \inf_{x_i \in U} [\max(\mu_A(x_i), 1 - \Pi_X(x_i))] \qquad \textbf{(EQ 3.57)}$$

We will leave the derivation of this formula as an exercise. By comparing this formula with the fuzzy subsethood defined in Equation 3.38, we can see that

$$Nec(A|\Pi_{\hat{x}}) = s_1(\Pi_{\hat{x}}, A) \qquad \text{(EQ 3.58)}$$

Rhis is consistent with our intuition that if the possibility distribution Π_x is a subset of A, the necessity measure of A given $\Pi_{\hat{x}}$ must be 1.

Consider the example shown in Figure 3.20. The possibility measure is the height of the intersection. Thus we have

$$Pos(A|\Pi_X) = 0.5$$

The necessity measure can also be calculated using Equation 3.55. The result is

$$Nec(A|\Pi_X) = 0$$

FIGURE 3.20   An Example of Possibility Measure

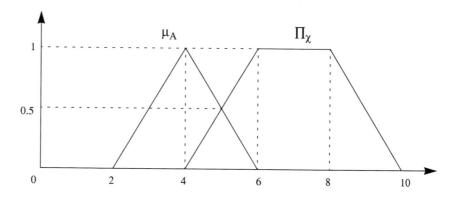

3.7 Summary

In this chapter we introduced fuzzy sets, different ways to represent them, their basic operations, and important properties about them. More specifically, we have discussed the following concepts:

- Different types of membership functions.
- The law of excluded middle and the law of contradiction in classical set theory no longer hold in fuzzy sets.
- Hedges as modifiers to fuzzy sets.
- The axioms of fuzzy conjunction (intersection) operators and fuzzy disjunction (union) operators.
- Cardinality and subsethood of fuzzy sets.
- Height, alpha-cuts, and resolution identity of fuzzy sets.
- Convexity of fuzzy sets.

- A geometric interpretation of fuzzy sets.
- Possibility measures and necessity measures for determining how well a possibility distribution of a variable matches a fuzzy condition about the variable.

These concepts will be used in subsequent chapters.

Bibliographical and Historical Notes

Almost all of the concepts and techniques in this chapter were introduced in L.A. Zadeh's seminal paper on fuzzy sets published in 1965 [708]. The concept of linguistic variable was first introduced by Zadeh in a frequently cited 1973 paper [703], and fully developed later in [699]. A more detailed discussion on the foundation of Boolean algebra discussed in Section 3.1.1 can be found in [635]. Bart Kosko introduced the geometric interpretation of fuzzy sets and investigated issues such as fuzzy subsethood based on the interpretation in the late 80s [332, 335].

Exercises

3.1   Using the axioms of t-norms, show that all t-norms are bounded above by min, and bounded below by drastic product, i.e.,

$t_i(x, y) \le t(x, y) \le min(x, y)$

3.2   Using the axioms of t-conorms, show that all t-conorms are bounded above by drastic sum, and bounded below by max, i.e.,

$max(x, y) \le s(x, y) \le s_1(x, y)$

3.3   How does a linguistic variable differ from a symbolic variable in a conventional AI system?

3.4   Let x be a linguistic variable that measures a university's academic excellence, which takes values from the universe of discourse $U = \{1, 2, 3, 4, 5, 6, 7, 8, 9, 10\}$. Suppose the term set of x includes *Excellent, Good, Fair,* and *Bad.*
The membership functions of these linguistic labels are listed below:

- μ *Excellent* = {(8, 0.2) (9, 0.6) (10, 1)}
- μ *Good* = {(6, 0.1) (7, 0.5) (8, 0.9) (9, 1) (10, 1)}
- μ *Fair* = {(2, 0.3) (3, 0.6) (4, 0.9) (5, 1) (6, 0.9) (7, 0.5) (8, 0.1)}
- μ *Bad* = {(1, 1) (2, 0.7) (3, 0.4) (4, 0.1)}

Construct the membership functions of the following compound sets:

- *Not Bad but Not Very Good*
- *Good but Not Excellent*

3.5 Give a formal definition of the syntactic rule as well as the semantics of the possible values V of a linguistic variable. You may assume that T is a set of primitive linguistic terms, and the meaning of a term t in T is defined by their membership function μ_t. Syntactic rules can be defined recursively, e.g.,

- $V = t$
- $V = (\neg V)$

Each syntactic rule should be associated with a semantic rule to define the membership function of the linguistic expression constructed, e.g.,

- $\mu_{\neg V} = 1 - \mu_V$

3.6 Consider the statement, *steam pressure is dangerously High.* Define the term *dangerously High.* Make sure you choose an appropriate universe of discourse. Can you define a context-free definition for *dangerously* in a manner that can be used as an *operator* on any given instance of the term *high*? If so, can you extend this definition to make it applicable to linguistic terms other than *high*? If you find this impossible, can you explain why?
Hint: Consider the way the term *very* operates in a similar situation.

3.7 Define a simple, that is, noncompound, term that can be represented as a subnormal fuzzy subset.

3.8 Compute, using operations on fuzzy sets, the meaning of the term, *very hard but not impossible homework.* Make sure you define an appropriate universe of discourse.

3.9 Define the term, *short and sweet*

3.10 Let A be a fuzzy subset of the universe U α_1 and α_2 are two numbers such that $0 \le \alpha_1 < \alpha_2 \le 1$. Prove that $A_{\alpha_1} \supseteq A_{\alpha_2}$ where $\supseteq$ denotes a crisp superset relation.

3.11 Suppose you encounter a critic of fuzzy logic who challenges you as follows:

The meaning of the concept "tall person" is different for Americans and for Japanese. However, you can represent a fuzzy concept like "tall person" using only one membership function. This creates a problem for fuzzy set theory.

What is your response?

3.12 Let A be a fuzzy set in the universe of discourse U. Consider the geometric interpretation of fuzzy sets described in Section 3.5. Show that the cardinality of A is the Hamming distance between A and the origin of the cube in the geometric interpretation, i.e., $|A| = d_H(A, \phi)$ where d_H denotes the Hamming distance defined in Equation 3.52.

3.13 Use a Venn diagram like Figure 3.19 to show that "no possibility implies no necessity".

3.14 Use a Venn diagram like Figure 3.19 to show that "a variable is possible to be NOT A if and only if it is not necessarily A."

3.15   Derive a necessity measure (Equation 3.57) from the formula of possibility measure in Equation 3.56 and the relationship between the two measures in Equation 3.54.

3.16   Calculate the possibility and the necessity measures for x to be B given the possibility distribution Π_X shown in Figure 3.21. Let x be a variable whose universe of discourse U is the set of integers between 0 and 100. B is a fuzzy subset of U whose membership function is shown in Figure 3.21.

FIGURE 3.21   Exercise for Possibility Measures and Necessity Measures

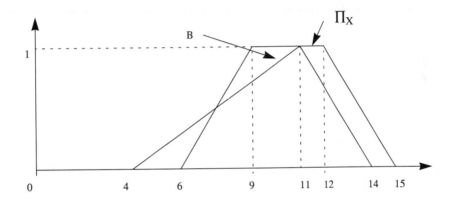

4 FUZZY RELATIONS, FUZZY GRAPHS, AND FUZZY ARITHMETIC

In this chapter, we introduce three important concepts in fuzzy logic: fuzzy relations, fuzzy graphs, and the extension principle. The former two together form the foundation of fuzzy rules, while the last is the basis of fuzzy arithmetic.

4.1 Fuzzy Relations

A fuzzy relation generalizes the notion of a classical black-and-white relation into one that allows partial membership. For example, a binary relation "friend" will classify all human relationships into either being friend or not being friend. A fuzzy relation "Friend," in contrast, can describe the degree of friendship between two persons.

The classical notion of relation describes a relationship that holds between two or more objects. A relationship between two objects is represented by a *binary relation*, a relation with two arguments. For example, the parent relationship between a parent and his/her child can be represented as a binary relation. More generally, we can use an n-ary relation, a relation with n arguments, to describe a relationship between n objects. For instance, we can use an n-ary relation to describe that student X took course Y during semester Z in year W. This relation has four arguments such as took_course (student, course, semester, year). An n-ary relation can be formally defined as set of ordered lists of n objects. Each list describes a case in which the relation holds.

A binary relation on variables x and y, whose domains are X and Y respectively, can be defined as a set of ordered pairs in $X \times Y$. For instance, the binary relation "less than" between two real numbers can be formally defined as

$$R = \{(x, y) | x < y, x, y \in R\}$$

It is easy to see that the relation is a subset of $X \times Y$. In general, an n-ary relation on $x_1, x_2, ..., x_n$ whose domains are $X_1, X_2, ..., X_n$ is a subset of $X_1 \times X_2 \times ... \times X_n$.

Since a relation can be viewed as a set, we can easily generalize the classical notion of a relation using fuzzy scts. We will show how to generalize binary relations. First, we can represent a binary relation R on x, y with domains X, Y as a function that maps an ordered pair (x,y) in $X \times Y$ to 0 (i.e., the relation does not hold between x and y) or 1 (i.e., the relation holds between x and y), i.e., $R = X \times Y \rightarrow \{0, 1\}$

A fuzzy relation generalizes the classical notion of relation into a matter of degree. As we mentioned earlier, the fuzzy relation *Friend* describes the degree of friendship between two persons. Similarly, a fuzzy relation *Petite* between height and weight of a person describes the degree by which a person with a specific height and weight is considered petite. Formally, fuzzy relation R between variables x and y, whose domains *are X* and *Y,* respectively, is defined by a function that maps ordered pairs in $X \times Y$ to their degree in the relation, which is a number between 0 and 1, i.e., $R: X \times Y \rightarrow [0, 1]$.

More generally, a fuzzy n-ary relation R in $x_1, x_2, ..., x_n$, whose domains are $X_1, X_2, ..., X_n$, respectively, is defined by a function that maps an n-tuple $<x_1, x_2, ..., x_n>$ in $X_1 \times X_2 \times ... \times X_n$ to a number in the interval, i.e., $R: X_1 \times X_2 \times ... \times X_n \rightarrow [0, 1]$. Just as a classical relation can be viewed as a set, a fuzzy relation can be viewed as a fuzzy subset. From this perspective, the mapping above is equivalent to the membership function of a multidimensional fuzzy set.

If the possible values of x and y are discrete, we can express a fuzzy relation in a matrix form. For example, suppose we wish to express a fuzzy relation *Petite* in terms of the height and the weight of a female. Suppose the range of the height and the weight of interest to us are {5', 5'1", 5'2", 5',3", 5'4", 5'5", 5'6"}, denoted h, and {90, 95, 100, 105, 110, 115, 120, 125} (in lb,) denoted w, respectively. We can express the fuzzy relation in a matrix form as shown below:

	90	95	100	105	110	115	120	125
5'	1	1	1	1	1	1	0.5	0.2
5'1"	1	1	1	1	1	0.9	0.3	0.1
5'2"	1	1	1	1	1	0.7	0.1	0
5'3"	1	1	1	1	0.5	0.3	0	0
5'4"	0.8	0.6	0.4	0.2	0	0	0	0
5'5"	0.6	0.4	0.2	0	0	0	0	0
5'6"	0	0	0	0	0	0	0	0

Each entry in the matrix indicates the degree a female with the corresponding height (i.e., the row heading) and weight (i.e., the column heading) is considered to be petite. For instance, the entry corresponding to a height of 5'3" and a weight of 115 lb. has a value 0.3, which is the degree to which such a female person will be considered a *petite* person; i.e., *Petite(5'3", 115 lb) = 0.3*.

Once we define the *Petite* fuzzy relation, we can answer two kinds of questions:

- What is the degree that a female with a specific height and a specific weight is considered to be petite?
- What is the possibility that a petite person has a specific pair of height and weight measures?

In answering the first question, the fuzzy relation is equivalent to the membership function of a multidimensional fuzzy set. In the second case, the fuzzy relation becomes a possibility distribution assigned to a petite person whose actual height and weight are unknown.

4.2 The Composition of Fuzzy Relations

We have shown that once we have a fuzzy relation, we can directly answer two kinds of interesting questions. We can view these question-answer capabilities as inferences (e.g., inferring the possible weight-height combinations knowing that a person is petite). An even more useful kind of inference is the following:

- Given a two-dimensional fuzzy relation and the possible values of one variable, infer the possible values of the other variable.

For instance, we may wish to know the possible weight of a petite female called Michelle who is *about 5'4"* tall where *"about"* indicates impression. The answer to this question can be obtained through the *composition of fuzzy relations*, which is also referred to as the *compositional rule of inference*. To find out the answer, we will then consider whether it is possible for such a person to weigh 90 lb, 95 lb, ... , 120 lb. Because we are interested in finding all possible weights of the person, we may get multiple positive answers. Furthermore, the answer should indicate the degree of possibility for each weight.

How do we find out whether it is possible for Michelle to have a specific weight, say 110 lb? If we think about this for a moment, it is not difficult to realize that we need to consider all possible heights of the person and see if a petite person with such a height can weigh 110 lb. For example, we may start with 5 feet and check if the following two conditions are both true:

- What is the possibility that Michelle's height is 5 feet given that she is about 5'4" tall?
- What is the possibility that a petite person is 5 feet tall and weighs 110 lb?

If the answers to both questions are positive, we infer that 110 lb is a possible weight of Michelle. We can ask two similar questions for the next height, which is 5 feet 2 inches.

- What is the possibility that Michelle's height is 5 feet 2 inches given that she is about 5' 4" tall?
- What is the possibility that a petite person is 5 feet 2 inches tall and weighs 110 lb?

The procedure can be repeated for all the remaining height measures (i.e., 5'4", 5' 6", ... , 6'). We can represent such a procedure in predicate logic. Predicate logic is a logic system that uses "predicate," as building blocks for constructing logic formulae. A predicate is like a function that maps its arguments to either true or false. The arguments of a predicate are usually variables or constants. Predicates are often connected into a logic formula using and (denoted $\wedge$), or (denoted $\vee$), and imply (denoted $\rightarrow$). A bidirectional implication is denoted by ↔. Variables are quantified by a "for all" quantifier ($\forall$) or a "there exists" quantifier ($\exists$). These notations are summarized in Table 4.1. To help you get familiar with these notations, we will use them in the discussion below.

We will give a more detailed introduction to predicate logic in Chapter 6. We use

TABLE 4.1   Predicate Logic Notations

Symbols	Meaning
∧	and
∨	or
∀	for all
∃	there exists
↔	if and only if
→	imply

three predicates to describe the reasoning process above in logic:

Possible-height (h_i): The predicate is true if h_i is a possible height of a person.
Possible-weight (w_j): The predicate is true if w_j is a possible weight of the person.
Petite(h_i, w_j): the predicate is true if a person with height h_i and w_j weight is petite.

Notice that the first two predicates represent crisp (i.e., nonfuzzy) constraints on possible heights and weights, whereas the last predicate represents a crisp binary relation. Therefore, they cannot truly represent the possibility distribution of *About 5' 4"*, or the fuzzy relation *Petite*. However, we will use these predicates to show how the inference procedure above can be expressed in classical logic. We will then generalize such a logic expression to obtain the compositional rule of inference in fuzzy logic. We first represent the inference procedure described earlier in logic as follows:

[(*Possible-height (5')*   ∧ *Petite (5', 90))* ∨
    (*Possible-height (5'2")* ∧ *Petite (5', 90))* ∨

.

.

.

    (*Possible-height (6')*   ∧ *Petite (6', 90))*] ⇔ *Possible-weight(90)*

Since the same procedure can be used to determine whether it is possible for a person to weigh 95, 100, 105, ..., 125 pounds, we can use a more general expression in logic as follows:

$\forall$ w_j [(*Possible-height (5')*   ∧ *Petite (5', w_j))* ∨
    (*Possible-height (5'2")* ∧ *Petite (5'2", w_j))* ∨

.

.

.

    (*Possible-height (6')* ∧ *Petite (6', w_j))*] ⇔ *Possible-weight (w_j)*
This can be rewritten into a more compact form:

$$\forall w_j \left[Possible\text{-}weight(w_j) \leftrightarrow \underset{h_i}{\vee} \; Possible\text{-}height(h_i) \wedge Petite(h_i, w_j) \right] \quad \textbf{(EQ 4.1)}$$

In fact, this is a special case (i.e., the binary case) of the composition of a fuzzy relation. In fuzzy logic, as we already know, both the notion of possibility and the notion of relation become matters of degree. Let us return to our original question, "what are the possible weights of a petite female called Michelle who is about 5'4" tall?" Suppose the meaning of petite is expressed by the fuzzy relation *Petite* in the previous section. We introduce the following notations to represent related possibility distributions:

- $\Pi_{\text{height}(x)}(h_i)$:The possibility degree for a person's height to be h_i.
- $\Pi_{\text{petite}}(h_i, w_j)$: The possibility degree for a petite person to have a height of h_i, and a weight of w_j.

Furthermore, the conjunctions ($\wedge$) and disjunctions ($\vee$) in Equation 4.1 in binary logic need to be generalized to a fuzzy conjunction ($\otimes$) and a fuzzy disjunction ($\oplus$) operator. Therefore, we can infer Michelle's possible weight from her possible height and also the fact that she is petite using a generalized version of Equation 4.1, which is given below:

$$\Pi_{\text{weight}(x)}(w_j) = \underset{h_i}{\oplus}(\Pi_{Height(x)}(h_i) \otimes \Pi_{Petite}(h_i, w_j)) \quad \textbf{(EQ 4.2)}$$

This is an application of the compositional rule of inference to our example. We define the compositional rule of inference more formally below.

DEFINITION 6   Let X and Y be the universes of discourse for variables x and y, respectively, and x_i and y_j be elements of X and Y. Let R be a fuzzy relation that maps $X \times Y$ to [0, 1] and the possibility distribution of X is known to be $\Pi_x(x_i)$. The compositional rule of inference infers the possibility distribution of Y as follows:

$$\Pi_Y(y_j) = \underset{x_i}{\oplus}(\Pi_X(x_i) \otimes \Pi_R(x_i, y_j)) \quad \textbf{(EQ 4.3)}$$

The computation steps for the compositional rule of inference are similar to those of matrix multiplication. The fuzzy conjunction and disjunction operations in Equation 4.3 correspond respectively to the multiplication and summation steps in matrix multiplication.

The compositional rule of inference is not uniquely defined. By choosing different fuzzy conjunction and fuzzy disjunction operators, we get different compositional rules of inference. We list two that are commonly used in practice:

1. max-min composition:

$$\Pi_Y(y_j) = \underset{x_i}{\max}(min(\Pi_X(x_i), \Pi_R(x_i, y_j))) \quad \textbf{(EQ 4.4)}$$

2. max-product composition:

$$\Pi_Y(y_j) = \underset{x_i}{\max}(\Pi_X(x_i) \times \Pi_R(x_i, y_j)) \quad \textbf{(EQ 4.5)}$$

Returning to our example, let us assume *About-5'4"* is defined as

$$About\text{-}5'4" = \{0/5', 0/5'1", 0.4/5'2", 0.8/5'3", 1/5'4", 0.8/5'5", 0.4/5'6"\} \quad \text{(EQ 4.6)}$$

Using the max-min compositional rule of inference, we can compute the weight possibility distribution of a petite person about 5'4" tall:

$$\Pi_{weight}(90) = (0 \wedge 1) \vee (0 \wedge 1) \vee (0.4 \wedge 1) \vee (0.8 \wedge 1)$$
$$\vee (1 \wedge 0.8) \vee (0.8 \wedge 0.6) \vee (0.4 \wedge 0) = 0.8 \quad \text{(EQ 4.7)}$$

Similarly, we can compute the possibility degree for other weights. The final result is

$$\Pi_{weight} = \{0.8/90, 0.8/95, 0.8/100, 0.8/105, 0.5/110,$$
$$0.4/115, 0.1/120, 0/125\}$$

Mathematically, the composition of fuzzy relations can be defined using three basic operations: *projection, cylindrical extension*, and *intersection*. We introduce the first two concepts before we give a formal definition of the composition of fuzzy relations.

4.2.1 Cylindrical Extension

Cylindrical extension and projection are dual operations. The former extends the dimension of a fuzzy relation while the latter reduces the dimension of a fuzzy relation.

Typically, we apply cylindrical extension to two fuzzy relations so that they have the same dimensionality in order to apply set operations (i.e., intersections and union) to them. For example, we can apply cylindrical extension to the fuzzy set *About 5'4"* to $H \times W$. The result is

$$About\text{-}5'4" = \begin{array}{c} \\ 5' \\ 5'1" \\ 5'2" \\ 5'3" \\ 5'4" \\ 5'5" \\ 5'6" \end{array} \begin{array}{c} 90 \;\; 95 \;\; 100 \;\; 105 \;\; 110 \;\; 115 \;\; 120 \;\; 125 \\ \begin{bmatrix} 0 & 0 & 0 & 0 & 0 & 0 & 0 & 0 \\ 0 & 0 & 0 & 0 & 0 & 0 & 0 & 0 \\ 0.4 & 0.4 & 0.4 & 0.4 & 0.4 & 0.4 & 0.4 & 0.4 \\ 0.8 & 0.8 & 0.8 & 0.8 & 0.8 & 0.8 & 0.8 & 0.8 \\ 1 & 1 & 1 & 1 & 1 & 1 & 1 & 1 \\ 0.8 & 0.8 & 0.8 & 0.8 & 0.8 & 0.8 & 0.8 & 0.8 \\ 0.4 & 0.4 & 0.4 & 0.4 & 0.4 & 0.4 & 0.4 & 0.4 \end{bmatrix} \end{array} \quad \text{(EQ 4.8)}$$

Fig 4.1 shows graphically the effect of applying cylindrical extension to a fuzzy subset A of U to the extended dimension $U \times V$. As shown in the figure, the membership function of the extended fuzzy set has a cylinder shape from which the operation gets its name. We formally define cylindrical extension below.

DEFINITION 7   Let R be a fuzzy subset of $U_{i1} \times U_{i2} \times \ldots \times U_{ik}$, where $(i_1, i_2, \ldots, i_k)$, is a subsequence of $(1, 2, \ldots, n)$. The *cylindrical extension* of R in $U_1 \times U_2 \times \ldots \times U_n$ is a fuzzy subset of $U_1 \times U_2 \times \ldots \times U_n$, denoted as $\bar{R}$, whose membership function is defined as

$$\mu_{\bar{R}}(U_1, U_2, \ldots, U_n) = \mu_R(U_{i1}, U_{i2}, \ldots, U_{ik}) \qquad \textbf{(EQ 4.9)}$$

FIGURE 4.1   An Example of Cylindrical Extension

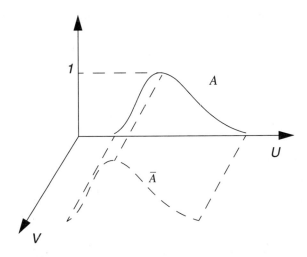

4.2.2 Projection

The projection operation, as the name suggests, projects a fuzzy relation to a subset of selected dimensions. This operation is often used to extract the possibility distribution of a few selected variables from a given fuzzy relation. For instance, the projection of the *Petite* fuzzy relation to the universe of weight gives us a possibility distribution of a petite female person's weight. Projection to a fuzzy relation is analogous to computing the marginal probability on a joint probability distribution. Instead of adding the probability across a row or a column in a joint probability distribution, we perform fuzzy disjunction across a row or a column of a fuzzy relation. Hence, the result of projecting the *Petite* relation to weight is

$$\mu_{Proj_w \; Petite} = \{1/90, 1/95, 1/100, 1/105, 1/110, 1/115, 0.5/120, 0.2/125\} \quad \textbf{(EQ 4.10)}$$

Similarly, we can compute the projection of the relation to height as well. We will leave this as an exercise. Formally, we define the projection as follows.

DEFINITION 8   Let R be a fuzzy relation in $U_1 \times U_2 \times \ldots \times U_n$, and $(i_1, i_2, \ldots, i_k)$ be a subsequence of $(1, 2, \ldots, n)$. The *projection* of R on $U_{i1} \times U_{i2} \times \ldots \times U_{ik}$, denoted as *Proj R*, is defined as

$$\mu_{\underset{U_{i1}, \dots, U_{ik}}{Proj\ R}}(u_{i1}, \dots, u_{in}) = \underset{U_{j1} \times U_{j2} \times \dots \times U_{jl}}{\oplus}\mu_R(u_1, \dots, u_n) \quad \textbf{(EQ 4.11)}$$

where $(j_1, j_2, \dots, j_l)$ is the sequence complementary to $(i_1, \dots, i_k)$, [e.g., if R is a fuzzy relation involving six variables, (1, 2, 4) is complementary to (3, 5, 6)], u_i denotes the variable whose universe of discourse is U_i, and $U_{j1} \times U_{j2} \times \dots \times U_{jl}$ $\oplus$ denotes applying fuzzy disjunction to every point in $U_{j1} \times \dots \times U_{jl}$.

4.2.3 A Formal Definition of the Composition of Fuzzy Relation

Now we are ready to introduce the formal definition of composition. A composition of two fuzzy relations is the result of three operations: (1) cylindrically extending each relation so that their dimensions are identical, (2) intersecting the two extended relations, and (3) projecting the intersection to the dimensions not shared by the two original relations. This is formally stated below for the composition of binary fuzzy relations.

DEFINITION 9   Let R and S be two binary fuzzy relations in $U_1 \times U_2$ and $U_2 \times U_3$ respectively. The *composition* of the two relations, denoted as $R \circ S$, is

$$R \circ S = \underset{U_1, U_3}{Proj}(\bar{R} \cap \bar{S}) \quad \textbf{(EQ 4.12)}$$

where $\bar{R}$ and $\bar{S}$ are cylindrical extensions of R and S in $U_1 \times U_2 \times U_3$.

4.3 Fuzzy Graphs

A fuzzy relation may not have a meaningful linguistic label such as the petite relation does. In fact, most fuzzy relations used in real-world applications do not represent a concept, rather they represent a *functional mapping* from a set of input variables to one or more output variables. Often, a set of fuzzy rules used in a fuzzy logic controller describes a fuzzy relation from the observed state variables to a control decision. In other words, a fuzzy relation underlying a fuzzy logic controller can be constructed by a set of if-then fuzzy rules. How is this achieved? To answer this question, we need to introduce *fuzzy graphs*.

The term "fuzzy graph" has been used for two different concepts in the literature, one introduced by Zadeh, and the other introduced by Rosenfeld. Rosenfeld's fuzzy graph is a generalization of conventional graph theory. In this book, we will discuss only Zadeh's fuzzy graph because it is much more relevant to fuzzy logic control and its industrial applications.

A fuzzy graph describes a functional mapping between a set of input linguistic variables and an output linguistic variable. Assume that a function $f: U \rightarrow V, X \in U, Y \in V$, is approximated by the following fuzzy IF-THEN rules:

$f:$ IF X is small THEN Y is small.
     IF X is medium THEN Y is large.
     IF X is large THEN Y is small.

which form a *fuzzy graph f** (Fig 4.2), where

$$f^* = small \times small + medium \times \text{large} + \text{large} \times small$$

FIGURE 4.2 Fuzzy Graph Approximation by a Disjunction of Cartesian products

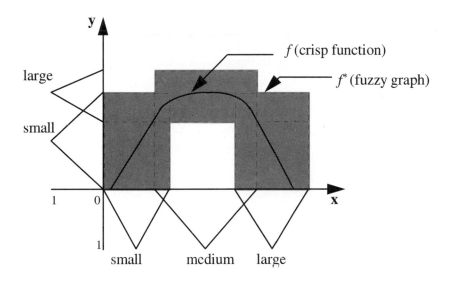

FIGURE 4.3 A Fuzzy Relation Formed by a Cartesian Product

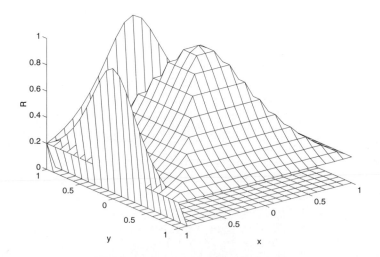

In f^*, + and $\times$ denote, respectively, the disjunction and Cartesian product. An expression of the form $A \times B$ where A and B are *words* (fuzzy sets) is referred as a Cartesian granule [680].

Suppose that x and y are two variables with universes of discourse X and Y respectively, and A and B are two fuzzy subsets of X and Y. The Cartesian product of A and B, denoted by $A \times B$, is defined as

$$\mu_{A \times B}(u, v) = \mu_A(u) \otimes \mu_B(v) \qquad \textbf{(EQ 4.13)}$$

Fig 4.3 shows a fuzzy relation formed by a Cartesian product using min as the fuzzy conjunction operator. A fuzzy graph f^* from X to Y is thus a union of Cartesian products involving linguistic input-output associations (i.e., pairs of "x is A_i" and "y is B_i") i.e.,

$$f^* = \bigcup_i A_i \times B_i \qquad \textbf{(EQ 4.14)}$$

The resulting fuzzy graph is basically a fuzzy relation. We will discuss the role of fuzzy graph and fuzzy relations within the context of fuzzy rule-based reasoning in Chapter 5. We should point out though that Cartesian products are not the only way to form a fuzzy relation. For instance, we will see in Chapter 6 how fuzzy implications can be used to form another type of fuzzy relations (i.e., implication relations).

4.4 Fuzzy Numbers

A fuzzy number is a fuzzy subset of the universe of a numerical number. For instance, a fuzzy real number is a fuzzy subset of the domain of real numbers. A fuzzy integer is a fuzzy subset of the domain of integers. An example of a fuzzy real number, *About-10*, is shown in Fig 4.4.

FIGURE 4.4   A Fuzzy Number

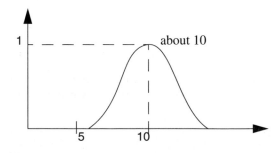

4.5 Function with Fuzzy Arguments

If we are given a precise function and would like to apply the function to fuzzy numbers, we need to use a technique in fuzzy logic called *the extension principle*. A fuzzy argument describes a possibility distribution of the argument. In computing the functional image of a fuzzy argument, we not only need to find out the functional image for each possible value, but also the possibility of this image. Furthermore, because different input values can map to the same output value, we need to determine the possibility of such an output value by combining the possibility degree of all inputs that map to the same output value. Applying a function to fuzzy arguments generalizes the notion of applying a function to intervals. Hence, we first review how to compute the functional image of an interval.

We can apply a function f to an interval I by mapping all elements in the interval to their image: $f(I) = \{y | y = f(x) \ x \in I\}$. If f is a continous function, $F(I)$ is also an interval, as illustrated in Fig 4.5.

As we have mentioned in Chapter 2, an interval is a special kind of possibility distribution whose possibility degree is 1 for points inside the interval, and 0 for points outside the interval. To generalize the mapping of an interval through a function to the mapping of any possibility distribution, we need to somehow "map" the degree of possibility through the function. We will discuss this by first considering monotonic functions, and then the general case of nonmonotonic functions.

FIGURE 4.5   An example of Applying a Function to an Interval

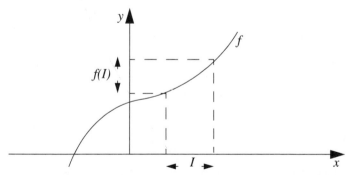

Supposing f is a monotonic continous function, the situation is simple. For each point that is mapped, the corresponding possibility degree is mapped along with it, as shown in Fig 4.6.

FIGURE 4.6   Mapping a Possibility Distribution to a Monotonic Function

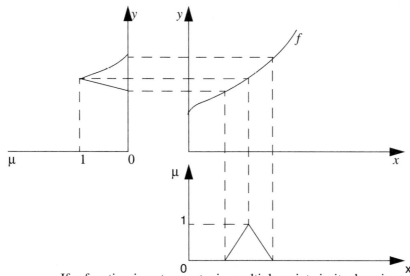

If a function is not monotonic, multiple points in its domain may map to the same point in its range. For instance, let A be a fuzzy number of variable x, f be a second-order polynomial function shown, in Fig 4.7. Points a and b map to the same point c as shown in the figure. Since $f(x) = c$ if either $x=a$ **or** $x=b$, the possibility degree for $f(A)$ to be c is the possibility degree for $x = a$ or $x=b$, i.e.,

$$\Pi_{f(A)} = \Pi_A(a) \oplus \Pi_A(b)$$ **(EQ 4.15)**

If we use the max disjunction operator, we can construct the possibility distribution of $f(A)$, which is shown by thick curves in Fig 4.7. We have introduced the two major concepts in extension principle through this example. We summarize them below:

- The possibility of an input value is directly propagated to the possibility of its image.
- When multiple input combinations map to the same output, the possibility of the output is obtained by combining the possibility of these inputs through fuzzy disjunction.

We give a concrete example below that demonstrates these two concepts.

FIGURE 4.7   Mapping a Possibility Distribution to a Nonmonotonic Function

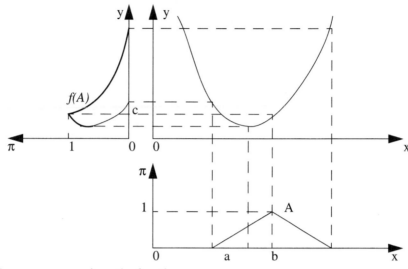

EXAMPLE 3   Suppose we are given the function

$$y = f(x) = (x-3)^2 + 2 = x^2 - 6x + 11 \qquad \text{(EQ 4.16)}$$

and a fuzzy integer number *Around-4* for x as shown in Fig 4.8:

$$Around\text{-}4 = 0.3/2 + 0.6/3 + 1/4 + 0.6/5 + 0.3/6 \qquad \text{(EQ 4.17)}$$

where + denotes union. Applying the extension principle, we get

$$f(Around\text{-}4) = 0.3/f(2) + 0.6/f(3) + 1/f(4) + 0.6/f(5) + 0.3/f(6)$$
$$= 0.3/3 + 0.6/2 + 1/3 + 0.6/6 + 0.3/11 \qquad \text{(EQ 4.18)}$$

Notice that $f(2) = f(4) = 3$. The possibility of the image $y = 3$ is thus a fuzzy disjunction of the possibility of $x = 2$ and $x = 4$. Therefore, we get

$$f(Around\text{-}4) = 0.6/2 + (0.3 \oplus 1)/3 + 0.6/6 + 0.3/11 \qquad \text{(EQ 4.19)}$$

If we choose the "max" fuzzy disjunction operator, we obtain the following final result:

$$f(Around\text{-}4) = 0.6/2 + 1/3 + 0.6/6 + 0.3/11 \qquad \text{(EQ 4.20)}$$

FIGURE 4.8   An Example of the Extension Principle

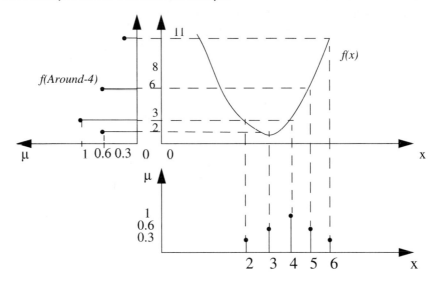

We now are ready to give a formal definition of the extension principle.

DEFINITION 10   Suppose f is a function with n arguments that maps a point in $U_1 \times U_2 \times \ldots \times U_n$ to a point in V. Let A be a fuzzy subset of $U_1 \times U_2 \times \ldots \times U_n$. The *extension principle* states that the image of A under f is a fuzzy subset of V with the following membership function (i.e., possibility distribution):

$$\mu_{f(A)}(y) = \underset{\substack{x_1, x_2, \ldots x_n \\ f(x_1, x_2, \ldots x_n) \,=\, y}}{\oplus} \mu_A(x_1, x_2, \ldots x_n) \qquad \textbf{(EQ 4.21)}$$

In the definition above, we assumed that we use a single fuzzy set to describe the possibility distribution of all arguments. However, fuzzy arguments to a function can each be a possibility distribution (e.g., $x_1 = A_1, x_2 = A_2, \ldots, x_n = A_n$). Under such a circumstance, Equation 4.21 in Definition 10 needs to be modified so that $\mu_A(x_1, x_2, \ldots, x_n)$ in the right hand side of the equation is replaced by a fuzzy conjunction of $\mu_{A_1}(x_1), \mu_{A_2}(x_2), \ldots, \mu_{A_n}(x_n)$. Equation 4.21 thus becomes

$$\mu_{f(A)}(y) = \underset{\substack{x_1, x_2, \ldots, x_n \\ f(x_1, x_2, \ldots, x_n) \,=\, y}}{\oplus} \mu_{A_1}(x_1) \otimes \mu_{A_2}(x_2) \otimes \ldots \otimes \mu_{A_n}(x_n) \qquad \textbf{(EQ 4.22)}$$

In Example 3, the fuzzy argument to the function is a fuzzy integer. How do we handle a fuzzy real number? This is illustrated using the following example.

EXAMPLE 4   We are given the following function f:

$$y = f(x) = \sqrt{x} \qquad \text{(EQ 4.23)}$$

and x is a fuzzy number *Around-3* characterized by the following triangular membership function (see Figure 4.9):

$$\mu_A(x) = \begin{cases} \dfrac{x-1}{2} & 1 \le x \le 3 \\ \dfrac{5-x}{2} & 3 \le x \le 5 \end{cases} \qquad \text{(EQ 4.24)}$$

For convenience, we use B to denote the image of A under f (i.e., $f(A) = B$). We calculate the image for each interval in which μ_A was defined.
Case 1: $1 \le x \le 3$.
Because f is monotonic, the interval maps to $1 \le y \le \sqrt{3}$.
Taking the inverse of f, we get

$$x = f^{-1}(y) = y^2 \qquad \text{(EQ 4.25)}$$

Substituting Equation 4.25 into Equation 4.24, we get B's membership function corresponding to the interval $1 \le y \le \sqrt{3}$:

$$\mu_B(y) = \frac{y^2-1}{2} \qquad 1 \le y \le \sqrt{3} \qquad \text{(EQ 4.26)}$$

Case 2: $3 \le x \le 5$.
Similarly, we get

$$\mu_B(y) = \frac{5-y^2}{2} \qquad \sqrt{3} \le y \le \sqrt{5} \qquad \text{(EQ 4.27)}$$

Combining the two cases we have

$$\mu_B(y) = \begin{cases} \dfrac{y^2-1}{2} & 1 \le y \le \sqrt{3} \\ \dfrac{5-y^2}{2} & \sqrt{3} \le y \le \sqrt{5} \end{cases} \qquad \text{(EQ 4.28)}$$

As illustrated in this example, finding the image of fuzzy arguments to a real function involves finding the inverse of the function. More formally, we can rewrite the extension principle into the following form, which is useful for calculating the image of a fuzzy real argument if the inverse of f exists (i.e. f is monotonic).

$$\mu_{f(A)}(y) = \bigoplus_{\substack{x \\ f(x) = y}} \mu_A(x) \qquad \textbf{(EQ 4.29)}$$

$$= \bigoplus_{\substack{x \\ x = f^{-1}(y)}} \mu_A(f^{-1}(y))$$

If f is nonmonotonic, we apply the similar steps to segments of f that are monotonic.

FIGURE 4.9   The Image of a Fuzzy Real Number under the Function $f(x) = \sqrt{x}$

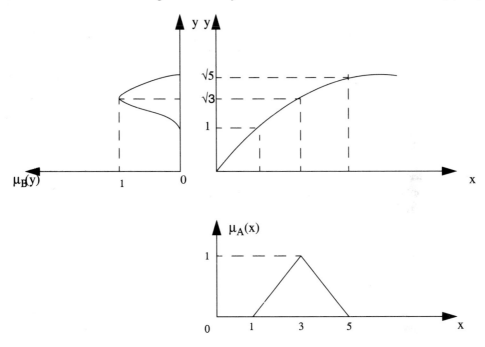

4.6   Arithmetic Operations on Fuzzy Numbers

Applying the extension principle to arithmetic operations (i.e., addition, subtraction, multiplication, and division), we have the following fuzzy arithmetic operations, where x and y are the operands, z is the result, and A and B denote the fuzzy values for x and y, respectively.

Fuzzy Addition:

$$\mu_{A+B}(z) = \bigoplus_{\substack{x, y \\ x + y = z}} \mu_A(x) \otimes \mu_B(y) \qquad \textbf{(EQ 4.30)}$$

Fuzzy Subtraction:

$$\mu_{A-B}(z) = \underset{x,\,y}{\oplus} \mu_A(x) \otimes \mu_B(y) \qquad \text{(EQ 4.31)}$$
$$x - y = z$$

Fuzzy Multiplication:

$$\mu_{A \times B}(z) = \underset{x,\,y}{\oplus} \mu_A(x) \wedge \mu_B(y) \qquad \text{(EQ 4.32)}$$
$$x \times y = z$$

Fuzzy Division:

$$\mu_{A/B}(z) = \underset{x,\,y}{\oplus} \mu_A(x) \wedge \mu_B(y) \qquad \text{(EQ 4.33)}$$
$$x/y = z$$

EXAMPLE 5   Let A and B be two fuzzy integers defined as
$$A = 0.3/1 + 0.6/2 + 1/3 + 0.7/4 + 0.2/5$$
$$B = 0.5/10 + 1/11 + 0.5/12$$
Suppose we want to calculate the sum of these two fuzzy integers. We apply Equation 4.29 and get
$$f(A+B) = 0.3/11 + 0.5/12 + 0.5/13 + 0.5/14 + 0.2/15 + 0.3/12 + 0.6/13$$
$$1/14 + 0.7/15 + 0.2/16 + 0.3/13 + 0.5/14 + 0.5/15 + 0.5/16 + 0.2/17$$
Now we will get the max of the duplicates, so we get
$$f(A+B) = (0.3/11) + 0.5/12 + 0.6/13 + 1/14 + 0.7/15 + 0.5/16 + 0.2/17$$

4.7 Summary

We summarize main ideas of this chapter below.

- A fuzzy relation is a multidimensional fuzzy set that describes the degree of an elastic relationship between two or more variables.

- Two fundamental operations on fuzzy relations are projection and cylindrical extension.

- A composition of two fuzzy relations is an important technique.

- A fuzzy graph is a fuzzy relation formed by pairs of Cartesian products of fuzzy sets.

- A fuzzy graph is the foundation of fuzzy mapping rules, as we will see in the next chapter.

- The extension principle is another important technique in fuzzy logic that allows a fuzzy set to be mapped through a function.

- Addition, subtraction, multiplication, and division of fuzzy numbers are all defined based on the extension principle.

Bibliographical and Historical Notes

Most of the concepts in this chapter were first introduced by L. A. Zadeh in his 1965 seminal paper [708]. They include fuzzy relation, projection, and the notion of composition of two fuzzy relations. The paper also introduced the technique for computing fuzzy sets induced by functional mappings, which is the essence of the extension principle. The name "extension principle" was later introduced by Zadeh in an important 1975 paper [699]. The notion of cylindrical extension was also formalized in the same paper. A more detailed discussion on fuzzy arithmetic can be found in an introductory book to the subject by A. Kaufmann and M. M. Gupta [305].

Exercises

4.1   Compute the projection of the *Petite* relation on the universe of a person's weight.

4.2   Let A and B be two fuzzy subsets of U and V, respectively. Let R be the Cartesian product of A and B. Is the projection of R on U identical to A? If so, prove it. If not, give a counterexample.

4.3   Let R and S be two fuzzy relations defined below:

$$R = \begin{array}{c} \\ x_1 \\ x_2 \end{array} \begin{array}{c} \begin{array}{ccc} y_1 & y_2 & y_3 \end{array} \\ \left[\begin{array}{ccc} 0.0 & 0.2 & 0.8 \\ 0.3 & 0.6 & 1.0 \end{array} \right] \end{array} \qquad \text{(EQ 4.34)}$$

$$S = \begin{array}{c} \\ y_1 \\ y_2 \\ y_3 \end{array} \begin{array}{c} \begin{array}{ccc} z_1 & z_2 & z_3 \end{array} \\ \left[\begin{array}{ccc} 0.3 & 0.7 & 1.0 \\ 0.5 & 1.0 & 0.6 \\ 1.0 & 0.2 & 0.0 \end{array} \right] \end{array} \qquad \text{(EQ 4.35)}$$

(a) Compute the result of $R \circ S$ using sup-min composition.
(b) Compute the result of $R \circ S$ using sup-product composition.

4.1   Use the extension principle to find the image of $\bar{A}$ for the function $f(x) = x^2(x-1)^2$ where $\bar{A}$ is defined by the triangular membership function below.

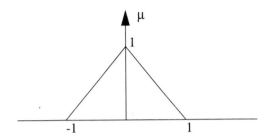

4.2   Compute the following fuzzy algebraic expressions:

- $\tilde{3} + \tilde{4}$
- $\tilde{3} \times \tilde{2}$

where each fuzzy numbers $\tilde{n}$ is characterized by a triangular membership function with a support of $[n - 0.5, n + 0.5]$ like the one shown below:

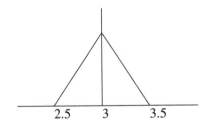

4.3   Compare $\tilde{2} + \tilde{1}$, $\tilde{2} - \tilde{1}$. How "fuzzy" are the two outcomes?

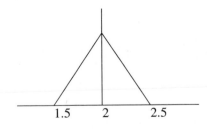

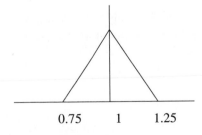

4.4 Compute the value of adding the following two fuzzy integers:

- $a1 = 0.2/2 + 0.5/3 + 1/4 + 0.6/5 + 0.3/6$
- $a2 = 0.3/0 + 0.7/1 + 1/2 + 0.8/3 + 0.4/4$

5   FUZZY IF-THEN RULES

5.1 Introduction

Among all the techniques developed using fuzzy sets, fuzzy if-then rules are by far the most visible due to their wide range of successful applications. Fuzzy if-then rules have been applied to many disciplines such as control systems, decision making, pattern recognition, and system modΓigeling. Fuzzy if-then rules also play a critical role in industrial applications ranging from consumer products, robotics, manufacturing, process control, medical imaging, to financial trading.

A fuzzy if-then rule associates a *condition* described using linguistic variables and fuzzy sets to a *conclusion*. From a knowledge representation viewpoint, a fuzzy if-then rule is a scheme for capturing knowledge that involves imprecision. The main feature of reasoning using these rules (i.e., fuzzy rule-based reasoning) is its *partial matching* capability, which enables an inference to be made from a fuzzy rule even when the rule's condition is only partially satisfied.

Learning fuzzy if-then rules is both easy and challenging. It is easy to understand the basics of fuzzy if-then rules and to start building your own application using a commercially available fuzzy logic development tool such as the Matlab Fuzzy Logic Tool Box. It is challenging, however, to master fuzzy if-then rules for two reasons. First, there are two different kinds of fuzzy rules: *fuzzy mapping rules* and *fuzzy implication rules*. Understanding the fundamental differences between these two is important for developing their successful application. Second, there are several fuzzy rule-based inference techniques. Understanding the theoretical foundation of these techniques and their relationship to suitable applications can be challenging.

We introduced in Chapter 2 the major inference steps of fuzzy rule-based reasoning. In this chapter, we will first discuss the differences between fuzzy mapping rules and fuzzy implication rules. The rest of the chapter is devoted to the foundation and major types of fuzzy mapping rules, while the next chapter will discuss fuzzy implication rules.

5.1.1 Basics of Fuzzy Rules

A fuzzy if-then rule is a knowledge representation scheme for capturing knowledge (typically human knowledge) that is imprecise and inexact by nature. This is achieved by using linguistic variables to describe elastic conditions (i.e., conditions that can be satisfied to a degree) in the "if part" of fuzzy rules.

The main feature of fuzzy rule-based inference is its capability to perform inference under partial matching. That is, it computes the degree the input data matches the condition of a rule. FigFig 5.1 illustrates one way to calculate the matching degree between a fuzzy input A' and a fuzzy condition A that was introduced in Chapter 2:

$$\text{matchingdegree}\,(A, A') = \sup_{x}\;\min(\mu_A(x), \mu_{A'}(x)) \qquad \textbf{(EQ 5.1)}$$

FIGURE 5.1   Matching a Fuzzy Input A' with a Fuzzy Condition A

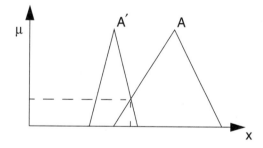

This matching degree is combined with the consequent (i.e., "then" part) of the rule to form a conclusion inferred by the fuzzy rule. The higher the matching degree, the closer the inferred conclusion to the rule's consequent.

Words play an important role in fuzzy rule-based systems. The elastic condition and the consequent of a fuzzy rule are often described by words (i.e., linguistic labels) whose meanings are imprecise. These words facilitate the extraction and the documentation of human knowledge in an explicit and easily comprehensible form, especially those that are imprecise by nature. Words in fuzzy rules differ from symbols in classical rules in Artificial Intelligence (AI) — the meaning of symbols for a numeric variable is often described using intervals, whereas the meaning of words in a fuzzy rule is characterized by membership functions that smooth the sharp boundaries of intervals. Because of the importance of these "words" in a fuzzy rule, fuzzy rules and, more generally, fuzzy logic are referred to as a paradigm called "computing with words" [680].

5.2 Two Types of Fuzzy Rules

There are two types of fuzzy rules: 1) *fuzzy mapping rules*, and 2) *fuzzy implication rules*. A fuzzy mapping rule describes a functional mapping relationship between inputs and an output using linguistic terms, while a fuzzy implication rule describes a generalized logic implication relationship between two logic formulas involving linguistic variables and imprecise linguistic terms. The foundation of a fuzzy mapping rule is a fuzzy graph, while the foundation of a fuzzy implication rule is the narrow sense of fuzzy logic. The two types of fuzzy rules are also related to different disciplines. Fuzzy mapping rules are related to other function approximation techniques in system identification and artificial neural networks, whereas fuzzy implication rules are related to classical two-valued logic and multivalued logic. We will discuss these two types of rules below.

5.2.1 Fuzzy Mapping Rules

In many real-world problems, we are interested in finding the functional relationship between a set of observable parameters and one or multiple parameters whose values we do not know. For example, a financial trading broker is interested in the relationship between the history of various financial indices and future stock values. A control engineer is interested in finding an appropriate control law, which can also be viewed as a functional relationship between observable state variables and controllable parameters. Indeed, fuzzy logic controllers use rules of this type to approximate a mapping (typically nonlinear) from the observed state to a desired control action. In fact, most industrial applications of fuzzy logic use fuzzy mapping rules.

The needs to approximate a function of interest is often due to one or more of the following reasons. First, the mathematical structure of the function is not precisely known. Were the structure known, one could use various parameter identification techniques to find the parameters. Second, the function is so complex that finding its precise mathematical form is either impossible or practically unfeasible due to its high cost. Third, even if finding the function is not impractical, implementing the function in its precise mathematical form in a product or service may be too costly. This is particularly important for low-cost high volume products (e.g., automobiles, cameras, and other consumer products).

Even though human insights about the function can often be captured using fuzzy rules, the focus of this type of fuzzy rule inference is on exploring the trade-off between accuracy and cost. This actually echoes the motivation for developing fuzzy logic technology that we have discussed in Chapter 1. Because each fuzzy rule approximates a small segment of the function, the entire function is approximated by a set of fuzzy mapping rules. We will refer to such a collection of fuzzy mapping rules as *fuzzy rule-based models* or simply *fuzzy models*. The inference (i.e., mapping) for this type of rule is always in the forward direction. The main difference between fuzzy rules and non-fuzzy rules for function approximation lies in their "interpolative reasoning" capability, which allows the output of multiple fuzzy rules to be fused for a given input. As we will see later, the concept of interpolative reasoning can be viewed as a kind of partition-based modeling in system identification.

Function approximation techniques can be broadly classified in three categories: global techniques, superimposition techniques, and partition-based techniques. The global techniques approximate a function globally using one mathematical structure (e.g., linear, second-order polynomial, etc.). A major issue of this technique lies in finding the suitable model structure for a given problem. The superimposition techniques approximate a function by superimposing functions of a given form (e.g., B-splines, Taylor expansions). The partition-based approximate techniques approximate a function by partitioning the input space of the function and approximate the function in each partitioned region separately (e.g., piecewise linear approximation). The latter two are also referred to as *local modeling* techniques, while the first is referred to as a *global modeling* technique.

Fuzzy rule-based function approximation is a partition-based technique. It generalizes classical partition by allowing a subregion to **partially overlap** with neighboring subregions. We will refer to this as a *fuzzy partition*. In the input space where subregions partially overlap with each other, the function is approximated using a kind of *interpolation* technique. We will elaborate on these concepts in Section 5.5.

5.2.2 Fuzzy Implication Rules

Fuzzy implication rules are a generalization of "implication" in two-valued logic. Their aim is to mimic human reasoning in its ability to reason with ideas or statements that are imprecise by nature. Even though it is obvious that human beings perform certain kinds of approximate reasoning, it is unclear how one can characterize such a reasoning process due to our present limited understanding of the human reasoning process to date. Consequently, there has not been posited a unique set of desired properties for fuzzy implication rules. Rather, several sets of desired properties have been developed, as we shall see later in the next chapter. Even though the set of desired properties is not unique, they are useful for comparing different reasoning schemes using fuzzy implication rules. We want to point out this "nonuniqueness" aspect of fuzzy implication, because it is in sharp contrast with many other techniques in science and engineering that were built on a set of well-defined axioms, properties, or principles.

The inference of fuzzy implications generalizes two kinds of logic inference using the implications in classical logic: *modus ponens* and *modus tollens*. While we will describe these inferences in detail in the next chapter, we can briefly illustrate here these logic inferences using the following implication:

IF a person's IQ is high THEN the person is smart.

Given the implication and the fact that "Jack's IQ is high", modus ponens enables us to infer that "Jack is smart." On the other hand, if we are given the implication and the fact that "Jack is not smart," modus tollens enables us to infer that "Jack's IQ is not high."

Even though classical logic can perform these two kinds of inferences, they are limited in two ways. First, these inferences insist on perfect matching. If Jack's IQ is more or less high, for instance, no modus ponens inference can be made from the implication above because "more or less smart" is not identical to "smart." However, our "common-sense reasoning" usually suggests that we can infer that "Jack is more or less smart." Second, these inferences cannot handle uncertainty. For instance, if Jack told us his IQ is high but cannot provide documents supporting the claim, we may be somewhat uncertain about the claim. Under such a circumstance, however, logic cannot reason about the uncertainty.

These limitations motivated L.A. Zadeh to develop a reasoning scheme that generalizes classical logic so that (1) it can conduct "common-sense" reasoning under partial matching, and (2) it can reason about the certainty degree of a statement. In particular, logic implications are generalized to allow partial matching. In our example, this means that we can infer "Jack is somewhat smart" from "Jack's IQ is somewhat high" and the fuzzy implication "If Jack's IQ is *High*, then Jack is *Smart*," as shown below.

Given: Jack's IQ is *High* $\rightarrow$ Jack is *Smart*.
Jack's IQ is *somewhat High*.

Infer: Jack is *somewhat Smart*.

The second limitation of logic (i.e., inability to deal with uncertainty) has motivated another extension to classical logic: multivalued logic. Multivalued logic extends the truth values in logic from binary (i.e., true and false) to multiple (i.e., more than two) so that different degrees of certainty are expressed by different truth values. Since fuzzy logic also generalizes the truth values in classical logic beyond true and false, it is related to multivalued logic, as we shall see later in the next chapter. However, fuzzy logic differs from multivalued logic in that it also addresses the first limitation of logic (i.e., restricted to a perfect match). The reason that a fuzzy implication can address the first problem is that it uses linguistic variables in its antecedent (i.e., if parts). Consequently, the statement in the antecedent describes an elastic condition that can be partially satisfied. In our example " A person's IQ" is a linguistic variable in the antecedent, and " A person's IQ is high" describes an elastic condition about a person's IQ. This enables a person with a somewhat high IQ to partially match the condition. Obviously, the degree to which the condition in a fuzzy implication is matched influences the inference that can be made from the fuzzy implication. We will discuss the foundation of this inference scheme by fuzzy implications in the next chapter.

Other approaches for reasoning under uncertainty include Bayesian probabilistic inference, the Dempster-Shafer theory, and nonmonotonic logic. While these approaches have their merits for reasoning under uncertainty, fuzzy logic is unique in that it addresses both the uncertainty management problem and the partial matching issue.

The distinctions between fuzzy implication rules and fuzzy mapping rules are subtle, yet important. Prof. Zadeh mentioned in this book's Forward that fuzzy logic has four principle facets: (1) the logical facet, (2) the set-theoretic facet, (3) the relational facet, and (4) the epistemic facet. Fuzzy implication rules are primarily in the logic facet, whereas fuzzy mapping rules are in the relational facet because functional mapping is a kind of relation. We summarize several major differences between these two types of fuzzy rules in Table 5.1.

TABLE 5.1   A Comparison of Two Types of Fuzzy Rules

	Fuzzy Implication Rules	Fuzzy Mapping Rules
Purpose	generalize implications for handling imprecision	approximate functional mappings
Desired Inference	generalized modus ponens and modus tollens	forward only
Application	diagnostics, high-level decisionmaking	control, system modeling, and signal processing
Related Disciplines	classical logic, multivalued logic (other extended logic systems	system ID, piecewise linear interpolation, nueral networks
Typical Design Approach	designed individually	designed as a rule set
Suitable Problem Domains	domains with continuous and discrete variables	continuous nonlinear domains

The remainder of this chapter will focus on fuzzy mapping rules, because they are central to fuzzy logic control and most industrial applications. We will discuss fuzzy implication rules in detail in the next chapter.

5.3 Fuzzy Rule-based Models for Function Approximation

For the convenience of our discussion, we will refer to a model that describes a functional mapping relationship using a set of fuzzy mapping rules as a *fuzzy rule-based model*, or simply *fuzzy model*. Even though the successful applications of this kind of model in control have created the popular image of fuzzy logic, we should point out that there are other kinds of models in fuzzy logic technology (e.g., fuzzy models for pattern recognition) that do not use rules. We will introduce those models in later chapters.

A fuzzy model describes a mapping (i.e., function) from a set of input variables to a set of output variables. A fuzzy model can be used in various ways. For example, a fuzzy model of the stock market can be used to predict future changes of The Dow Jones Average. A fuzzy model of a petrochemical process can be used to predict the future state of the process. Similarly, a fuzzy model of a subway control system can be used to predict the future state of a subway train.

What is a fuzzy model? A fuzzy model is a model that is obtained by *fusing* multiple *local models* that are associated with fuzzy subspaces of the given input space. A fuzzy subspace is a region whose boundary allows a gradual transition from "inside the region" to "outside the region." Hence, a fuzzy subspace usually partially overlaps with its neighboring fuzzy subspaces. A fuzzy model contains a set of fuzzy subspaces that form a fuzzy decomposition (i.e., also called *fuzzy partition*) of the input space. The result of fusing multiple local models is usually a fuzzy conclusion, which is converted to a crisp final output through a *defuzzification process*. The four major concepts in fuzzy rule-based models thus are (1) fuzzy partition, (2) mapping of fuzzy subregions to local models, (3) fusion of multiple local models, and (4) defuzzification. These four concepts together enable the construction of a complex global model using a set of simpler local models.

5.3.1 Fuzzy Partition

A classical partition of a space is a collection of disjoint subspaces whose union is the entire space. For example, Fig 5.2 shows an example of classical partition for an input space involving two variables: X_1 and X_2. The partition is constructed by first partitioning the range of X_1 into four intervals (i.e., I_1, I_2, I_3 and I_4), and the range of X_2 into three intervals (i.e., J_1, J_2 and J_3). Each pair of intervals I_k and J_l specifies a subspace A_{kl} that can be described as

$$X_1 \; \varepsilon \; I_k \text{ and } X_2 \; \varepsilon \; J_l$$

A total of 12 subspaces partition the entire input space in this example.

A fuzzy partition generalizes classical partition so that the transition from one subspace into a neighboring one is smooth. That is, the membership degree of a point X in one subspace A increases while its membership degree in a neighboring subspace B decreases as X gradually moves out of A and moves into B. As we shall see later, this generalization enables interpolative reasoning in fuzzy models.

FIGURE 5.2   An Example of Classical Partition

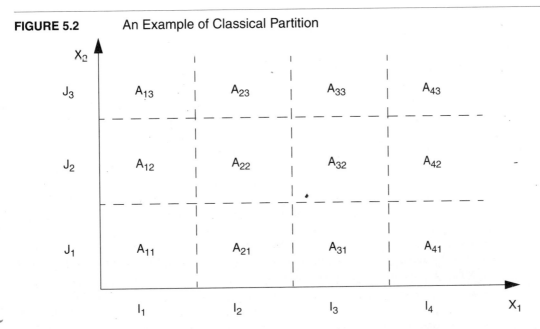

Fig 5.3 shows a fuzzy partition for a system that controls the contrast of a TV based on two inputs: (1) the distance between the viewer and the TV and (2) the brightness of the room. In this example, the distance between the viewer and the TV, denoted d, is classified

into three fuzzy sets: *Near, Medium,* and *Far.* Similarly, the brightness of the room, denoted by *b*, is classified into two fuzzy sets: *Bright* and *Dark.* A fuzzy partition of the entire input space is thus formed by six fuzzy subregions specified by pairs of fuzzy sets, (i.e., one for each variable). Fig 5.3, shows two of the six fuzzy subregions:

d is *Medium* AND *b* is *Dark.*

d is *Far* AND *b* is *Dark*

As shown in the figure, the to fuzzy subregions partially overlap with each other because fuzzy sets *Medium* and *Far* partially overlap. Obviously, the same observation can be made for other neighboring fuzzy regions.

FIGURE 5.3   An Example of Fuzzy Partition

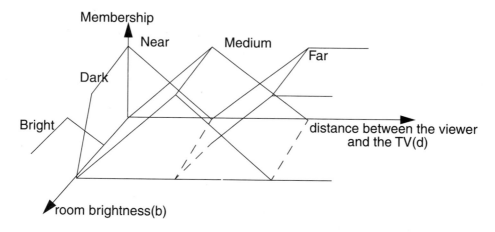

In a general sense, a fuzzy partition of a space is a collection of fuzzy subspaces whose boundaries partially overlap and whose union is the entire space. Even though there are different operators to choose from for the union operation in fuzzy sets (as discussed in Chapter 3), the one that is most suitable for defining fuzzy partition is the addition operator. Thus, we can formally define a *fuzzy partition* of a space *S* as a collection of fuzzy subspaces A_i of *S* that satisfies the following condition:

$$\sum_i \mu_{A_i}(x) = 1 \quad \forall\, x \in S \qquad \textbf{(EQ 5.2)}$$

That is, for any element of the space, its membership degree in all subspaces always adds up to 1. Assuming that we use the addition operator to compute the union, it is easy to show that the condition above is equivalent to

$$\bigcup_i A_i = S \qquad \textbf{(EQ 5.3)}$$

We leave this as an exercise. Even though many fuzzy systems developed in the real world satisfy Equation 5.2, not all researchers in the fuzzy logic community adopt the definition above for fuzzy partition. A somewhat relaxed definition of fuzzy partition is one that replaces the sum-to-one condition in Equation 5.2, with the condition that membership degrees in all subspaces sum to a number greater than 0, but not greater than 1. We call a collection of fuzzy subsets A_i of S a *weak fuzzy partition* of S if and only if it satisfies the following condition:

$$0 < \sum_i \mu_{A_i}(x) \le 1 \quad \forall x \in S \qquad \text{(EQ 5.4)}$$

The "greater than 0" condition requires each element in the space S to be covered by at least one fuzzy subspace in the partition. In an intuitive sense, this means that a fuzzy partition does not leave any "holes." The "sum to 1" condition of a fuzzy partition can be relaxed to the "sum to less or equal to 1" condition because the interpolative reasoning of fuzzy models includes a normalization step. It is obvious that a fuzzy partition is always a weak fuzzy partition, but not vice versa.

Research Note: By generalizing the interpolative reasoning capability in fuzzy logic to use both "distance" and "membership degree" (instead of using only "membership degrees"), fuzzy logic researchers have generalized the notion of "fuzzy partition" by removing the "greater than 0" condition, i.e., $\sum_i A_i(x) < 1$. This is because even if an element x is not covered by any fuzzy subspaces, distance-based interpolative reasoning can still be performed on x by using fuzzy subspaces in the partition that are close to x. The term "fuzzy partition," however, may no longer be appropriate to describe such a set of fuzzy subspaces because its analogy to the conventional notion of "partition" is very weak. A more appropriate term might be "fuzzy cases" or "fuzzy patches." It should be emphasized that "fuzzy partition" and "interpolative reasoning" are like two sides of a coin. Extending one usually requires extending the other. The two together makes it possible for fuzzy logic to exploit the trade-off between cost and precision we discussed in Chapter 1.

Research Note: It has been shown that $\sum_i A_i(x) = 1$ is a desirable property in a framework for analyzing the stability of fuzzy logic controllers.

5.3.2 Mapping a Fuzzy Subspace to a Local Model

A *local model* for a subspace of the entire input space describes the system's input-output mapping relationship in the small subspace. In contrast, a *global model* for an input space describes the system's input-output relationship for the entire input space. Because the scope of the local model is smaller than that of a global model, it is usually easier to develop a local model. In particular, a nonlinear global model (i.e., whose input-output mapping function is not linear) can often be approximated by a set of linear local models. This can be understood by remembering the well-known approximation technique called "piecewise linear approximation," which approximates an arbitrary nonlinear function using segments of lines. An example of such an approximation technique is shown in Fig 5.4, where the dotted line indicates the function being approximated.

Piecewise linear approximation has two major components:

1. **partitioning** the input space to crisp regions,

2. **mapping** each partitioned region to a linear local model.

The main difference between fuzzy modeling and piecewise linear approximation is that the transition from one local subregion to a neighboring one is gradual rather than abrupt.

FIGURE 5.4   An Example of Piecewise Linear Approximation

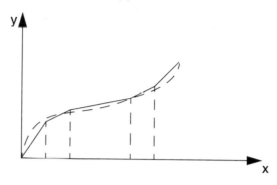

Generally, the mapping from a fuzzy subspace to a local model is represented as a fuzzy if-then rule in the form of

$$\text{IF } \grave{x} \text{ is in } FS_i \text{ THEN } y_j = LM_i(\grave{x})$$

where $\grave{x}$ and y_j denote the vector of input variables and an output variable respectively, FS_i and LM_i denote the ith. fuzzy subspace and the corresponding local model, respectively. The local model is, in general, a function of the input variables, even though they can be constant, as we will see in the next section.

Because fuzzy mapping rules describe a functional relationship between input and output, two rules that associate the same fuzzy subregion to two different local models are said to be *inconsistent*. For instance, the two rules below are inconsistent because they associate two different local models to a fuzzy subspace (i.e. x is *small*).

$$\text{IF } x \text{ is } \textit{Small} \text{ THEN } y \text{ is } \textit{Large}$$
$$\text{IF } x \text{ is } \textit{Small} \text{ THEN } y \text{ is } \textit{Small}.$$

The local model associated with each fuzzy subspace can be of four different types: (1) a crisp constant, (2) a fuzzy constant, (3) a linear model, or (4) a nonlinear model. The one that is most frequently used among fuzzy logic applications is the second type: fuzzy constant. We discuss each type of local model below.

1. *Crisp Constant:* This type of local model is simply a crisp (nonvisual) constant (e.g., 4.5), e.g.,

$$\text{IF } x_2 \text{ is Small THEN } y = 4.5.$$

2. *Fuzzy Constant:* A local model that is a fuzzy constant (e.g., SMALL) belong to this type. An example is given below:

$$\text{IF } x \text{ is Small THEN } y \text{ is Medium}.$$

3. *Linear Model:* This kind of local model describes the output as a linear function of the input variables, such as in the rule below:

$$\text{IF } x \text{ is small AND } x_2 \text{ is Large THEN } y = 2x_1 + 5x_2 + 3.$$

4. *Nonlinear Model:* Theoretically speaking, a local model can be more complex than a linear model. In practice, however, there is rarely such a need. A fuzzy model that

fuses local models of the first three types is often sufficient to approximate an unknown complex plant or system. Nonlinear local models have been introduced in a hybrid neuro-fuzzy system that uses neural networks to represent nonlinear local models associated with the rule [553].

Research Note: The integration of fuzzy models and nonlinear models is a promising research area that may provide solutions to problems that are more complex than those which fuzzy models alone have been able to solve to date. Ideally, such a hybrid modeling approach could benefit from the insights into the mathematical structure of the model as well as the flexibility of fuzzy models.

5.3.3 Fusion of Local Models through Interpolative Reasoning

Fuzzy models use *interpolative reasoning* to fuse multiple local models into a global model. The basic idea behind *interpolative reasoning* is analogous to drawing a conclusion from a panel of experts, each of whom is specialized in a subarea of the entire problem. Each expert's opinion is associated with a weight, which reflects the degree to which the current situation is in the expert's specialized subarea. These weighted opinions are then combined to form an overall opinion. In this analogy, an expert corresponds to a fuzzy if-then rule, the specialized subarea of the expert corresponds to the fuzzy subspace associated with the if-part of the rule. The weight of an expert's opinion is determined by the degree to which the current situation belongs to the subspace.

5.3.4 Defuzzification

We may interpret a possibility distribution either through *linguistic approximation*, or through *defuzzification*. The former gives a qualitative interpretation, while the latter gives a quantitative summary and is more commonly used in fuzzy logic control and many other industrial applications. Given a possibility distribution of a fuzzy model's output, defuzzification amounts to selecting a single representative value that captures the essential meaning of the given distribution. There are two common defuzzification techniques: mean of maximum and center of area.

5.3.4.1 Mean of Maximum (MOM)

The Mean of Maximum (MOM) defuzzification method calculates the average of those output values that have the highest possibility degrees. Suppose "y is A" is a fuzzy conclusion to be defuzzified. We can express the MOM defuzzification method using the following formula:

$$MOM(A) = \frac{\sum\limits_{y^* \in P} y^*}{|P|} \qquad \text{(EQ 5.5)}$$

where P is the set of output values y with highest possibility degree in A, i.e.,

$$P = \left\{ y^* \,\middle|\, \mu_A(y^*) = \sup_y \mu_A(y) \right\} \qquad \text{(EQ 5.6)}$$

If P is an interval, the result of MOM defuzzification is obviously the midpoint in that interval. An Example of MOM defuzzification is shown in Fig 5.5.

FIGURE 5.5   An Example of MOM Defuzzification

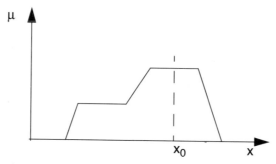

A major limitation of MOM defuzzification is that it does not take into account the overall shape of the possibility distribution. Two fuzzy conclusions with the same peak points, but otherwise different shapes, will yield the same deffuzzified result using the MOM method. This is counterintuitive. Fig 5.6 depicts an example of such results.

5.3.4.2   Center of Area (COA)

The Center-of-Area (COA) method (also referred to as the center-of-gravity, or centroid method in the literature) is the most popular defuzzification technique. Unlike MOM, the COA method takes into account the entire possibility distribution in calculating its representative point. The defuzzification method is similar to the formula for calculating the center of gravity in physics, if we view $\mu_A(x)$ as the density of mass at x. Alternatively, we can view the COA method as calculating a weighted average, where $\mu_A(x)$ serves as the weight for value x. If x is discrete, the defuzzification result of A is

$$COA(A) = \frac{\sum_x \mu_A(x) \times x}{\sum_x \mu_A(x)} \qquad \text{(EQ 5.7)}$$

Similarly, if x is continuous, the result is

$$COA(A) = \frac{\int \mu_A(x) x \, dx}{\int \mu_A(x) \, dx} \qquad \text{(EQ 5.8)}$$

FIGURE 5.6   A Counterintuitive Result of MOM Method

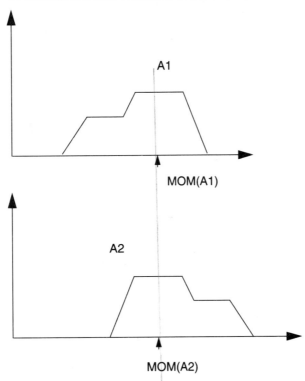

An example of COA defuzzification is shown in Fig 5.7. The main disadvantage of the COA method is its high computational cost. However, as we will see later in Section 5.8, the calculation can be simplified for some fuzzy models.

So far, we have assumed that defuzzification follows the fusion step described in the previous section. Even though this is the case for most fuzzy rule-based systems, there are situations in which defuzzification cannot be completely separated from the fusion step. We will discuss such a case in mobile robot navigation in Chapter 11.

FIGURE 5.7   An Example of COA Defuzzification

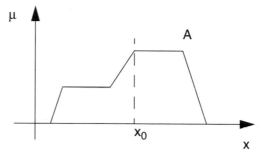

5.3.4.3 The Height Method

The third defuzzification technique is the *height method,* which can be viewed as a two-step procedure. First, we convert the consequent membership function C_i into crisp consequent $y = c_i$ where C_i is the center of gravity of C_i. The centroid defuzzification is then applied to the rules with crisp consequents, which gives us the following formula:

$$y = \frac{\displaystyle\sum_{i=1}^{M} w_i c_i}{\displaystyle\sum_{i=1}^{M} w_i} \tag{EQ 5.9}$$

where w_i is the degree to which the ith rule matches the input data. The main benefit of the height method is its simplicity. The calculation of C_i can be preformed during compilation. Consequently, the only computation required during run time is a normalized weighted sum. This not only reduces the computation cost required by the previous two defuzzification techniques but also facilitates the application of neural networks learning to fuzzy systems. Hence, many well-known neuro-fuzzy models [382, 449, 621, 619] use this type of defuzzification method. The main disadvantage of this method is that it is not well justified and is often considered an approximation to the centroid defuzzification.

5.4 A Theoretical Foundation of Fuzzy Mapping Rules

5.4.1 A Mathematical Representation of Fuzzy Mapping Rules

A fuzzy mapping rule imposes an elastic constraint on possible associations between input and output variables. The constraint is elastic because a fuzzy rule can describe input-output associations that are somewhat possible (i.e., the gray area between totally possible and totally impossible). The degree of possibility of an input-output association imposed by a rule R can be expressed as a possibility distribution, denoted by Π_R.

Since a fuzzy relation is a general way for describing a possibility distribution, it is natural to use it to represent the possibility distribution imposed by a fuzzy rule. Even though this idea seems appealing, a major question remains to be answered: How do you construct the fuzzy relation that represents fuzzy mapping rules? The answer is to use the concept of *Cartesian product* introduced in Chapter 4.

A fuzzy mapping rule is represented mathematically as *fuzzy relations* formed by the Cartesian product of the variables referred to in the rule's if-part and then-part. For example, the mapping rule

IF x *is A*, THEN *y is B*

is mathematically represented as a fuzzy relation R defined as $\mu_R(x, y) = \mu_{A \times B}(x, y)$. If we use the min operator for the Cartesian product, the fuzzy relation R becomes $\mu_R(x, y) = min\{\mu_A(x), \mu_B(y)\}$. The two-dimensional membership function of a fuzzy relation constructed this way is depicted in Fig 5.8

FIGURE 5.8   A Fuzzy Relation Formed by a Fuzzy Mapping Rule

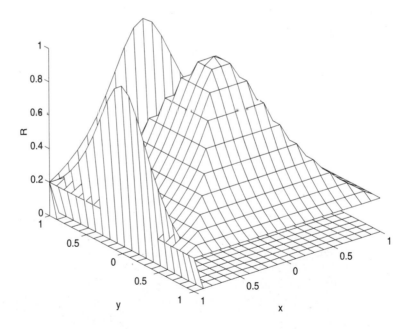

EXAMPLE 6   Let us consider the following fuzzy mapping rule from X to Y where $X = \{2, 3, 4, 5, 6, 7, 8, 9\}$ and $Y = \{1, 2, 3, 4, 5, 6\}$:

IF x *is Medium*, THEN *y is Small*

where *Medium* and *Small* are fuzzy subsets of X and Y characterized by the following membership functions:

$Medium \equiv 0.1/2 + 0.3/3 + 0.7/4 + 1/5 + 1/6 + 0.7/7 + 0.5/8 + 0.2/9$

$Small \equiv 1/1 + 1/2 + 0.9/3 + 0.6/4 + 0.3/5 + 0.1/6$

The fuzzy relationship R representing the rule is the Cartesian product of *Medium* and *Small*. If we use the min operator to construct the Cartesian product, we have

$$\mu_R(x, y) \;=\; \min\{\mu_{Medium}(x), \mu_{Small}(y)\} \qquad \textbf{(EQ 5.10)}$$

The resulting fuzzy relation representing the rule is

$$R \equiv \begin{bmatrix} 0.1 & 0.1 & 0.1 & 0.1 & 0.1 & 0.1 \\ 0.3 & 0.3 & 0.3 & 0.1 & 0.3 & 0.1 \\ 0.7 & 0.7 & 0.7 & 0.6 & 0.3 & 0.1 \\ 1 & 1 & 0.9 & 0.6 & 0.3 & 0.1 \\ 1 & 1 & 0.9 & 0.6 & 0.3 & 0.1 \\ 0.7 & 0.7 & 0.7 & 0.6 & 0.3 & 0.1 \\ 0.5 & 0.5 & 0.5 & 0.5 & 0.3 & 0.1 \\ 0.2 & 0.2 & 0.2 & 0.2 & 0.2 & 0.1 \end{bmatrix} . \qquad \textbf{(EQ 5.11)}$$

5.4.2   The Foundation of Fuzzy Rule-based Models: The Fuzzy Graph

The theoretical foundation of fuzzy mapping rules is a fuzzy graph and a compositional rule of inference, which were introduced in Chapter 4. A Fuzzy graph can be conveniently described by fuzzy rules in the form of

IF x is A THEN y is B.

Such a statement (or rule), as pointed out by Zadeh, generalizes the dependency relationship between variables in a lookup table such as

IF x is 5 THEN y is 10

IF x is 10 THEN y is 14.

A set of such dependencies forms a functional mapping from x to y. Generalizing these point-to-point mappings to a mapping from fuzzy sets to fuzzy sets introduces two benefits. (1) We can reduce the total number of point-to-point rules required for approximating a function. (2) Using words in fuzzy rules makes it easier to capture, understand, and communicate the underlying human knowledge.

Let f^* be a fuzzy graph described by a set of fuzzy mapping rules in the form of

IF x is A_i then y is B_i.

As we have discussed in Chapter 4, the fuzzy graph can be expressed mathematically as

$$f^* = \bigcup_i A_i \times B_i \qquad \textbf{(EQ 5.12)}$$

The inference (i.e., interpolative reasoning) of such a fuzzy rule-based model is based on the *compositional rule of inference*, which we also introduced in Chapter 4. The net effect is a possibility distribution over the domain of definition of the output variable. In particular,

$$B' = A' \circ f^* \tag{EQ 5.13}$$

where f^* represents the fuzzy graph of a given fuzzy model, A' is an input which can be fuzzy or crisp, and B' is the inferred output value before defuzzification. Using the definition of a compositional rule of inference, we can express this as

$$A' \circ f^* = \text{Proj}_Y(\overline{A'} \cap f^*) \tag{EQ 5.14}$$

$$= \text{Proj}_Y[\overline{A'} \cap (\bigcup_i A_i \times B_i)]$$

$$= \bigcup_{x \in X}[\overline{A'} \cap (\bigcup_i A_i \times B_i)]$$

where X and Y are the universe of discourse of x and y respectively, and $\overline{A'}$ denotes the cylindrical extension of A' to $X \times Y$

EXAMPLE 7   Consider the following rule in Example 6:

$$\text{IF } x \text{ is } Medium \text{ THEN } y \text{ is } Small$$

and the input data

$$x \text{ is } Small$$

where $Small$ for x is defined as

$$Small \equiv 1/1 + 0.9/2 + 0.6/3 + 0.3/4 + 0.1/5$$

To find out the possible values of y, we compose the possible values of x with the fuzzy relation R in Equation 5.11 using the sup-min composition:

$$small \circ R = \begin{bmatrix} 1 & 0.9 & 0.6 & 0.3 & 0.1 & 0 & 0 & 0 \end{bmatrix} \circ \begin{bmatrix} 0.1 & 0.1 & 0.1 & 0.1 & 0.1 & 0.1 \\ 0.3 & 0.3 & 0.3 & 0.1 & 0.3 & 0.1 \\ 0.7 & 0.7 & 0.7 & 0.6 & 0.3 & 0.1 \\ 1 & 1 & 0.9 & 0.6 & 0.3 & 0.1 \\ 1 & 1 & 0.9 & 0.6 & 0.3 & 0.1 \\ 0.7 & 0.7 & 0.7 & 0.6 & 0.3 & 0.1 \\ 0.5 & 0.5 & 0.5 & 0.5 & 0.3 & 0.1 \\ 0.2 & 0.2 & 0.2 & 0.2 & 0.2 & 0.1 \end{bmatrix} \tag{EQ 5.15}$$

$$= \begin{bmatrix} 0.6 & 0.6 & 0.6 & 0.6 & 0.3 & 0.1 \end{bmatrix} \tag{EQ 5.16}$$

Therefore, the result of the inference is

$$y = 0.6/2 + 0.6/3 + 0.6/4 + 0.6/5 + 0.3/6 + 0.1/7$$

In this example, we consider only one rule. However, a fuzzy model for function approximation is usually formed by a set of fuzzy mapping rules. In such a case, the fuzzy relation of the entire model (denoted FM) is constructed by forming the union of fuzzy relations of individual rules:[1]

$$\mu_{FM} = \mu_{R_1} \cup \mu_{R_2} \cup \ldots \cup \mu_{R_n} \qquad \text{(EQ 5.17)}$$

The compositional rule of inference is then applied to the model's input and the fuzzy relation representing the model to obtain the model's output. Different types of fuzzy models differ in their choice of compositional operators and their fuzzy disjunction (i.e., union) operators in combining multiple rules. We discuss these differences in the rest of this chapter.

5.5 Types of Fuzzy Rule-based Models

There are three types of fuzzy rule-based models for function approximation: (1) the Mamdani model, (2) the Takagi-Sugeno-Kang (TSK) model, and (3) Kosko's additive model (SAM). The inference scheme of SAM is similar to that of TSK model. Both of them use an inference analogous to the weighted sum to aggregate the conclusion of multiple rules into a final conclusion. Therefore, we refer to these rule models as additive rule models. In contrast, the Mamdani model combines inference results of rules using superimposition, not addition. Hence, it is a *nonadditive rule model*. Figure 5.9 shows the classification of these fuzzy rule-based models using a tree structure,

FIGURE 5.9   A Classification of Fuzzy Rule-based Models for Function Approximation

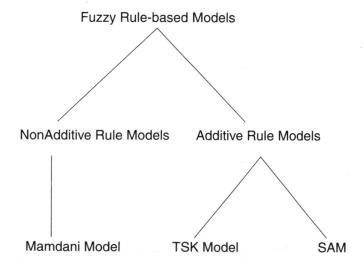

---

[1] Because these fuzzy mapping rules have the same antecedent variables and consequent variables, their fuzzy relations R_1, R_2, ..., R_n are defined in the same space.

The first rule-based model developed was the Mamdani model, which was named after E. H. Mamdani who developed the first fuzzy logic controller using the model. Most fuzzy control systems developed in the 80s use the Mamdani model. The Takagi-Sugeno-Kang (TSK) model was first introduced by T. Takagi and Prof. M. Sugeno around 1985. Another student of Sugeno, K. T. Kang, continued to work on applications and identification of the model. The TSK model has drawn much more attention in the 90's, both in the research community and in industry. One of the main advantages of the TSK model is that it can approximate a function using fewer rules.

The Mamdani model and SAM use rules whose consequent part is a fuzzy set:

$$R_i: If \quad x_1 \text{ is } A_{i1} \text{ and } x_2 \text{ is } A_{i2} \text{ and } ... \text{ and } x_s \text{ is } A_{is}$$
$$Then \quad y \text{ is } C_i, i = 1, 2, ..., M$$

(EQ 5.18)

where M is the number of fuzzy rules, $x_j \in U_j$ ($j = 1, 2, ..., s$) are the input variables, $y \in V$ is the output variable, and A_{ij} and C_i are fuzzy sets characterized by membership functions $\mu_{A_{ij}}(x_j)$ and $\mu_{C_i}(y)$, respectively.

These two fuzzy rule-based models differ in their inference schemes. The TSK model uses a rule whose then part is a linear model:

$$R_i: If \quad x_1 \text{ is } A_{i1} \text{ and } x_2 \text{ is } A_{i2} \text{ and } ... \text{ and } x_s \text{ is } A_{is}$$
$$Then \quad y = f_i(x_1, x_2, ..., x_s), \quad i = 1, 2, ..., M$$

where f_i is a linear function. We discuss each model as well as its foundation below.

5.6   The Mamdani Model

One of the most widely used fuzzy models in practice is the *Mamdani model* [401], which consists of the following linguistic rules that describe a mapping from $U_1 \times U_2 \times ... \times U_r$ to W.

$$R_i: \text{IF } x_1 \text{ is } A_{i1} \text{ and } ... \text{ and } x_r \text{ is } A_{ir} \text{ THEN } y \text{ is } C_i$$

where x_j ($j = 1, 2, ..., r$) are the input variables, y is the output variable, and A_{ij} and C_i are fuzzy sets for x_j and y respectively. Given inputs of the form:

$$x_1 \text{ is } A'_1, x_2 \text{ is } A'_2, ..., x_r \text{ is } A'_r$$

where A'_1, A'_2, ..., A'_r are fuzzy subsets of $U_1, U_2, ... U_r$ (e.g., fuzzy numbers), the contribution of rule R_i to a Mamdani model's output is a fuzzy set whose membership function is computed by

$$\mu_{C'_i}(y) = (\alpha_{i1} \wedge \alpha_{i2} \wedge ... \wedge \alpha_{in}) \wedge \mu_{C_i}(y) \qquad \textbf{(EQ 5.19)}$$

where α_i is the matching degree (i.e., firing strength) of rule R_i, and where α_{ij} is the matching degree between x_j and R_i's condition about x_j.

$$\alpha_{ij} = \sup_{x_j}(\mu_{A'_j}(x_j) \wedge \mu_{A_{ij}}(x_j)) \qquad \textbf{(EQ 5.20)}$$

and $\wedge$ denotes the "min" operator. This is the "clipping inference method" we introduced in Chapter 2.

The final output of the model is the aggregation of outputs from all rules using the max operator:

$$\mu_C(y) \;=\; \max\{\mu_{C'_1}(y), \mu_{C'_2}(y), ..., \mu_{C'_L}(y)\} \qquad \textbf{(EQ 5.21)}$$

Notice that the output C is a fuzzy set. This fuzzy output can be defuzzified into a crisp output using one of the defuzzification techniques introduced in Section 5.3.4.

Now that we have described the inference scheme of the Mamdani model, we will show below that the model can be derived from the foundation discussed in the previous section using the following operators:

- sup-min composition
- min for Cartesian products
- min for conjunctive conditions in rules
- max for aggregating multiple rules

To see this we first describe a lemma regarding an important property of min and max operators, and then we show that applying sup-min composition to a model is equivalent to first composing with individual rules and then aggregating the results.

LEMMA 1   Let $*$ be a monotonically nondecreasing operation (i.e., if a, b, c, d are real numbers, $a \geq b$, $c \geq d \rightarrow a*b \geq c*d$). Then, the operation $*$ distributes over $\wedge$ (min) and $\vee$ (max).

$$a*(b \wedge c) = (a*b) \wedge (a*c) \qquad \textbf{(EQ 5.22)}$$

$$a*(b \vee c) = (a*b) \vee (a*c) \qquad \textbf{(EQ 5.23)}$$

Proof: We will leave this proof as an exercise.

It is worthwhile to point out that all t-norms and t-conorms are monotonically nondecreasing operations. Therefore, they can be distributed over min and max. Obviously, min can distribute over max and max can distribute over min.

THEOREM 3   Suppose a fuzzy model that describes a fuzzy mapping from $U * V$ to W is described by n rules in the form of

IF x is A_i AND y is B_i THEN z is C_i

where A_i, B_i, and C_i are fuzzy subsets of U, V, and W, respectively. The model's fuzzy graph f^* is expressed as

$$f^* = \bigcup_{i=1}^{n} R_i = \bigcup_{i=1}^{n} (\overline{A}_i \cap \overline{B}_i) \times C_i \qquad \textbf{(EQ 5.24)}$$

where R_i is the fuzzy relation of ith. rule. Let A' and B' be fuzzy subsets of U and V respectively. If we compose A' and B' with f^* using **sup-min** composition and compute the union in Equation 5.24 using the **max** operator, then composing inputs with f^* (i.e., the entire fuzzy rule-based model) are equivalent to first composing inputs with individual rules in f^* and then aggregating their composition results.

$$(\bar{A}' \cap \bar{B}') \circ f^* = \bigcup_{i=1}^{n} (\bar{A}' \cap \bar{B}') \circ R_i \qquad \textbf{(EQ 5.25)}$$

Proof: Let us denote the result of composition as C', i.e. $(\bar{A}' \cap \bar{B}') \circ f^* = C'$. The membership function of C' is obtained by applying the definition of sup-min composition to the left-hand side of Equation 5.25:

$$\mu_{c'}(z) = \sup_{x,\, y}(\mu_{A'}(x) \wedge \mu_{B'}(y)) \wedge \left(\bigvee_{i=1}^{n} \mu_{R_i}(x, y, z) \right) \qquad \textbf{(EQ 5.26)}$$

Based on the Lemma 1, we can rewrite Equation 5.26 as

$$\mu_{C'}(z) = \sup_{x,\, y} \bigvee_{i} [(\mu_{A'}(x) \wedge \mu_{B'}(y)) \wedge \mu_{R_i}(x, y, z)] \qquad \textbf{(EQ 5.27)}$$

$$= \bigvee_{i} \sup_{x,\, y} [(\mu_{A'}(x) \wedge \mu_{B'}(y)) \wedge \mu_{R_i}(x, y, z)]$$

$$= \bigvee_{i=1} \mu_{(\bar{A}' \cap \bar{B}') \circ R_i}(z)$$

Therefore, we have proved

$$(\bar{A}' \cap \bar{B}') \circ f^* = \left((\bar{A}' \cap \bar{B}') \circ \bigcup_{i=1}^{n} R_i \right) \qquad \textbf{(EQ 5.28)}$$

$$= \bigcup_{i=1}^{n} (\bar{A}' \cap \bar{B}') \circ R_i \qquad \textbf{(EQ 5.29)}$$

Based on Theorem 3, we can now derive a Mamdani model from a sup-min composition [363, 138].

THEOREM 4   Suppose a fuzzy rule-based model maps $X \times Y$ to Z using a set of n rules in the form of
IF x is A_i AND y is B_i THEN z is C_i. $1 \leq i \leq n$
and receives inputs in the form of x is A' and y is B' where A' and B' are fuzzy subsets of U and V. Suppose the fuzzy inference of the model is based on a sup-min composition between inputs and a fuzzy graph that is defined using *max* and *min* for all fuzzy disjunctions and fuzzy conjunctions operations. Then, the output of the fuzzy model (before defuzzification), denoted by C', is characterized by the following membership function:

$$\mu_C(z) = \max_{i=1}^{n}(\alpha_i \wedge \mu_{C_i}(z)) \tag{EQ 5.30}$$

where $\alpha_i = \sup_x(\mu_{A'}(x) \wedge \mu_{A_i}(x)) \wedge \sup_y(\mu_{B'}(y) \wedge \mu_{B_i}(y))$ and $\wedge$ denotes the min operator.

Proof: Since we use the min operator for fuzzy conjunction, we have

$$\mu_{\bar{A}' \cap \bar{B}'}(x, y) = \mu_{A'}(x) \wedge \mu_{B'}(y) \tag{EQ 5.31}$$

Based on the definition of fuzzy graph, we have,

$$R_i = (\bar{A}_i \cap \bar{B}_i) \times C_i \tag{EQ 5.32}$$

Hence, the fuzzy relation of rule R_i is defined as

$$\mu_{R_i}(x, y, z) = \mu_{(\bar{A}_i \cap \bar{B}_i) \times C_i}(x, y, z) = (\mu_{A_i}(x) \wedge \mu_{B_i}(y)) \wedge \mu_{C_i}(z) \tag{EQ 5.33}$$

where the first fuzzy conjunction corresponds to "AND" in the if-part of the rules, and the second fuzzy conjunction corresponds to the Cartesian product.

The output of a fuzzy model, in general, is

$$C' = (\bar{A}' \cap \bar{B}') \circ \bigcup_{i=1}^{n} R_i \tag{EQ 5.34}$$

Since the model uses sup-min composition and max for rule aggregation, we can apply Theorem 3 to the equation above to get

$$C' = \left(\bigcup_{i=1}^{n} (\bar{A}' \cap \bar{B}') \right) \circ R_i \tag{EQ 5.35}$$

From Equation 5.33 and the definition of sup-min composition, we obtain the membership function of C' as follows:

$$\mu_C(z) = \max_{i=1}^{n} \sup_{x,y} \left[(\mu_{A'}(x) \wedge \mu_{B'}(y)) \wedge \left(\mu_{A_i}(x) \wedge \mu_{B_i}(y) \wedge \mu_{C_i}(z) \right) \right] \tag{EQ 5.36}$$

$$= \max_{i=1}^{n} \sup_{x,y} \left[\left(\mu_{A'}(x) \wedge \mu_{A_i}(x) \right) \wedge (\mu_{B'}(y) \wedge \mu_{B_i}(y)) \wedge \mu_{C_i}(z) \right] \tag{EQ 5.37}$$

$$= \max_{i=1}^{n} \sup_{x,y} \left[\left(\mu_{A'}(x) \wedge \mu_{A_i}(x) \right) \wedge (\mu_{B'}(y) \wedge \mu_{B_i}(y)) \right] \wedge \mu_{C_i}(z) \tag{EQ 5.38}$$

$$= \max_{i=1}^{n} \left[\sup_{x}(\mu_{A'}(x) \wedge \mu_{A_i}(x)) \right] \wedge \left[\sup_{y}(\mu_{B'}(y) \wedge \mu_{B_i}(y)) \right] \wedge \mu_{C_i}(z) \qquad \textbf{(EQ 5.39)}$$

$$= \max_{i=1}^{n} \alpha_i \wedge \mu_{C_i}(z) \qquad \textbf{(EQ 5.40)}$$

Thus, the theorem is proved.

5.7   The TSK Model

The Takagi-Sugeno-Kang (TSK) model was introduced by T. Takagi and M. Sugeno in 1984, about one decade after the Mamdani model. Later M. Sugeno and K.T. Kang also worked on the identification of this type of fuzzy models.

The main motivation for developing this model is to reduce the number of rules required by the Mamdani model, especially for complex and high-dimensional problems. To achieve this goal, the TSK model replaces the fuzzy sets in the consequent (then-part) of the Mamdani rule with a linear equation of the input variables. For example a two-input one-output TSK model consists of rules in the form of

IF x is A_i and y is B_j
THEN $z = ax + by + c$

where a, b, c are numerical constants. In general, rules in a TSK model have the form

IF x_1 is A_{i1} and ... and x_r is A_{ir}
THEN $y = f_i(x_1, x_2, ..., x_r) = b_{i0} + b_{i1}x_1 + ... + b_{ir}x_r$

where f_i is the linear model, and b_{ij} ($j = 0, 1, ..., r$) are real-valued parameters.

The inference performed by the TSK model is an interpolation of all the relevant linear models. The degree of relevance of a linear model is determined by the degree the input data belong to the fuzzy subspace associated with the linear model. These degrees of relevance become the weight in the interpolation process.

The total output of the model is given by the equation below where α_i is the matching degree of rule R_i, which is analogous to the matching degree computed by Equation 5.19 of the Mamdani model.

$$y = \frac{\displaystyle\sum_{i=1}^{L} \alpha_i f_i(x_1, x_2, ..., x_r)}{\displaystyle\sum_{i=1}^{L} \alpha_i} = \frac{\displaystyle\sum_{i=1}^{L} \alpha_i(b_{i0} + b_{i1}x_1 + ... + b_{ir}x_r)}{\displaystyle\sum_{i=1}^{L} \alpha_i} \qquad \textbf{(EQ 5.41)}$$

The inputs to a TSK model are crisp (nonfuzzy) numbers. Therefore, the degree the input $x_1 = a_1$, $x_2 = a_2$, ..., $x_r = a_{1r}$ matches ith. rule is typically computed using the min operator:

$$\alpha_i = \min(\mu_{A_{i1}}(a_1), \mu_{A_{i2}}(a_2), ..., \mu_{A_{ir}}(a_r)) \qquad \textbf{(EQ 5.42)}$$

However, the product operator can also be used:

$$\alpha_i = \mu_{A_{i1}}(a_1) \times \mu_{A_{i2}}(a_2) \times \dots \times \mu_{A_{ir}}(a_r)$$ **(EQ 5.43)**

Let us consider a TSK model consisting of the following three rules:

IF x is *Small*   THEN $y = L1(x)$.
IF x is *Medium*   THEN $y = L2(x)$.
IF x is *Large*   THEN $y = L3(x)$.

The output of such a model is

$$y = \frac{\mu_{small}(x) \times L1(x) + \mu_{medium}(x) \times L2(1)(x) + \mu_{large}(x) \times L3(x)}{\mu_{small}(x) + \mu_{medium}(x) + \mu_{large}(x)}$$

The resulting model is shown in Fig 5.10.

FIGURE 5.10   An Example of TSK Model

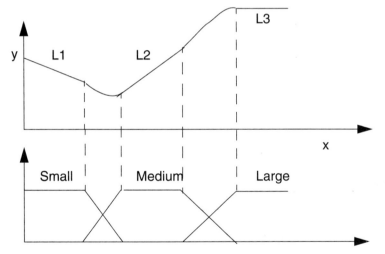

The TSK model provides a powerful tool for modeling complex systems. It can express a highly nonlinear functional relation using a small number of rules. The potential application of TSK Models, hence, is very broad.

The great advantage of the TSK model is its representative power; it can describe a highly nonlinear system using a small number of rules. Moreover, due to the explicit functional representation form, it is convenient to identify its parameters using some learning algorithms. Several commonly used neuro-fuzzy systems such as ANFIS [279] and the one in [575] are constructed on the basis of the TSK model.

Research Note: In principle, f_i can be an arbitrary polynomial function. In practice, however, it is usually assumed that f_i is linear, i.e., $f_i(x_1, x_2, \dots, x_s) = b_{i0} + b_{i1}x_1 + \dots + b_{is}x_s$. An extreme case is that f_i is taken as a constant, i.e., $f_i = b_{i0}$. In this case the output of the model can be expressed in a similar form to height defuzzification (Equation 5.9), where c_i are replaced by b_{i0}.

5.8 Standard Additive Model

The Standard Additive Model (SAM) was introduced by B. Kosko in 1996 [330]. The structure of fuzzy rules in SAM is identical to that of the Mamdani model. Nevertheless, there are four differences between the inference scheme of these two models:

(1) SAM assumes the inputs are crisp, while the Mamdani model handles both crisp and fuzzy inputs.

(2) SAM uses the scaling inference method introduced in Chapter 2, while the Mamdani model uses the clipping method. As we shall see later this difference is due to the choice of composition operators used by the two models.

(3) SAM uses addition to combine the conclusions of fuzzy rules, while the Mamdani model uses max.

(4) SAM includes the centroid defuzzification technique, while the Mamdani model does not insist on a specific defuzzification method.

As in Section 5.6, we will focus our discussion on a two-input one-output model. The generalization to a model with arbitrary inputs is straightforward. Let us consider a standard additive model consisting of rules of the form of

$$\text{IF } x \text{ is } A_i \text{ AND } y \text{ is } B_i \text{ THEN } z \text{ is } C_i. \qquad \textbf{(EQ 5.44)}$$

Given crisp inputs $x = x_0, y = y_0$, the output of the model[2] is

$$z = Centroid\left(\sum_i \mu_{A_i}(x_0) \times \mu_{B_i}(y_0) \times \mu_{C_i}(z)\right) \qquad \textbf{(EQ 5.45)}$$

where centroid is the function that preforms the centroid defuzzification. Equation 5.45 can be transformed into another form that is efficient to compute. This is formally stated in the Theorems below.

THEOREM 5   Suppose a SAM model that describes a mapping from $U \times V$ to W contains rules in the form of

IF x is A_i AND y is B_i THEN z is C_i.

Then the model's output for inputs $x = x_0; y = y_0$ is

$$z = \frac{\displaystyle\sum_{i=1}^{n}(\mu_{A_i}(x_0) \times \mu_{B_i}(y_0)) \times A_i \times g_i}{\displaystyle\sum_{i=1}^{n}(\mu_{A_i}(x_0) \times \mu_{B_i}(y_0)) \times A_i} \qquad \textbf{(EQ 5.46)}$$

---

[2]. The most general form of SAM model allows each rule to be associated with a weight. We avoid the complex SAM model here for both clarity and ease of comparison with other models.

where A_i is the area under the ith. rule's conclusion C_i and g_i is the centroid of C_i (i.e., center of gravity):

$$A_i = \int \mu_{C_i}(z)dz \qquad \text{(EQ 5.47)}$$

$$g_i = \frac{\int z \times \mu_{C_i}(z)dz}{\int \mu_{C_i}(z)dz} \qquad \text{(EQ 5.48)}$$

if z is a continuous variable. Should z be discrete, we simply replace the integral with a summation.

Proof: We will leave this proof as an exercise.

The main advantage of the standard additive model is the efficiency of its computation, because both A_i and g_i in the theorem above can be precomputed (i.e., they are constants once the rules are defined). The foundation of the standard additive model is a fuzzy graph, the sup-product composition, and the use of "addition" as a rule aggregation operator. We formally state this as a theorem below.

THEOREM 6 Suppose $f*$ is a fuzzy graph consisting of rules in the form of

IF x is A_i AND y is B_i THEN z is C_i.

If the inference of the model uses sup-product composition, "product" for all fuzzy conjunction, and "addition" for rule aggregation, and centroid defuzzification, then the model's output for crisp inputs $x = x_0$; $y = y_0$ is

$$z = Centroid\left(\sum_{i=1}^{n} \mu_{A_i}(x_0) \times \mu_{B_i}(y_0) \times C_i\right) \qquad \text{(EQ 5.49)}$$

Proof: We will use A_0 and B_0 to denote fuzzy sets that are equivalent to the crisp inputs, i.e.,

$$\mu_{A_0}(x) = \begin{cases} 1 & x = x_0 \\ 0 & x \neq x_0 \end{cases}$$

$$\mu_{B_0}(x) = \begin{cases} 1 & y = y_0 \\ 0 & y \neq y_0 \end{cases}$$

The output of the model is

$$z = Centroid\left[(\bar{A}_0 \cap \bar{B}_0) \circ \bigcup_{i=1}^{n} (A_i \cap B_i) \times C_i\right] \qquad \text{(EQ 5.50)}$$

Let us denote the inference result before defuzzification as C'. Based on the definition of sup-product composition, we have

$$\mu_C(z) = \sup_{x,y}(\mu_{A_0}(x_0) \times \mu_{B_0}(y)) \times \left(\sum_{i=1}^{n} \mu_{A_i}(x) \times \mu_{B_i}(y) \times \mu_{C_i}(z) \right) \quad \textbf{(EQ 5.51)}$$

Because A_0 and B_0 are singletons, we can simplify the equation to

$$\mu_C(z) = \sum_{i=1}^{n} \mu_{A_i}(x_0) \times \mu_{B_i}(y_0) \times \mu_{C_i}(z) \qquad \textbf{(EQ 5.52)}$$

The defuzzified output is thus

$$z = Centroid(C) \qquad \textbf{(EQ 5.53)}$$

$$= Centroid\left(\sum_{i=1}^{n} \mu_{A_i}(x_0) \times \mu_{B_i}(y_0) \times \mu_{C_i}(z) \right)$$

The theorem is thus proved.

Combining the outputs of the individual fuzzy rules and then defuzzifying the combined outcome to obtain a crisp value is the standard procedure, particularly in early fuzzy modeling practice. A different procedure was proposed recently in Sugeno and Yasukawa (1993) under the framework of *qualitative modeling*.

Instead of defuzzifying the combined fuzzy output, this method first defuzzifies the individual consequent constituents C_i in Equation 5.18 and then takes the weighted average of the individual defuzzified constituents with respect to the matching degree α_i. Let d_i be the defuzzified constituent of the ith rule and computed by the centroid defuzzifier, i.e.,

$$d_i = \frac{\int_{S_i} \mu_{C_i}(y) y \, dy}{\int_{S_i} \mu_{C_i}(y) \, dy} \qquad \textbf{(EQ 5.54)}$$

where S_i denotes the support of $\mu_{C_i}(y)$. Then the crisp output of the model is given by

$$y = \frac{\displaystyle\sum_{i=1}^{M} w_i d_i}{\displaystyle\sum_{i=1}^{M} w_i} \qquad \textbf{(EQ 5.55)}$$

Comparing Equations 5.46 and 5.55, we can see the great similarity between Sugeno-Yasukawa's procedure and Kosko's procedure: both try to combine the crisp outputs of all M rules in an *additive* fashion, and both try to perform the defuzzification step by working on the consequent constituents C_i (Kosko's procedure applied the defuzzification to the combined output, but the final implementation was actually done based on the parameters of C_i, i.e., A_i and g_i).

Since C_i is a fuzzy set that has a preestablished formation, defuzzifying it is much easier than defuzzifying *the combined fuzzy output.*Moreover, since the parameters of C_i have been explicitly included in the defuzzified constituent d_i, it is possible to determine them using some learning procedure.

Research Note: In Sugeno and Yasukawa's original paper, the defuzzified constituents d_i are computed by the centroid defuzzifier 5.7. Alternatively, they can also be computed using the *height defuzzifier* [162, 410] described in Section 5.3.4.3.

5.9 Summary

This is perhaps one of the most important chapters in this book because of the wide applications of fuzzy rule-based models as well as some of the confusion about them in the literature. The most important ideas of this chapter are the following.

- There are two different kinds of fuzzy if-then rules: fuzzy implication rules and fuzzy mapping rules. The former are for describing a logic relationship between two logic sentences, while the latter are for describing a functional mapping relationship between inputs and outputs of a model.

- The foundation of a fuzzy mapping rule is a fuzzy graph and a compositional rule of inference introduced in Chapter 4.

- A set of fuzzy mapping rules form a fuzzy rule-based model.

- The major concepts of fuzzy rule-based models are (1) fuzzy partitioning of the input space, (2) mapping each fuzzy subregion to a local model using a fuzzy mapping rule, (3) fusing the output of multiple rules through interpolative reasoning, and (4) defuzzifying.

- There are three major types of fuzzy models: Mamdani, TSK, and SAM. The latter two are additive, while the first is not additive.

- The fundamental difference between the Mamdani model and SAM lies in the choice of composition, conjunction, and disjunction operators in their reasoning.

- The difference between the TSK model and SAM is the structure of their local models (i.e., then parts of rules). TSK model uses a linear local model, whereas SAM uses a fuzzy constant as its rule's local model.

Bibliographical and Historical Notes

The distinction between fuzzy implication rules and fuzzy mapping rules has not been clear in the literature. Until the early 1990s, fuzzy mapping rules in control systems were been viewed as a special kind of fuzzy implication rule. This view is most clearly expressed in a survey paper by C. C. Lee [363]. However, it is difficult to explain the use of a conjunction operator in forming the "fuzzy implication relation" of rules, because implication and conjunction are not logically equivalent. This difficulty gradually leads to the crystallization of the fundamental differences between the two types of rules. Zadeh, Kosko, and many others have contributed to this process through keynote speeches (Zadeh), books (Kosko), conferences, and journal publications.

The concept of the fuzzy graph was initially introduced in 1971 [704], and in a more explicit form in 1974 [700, 696]. It was not until the 1990s that Zadeh clearly pointed out the connection between fuzzy graph and fuzzy mapping rules. This development is now widely used in fuzzy logic control applications [682].

Motivated by Zadeh's work on fuzzy algorithms [707] and linguistic analysis [701], E. H. Mamdani introduced a model for a fuzzy logic controller using fuzzy if-then rules in 1974 [400]. The Mamdani model was not only the first fuzzy logic controller, but also the first application of fuzzy mapping rules. Not surprisingly, the historical development of fuzzy mapping rules is tightly linked to that of fuzzy logic control. A more detailed account of the latter will be given in Chapter 8. After T. Takagi graduated, Prof. M. Sugeno continued this line of research with another student, K. T. Kang [559]. Since then the TSK model has become increasingly popular. Most of the research regarding the TSK model, however, is not on modifying the model itself, but on the identification of the model. Hence, we will defer discussing this until Chapter 14. Original proofs of Theorem 3 and 4 were given by C. C. Lee [363] and by V. Cross and T. Sudkamp [138].

Kosko gave a thorough treatment of the standard additive model in his 1997 book [330]. The idea of defuzzifying each rule's fuzzy consequent before processing the input has also been presented by M. Sugeno and T. Yasukawa [562].

Additive fuzzy models (including the TSK model and SAM) are also related to radial basis functions, as shown by Wang and Mendel [620]. We will discuss the relationship between additive fuzzy models and other similar models in system identification (e.g., B-splines) in Chapter 14.

Exercises

5.1   Give an example of a fuzzy mapping rule and a fuzzy implication rule each related to your daily life.

5.2   Let a Mamdani model contains the following three rules:

(a) IF x is Small   THEN y is Medium.
(b) IF x is Medium THEN y is Large.
(c) IF x is Large   THEN y is Small.

where Small, Medium, and Large are fuzzy sets defined as
Small $= 1/1 + 0.5/2$
Medium $= 0.5/2 + 1/3 + 0.6/4 + 0.3/5$
Large $= 0.4/4 + 0.7/5 + 1/6$

5.3   Let $X = 6.3/3 + 1/4 + 0.5/6$.

(a) Using a compositional rule of inference, compute the output of the Mamdani model before defuzzification. Show the steps and intermediate results.

(b) Using centroid defuzzification, compute the defuzzified result.

(c) Using the SAM model, compute the defuzzified output.

5.4   Suppose you were to rank the graduate programs in your area of study. To do so, you may use as criteria, among other things, the number of Ph.D. graduates in a given year, the dollar amount of research funding, the selectivity of the admissions process, quality and reputation of the faculty. Consider the first three criteria and decide whether a number of the programs considered rank *high* or *moderately high*. We are not concerned with lower rankings here

(a) State the rules that you would consider most appropriate for this process. Your answer must be in the form of *if-then* rules and must be based on the following term sets:

- Number of Ph.D. graduates: *high(H), medium(M),* or *low(L),*
- Research funding: *high(H), moderately high(MH),* or *average(AVG),*
- Admissions process: *highly selective(HS), selective(S), not very selective(NVS).*

You may use conjunctions or adjectives in your rules, or use the term *ANY*. Use normalized scales to define the appropriate universe of discourse.

(b) Suppose some graduate program has the following description:

- Number of Ph.D. graduates: *medium high,*
- Research funding: *slightly above average,*
- Admissions process: *rather selective.*

Define the given linguistic terms appropriately. You need only show your answer schematically on the same graphs as used in part a.

(c) Determine the ranking of the given program using your rule set by sketching the fuzzy set that represents the conclusion of the inference process.

(d) Find a defuzzified conclusion using MOM and COA methods. Show your results schematically, that is, you need not accurately and algebraically determine the value.

5.5   Derive the scaling inference method described in Chapter 2 for the case of crisp input from the foundation of fuzzy mapping rules. You may assume that the rules are in the form of
R_i: IF x is A_i THEN y is B_i.
and the input value for x is a (i.e., $x = a$).

5.6   Can you derive the scaling inference method for the case of fuzzy input from the foundation of fuzzy mapping rules? If so, show it. If not, explain the difficulties.

5.7   Let x and y be two variables taking values from U and V respectively, where $U = \{1, 2, 3, 4, 5\}$ and $V = \{0, 10, 20\}$. The fuzzy subsets of the universe of discourse U are defined as follows:
Large = {(3, 0.3) (4, 0.7) (5, 1)}
Small = {(1, 1) (2, 1) (3, 0.7) (4, 0.3)}
Consider the following TSK model with two rules:
R1: If x is Small then $y = x + 3$.
R2: If x is Large then $y = 10 - x$.
What is the model's output for x = 2, 3, 4, 5?

5.8   Prove Lemma 1.

5.9   Prove Theorem 5.

5.10   What are the choices of the following operators for the Mamdani model and for the SAM model respectively?

- Cartesian product
- AND in rule condition
- Compositional rule of inference
- Aggregation of rules

6 FUZZY IMPLICATIONS AND APPROXIMATE REASONING

Even though we have used the term "fuzzy logic" frequently in previous chapters, we used the term in the general sense (i.e., referring to a collection of techniques based on fuzzy sets). In this chapter, we will discuss topics related to "fuzzy logic" in the narrow sense (i.e., referring to a generalization of classical two-valued logic systems). Classical logic offers a formal framework for reasoning. A logic system has two major components: (1) a formal language for constructing statements about the world, and (2) a set of inference mechanisms for inferring additional statements about the world from those already given. Two of the most commonly used logic systems are (1) propositional logic, and (2) first-order predicate logic.

After a brief review of these two bivalent logic systems, we introduce fuzzy logic with an emphasis on fuzzy implications and its reasoning. We chose to focus on fuzzy implications for two reasons. First, it is the most commonly used reasoning scheme in applications of fuzzy logic (narrow sense). An important application of fuzzy implications is fuzzy expert systems, which we will discuss in Chapter 11.

The second reason we elaborate on fuzzy implications is the subject is complicated by the fact that there isn't a unique definition of fuzzy implications. In fact, there are three major ways to define fuzzy implication, which lead to three families of fuzzy implications. We will discuss the rationales of these different approaches and compare them based on some desired criteria of their inference.

6.1 Propositional Logic

In propositional logic, a statement about the world is constructed by (1) representing simple sentences as basic units called *propositions*, and (2) connecting propositions with the following connectors to form complex sentences: $\neg$ (not), $\wedge$ (and), $\vee$ (or), $\Rightarrow$ (implies). For example, a statement such as

If today is a weekday and the current time is rush hour,
then the traffic is congested.

can be represented by first defining the following three propositions:

P: today is a weekday
Q: the current time is during rush hour
R: the traffic is congested

then connecting them using "and" and "implication" we would have:

$$(P \wedge Q) \Rightarrow R \qquad \text{(EQ 6.1)}$$

A classical logic statement can have two possible truth values: *true* or *false*. Since a logic connective is applied to a subformula or a pair of subformula, its meaning can be defined by listing the truth value of the resulted (compound) formula for all possible truth value combinations of the subformula. A table that provides such a listing is called *a truth table*. The truth table of logic connectives mentioned above are shown in Table 6.1.

TABLE 6.1   A Truth Table of Logic Connectives

α	β	$\neg\alpha$	$\alpha \wedge \beta$	$\alpha \vee \beta$	$\alpha \Rightarrow \beta$
False	False	True	False	False	True
False	True	True	False	True	True
True	False	False	False	True	False
True	True	False	True	True	True

The implication connective is especially important, because it is the basis of fuzzy implication rules. An implication has two parts: a premise (the "if-part" preceding the implication connective) and a conclusion (the "then-part" following the implication connective). In the previous example (Equation 6.1), for instance, $P \wedge Q$ is the premise and R is the conclusion.

An implication $\alpha \Rightarrow \beta$ is logically equivalent to $\neg\alpha \vee \beta$. That is, an implication is true if either its premise is false or its conclusion is true. The rationale of this definition is that the implication should not tell us anything regarding whether β is true when the premise is false. This can easily be verified using the truth table above. To find out the truth value of β when α is false and $\alpha \Rightarrow \beta$ is true, we identify those rows in the truth table that have the correct truth value assignments under the columns of α and $\alpha \Rightarrow \beta$ (i.e., the first and second rows). The truth values of β in these two rows are "False" and "True," respectively. Therefore, nothing can be inferred about the truth value of β (i.e., it is unknown).

One of the most important inference schemes in propositional logic is *modus ponens*. Given that an implication and its premise are true, modus ponens enables us to deduce that the consequent is true.

$$\alpha \Rightarrow \beta$$
$$\alpha$$

$$\beta$$

For instance, assuming that we know today is a weekday and the time is during rush hour, we can deduce that the traffic is congested using the implication in Equation 6.1:

$$(P \wedge Q) \Rightarrow R$$
$$P$$
$$Q$$

$$R$$

Another inference scheme involving implication is *modus tolens*. From an implication and the negation of its conclusion, we can deduce the negation of its premise:

$$\alpha \Rightarrow \beta$$
$$\neg \beta$$

$$\neg \alpha$$

One of the main limitations of propositional logic is that it cannot easily describe knowledge that applies to a class of objects. To do this, we need first-order predicate calculus.

6.2 First-Order Predicate Calculus

First-Order Predicate Calculus (FOPC) offers a formal language that is more powerful than that of propositional logic. More specifically, it allows the use of variables in a logic statement. A variable in FOPC is associated with one of the two quantifiers: the *universal quantifier* $\forall$ (read "for all") and the *existential quantifier* $\exists$ (read "there exists"). The former is used to describe a statement that is true for **all** possible objects, while the latter is used to describe a statement that is true for **at least one** object.

Unlike propositional logic, FOPC uses *predicates* to describe simple sentences. A predicate represents a set of objects (e.g., the predicate *Aggies* represents the set of Aggies) or a relation (e.g., the predicate *Friend* represents a friendship relation between two persons). A predicate has a number of arguments, which can be variables or constants. The following are some examples of predicates with constant arguments:

Aggies(John) *John is an Aggie.*
RushHour(5pm): *Five o'clock in the afternoon is during rush hour.*
Friend(Jack, Jill): *Jack and Jill are friends.*

Like propositional logic, FOPC combines simple logic expressions into complex ones using logic connectives. For instance, a statement such as "all Aggies like Aggie traditions" can be described in a logic system called first-order predicate logic as follows:

$$\forall x,y \quad \text{Aggies}(x) \wedge \text{AggieTradition}(y) \Rightarrow \text{Like}(x,y) \qquad \textbf{(EQ 6.2)}$$

where x and y are variables; Aggies, Aggie Tradition, and Like are *predicates*.

All inference rules in propositional logic can be extended to FOPC by finding proper substitutions of variables using a *unification* algorithm. For example, given the following facts

> Aggie (John)
> AggieTradition(Bonfire)

modus ponens can be applied to Equation 6.2 by substituting variables x and y with John and Bonfire respectively, denoted as $\{\text{John}/x, \text{Bonfire}/y\}$.

$$\forall x,y \quad \text{Aggies}(x) \wedge \text{AggieTradition}(y) \Rightarrow \text{Like}(x,y)$$
$$\text{Aggie}(\text{John})$$
$$\text{AggieTradition}(\text{Bonfire})$$

Like(John, Bonfire) $\phi = \{\text{John}/x, \ \text{Bonfire}/y\}$

where ϕ denotes a variable substitution. Such variable substitutions are called *unifiers*. The algorithms for finding these variable substitutions are called unification algorithms. The detail of the algorithm and other reasoning schemes in first-order logic (e.g., resolution principle) can be found in a textbook on artificial intelligence by Russell and Norvig [517].

One of the main limitations of classical logic is that it cannot easily represent and reason about knowledge that is uncertain. Fuzzy logic aims to generalize classical logic for reasoning under uncertainty.

In addition to fuzzy logic, there are other approaches to extend the classical logic framework for reasoning under uncertainty. These include *multivalued logic*, *nonmonotonic logic* developed in the artificial intelligence community over the past two decades, and *probabilistic logic* in its infancy.

6.3 Fuzzy Logic

Fuzzy logic generalizes the notion of truth values in classical logic (i.e., true or false) into a matter of degree. A statement in fuzzy logic, thus, can be partially true. A truth value is a number between 0 (false) and 1 (true). For example, the fact that one of the authors is only somewhat bald, but not entirely bald yet, can be represented by assigning a small truth value (say 0.4) to the predicate Bald(John Yen).

An important goal of fuzzy logic is to be able to make reasonable inference even when the condition of an implication rule is partially satisfied. This capability is sometimes referred to as *approximate reasoning*. As we have mentioned before, this is achieved in fuzzy logic by two related techniques: (1) representing the meaning of a fuzzy implication rule using a fuzzy relation, and (2) obtaining an inferred conclusion by applying the compositional rule of inference to the fuzzy implication relation.

Even though we have introduced the compositional rule of inference, we have not discussed how to construct a fuzzy relation that provides a formal representation of the semantics of a fuzzy implication rule. This will be the topic of this section.

6.3.1 Fuzzy Implication

We mentioned in Chapter 5 that a fuzzy mapping rule can be represented as a fuzzy relation between antecedent variables and consequent variables. Similarly, we can represent a fuzzy implication rule using a fuzzy relation. However, the content of the fuzzy relations for these two types of rules are very different due to the difference in their semantics. A fuzzy mapping rule, as we mentioned in Chapter 5, describes an association; therefore, its fuzzy relation is constructed from the Cartesian product of its antecedent fuzzy condition and its consequent fuzzy conclusion. A fuzzy implication rule, however, describes a generalized logic implication; therefore, its fuzzy relation needs to be constructed from the semantics of a generalization to implication in two-valued logic.

The difference between the semantics of fuzzy mapping rules and fuzzy implication rules can be seen from the difference in their inference behavior. Even though these two types of rules behave the same when their antecedents are satisfied, they behave differently when their antecedents are not satisfied. We will illustrate this using an example. Suppose x and y are two integer variables taking values from the interval $[0, 10]$. Suppose we know that if x is between 1 and 3, then y is either 7 or 8. This knowledge can be represented in at least two ways: (1) as a logic implication, and (2) as a conditional statement in a procedural programming language (e.g., C, Visual Basic, etc.). The two representations are different, however. Assuming that we also know the value of x is 5, the logic representation will include that y is unknown (i.e., y can be any integer in the interval $[0, 10]$), but the procedural representation will not make any conclusion about the value of y, since the "else" component of the if-then-else statement is missing. For ease of comparison, we summarize these two different results below:

Implication Rule (Logic Representation)

Given: $x \in [1, 3] \rightarrow y \in [7, 8]$

$\underline{\quad x = 5 \quad\qquad\qquad\qquad\qquad}$

Infer: y is unknown ($y \in [0, 10]$)

Mapping Rule (Procedural Representation)

Statement: IF $x \in [1, 3]$ THEN $y \in [7, 8]$
Variable Value: $\underline{x = 5 \qquad\qquad\qquad\qquad}$

Execution Result: no action

The two representations correspond to the two types of fuzzy rules: the logic representation is the basis of fuzzy implication rules, while the procedural representation is the essence of fuzzy mapping rules.

To establish the foundation of representing a fuzzy implication using a fuzzy relation, we will first show how an implication involving classical sets can be represented as a binary (i.e., two-valued) possibility relation. To see this, let us first consider the following example:

$$x \in \{b, c, d\} \Rightarrow y \in \{s, t\} \qquad \text{(EQ 6.3)}$$

where the universe of discourse of x and y are $U = \{a, b, c, d, e, f\}$ and $V = \{r, s, t, u, v\}$, respectively.

Such a set-to-set implication actually specifies a set of *possible* implications such as

$x = b \rightarrow y = s$
$x = b \rightarrow y = t$
$x = c \rightarrow y = s$

and a set of **impossible implications** such as

$x = b \rightarrow y = r$
$x = b \rightarrow y = u$
$x = b \rightarrow y = v$

Furthermore, if the antecedent is known to be false, the implication is true regardless of y's value. Therefore, the following implications are all possible:

$x = a \rightarrow y = r$
$x = a \rightarrow y = s$
.
$x = e \rightarrow y = r$
.
$x = f \rightarrow y = u$
$x = f \rightarrow y = v$

Hence, we can represent the meaning of the set-to-set implication using the following possibility relation:

$$R(x_i, y_j) = \begin{array}{c} \\ a \\ b \\ c \\ d \\ e \\ f \end{array} \begin{array}{c} r\ s\ t\ u\ v \\ \begin{bmatrix} 1 & 1 & 1 & 1 & 1 \\ 0 & 1 & 1 & 0 & 0 \\ 0 & 1 & 1 & 0 & 0 \\ 0 & 1 & 1 & 0 & 0 \\ 1 & 1 & 1 & 1 & 1 \\ 1 & 1 & 1 & 1 & 1 \end{bmatrix} \end{array} \qquad \text{(EQ 6.4)}$$

where an entry in the relation $R(x_i, y_j)$ represents whether

$$(x = x_i) \rightarrow (y = y_j) \qquad \text{(EQ 6.5)}$$

is possible. We will call such a possibility relation an *implication relation*. Like other possibility distributions, "1" means possible, and "0" means impossible. It is also easy to see that such a relation can be constructed by replacing x and y in Equation 6.3 with pairs of x_i and y_j and determine whether the resulting (i.e., instantiated) implication is true or false, i.e.,

$$R(x_i, y_j) = \begin{cases} 1 & \text{if } ((x_i \in \{b, c, d\}) \rightarrow (y_j \in \{s, t\})) \\ 0 & \text{if } \neg((x_i \in \{b, c, d\}) \rightarrow (y_j \in \{s, t\})) \end{cases} \qquad \text{(EQ 6.6)}$$

For instance, $R(a, r)$ is 1 because the antecedent $a \in [b, c, d]$ is false.

It may be worthwhile now to point out the difference between the possibility relation of a mapping rule and that of an implication rule. Fig 6.1 depicts such a difference for the rule "If x is A then y is B." The points in shaded areas are possible (i.e., possibility is 1) while those in the white area are not possible (i.e., possibility is 0). The shaded area in Fig 6.1 corresponds to the "1"entries in the implications relation in Equation 6.4.[1] Fig 6.1 echoes a point we made earlier — implication rules and mapping rules differ in their treatment of the situation that fails to satisfy the if-condition. In Fig 6.1, these situations are in the two regions outside of the rectangular area corresponding to "x is A."

FIGURE 6.1   A Pictorial View of the Possibility Relation for (a) an Implication Rule and (b) a Corresponding Mapping Rule

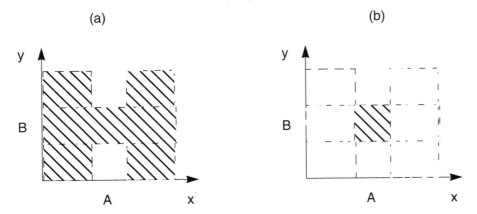

Having discussed the meaning of a set-to-set binary implication, we can now consider an implication involving fuzzy sets (i.e., fuzzy implication):

$$(x \text{ is } A) \rightarrow (y \text{ is } B) \qquad \textbf{(EQ 6.7)}$$

where A and B are fuzzy subsets of U and V, respectively. As in the previous example, this implication also specifies the possibility of various point-to-point implications. The main difference here is that the possibilities are no longer binary. Rather, they become a matter of degree. Therefore, the meaning of the fuzzy implication can be represented by an implication relation R defined as

$$R_I(x_i, y_j) = \Pi_I((x = x_i) \rightarrow (y = y_j)) \qquad \textbf{(EQ 6.8)}$$

where Π_I denotes the possibility distribution imposed by the implication. In fuzzy logic, this possibility distribution is constructed from the truth values of the instantiated (i.e.,

---

[1.] Because the way a fuzzy implication is constructed, we need to turn it counter clockwise 90 degrees to see how Fig 6.1 (a) is its missor image.

grounded) implications obtained by replacing variables in the implication (i.e., x and y) with pairs of their possible values (i.e., x_i, y_j):

$$\Pi((x = x_i) \to (y = y_j)) = t((x_i \text{ is A}) \to (y_j \text{ is B})) \qquad \textbf{(EQ 6.9)}$$

where t denotes the truth value of a proposition.

The equation above establishes an important relationship between fuzzy implication and multivalued logic. The truth value of the implication "x_i is $A \to y_j$ is B" is defined in terms of the truth value of the proposition "x_i is A" and the truth value of the proposition "y_j is B." For the convenience of our discussion, we will refer to these truth values as α_i and β_j, respectively, i.e.,

$t\ (x_i \text{ is A}) = \alpha_i$
$t\ (y_j \text{ is B}) = \beta_j$

The truth value of the implication (x_i is $A \to y_j$ is B) is thus a function I of α_i and β_j:

$t\ (x_i\ A \to Y_j \text{ is } B) = I\ (\alpha_i, \beta_j)$

We call the function I an "*implication function.*"

There isn't a unique definition for implication function. Different implication functions lead to different fuzzy implication relations. However, all implication function should, at the minimum, be consistent with the truth table of implication in propositional logic:

$I\ (0, \beta_j) = 1$
$I\ (\alpha_i, 1) = 1$

Various definitions of implication functions have been developed from both the fuzzy logic and multivalued logic research communities. Before we introduce them, however, we will describe several intuitive criteria of desired inference results of fuzzy implications in Section 6.3.3. These criteria will thus form the basis for evaluating and comparing different fuzzy implication functions in Section 6.3.4.

6.3.2 Approximate Reasoning

Given a possibility distribution of the variable X and the implication possibility from X to Y, we infer the possibility distribution of Y. This section shows how such an inference scheme is obtained.

Given: $x = x_i$ is possible   AND
$x = x_i \to y = y_j$ *is possible.*

Infer: $Y = y_j$ *is possible*

More generally, we have

Given: $\Pi(X = x_i) = a$ AND

$\Pi(X = x_i \rightarrow Y = y_j) = b$

Infer: $\Pi(Y = y_j) = a \otimes b$

where $\otimes$ is a fuzzy conjunction operator.

When different values of X imply an identical value of Y say y_j with potentially varying possibility degrees, these inferred possibilities about $Y = y_j$ need to be combined using fuzzy disjunction. Hence, the complete formula for computing the inferred possibility distribution of Y is

$$\Pi(Y = y_j) = \bigoplus_{x_i}(\Pi(X = x_i) \otimes \Pi((X = x_i) \rightarrow (Y = y_j))) \quad \textbf{(EQ 6.10)}$$

which is the compositional rule of inference.

Even though both fuzzy implications and fuzzy mapping rules use the compositional rule of inference to compute their inference results, their usages differ in two ways. First, the compositional rule of inference is applied to individual implication rules, while composition is applied to a set of fuzzy mapping rules that approximate a functional mapping. Second, the fuzzy relation of a fuzzy mapping rule is a Cartesian product of the rule's antecedent and its consequent part. An entry in the fuzzy implication relation, however, is the possibility that a particular input value implies a particular output value.

6.3.3 Criteria of Fuzzy Implications

The criteria of desired inference involving fuzzy implication results can be grouped into six groups:

    (1)   the basic criterion of modus ponens,

    (2)   the generalized criterion of modus ponens involving hedges,

    (3)   the mismatch criterion,

    (4)   the basic criterion of modus tolens,

    (5)   the generalized criterion of modus tolens involving hedges, and

    (6)   the chaining criterion of implications.

All but the last criterion were introduced by Fukami, Mizumoto and Tanaks [194]. The last criterion was proposed by Zadeh [700]. We discuss these criteria below.

6.3.3.1 Fundamental Modus Ponens

The first criterion states that if the given value of x is exactly the same as the antecedent of the implication, it is desirable to infer that the consequent of the implication is true.

Criterion I

    Given: x is $A \rightarrow y$ is B

$$x \text{ is } A$$

Infer: y is B

where x and y are linguistic variables whose universe of discourse are U and V, respectively; A and B are fuzzy subsets of U and V.

6.3.3.2   Generalized Modus Ponens Involving Hedges

The criterion in this group describes desired inference for the situation that the given value x is similar to the antecedent except that it is modified by a hedge. More specifically, the known proposition regarding x is in the form of " x is very A" or "x is somewhat A."

Unfortunately, there isn't a consensus on the "desired inference" for the case "x is very A." There are two "reasonable" criteria. One propagates the hedge to the conclusion, while the other doesn't.

Criterion II-1

Given: x is $A \rightarrow y$ is B

x is very A

Infer: y is very B

An example is given below for which such a criterion is intuitive.

IF the color of a tomato is red, THEN the tomato is ripe.
The color of this tomato is very red

This tomato is very ripe

An alternative criterion for the same situation is to expect the inference to be the same as the consequent of the implication:

Criterion II-2

Given: x is $A \rightarrow y$ is B

x is very A

Infer: y is B

A more general version of this criterion is to state that the inference result is desired to be the consequent whenever the given fact about x is a subset of A.

Criterion II-2*

Given: x is $A \rightarrow y$ is B

x is A'

$A' \subset A$

Infer: y is B

We can also establish a criterion similar to Criterion I-1 and Criterion II-2 for the situation where x is known to be more or less A.

Criterion III-1

Given: x is $A \rightarrow y$ is B

x is more or less A

Infer: y is more or less B

Criterion III-2

Given: x is $A \rightarrow y$ is B

x is more or less A

Infer: y is B

The example relating the color of a tomato to its ripeness can also be used to illustrate this criterion.

IF the color of a tomato is red, THEN the tomato is ripe.
The color of this tomato is more or less red

This tomato is ripe

6.3.3.3 Mismatch

In two-valued logic, if we know the premise of an implication is false (i.e., its negation is true), we cannot infer if the consequent is true or false. Hence, it is considered desirable for fuzzy inference to have the following criterion:

Criterion IV

Given: x is $A \rightarrow y$ is B

x is not A

Infer: y is V (unkown)

The fact that "y is V" represents "y is unknown" needs some explanation. We first recall that fuzzy implication is used to infer **possibility distribution** of the consequent var y. Therefore, when we do not have any information about the possible values of y, we can simply say $y \in V$, where V is the universe of discourse of y. Using the notation in possibility theory, this can be expressed as

$$\Pi_y(v) = 1 \qquad \forall v \in V$$

which means that every element of V is entirely possible to be y's value. This is equivalent to assigning the universe V to y, i.e., y is V.

6.3.3.4 Generalized Modus Tolens

The modus tolens in two-valued logic states that we can deduce that the antecedent is false if the consequent is known to be false, i.e.,

Given: $P \rightarrow Q$

$\qquad \neg Q$

Infer: $\neg P$

Hence, similar inference capability is desirable for fuzzy implications: This gives us the fifth criterion:

Criterion V

Given: x is $A \rightarrow y$ is B

$\qquad y$ is not B

Infer: x is not B

6.3.3.5 Generalized Modus Tolens Involving Hedges

We can generalize the basic criterion of modus tollens for dealing with situations that y is not very B or y is not more or less B. Suppose we know the following about tomato:

Given: IF the color of a tomato is red, THEN the tomato is ripe.

$\qquad$ This tomato is not very ripe

Our common sense reasoning usually infers from these two sentences that this tomato is not very red. A similar example can be constructed for the hedge "more or less". These examples inspired the following criterion for fuzzy implication.

Criterion VI

Given: x is $A \rightarrow y$ is B

$\qquad y$ is not (very B)

Infer: x is not (very A)

Criterion VII
Given: x is $A \rightarrow y$ is B
y is not (more or less B)

Infer: x is not (more or less A)

In two-valued logic, if we know that the conclusion of an implication is true, then we can not make any further inference about whether the premise is true. Therefore, we establish the following criterion for fuzzy implication:

Criterion VIII
Given: x is $A \rightarrow y$ is B
y is B

Infer: x is U (unkown)

Similar to Criterion IV, "x is U" represents that all elements of U are possible values of x. In other words, the true value of x is completely unknown.

6.3.3.6 Chaining

In propositional logic, the following inference is valid for any proposition p, q, r:

Given: $p \rightarrow q$
$q \rightarrow r$

Infer: $p \rightarrow r$

The following chaining criterion for fuzzy implications is thus established.

Criterion IX
Given: x is $A \rightarrow y$ is B
y is $B \rightarrow z$ is C

Infer: x is $A \rightarrow z$ is C

We summarized the criteria above in Table 6.2.

TABLE 6.2   Intuitive Criteria for Involving Fuzzy Implication x is $A \rightarrow y$ is B

Criterion	Given	Infer
I	x is A	y is B
II-1	x is very A	y is very B
II-2	x is very A	y is B
II-2*	x is A' and $A' \subset A$	y is B
III-1	x is more or less A	y is more or less B
III-2	x is more or less A	y is B
IV	x is not A	y is V (unknown)
V	y is not B	x is not A
VI	y is not (very B)	x is not (very A)
VII	y is not (more or less B)	x is not (more or less A)
VIII	y is B	x is U (unknown)
IX	$\dot{y}$ is $B \rightarrow z$ is C	x is $A \rightarrow z$ is C

6.3.4   Families of Fuzzy Implications

Fuzzy implications can be classified into three families. Each family extends a particular logic formulation of implication in propositional logic. Even though these formulations are equivalent in classical logic, they are not equivalent in fuzzy logic because the law of excluded middle no longer holds in fuzzy logic (or in any many-valued logic system). Fuzzy implications within a family differ on their choice of the fuzzy conjunction and fuzzy disjunction operators. We briefly describe each family below.

The first family of fuzzy implication is obtained by generalizing material implications in two-valued logic to fuzzy logic. A material implication $p \rightarrow q$ is defined as $\neg p \vee q$. Generalizing this to fuzzy logic gives us $t(p \rightarrow q) = t(\neg p \vee q)$. More specifically, fuzzy implications in this family can be generically defined as:

$$t(x_i \text{ is } A \rightarrow y_j \text{ is } B) = t(\neg(x_i \text{ is } A) \vee (y_j \text{ is } B))$$
$$= ((1 - \mu_A(x_i)) \oplus \mu_B(y_j)) \qquad \textbf{(EQ 6.11)}$$

The second family of fuzzy implication is based on logic equivalence between implications $p \rightarrow q$ and $\neg p \vee (p \wedge q)$. We will leave the proof of this equivalence as an exercise. Fuzzy implications in this family thus have the following form:

$$t(x_l \text{ is } A \rightarrow y_j \text{ is } B) = t(\neg(x_i \text{ is } A) \vee [(x_i \text{ is } A) \wedge y_j \text{ is } B])$$
$$= (1 - \mu_A(x_i)) \oplus (\mu_A(x_i) \otimes \mu_B(y_j)) \qquad \textbf{(EQ 6.12)}$$

The third family of fuzzy implication generalizes the "standard sequence" of many-valued logic and its variants. The implication in these logic systems is defined to be true whenever the consequent is as true or truer than the antecedent, i.e., $t(p \rightarrow q) = 1$ whenever $t(p) \leq t(q)$. This is an important property of many multivalued logic systems because it allows the following tautology (i.e., a logic formula that is always true) in two-valued logic to be maintained in multi-valued logic: $f \rightarrow f$ where f is any formula. In other words, a logic formula always implies itself, regardless of its truth value. The fuzzy implication function in this family can all be described in the following form:

$$t(x_i \ is \ A \rightarrow y_j \ is \ B) = \sup\{\alpha | \alpha \in [0,1], \ \alpha \otimes t(x_i \ is \ A) \leq t(y_i \ is \ B)\}$$

$$= \sup\{\alpha | \alpha \in [0,1], \ \alpha \otimes \mu_A(x_i) \leq \mu_B(y_j)\} \textbf{(EQ 6.13)}$$

6.3.5 Major Fuzzy Implication Functions

We introduce below several major implication functions in these families.

1. Zadeh's arithmetic fuzzy implication (Family 1)

$$t(x_i \ is \ A \rightarrow y_j \ is \ B) = 1 \wedge (1 - (\mu_A(x_i) + (\mu_B(y_j)) \qquad \textbf{(EQ 6.14)}$$

2. Zadeh's maximum fuzzy implication (Family 2)

$$t(x_i \ is \ A \rightarrow y_j \ is \ B) = (1 - \mu_A(x_i)) \vee (\mu_A(x_i) \wedge \mu_B(y_j)) \qquad \textbf{(EQ 6.15)}$$

Arithmetic fuzzy implication is obtained from the first family by using a bounded sum operator for fuzzy disjunction. Maximum fuzzy implication is from the second family by using min for fuzzy conjunction and max for fuzzy disjunction. Both fuzzy implications were proposed by Zadeh.

3. Standard sequence fuzzy implication (Family 3)

$$t(x_i \ is \ A \rightarrow y_j \ is \ B) = \begin{cases} 1 & \mu_A(x_i) \leq \mu_B(y_j) \\ 0 & \mu_A(x_i) > \mu_B(y_j) \end{cases} \qquad \textbf{(EQ 6.16)}$$

4. Godelian sequence fuzzy implication (Family 3)

$$t(x_i \ is \ A \rightarrow y_j \ is \ B) = \begin{cases} 1 & \mu_A(x_i) \leq \mu_B(y_j) \\ \mu_B(y_j) & \mu_A(x_i) > \mu_B(y_j) \end{cases} \qquad \textbf{(EQ 6.17)}$$

5. Goguen's fuzzy implication (Family 3)

$$t(x_i \ is \ A \rightarrow y_j \ is \ B) = \begin{cases} 1 & \mu_A(x_i) \leq \mu_B(y_j) \\ \dfrac{\mu_B(y_j)}{\mu_A(x_i)} & \mu_A(x_i) > \mu_B(y_j) \end{cases} \qquad \textbf{(EQ 6.18)}$$

All the latter three fuzzy implications came from multivalued logic systems. They came from, respectively, the standard sequence many-valued logic system (denoted S_n in the literature), a many-valued logic system proposed by Kurt Godel (denoted G_n in the literature), and a many-valued logic system J. A. Goguen introduced in 1969. Fig 6.2 depicts the five fuzzy implication functions we introduced above.

Even though implication functions in multivalued logic systems can be used for constructing fuzzy implication relations, approximate reasoning in fuzzy logic is fundamentally different from logic inference in multi-valued logic — approximate reasoning infers possible values of a variable, whereas multivalued logic infers the truth values of propositions. The connection between the two was established by Equation 6.9. Even if we choose to use a fuzzy implication function originated in a multivalued logic system (e.g., standard sequence, Godelian implication, or Goguen's implication), approximate reasoning still benefits from the following important concepts and techniques in fuzzy logic: the compositional rule of inference, fuzzy relations, and possibility distributions. Without them, approximate reasoning would not have been possible.

Fig 6.2 shows graphically the function surface of the five fuzzy implication functions we discussed.

FIGURE 6.2 Five Fuzzy Implication Functions: (a) Zadeh's Arithmetic Fuzzy Implication, (b) Zadeh's Maxmin Fuzzy Implication, (c) Standard Sequence Fuzzy Implication, (d) Godelion Fuzzy Implication, and (e) Goguen Fuzzy Implication

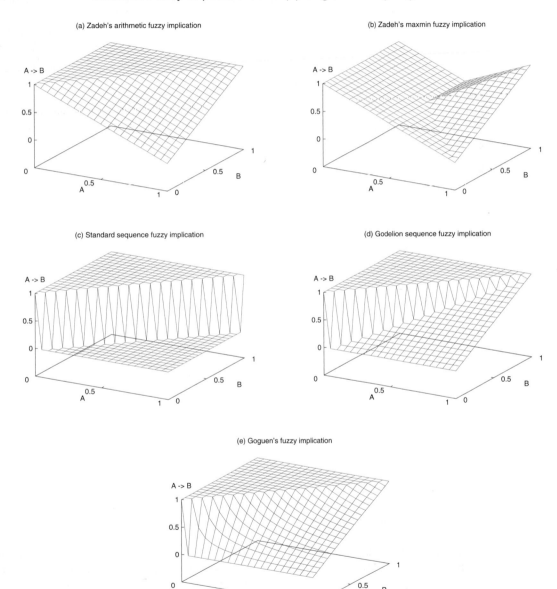

Table 6.3 summaries how the criteria introduced in the previous section are satisfied by the five fuzzy implication functions based on sup-min composition (except that the sup-product composition is applied to Goguen's fuzzy implication). We applied the sup-product composition to Goguen's fuzzy implication because the conjunction operator in Goguen's multivalued logic system uses product as its conjunction operator. In other words, the choice of implication function and the choice of disjunction/conjunction operator in the compositional rule of inference are not unrelated. We have to be careful not to mix a fuzzy implication function with an inappropriate operator in the compositional rule of inference.

TABLE 6.3   Satisfaction of Fuzzy Inference Criteria by Fuzzy Implication Functions ($I(x,y)$)
X: denotes that a criterion is not supported by $I(x,y)$
and O denotes that a criterion is supported by $I(x,y)$

	Arithmetic	Maximum	Standard	Godel	Goguen
I	X	X	O	O	O
II-1	X	X	O	X	X
II-2	X	X	X	O	O
III-1	X	X	O	O	X
III-2	X	X	X	X	X
IV	O	O	O	O	O
V	X	X	O	X	X
VI	X	X	O	X	X
VII	X	X	O	X	X
VIII	O	X	O	O	O
IX	X	X	O	O	O

As shown in this table, the only criterion satisfied by all five implication functions is Criterion IV, which is restated below:

Given: x is $A \rightarrow y$ is B
x is not A

Infer: y is V (unkown)

The table also indicates that standard sequence fuzzy implication satisfies most criteria. It is certainly desirable to state general conditions regarding the satisfaction of certain criteria. The theorem below states that the first criterion is always satisfied as long as three conditions hold. The first condition is the property of fuzzy implication functions in the third family. The second condition is related to the choice of the compositional rule of inference. The third condition is about the content of the implication rule itself. This condition requires that for every possible values x_i of x, there is at least one y_j such that $\mu_A(x_i) = \mu_B(y_j)$. We formally state the theorem below.

THEOREM 7   Let I be a fuzzy implication function. Let x and y be two linguistic variables with universe of discourse U and V, respectively, A and B are fuzzy subsets of U and V. Using a compositional rule of inference and the implication function I to perform reasoning using the fuzzy implication $(x\ is\ A) \rightarrow (y\ is\ B)$ satisfies Criterion I if the following three conditions are satisfied:

(1) $t_I(x\ is\ A \rightarrow y\ is\ B) = 1$ if $\mu_A(x_i) \le \mu_B(x_j)$

(2) $\mu_A(x_i) \otimes t_I(x\ is\ A \rightarrow y\ is\ B) \le \mu_B(y_j)\ \ \forall x_i \in U, \forall y_j \in V$

 where $\otimes$ denotes the fuzzy conjunction used in the compositional rule of inference.

(3) $\forall y_j \in V, \exists x_i \in U$ such that $\mu_A(x_i) = \mu_B(y_j)$

   Proof: Based on the compositional rule of inference, we can express the result of fuzzy inference related to Criterion I as

$$\Pi_Y(y_j) = \sup_{x_i}(\Pi_X(x_i) \otimes \Pi(x= x_i \rightarrow y= y_j))$$

$$= \sup_{x_i}((\mu_A)(x_i) \otimes t_I(x\ is\ A \rightarrow y\ is\ B))$$

$$= \sup_{x_i}(\mu_A(x_i) \otimes I(\mu_A(x_i), \mu_B(y_j))) \qquad \textbf{(EQ 6.19)}$$

For each y_j in V, we partition x_i in U into two groups:

 $G1 = \{x_i | x_i \in U, \mu_A(x_i) \le \mu_B(y_j)\}$

 $G2 = \{x_i | x_i \in U, \mu_A(x_i) > \mu_B(y_j)\}$

We consider each group separately below:

Group 1: $\mu_A(x_i) \le \mu_B(y_j)$

From the first condition and the definition of G1, we know that

$\forall x_i \in G1\ \ \mu_A(x_i) \otimes t_I(x\ is\ A \rightarrow y\ is\ B) = \mu_A(x_i) \otimes 1 = \mu_A(x_i)$

From the third condition, we have

$$\sup_{x_i \in G1} \mu_A(x_i) \otimes t_I(x\ is\ A \rightarrow y\ is\ B) = \mu_B(y_j) \qquad \textbf{(EQ 6.20)}$$

Group 2: $\mu_A(x_i) > \mu_B(y_j)$

From the second condition, we have

$$\sup_{x_i \in G2} \mu_A(x_i) \otimes t_I(x\ is\ A \rightarrow y\ is\ B) \le \mu_B(y_j) \qquad \textbf{(EQ 6.21)}$$

Combining Equations 6.20 and 6.21 we get

$$\sup_{x_i \in U} \mu_A(x_i) \otimes t_I(x\ is\ A \rightarrow y\ is\ B) = \mu_B(y_j) \qquad \textbf{(EQ 6.22)}$$

That is, the inferred possibility distribution of Y is B:

$$\Pi_Y(y_j) = \mu_B(y_j)$$

Thus, Criterion I is satisfied. The theorem is thus proved.

Based on this theorem, the following can easily be proved:

- The Godel implication satisfies Criterion I when sup-min composition is used.
- The standard sequence implication satisfies Criterion I for any sup-$\otimes$ composition where $\otimes$ is a fuzzy conjunction operator.
- Goguen's fuzzy implication satisfies Criterion I when it is used with sup-product composition.

We shall leave the proof of these as an exercise. Using sup-product composition for Goguen's fuzzy implication is not surprising, because Goguen's implication function is defined by using product as its fuzzy conjunction operator.

EXAMPLE 8   Let U and V be two universes representing numeric ratings (from 1 to 10) of redness and ripeness of tomatoes, respectively. We denote the variable of these two ratings as x and y, respectively. Let Red be a fuzzy subset of U defined as

$$Red \ = \ 0.25/6 + 0.5/7 + 0.75/8 + 1/9 + 1/10$$

and $Ripe$ be a fuzzy subset of V defined as

$$Ripe \ = \ 0.25/7 + 0.5/8 + 0.75/9 + 1/10$$

We are also given the fuzzy implication

$$tomato \ \ is \ \ Red \rightarrow tomato \ \ is \ \ Ripe$$

Using the standard sequence implication, Godel's implication, and Goguen's implication, we get the following three fuzzy implication relations, denoted as R_s, R_{gd}, and R_{gg}, respectively.

$$
R_s = \begin{array}{c} \\ 1 \\ 2 \\ 3 \\ 4 \\ 5 \\ 6 \\ 7 \\ 8 \\ 9 \\ 10 \end{array}
\begin{array}{c} 1\ 2\ 3\ 4\ 5\ 6\ 7\ 8\ 9\ 10 \\
\left[\begin{array}{cccccccccc}
1 & 1 & 1 & 1 & 1 & 1 & 1 & 1 & 1 & 1 \\
1 & 1 & 1 & 1 & 1 & 1 & 1 & 1 & 1 & 1 \\
1 & 1 & 1 & 1 & 1 & 1 & 1 & 1 & 1 & 1 \\
1 & 1 & 1 & 1 & 1 & 1 & 1 & 1 & 1 & 1 \\
1 & 1 & 1 & 1 & 1 & 1 & 1 & 1 & 1 & 1 \\
0 & 0 & 0 & 0 & 0 & 0 & 1 & 1 & 1 & 1 \\
0 & 0 & 0 & 0 & 0 & 0 & 0 & 1 & 1 & 1 \\
0 & 0 & 0 & 0 & 0 & 0 & 0 & 0 & 1 & 1 \\
0 & 0 & 0 & 0 & 0 & 0 & 0 & 0 & 0 & 1 \\
0 & 0 & 0 & 0 & 0 & 0 & 0 & 0 & 0 & 1
\end{array}\right]
\end{array}
\qquad \text{(EQ 6.23)}
$$

$$
R_{gd} =
\begin{array}{c|cccccccccc}
 & 1 & 2 & 3 & 4 & 5 & 6 & 7 & 8 & 9 & 10 \\
\hline
1 & 1 & 1 & 1 & 1 & 1 & 1 & 1 & 1 & 1 & 1 \\
2 & 1 & 1 & 1 & 1 & 1 & 1 & 1 & 1 & 1 & 1 \\
3 & 1 & 1 & 1 & 1 & 1 & 1 & 1 & 1 & 1 & 1 \\
4 & 1 & 1 & 1 & 1 & 1 & 1 & 1 & 1 & 1 & 1 \\
5 & 1 & 1 & 1 & 1 & 1 & 1 & 1 & 1 & 1 & 1 \\
6 & 0 & 0 & 0 & 0 & 0 & 0 & 1 & 1 & 1 & 1 \\
7 & 0 & 0 & 0 & 0 & 0 & 0 & 0.25 & 1 & 1 & 1 \\
8 & 0 & 0 & 0 & 0 & 0 & 0 & 0.25 & 0.5 & 1 & 1 \\
9 & 0 & 0 & 0 & 0 & 0 & 0 & 0.25 & 0.5 & 0.75 & 1 \\
10 & 0 & 0 & 0 & 0 & 0 & 0 & 0.25 & 0.5 & 0.75 & 1
\end{array}
$$

(EQ 6.24)

$$
R_{gg} =
\begin{array}{c|cccccccccc}
 & 1 & 2 & 3 & 4 & 5 & 6 & 7 & 8 & 9 & 10 \\
\hline
1 & 1 & 1 & 1 & 1 & 1 & 1 & 1 & 1 & 1 & 1 \\
2 & 1 & 1 & 1 & 1 & 1 & 1 & 1 & 1 & 1 & 1 \\
3 & 1 & 1 & 1 & 1 & 1 & 1 & 1 & 1 & 1 & 1 \\
4 & 1 & 1 & 1 & 1 & 1 & 1 & 1 & 1 & 1 & 1 \\
5 & 1 & 1 & 1 & 1 & 1 & 1 & 1 & 1 & 1 & 1 \\
6 & 0 & 0 & 0 & 0 & 0 & 0 & 1 & 1 & 1 & 1 \\
7 & 0 & 0 & 0 & 0 & 0 & 0 & 1/2 & 1 & 1 & 1 \\
8 & 0 & 0 & 0 & 0 & 0 & 0 & 1/3 & 2/3 & 1 & 1 \\
9 & 0 & 0 & 0 & 0 & 0 & 0 & 1/4 & 1/2 & 3/4 & 1 \\
10 & 0 & 0 & 0 & 0 & 0 & 0 & 1/4 & 1/2 & 3/4 & 1
\end{array}
$$

(EQ 6.25)

Suppose we know that a specific tomato is very red. From the definition of hedge VERY, we can construct the following membership function:

$VERY \ \ RED \ = \ 0.0625/6 + 0.25/7 + 0.5625/8 + 1/9 + 1/10$

Part A:

Applying the sup-min composition to the standard sequence implication, we obtain the following possibility distribution about the ripeness of the tomato:

$\Pi_{ripeness} = \Pi_{redness} \circ R_s$

$$= \begin{bmatrix} 0 & 0 & 0 & 0 & 0 & 0.065 & 0.25 & 0.5625 & 1 & 1 \end{bmatrix} \circ \begin{bmatrix} 1 & 1 & 1 & 1 & 1 & 1 & 1 & 1 & 1 & 1 \\ 1 & 1 & 1 & 1 & 1 & 1 & 1 & 1 & 1 & 1 \\ 1 & 1 & 1 & 1 & 1 & 1 & 1 & 1 & 1 & 1 \\ 1 & 1 & 1 & 1 & 1 & 1 & 1 & 1 & 1 & 1 \\ 1 & 1 & 1 & 1 & 1 & 1 & 1 & 1 & 1 & 1 \\ 0 & 0 & 0 & 0 & 0 & 0 & 1 & 1 & 1 & 1 \\ 0 & 0 & 0 & 0 & 0 & 0 & 0 & 1 & 1 & 1 \\ 0 & 0 & 0 & 0 & 0 & 0 & 0 & 0 & 1 & 1 \\ 0 & 0 & 0 & 0 & 0 & 0 & 0 & 0 & 0 & 1 \\ 0 & 0 & 0 & 0 & 0 & 0 & 0 & 0 & 0 & 1 \end{bmatrix} \quad \textbf{(EQ 6.26)}$$

$$= \begin{bmatrix} 0 & 0 & 0 & 0 & 0 & 0 & 0.065 & 0.25 & 0.5625 & 1 \end{bmatrix} \quad \textbf{(EQ 6.27)}$$

The inferred possibility distribution turns out to be VERY RIPE.

Part B:

If we apply sup-min composition to Godel's fuzzy implication for the same problem, the possibility distribution about the ripeness of the tomato is inferred as follows:

$$\Pi_{ripeness} = \Pi_{redness} \circ R_{gd}$$

$$= \begin{bmatrix} 0 & 0 & 0 & 0 & 0 & 0.065 & 0.25 & 0.5625 & 1 & 1 \end{bmatrix} \circ \begin{bmatrix} 1 & 1 & 1 & 1 & 1 & 1 & 1 & 1 & 1 & 1 \\ 1 & 1 & 1 & 1 & 1 & 1 & 1 & 1 & 1 & 1 \\ 1 & 1 & 1 & 1 & 1 & 1 & 1 & 1 & 1 & 1 \\ 1 & 1 & 1 & 1 & 1 & 1 & 1 & 1 & 1 & 1 \\ 1 & 1 & 1 & 1 & 1 & 1 & 1 & 1 & 1 & 1 \\ 0 & 0 & 0 & 0 & 0 & 0 & 1 & 1 & 1 & 1 \\ 0 & 0 & 0 & 0 & 0 & 0 & 0.25 & 1 & 1 & 1 \\ 0 & 0 & 0 & 0 & 0 & 0 & 0.25 & 0.5 & 1 & 1 \\ 0 & 0 & 0 & 0 & 0 & 0 & 0.25 & 0.5 & 0.75 & 1 \\ 0 & 0 & 0 & 0 & 0 & 0 & 0.25 & 0.5 & 0.75 & 1 \end{bmatrix}$$

$$= \begin{bmatrix} 0 & 0 & 0 & 0 & 0 & 0 & 0.25 & 0.5 & 0.75 & 1 \end{bmatrix} \quad \textbf{(EQ 6.28)}$$

The inferred possibility distribution of the tomato's ripeness is thus RIPE.

Part C:

Finally, we apply sup-product composition to Goguen's implication for the same problem. As we mentioned earlier, we chose sup-product rather than sup-min because Goguen's implication is defined by using product as its fuzzy conjunction operator. We can compute the possibility distribution of the tomato's ripeness in a way similar to the previous two examples:

$$\Pi_{ripeness} = \Pi_{redness} \circ R_{gg}$$

$$= \begin{bmatrix} 0 & 0 & 0 & 0 & 0 & \dfrac{1}{16} & \dfrac{1}{4} & \dfrac{9}{16} & 1 & 1 \end{bmatrix} \circ \begin{bmatrix} 1 & 1 & 1 & 1 & 1 & 1 & 1 & 1 & 1 & 1 \\ 1 & 1 & 1 & 1 & 1 & 1 & 1 & 1 & 1 & 1 \\ 1 & 1 & 1 & 1 & 1 & 1 & 1 & 1 & 1 & 1 \\ 1 & 1 & 1 & 1 & 1 & 1 & 1 & 1 & 1 & 1 \\ 1 & 1 & 1 & 1 & 1 & 1 & 1 & 1 & 1 & 1 \\ 0 & 0 & 0 & 0 & 0 & 0 & 1 & 1 & 1 & 1 \\ 0 & 0 & 0 & 0 & 0 & 0 & 1/2 & 1 & 1 & 1 \\ 0 & 0 & 0 & 0 & 0 & 0 & 1/3 & 2/3 & 1 & 1 \\ 0 & 0 & 0 & 0 & 0 & 0 & 1/4 & 1/2 & 3/4 & 1 \\ 0 & 0 & 0 & 0 & 0 & 0 & 1/4 & 1/2 & 3/4 & 1 \end{bmatrix}$$

$$= \begin{bmatrix} 0 & 0 & 0 & 0 & 0 & 0 & \dfrac{1}{4} & \dfrac{1}{2} & \dfrac{3}{4} & 1 \end{bmatrix} \qquad \textbf{(EQ 6.29)}$$

The inferred possibility distribution of the tomato's ripeness is thus RIPE. The example above illustrates that standard sequence implication can satisfy Criterion II-1, while Godel's implication and Goguen's implication can satisfy Criterion II-2. However, we should point out that these results require some conditions about the membership functions in the implication rules, just as Theorem 7 needs the third condition. In addition to the third condition in Theorem 7, some of the checked entries in Table 6.3 require that fuzzy sets A and B be normal. Even though these assumptions are often satisfied in practice, it is important to understand them so that we can have a firm grasp of the foundation of approximate reasoning. We will leave a further investigation of these assumptions as an exercise.

6.4 Summary

In this chapter, we discussed the foundation of fuzzy implication rules, the desirable criteria of their inference, and several major types of fuzzy implication rules. We summarize several key points below.

The content of a fuzzy implication relation is determined by implication functions, which return the truth value of an implication based on the truth value of its antecedent and the truth value of its consequent.

- The inference involving a fuzzy implication rule is achieved by composing the possibility distribution of antecedent variables with the rule's fuzzy implication relation.

- A fuzzy implication rule is formally represented as a fuzzy relation of the implication possibility between values of the rule's antecedent variables and the consequent variable. Such a fuzzy relation is called an implication relation.

- Desirable criteria of inference using fuzzy implication are generalizations of modus ponens and modus tollens in two-valued logic. These criteria also involve hedges.

- Fuzzy implications functions can be classified into three families. One of these families is based on the definition of implication in multivalued logic systems.

- We introduced five fuzzy implication functions: (1) Zadeh's arithmetic fuzzy implication, (2) Zadeh's maxmin fuzzy implication, (3) standard sequence fuzzy implication, (4) Godelion fuzzy implication, and (5) Goguen's fuzzy implication.

- Among the five fuzzy implication functions, the standard sequence fuzzy implication satisfies most criteria.

- Sufficient conditions for satisfying certain inference criteria can be established for some families of fuzzy implications.

- The satisfaction of an inference criterion by a fuzzy implication function often requires some simple assumptions about the membership functions in the fuzzy implication.

Bibliographical and Historical Notes

Fuzzy implication rules and generalized modus ponens were first introduced by Zadeh in 1975 [699, 697]. This area soon became one of the most active research topics for fuzzy logic researchers. B. Gains discussed a wide range of issues regarding the foundation of fuzzy logic reasoning (including fuzzy implications and their relationships to implications in multivalued logic systems) in 1976 [200].

S. Fukami, M. Mizumoto, and K. Tanaka proposed a set of intuitive criteria for comparing fuzzy implication functions [194]. All but the last criterion introduced in this chapter came from their work. They also gave a comprehensive and formal discussion on how various fuzzy implication functions satisfy these criteria, which laid the foundation of theorem 7. The last criterion (i.e. chaining) was proposed by Zadeh and was referred to as "fuzzy syllogism" [700]. J. F. Baldwin and B. W. Pilsworth also discussed a similar set of intuitively desired properties of fuzzy implications and gave a less formal discussion on what fuzzy implication functions have these properties [23].

Trillas and Valverde described two ways to define fuzzy implication functions: the S-rules and the R-rules, which correspond to the first and the third fuzzy implication families in this chapter [600]. They also presented a scheme called modus ponens generating functions, which are similar to inference in multivalues logic systems, for reasoning about the truth value of a fuzzy proposition from the truth value of a relevant fuzzy implication [601]. Even though most of the research in this area focused on the choice of fuzzy implication functions, alternative schemes for approximate reasoning to replace the compositional rule of inference have also been proposed by P. Magrez and P. Smets [397].

Fuzzy implications have also been used in areas beyond approximate reasoning. In particular, W. Bandler and L. Kohout proposed an approach for defining subsethood and setequivalence using fuzzy implications [24]. B. Bouchon discussed justifications for viewing fuzzy implications as conditional possibility distributions [77].

Other fuzzy logic extensions to two-valued logic systems have also been made, even though their applications have been somewhat limited. R. C. T. Lee extended the resolution principle, which is the basis for automated theorem proving in Artificial Intelligence (AI) [368]. Based on this extension, M. Mukaidono, Z. Shen, and L. Ding developed the fundamentals of fuzzy Prolog, which generalizes Prolog — an AI programming language based on logic [430]. Based on Zadeh's possibility theory, D. Dubois and H. Parade have developed a framework for reasoning about interval truth values known as *possibilistic logic* [178, 169, 168].

Exercises

6.1   Prove that $p \rightarrow q$ is logically equivalent to $\neg p \vee (p \wedge q)$.

6.2   Let x and y be two variables taking values from U and V respectively, where $U = \{1, 2, 3, 4\}$ and $V = \{0, 10, 20\}$. A and B are fuzzy subsets of U and V, respectively, defined as follows:

$$A = 1/1 + 0.7/2 + 0.3/3$$
$$\mathbf{B} = 0.5/10 + 1/20$$

Consider the following fuzzy implication rule:
   If **x** is A then y is B

(a) Using the Godelian sequence fuzzy implication operator, construct the fuzzy implication relation for $R1$.

(b) Suppose that the value of x is VERY A. Use the sup-min compositional operator to infer the value of y.

6.3   Assume that Ripe in Example 7 is now defined as
   $$Ripe = 0.3/7 + 0.6/8 + 0.9/9 + 1/10$$
What is the result of fuzzy inference using the standard sequence implication and Godel's implication? How do you explain the inference result?

6.4   Prove that approximate reasoning using the standard sequence implication function and sup-min compositional rule of inference satisfies Criteria II-1, III-1, IV, V, VI, VII, VIII, and IX.

6.5   Under what condition will the Godel implication satisfy Criteria II-2? Prove this as a theorem.

6.6   Based on Theorem 7, prove the following:

(a) The standard sequence and the Godel implication satisfy Criterion I when sup-min composition is used.

(b) The sFigtandard sequence implication satisfies Criterion I for any sup-$\otimes$ composition where $\otimes$ is a fuzzy conjunction operator.

(c) Goguen's fuzzy implication satifies Criterion I when it is used with sup-product composition.

7 FUZZY LOGIC AND PROBABILITY THEORY

7.1 Introduction

Much of the criticism of fuzzy logic has come from the probability theory community. This is often due to some confusion about the difference between fuzzy logic and probability theory. This confusion is caused by a complicated relationship between fuzzy logic and probability. They are similar in certain perspectives, yet different in some other perspectives. In this chapter, we will reveal this mysterious relationship between fuzzy logic and probability. It is important that we understand their relationship clearly so that we can not only clarify confusion such as "fuzzy logic is a clever disguise of probability theory" but also learn to use both technologies appropriately such that their benefits are maximized.

Before we engage in a technical discussion, however, we would like to point out that the most important obstacle to understanding these two technologies is to *adopt a particular technology as a "religion."* Once a person adopts a technology as a religion, he orshe is likely to be convinced that the technology is the only true solution to a class of problems, just as a religion offers the true answer to the creation of universe, life, and people. Such a danger exists for both sides — whether you feel more comfortable with probability or fuzzy logic.

Fuzzy logic and probability theory are two different tools, like screw drivers and hammers. They were designed for different tasks. One could conceivably use a hammer to hammer a screw, yet it is much more effective to use a screw driver for the job. Only after we fully understand the tasks these tools are designed for can we use them properly. Too often, unfortunately, we have a tendency to avoid learning how to use a new tool, simply because we are useing to use an old tool that we are familiar with. Professor Zadeh once said, "if the only tool you know is a hammer, everything looks like a nail".   We hope this chapter helps you to understand how to use both screw driver and hammer wisely.

7.2 Possibility versus Probability

Among all concepts and techniques in fuzzy logic, *possibility theory* is the one that is most often confused with probability theory. For this reason, we will first clarify the differences between them. Possibility and probability measure two different kinds of uncertainty.   As we have mentioned in Chapter 2, the best way to introduce possibility theory is to start with the notion of interval values.

Imagine that you were a witness of a bank robbery. When the police asked you about the height of a male suspect, your answer may be one of the following :

- He is between 5 feet and 6 feet.

- He is somewhat tall.

Both of these answers impose a constraint (based on your impression) on the possible heights of the suspect. The first answer imposes a crisp constraint. That is, the suspect can't possibly be shorter than 5 feet, nor can he be taller than 6 feet. The second answer, on the other hand, typically imposes an imprecise constraint on the suspect's height. Unlike the interval-value answer, the second response does not have a well-defined sharp boundary between the possible heights and the impossible ones. Rather, it suggests that a certain height is more possible than another one.

In general, when we assign a fuzzy set, a variable whose value we don't know exactly, the assignment introduces an imprecise constraint on the variable's value. We call such a constraint a *possibility distribution*, because it specifies the degree of possibility for the variable to take a certain value.   The constraint imposed by an internal-value assignment is a special kind of possibility distribution. In other words, the notion of possibility distribution generalizes the notion of intervalvalue to smooth the boundary between the possible values and the impossible values so that possibility becomes a matter of degree. We use $\Pi(x)$ to denote the possibility distribution of a variable x.

Possibility measures *the degree of ease* for a variable to take a value, while probability measures the *likelihood* for a variable to take a value. Therefore, they deal with two different types of uncertainty. Possibility theory handles imprecision, and probability theory handles likelihood of occurrence. This fundamental difference leads to different mathematical properties of their distributions. Being a measure of occurrence, the probability distribution of a variable must add up to exactly 1. The possibility distribution, however, is not subject to this restriction since a variable can have multiple possible values. Returning to our example about the robbery suspect, we may use the following normal distribution with $\mu = 5$ ft, $\sigma = 0.2$ ft to describe the probabilities of the suspect's height:

$$P(h) \;=\; \frac{1}{0.2\sqrt{2\pi}} e^{-\frac{1}{2}\left(\frac{h-5}{0.2}\right)^2} \qquad\qquad \textbf{(EQ 7.1)}$$

It is well known that the area under a normal distribution is 1.

7.2.1 Noninteractiveness versus Independence

The notion of noninteractiveness in possibility theory is analogous to the notion of independence in probability theory. If two random variables x and y are independent, their joint probability distribution is the product of their individual distributions (also called marginal probabilities). Similarly, if two linguistic variables are noninteractive, their joint possibility distribution is formed by combining their individual possibility distributions through a fuzzy conjunction operator. Conceptually, two variables are noninteractive if they don't have an impact on each other. For example, the heights of two suspects for a bank robbery are noninteractive. One suspect's being tall does not suggest the other suspect is tall (or short). On the other hand, the height and the weight of a suspect are interactive. For instance, a short suspect can't be too heavy.

Even though possibility and probability are clearly different, they are related in a way. If it's impossible for a variable x to take a value x_i, it is also improbable for x to have the value x_i, i.e.,

$$\Pi_X(x_i) = 0 \implies P_X(x_i) = 0 \qquad \textbf{(EQ 7.2)}$$

In general, possibility can serve as an upper bound on probability: $P_x((x_i) \leq \Pi_x(x_i)$ We compare some properties of probability distribution and possibility distribution in Table 7.1.

TABLE 7.1   Probability versus Possibility

Probability Distribution	Possibility Distribution
0<P(x)<1	0<π(x)<1
$p(x,y)=p(x)*p(y)$ if x and y are independent, $\int p(x)dx=1$	$\pi(x,y)=\pi(x) \lor \pi(y)$ if x and y are notinteracting

Possibilities and probabilities can be combined to deal with problems in which both kinds of uncertainty (i.e., imprecision and likelihood) exist. We discuss two examples of such combination below.

7.3 Probability of a Fuzzy Event

An event is a subset of the sample space that shares a common characteristic that we are interested in. A day in Houston with a temperature between 70°F and 80°F is an event. The relevant sample space is all possible temperatures in Houston. Having more heads than tails in 20 tossings of a coin, is a classical event in statistics textbooks. The relevant sample space includes all 21 possible outcomes (ranging from no head to 20 heads). There are situations when the event we are interested in does not have well-defined sharp boundaries. We call them fuzzy events. Formally, a fuzzy event is a fuzzy subset of the sampling space. For instance, " a day with a warm temperature" is a fuzzy event. Another example of a fuzzy event is "in 20 tosses of a coin, there are *several more* heads than tails."

The formula for calculating the probability of a fuzzy event A is a generalization of the probability theory:

$$P(A) = \int \mu_A(x) \times P_X(x)\, dx \qquad \textbf{(EQ 7.3)}$$

if X is continuous, or

$$P(A) = \sum_i \mu_A(x_i) \times P_X(x_i) \qquad \textbf{(EQ 7.4)}$$

if X is discrete where P_X denotes the probability distribution function of X.

EXAMPLE 9   Let X be the random variable that counts the total number of heads in 10 coin tossings. Let the meaning of the fuzzy event A "several more heads than tails" be characterized by the membership function below:

$$\mu_A(x_i) = \begin{cases} 0 & 0 \le x_i \le 5 \\ \frac{1}{3}(x_i - 5) & 5 \le x_i \le 8 \\ 1 & 9 \le x_i \le 10 \end{cases} \qquad \textbf{(EQ 7.5)}$$

What is the probability of the fuzzy event A?.

Solution:

From probability theory, we first obtain the following probability distribution for X:

$$P_X(x_i) = \frac{20!}{i!(20 - i)!} \qquad \textbf{(EQ 7.6)}$$

Using Equation 7.4, we get

$$P(A) = \sum_{i=0}^{20} \mu_A(x_i) P_X(x_i) \qquad \textbf{(EQ 7.7)}$$

The probability of the fuzzy event A can be calculated by substituting into this formula the memberships function of A (i.e., Equation 7.5) and the probability density function for X (i.e. Equation 7.7). The result is

$$P(A) = \frac{1}{3} \times P_X(6) + \frac{2}{3} \times P_X(7) + P(8) + P(9) + P(10) \qquad \textbf{(EQ 7.8)}$$

7.3.1 Properties about Probability of Fuzzy Events

For any two events A and B of a sample space, we know from probability theory that

$$P(A \cup B) = P(A) + P(B) - P(A \cap B) \qquad \textbf{(EQ 7.9)}$$

Does this relationship hold for fuzzy events? The answer to this question depends on the choice of the fuzzy union (i.e., fuzzy disjunction) and fuzzy intersection (i.e., fuzzy conjunction) operators. In fact, it is straightforward to show that Equation 7.9 holds for the following two choices :

1. Union: max, intersection: min
2. Union: algerbraic sum, intersection: algebraic product (i.e, multiplication).

We will leave this as an exercise.

7.3.2 Conditional Probability of Fuzzy Events.

Suppose A and B are two events of a sample space the conditional probability of A given B is defined in probability theory as

$$P(A/B) = \frac{P(A \cap B)}{P(B)}$$ **(EQ 7.10)**

It is easy to show that the conditional probability of A given B and the conditional probability of A's complement given B adds to 1, i.e.,

$$P(A/B) + P(A^C/B) = 1$$ **(EQ 7.11)**

The conditional probability of fuzzy events can be defined in a similar way using Equation 7.10. However, Equation 7.11 holds only if we choose multiplication as the fuzzy intersection operator. For this reason, we will define the intersection and union of fuzzy events using algebraic product and algebraic sum operators, respectively.

DEFINITION 11   If A and B are any two fuzzy events in a sample space S, the probability of both fuzzy events occurring is defined as $P(A \cap B) = \sum_{x \in S} \mu_A(x) \times \mu_B(x) \times P(x)$.

Based on this definition, we can now formally define the conditional probability of fuzzy events.

DEFINITION 12   If A and B are any two fuzzy events in a sample space S and $P(A) \neq 0$, the **conditional probability** of A given B is $P(A/B) = \frac{P(A \cap B)}{P(B)}$. From this definition we get

$$P(A \cap B) = P(A/B) \times P(B)$$ **(EQ 7.12)**

7.3.3 Independent Fuzzy Events

In probability theory, two events A and B are independent if the occurrence or non-occurrence of either one does not affect the occurrence of the other one, i.e., $P(A/B) = P(A)$ and $P(B/A) = P(B)$. If we substitute $P(A)$ for $P(A/B)$ into Equation 7.12, we have $P(A \cap B) = P(A) \times P(B)$, which we will use to formally define the independence of fuzzy events.

DEFINITION 13   Two fuzzy events A and B are **independent** if and only if $P(A \cap B) = P(A) \times P(B)$.

7.3.4 Bayes' Theorem for Fuzzy Events.

Bayes' theorem is the basis for many probabilistic inference schemes (e.g., Bayes' networks, influence diagrams, etc.) used in intelligent systems. If $A_1, A_2, \ldots, A_k$ forms a partition of the sample space S (i.e., they are mutually exclusive and exhaustive) and $P(A_i) \neq 0$ for $i = 1, 2, \ldots, k$, then for any event B in S such that $P(B) \neq 0$, Bayes' theorem enables us to infer $P(A_j / B)$ from $P(A_i)$ and $P(B / A_i)$:

$$P(A_j / B) = \frac{P(A_j) \times P(B / A_j)}{\sum_{i=1}^{k} P(A_i) \times P(B / A_i)} \qquad \text{(EQ 7.13)}$$

for $j = 1, 2, \ldots, k$.

If A_i and B are fuzzy events, we need to generalize the condition in Bayes' theorem to allow fuzzy events $A_1, \ldots A_k$ to form a fuzzy partition of S such that

$$\sum_{i=1}^{k} \mu_{A_i}(x) = 1 \qquad \text{for } x \in S$$

We formally state the theorem below

THEOREM 8   If $A_1, A_2, \ldots, A_k$ are fuzzy events of S, such that

$$\sum_{i=1}^{k} \mu_{A_i}(x) = 1 \qquad \text{for } x \in S \qquad \text{(EQ 7.14)}$$

$P(A_i) \neq 0$ for $i = 1, 2, \ldots, k$, and B is a fuzzy event of S such that $P(B) \neq 0$,

$$P(A_j / B) = \frac{P(B / A_j) \times P(A_j)}{\sum_{i=1}^{k} P(B / A_i) \times P(A_i)} \qquad \text{(EQ 7.15)}$$

for $j = 1, 2, \ldots, k$.

Proof: From the definition of conditional probability for fuzzy events, we have

$$P(A_j / B) = \frac{P(A_j \cap B)}{P(B)} \qquad \text{(EQ 7.16)}$$

In order to show that the right-hand side of Equation 7.15 equals the ratio above, we expand the denominator of Equation 7.15:

$$\sum_{i=1}^{k} P(B / A_i) \times P(A_i)$$

$$= \sum_{i=1}^{k} P(A_i \cap B)$$

$$= \sum_{i=1}^{k} \sum_{x_k} \mu_{A_i}(x_k) \times \mu_B(x_k) \times P(x_k)$$

$$= \sum_{x_k} \mu_B(x_k) \times P(x_k) \times \left(\sum_{i=1}^{k} \mu_{A_i}(x_k) \right)$$

$$= \sum_{x_k} \mu_B(x_k) \times P(x_k)$$

$$= P(B)$$

Since, $P(B/A_j) \times P(A_j) = P(A_j \cap B)$, we can substitute the numerator and denominator of Equation 7.16 to get Equation 7.15. The theorem is thus proved.

7.4 Fuzzy Probability

In classical probability theory, a probability measure is a number between 0 and 1. There are cases, however, when such a precise probability measure is difficult to obtain. This motivated the development of interval probability theories, which can represent and reason about the lower bound and the upper bound of the probability measure. The interval formed by these two bounds contains the true probability value, which we do not know precisely. In other words, interval probabilities allow us to describe the *imprecision* of our probability estimates. The width of the interval reflects the degree of imprecision. The wider the interval, the less precise our probability estimate. In the literature of interval probability theory, this interval is sometimes referred to as *the degree of ignorance*.

One of the best known interval probability theories is the Dempster-Shafer (DS) theory of evidence, which is also called the theory of belief Functions. The Theory was first developed in a seminal paper by A.P. Dempster in 1967, and later extended by G. Shafer in 1976. In the DS theory, the lower bound on a probability is called its *belief measure*, while the upper bound is called a *plausibility measure*. These bounds are calculated from underlying constraints on the probability distribution. These constraints are expressed by restricting a certain amount of "probability mass" to be within a set of events. The constraints are often called the *basic probability assignment* (bpa). The notation $m(A)$ represents the amount of probability mass constrained within the set A (i.e., the bpa of A).

The notion of probability mass can often be confused with the concept of probability. Since the DS theory was developed to deal with subjective probabilities, it may be easier to explain these concepts within the context of subjective probabilities, in which a probability measures our *degree of belief* in a statement or a hypothesis. The probability mass of A contributes to A's probability, but it is often just a part of A's total probability. Other parts of A's probability come from those probability masses constrained to

(1) subsets of A (and therefore to A as well), and

(2) those sets partially overlaping with A.

The former contributes to A's lower probability because they can not escape A, while the latter contributes only to A's upper probability because they can be allocated.

Fuzzy probability is a generalization of interval probability in which the probability value is bounded by a fuzzy set. More accurately, a fuzzy probability uses a possibility distribution to constrain an unknown probability measure. For example, suppose the meaning of high probability is as depicted in Fig 7.1. The statement " the probability that it will rain tomorrow is high" imposes a possibilistic constraint on the probability of the event that it will rain tomorrow. For example, the probability is entirely possible to be 1, yet impossible to be less than 0.5. The degree of possibility for the probability to be a value in between is its membership degree in "high probability."

Even though fuzzy probabilities have not been used often in practice due to the complexity of their reasoning, they illustrate one way to combine possibility theory and probability theory. We should also point out that fuzzy probabilities are related to the concept of higher-order probabilities.

In interval probability theories, we have

$$P^*(\neg h) = 1 - P^*(h)$$
$$P^*(\neg h) = 1 - P_*(h)$$

where P_* and P^* denote lower probability and upper probability, respectively. How can we generalize these so that we can obtain the fuzzy probability of $\neg h$ from the fuzzy probability of h? We will leave this as an exercise.

FIGURE 7.1   Examples of Fuzzy Probabilities

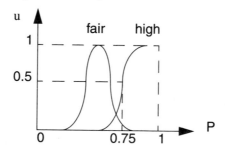

7.5 Probabilistic Interpretation of Fuzzy Sets

It is possible to give fuzzy sets a probabilistic interpretation — viewing membership degree of x in a fuzzy set A as the conditional probability of A given x, i.e.,

$$\mu_A(x) = P(A|x) \tag{EQ 7.17}$$

Such a probabilistic interpretation is consistent with the complement operation on fuzzy sets, because

$$\mu_{A^c}(x) = P(A^c|x) = 1 - P(A|x) = 1 - \mu_A$$

However, the probabilistic interpretation does not help us in calculating the intersection or the union of fuzzy sets.

$$\mu_{A \cap B}(x) = P(A \cap B | x) \tag{EQ 7.18}$$

$$\mu_{A \cup B}(x) = P(A \cup B | x) \tag{EQ 7.19}$$

From probability theory, we know

$$P(A \cup B | x) = P(A | x) + P(B | x) - P(A \cap B | x) \tag{EQ 7.20}$$

Therefore, should we adopt the probabilistic interpretation (Equation 7.15), we should get

$$\mu_{A \cup B}(x) = \mu_A(x) + \mu_B(x) - \mu_{A \cap B}(x) \tag{EQ 7.21}$$

However, this only establishes a constraint between union and intersection. If A and B are conditionally independent of x [i.e., $P(A | x, B) = P(A | x)$, $P(B | x, A) = P(B | x)$], we get $P(A \cap B | x) = P(A | x) \times P(B | x)$.
If one is a subset of another

$$P(A \cap B | x) = \begin{cases} P(A | x) & \text{if } A \subseteq B \\ P(B | x) & \text{if } B \subseteq A \end{cases} \tag{EQ 7.22}$$

or equivalently

$$P(A \cap B | x) = \min\{P(A | x), P(B | x)\} \tag{EQ 7.23}$$

As we can see from the two cases above, the probability $P(A \cap B | x)$ depends not only on $P(A | x)$, $P(B | x)$, but also the relationship between A and B. Hence finding a general intersection operator t such that

$$P(A \cap B | x) = t(P(A | x), P(B | x)) \tag{EQ 7.24}$$

is impossible. The same argument can be made about the union operator.

The above discussion points out a major limitation of the probabilistic interpretation of fuzzy sets; it suggests an uncertain reasoning scheme that is not *truth functional*. A scheme for reasoning under uncertainty is *truth functional* if the certainty degree of a conjunctive/disjunctive expression is a function of the certainty degree of its individual conjuncts/disjuncts.

7.6 Fuzzy Measure

Finally, it may be worthwhile to point out that both possibility measure and probability measure belong to a more general class of measures called fuzzy measure.

DEFINITION 14   A probability P over a set of events S is a function from 2^S to $[0,1]$ such that

1. $P(\phi) = 0$

2. $P(S) = 1$

3. If $A \cap B = 0$

   then $P(A \cup B) = P(A) + P(B)$

   where A, B are subsets of S.

4. $P(B|A) = \dfrac{P(A \cap B)}{P(A)}$ if $P(A) \neq 0$

It is easy to show that we get the following properties about probabilities from these axioms:

- $P(A) + P(A^c) = 1$
- If $A \subset B$, then $P(B) \geq P(A)$

FIGURE 7.2   Relationship between Fuzzy Measure, Probability, and Possibility

Fuzzy Measure

Possibility Measure   Probability Measure

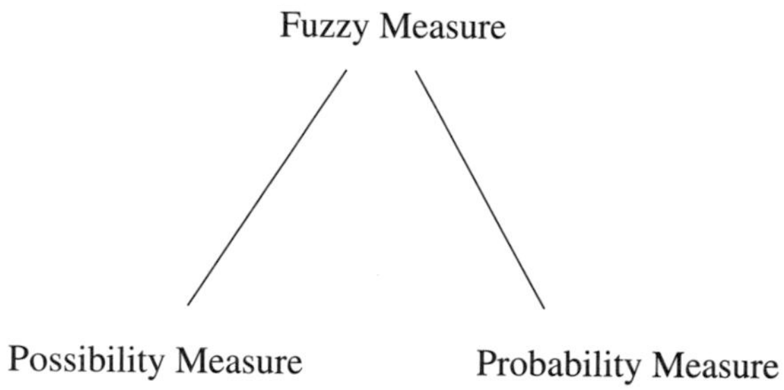

DEFINITION 15   A possibility measure Π over a set of events S is a function from 2^S to $[0,1]$ such that
1. $\Pi(\phi) = 0$
2. $\pi(S) = 1$
3. if $A \subset B$, then $\pi((A) \leq \pi(B))$
4. $\pi(A \cup B) = \max\{\pi(A), \pi(B)\}$

DEFINITION 16   A fuzzy measure g over a set of events S is a function from s to $[0,1]$ such that
1. $g(\phi) = 0$
2. $g(S) = 1$
3. if $A \subset B$, then $P(B) \geq P(A)$

Since the set of axioms for fuzzy measure is a subset of the axioms of probability and possibility measure above, it is more general than both of them.

7.7 Summary

In this chapter, we attempted to clarify the distinction between fuzzy logic and probability. In particular, we discussed the following topics.

- Possibility and probability are different just as interval-valued assignment and probability are different.

- Possibility and probability are related — if something is impossible, it is improbable. In general, a possibility distribution on a variable could serve as an upper bound on the probability distribution of the same variable.

- Fuzzy sets can be used to generalize the notions of "events" and "probabilities" in probability theory into the notions of "fuzzy events" and "fuzzy probabilities," respectively.

- Even though there is a probabilistic interpretation of fuzzy sets, it is difficult to use the interpretation to define operations (e.g., intersections and unions) of fuzzy sets.

- Both probability measure and possibility measure belong to a more general class of fuzzy measures.

Bibliographical and Historical Notes

Like most other important ideas in fuzzy logic, the notion of fuzzy events and their probability measures were first introduced by Zadeh [706]. Even though this work was published in 1968, it was not until 1981 that he formally introduced fuzzy probabilities [687, 684]. From 1978 to 1982, Zadeh published a series of papers on possibility theory and its connection to information analysis, uncertain data management, information processing, knowledge and meaning representation, human communication, and data analysis [692, 694, 686, 691, 690, 688]. Realizing the potential confusion between possibility theory and probability theory, he wrote a paper entitled "Is Possibility Different from Probability?" in 1983 [685]. He explicitly reasserted his position to this question in the title of a 1995 article — "Probability Theory and Fuzzy Logic are Complementary rather than Competitive" [681].

Other researchers have also made important contributions to the connection between fuzzy logic and probability theory. M. Sugeno introduced fuzzy measures in early 70's [35, 557]. B. R. Gaines wrote a paper on fuzzy logic and probabilistic logic in 1978 [198]. Several formal treatments of fuzzy random variables (i.e., a random variable whose outcome is fuzzy rather than precise) have been proposed by H. T. Nguyen [448], R. Kruse [340], and M. L. Puri and D. A. Ralescu [485]. The issue of consistency between the possibility distribution and the probability distribution of a variable was discussed by M. Delgado and S. Moral [152].

The probabilistic interpretation of fuzzy sets was given by B. Natvig in 1983 [438]. A few years later, E. Hisdal attempted to justify the fuzzy set operations using a model in which fuzzy sets are given a probabilistic interpretation [250, 253].

Many criticisms to fuzzy logic have been generated from statisticians and believers of Bayesian probability. Two of the most common attacks are

- "Anything that can be done with fuzzy logic can be better done with probability."
- "Probability is the only satisfactory measure of uncertainty."

B. Y. Lindley is probably the best known statistician to make these claims in 1982 [385]. Three years later, P. Cheeseman made similar arguments within the artificial intelligence community [114, 115]. Responding to this change, Zadeh wrote a paper to argue that probability theory alone is not sufficient for dealing with uncertainty in AI [683]. These debates continued into the 90s with a different twist. B. Kosko, who introduced an axiomatic foundation to fuzzy set theory, launched an attack on probability theory by arguing that "randomness is a special case of fuzziness" in 1990 [333]. A special issue of *IEEE Transactions on Fuzzy Systems* was dedicated to the debate on fuzziness vs. probability in 1994 [39], and contains a historical remark on the debate from the editor J. Bezdek and a paper by M. Laviolett and J. W. Seaman that challenges fuzzy logic again from a probability viewpoint [358]. The paper has been followed by responses from fuzzy logic researchers such as D. Dubois, H. Prade, E. Hisdal, G. J. Klir, B. Kosko, N. Wilson as well as a supporting reply from statistician D. V. Lindley.

While the debate is likely to continue for years to come, just like the debate between different groups of probabilists (the frequentist vs. the subjective, Bayesian vs. non-Bayesian) has never ended, I hope this chapter helps you to develop a better understanding about the *complementary* nature of the two theories as well as the subtlety of their different focus for handling two different kinds of uncertainty. Furthermore, the two claims mentioned earlier may have their theoretical significance. However, they are not as useful as the following question:

Given a problem, which technique offers the most cost-effective solution?
Depending on the problem, the answer may be fuzzy logic, probability, neither, or both (i.e., a combination of the two). When you become confident about responding to this question for the problems you are interested in, you are then truly a master of both tools.

Exercises

7.1   Let A and B be two fuzzy events of a sample space S. Prove that

$$P(A \cup B) = P(A) + P(B) - P(A \cap B)$$

if union and intersection of fuzzy events and defined as

i) $\mu_{A \cup B}(x) = \max \{\mu_A(x), \mu_B(x)\}$

  $\mu_{A \cap B}(x) = \min \{\mu_A(x), \mu_B(x)\}$

ii) $\mu_{A \cup B}(x) = \mu_A(x) + \mu_B(x) - \mu_A(x) \times \mu_B(x)$

  $\mu_{A \cap B}(x) = \mu_A(x) \times \mu_B(x)$

7.2   If A and B are two fuzzy events of a sample space S, prove that

$$P(A/B) + P(A^C/B) = 1$$

7.3   Suppose we change the definition of intersection of fuzzy events to use the min operator:

$$\mu_{A \cap B}(x) = \min\{\mu_A(x), \mu_B(x)\}$$

Will the following equality still hold?

$$P(A/B) + P(A^C/B) = 1$$

PART II

FUZZY LOGIC CONTROL

8   FUZZY LOGIC IN CONTROL ENGINEERING

8.1   Introduction

This chapter introduces fuzzy logic control and its application in control engineering. To this end a brief overview of control engineering is presented and certain fundamental concepts of this discipline are discussed. In particular, the concept of *control problem* and the manner in which it is formulated and ultimately resolved are outlined. In this connection it is crucial to recognize the role of *informal* means in both formulation and resolution of control problems. From this standpoint, one may view fuzzy logic control as extending the realm of application of control engineering by providing a mechanism to incorporate informal knowledge of the operation of a system in terms of a control algorithm, either independently or in conjunction with conventional control techniques.

With these remarks in mind, we start with a brief overview of the role of formal and informal means in the development of solution strategies to control engineering problems.

8.2   Fundamental Issues in Control Engineering

The term *control engineering* refers to a discipline whose main concern is with problems of regulating and generally controlling the behavior of physical systems. The term *physical system* here refers to

an interconnected collection of physical objects or entities that together serve a specific purpose or function predictable in accordance with physical laws.

For instance in the context of an automotive cruise control problem, illustrated in Fig 8.1, the collection of vehicle body, chassis, wheel assembly, engine,Fig and power train forms a physical system whose function is to move passengers or cargo in a prescribed manner or more specifically at a prescribed speed in spite of variations in the road grade and/or similar factors. This objective is realized, generally speaking, via a *feedback loop* as shown in the figure where comparison of the actual vehicle speed, v, with its prescribed or *reference* value, v_d, is used to adjust the throttle position, θ, so as to minimize or ideally eliminate the error, $e = v_d - v$.

The architecture shown in Fig 8.1 depicts a common means of representing a feedback control system via blocks representing its constituent elements, namely the plant (the physical system) and the controller. The connecting lines represent the flow of information among these elements. The terminology used in this context such as *plant, controller, reference, error, input, output,* and *disturbance* arc likewise standard means of referring to significant aspects of a feedback control system (Table 8.1.)

TABLE 8.1　　　　Control Engineering Terminology

Term	Standard Notation[a]	Meaning	Example
Plant	P	Physical system to be controlled	Vehicle
Controller	C	The computational device and/or the algorithm used to control the physical system	See Section 8.3.4
Input	u	Controllable attribute(s) of the plant	Throttle position, θ
Output	y	Observable attribute(s) of the plant	Vehicle speed, v
Reference	y_d or r	Desired value or setpoint for the plant output(s)	Desired speed, v_d
Error	e	Difference between reference and output	$e = v_d - v$
Disturbance	d	Uncontrollable input(s) of the plant	Gravity force, F_g

[a] These are generic notations. It is common to substitute physically meaningful variables when sensible as is done in Fig 8.1.

A key component of a feedback control system is the controller, whose purpose is to accomplish the performance objectives one states at the outset of the formulation of the given control problem. We will refer to these as the *external performance objectives*, which in the case of the automotive cruise control problem is to maintain the speed of the given vehicle at or "near" its prescribed or reference value in spite of variations in the road grade (disturbances) and to do so "quickly," "smoothly," and "efficiently." The task of controller design thus amounts to selection of the appropriate structure and parameters for the controller to accomplish the given external performance objectives.

FIGURE 8.1   An Automotive Cruise Control System

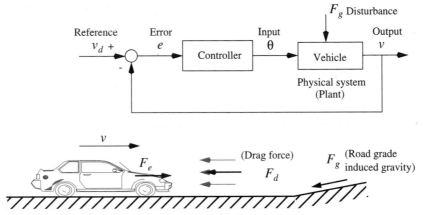

8.3 Control Design Process

The design process in control engineering is generally a multistage process involving *(i)* selection of control design technique or methodology, *(ii)* determination of technical design objectives, *(iii)* development of the plant model, and *(iv)* selection of controller structure and parameters.

8.3.1 Selection of Design Methodology

Generally design methodologies in control engineering are categorized as either *time* or *frequency-domain*-based. Frequency-domain design methods range from the classical "loop shaping" to the modern H_∞-based design techniques while time-domain design methods range from the simple PID design to linear quadratic optimal control. We will offer the reader a limited view of these techniques in the subsequent sections.

8.3.2 Determination of Technical Design Objectives

This task requires interpreting, refining, and quantifying the given external performance objectives into a set of *technical design objectives* compatible with the given design methodology. In the context of the automotive cruise control problem, for instance, one must translate such notions as "near prescribed speed" or "quickly," "smoothly," and "efficiently" into appropriate technical design objectives such as steady-state error, rise time/settling time, and maximum overshoot, or in the case of optimal control design as *performance weights* associated with time variation of error and the expenditure of control input.

8.3.3 Development of the Plant Model

In order to properly design a control strategy, we must have a *predictive model* of the plant. In general such a model is a mathematical description of the behavior of the given physical system and is derived according to applicable physical laws. In the case of the automotive cruise control problem, Newton's second law can be used to derive a model of the vehicle that predicts the variation of vehicle speed with time as a function of the applied forces: F_e (engine force, a function of the throttle position), F_d (drag force, a function of vehicle velocity), and F_g (gravity induced force, a function of the road grade) as

$$m\frac{dv}{dt} = F_e(\theta) - F_d(v) - F_g$$

$$\tau_e\frac{dF_e}{dt} = -F_e + F_{e1}$$

(EQ 8.1)

where m is the vehicle mass and τ_e is the engine response *time constant*. Typical functional forms of F_{e1} and F_d are as follows:

$$F_{e1}(\theta) = F_i + \gamma\sqrt{\theta}$$

$$F_d(v) = \alpha v^2 \mathrm{sgn}\, v$$

(EQ 8.2)

where F_i is the engine idle force, γ and α are positive constants, and

$$\mathrm{sgn}(v) = \begin{cases} -1 & v < 0 \\ 0 & v = 0 \\ 1 & v > 0 \end{cases}$$

(EQ 8.3)

Specific numerical values of the above parameters, used in the subsequent simulations, are listed in Table 8.2. (A Matlab/Simulink model that represents this vehicle model is given in the appendix to this chapter.)

TABLE 8.2   Values of the Numerical Constants Used in Conjunction with the Vehicle Model

Constant	Notation	Value(SI Units)
Vehicle mass	m	1000 kg
Drag coefficient	α	4 N/(m/s)^2
Engine force coefficient	γ	12,500 N
Engine idle force	F_i	6,400 N
Engine time constant	τ_e	0.1 to 1 second
Maximum throttle position	θ_{max}	30 to 60 degrees

8.3.3.1   Plant Model Simplification and Linearization

An essential step in the control design process is simplification and *linearization* of the plant model near an equilibrium or, more generally, an *operating point*. For this purpose we assume that the engine is relatively "fast" compared to the vehicle; i.e. we assume that the time constant, τ_e, is negligible and thus simplify the pair of equations in Equation 8.1 into

$$m\frac{dv}{dt} = F_e(\theta) - F_d(v) - F_g$$

(EQ 8.4)

$$0 = -F_e + F_{e1}$$

which can be combined into a single ordinary differential equation as

$$m\frac{dv}{dt} = F_{e1}(\theta) - F_d(v) - F_g$$

(EQ 8.5)

where F_{e1} is given for instance by Equation 8.2.

Next we assume that the vehicle will mostly be operated around some nominal velocity, v_0, and more or less on a flat road and proceed to linearize the above nonlinear differential equation. We note that the linearized model will be applicable to situations where the vehicle velocity differs from the nominal value of v_0; we will be able to accommodate reasonable, but not large, road grade changes via this linearized model.

We have, at this operating point,

$$m\frac{dv_0}{dt} = F_{e1}(\theta_0) - F_d(v_0),$$

(EQ 8.6)

from which we can determine the corresponding throttle position, θ_0. This task in principle involves solving a nonlinear ordinary differential equation. However with v_0 constant Equation 8.6 resolves into

$$0 = F_{e1}(\theta_0) - F_d(v_0),$$

(EQ 8.7)

In other words, for the vehicle to maintain its speed at v_0 on a flat road, the force produced by the engine should be sufficient to compensate for the wind resistance. We may use this fact to *graphically* determine the corresponding θ_0 as shown in Fig 8.2. We note that we can also determine θ_0 algebraically as follows:

$$\theta_0 = F_{e1}^{-1}(F_d(v_0)),$$

(EQ 8.8)

where $F_{e1}^{-1}(.)$ is the inverse of $F_e(.)$. The above equation, with the specific form of F_{e1} and F_d given earlier, resolves into

$$\theta_0 = \left(\frac{\alpha v_0^2 - F_i}{\gamma}\right)^2, ,$$

(EQ 8.9)

which for numeric values given in Table 8.2 and $v_0 = 40\text{m/s}$ results in $\theta_0 = 0.186\text{rad/s}$ or $10.7°$.

FIGURE 8.2   Nonlinear Description of the Engine and Drag Forces

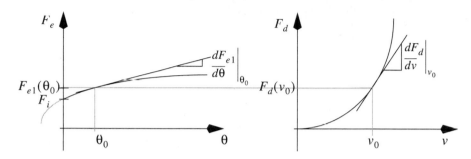

We further derive a relationship between the throttle position, θ, and the driving force, F_{e1}, for values of θ near θ_0 in terms of the Taylor series expansion of F_{e1} as follows:

$$F_{e1}(\theta) \cong F_{e1}(\theta_0) + \left.\frac{dF_{e1}}{d\theta}\right|_{\theta_0} (\theta - \theta_0) = F_{e1}(\theta_0) + k_\theta \delta\theta \qquad \textbf{(EQ 8.10)}$$

where $\delta\theta = \theta - \theta_0$ and we have denoted the gradient of F_{e1}, i.e., $\left.\frac{dF_{e1}}{d\theta}\right|_{\theta_0}$, by k_θ. Likewise we approximate F_d around v_0 via a similar (Taylor series) expansion as follows:

$$F_d(v) \cong F_d(v_0) + \left.\frac{dF_d}{dv}\right|_{v_0} (v - v_0) = F_d(v_0) + k_v \delta v \qquad \textbf{(EQ 8.11)}$$

where $\delta v = v - v_0$ and we denote $\left.\frac{dF_d}{dv}\right|_{v_0}$ by k_v. Now we can linearize Equation 8.1 as follows. Substituting for v its equivalent expression, $v_0 + \delta v$ and using the approximations given by Equations 8.10 and 8.11 in Equation 8.1 we have

$$m\frac{d}{dt}(v_0 + \delta v) = (F_{e1}(\theta_0) + k_\theta \delta\theta) - (F_d(v_0) + k_v \delta v) - F_g \qquad \textbf{(EQ 8.12)}$$

Expanding terms and taking Equation 8.7 into account, we have

$$m\delta\dot{v} = k_\theta \delta\theta - k_v \delta v - F_g \qquad \textbf{(EQ 8.13)}$$

Equation 8.13 is a first-order *linear ordinary differential equation* that can be written in standard form as

$$\tau\delta\dot{v} \ = \ -\delta v + k_s\delta\theta - \frac{1}{k_v}F_g \qquad\qquad \textbf{(EQ 8.14)}$$

where $k_s \ = \ k_\theta/k_v$ is the so-called *steady-state gain* of the plant and $\tau \ = \ m/k_v$ is the plant *time constant*. This equation can be analyzed rather easily and further used as the basis for control design as we shall see shortly. Typical step response of the vehicle to a 0.1-rad step change in the throttle position $(\delta\theta \ = \ 0.1\,\text{rad})$ is shown in Fig 8.3; $\tau \ = \ 3.125$ sec and $k_s \ = \ 45.2$ are based on values of m, α, and γ given in Table 8.2. The operating point and the initial condition are 40 m/s. (These values are used in the subsequent simulations as well.)

FIGURE 8.3   Vehicle Step Response with Initial Velocity of 40 m/s on a Flat Road

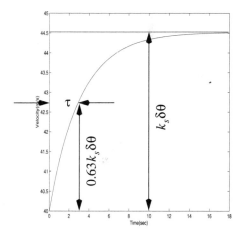

8.3.4   Control Design

The process of control design in the situation at hand amounts to selection of a *control algorithm* that transforms the deviation in speed $v_d - v$ into an appropriate action in terms of throttle position, θ, and in view of the stated performance objectives.

Since our emphasis in this chapter is on the potential use of formal design techniques, we define an *optimal control* problem wherein a cost associated with the deviation of the vehicle speed from its desired value and expenditure of control action is used to devise a control law that takes the form of a linear feedback strategy. In the particular case considered, such an optimization criterion or cost functional takes the form

$$J \ = \ \int_0^\infty (q(v_d - v)^2 + r\delta\theta^2)dt \qquad\qquad \textbf{(EQ 8.15)}$$

where q and r are *performance weights* discussed in Section 8.3.2 and, in effect, determine the relative significance of accuracy versus control cost in the optimization strategy; larger q implies better tracking and larger r implies less use of actuation power. Fig 8.4 illustrates this process.

FIGURE 8.4   Schematic of the Optimal Control Design Process for the Vehicle.

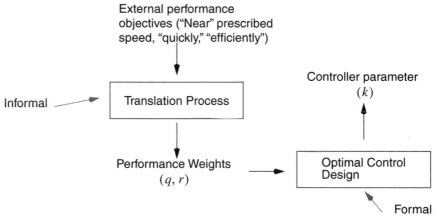

As we pointed out earlier, it is not generally possible to uniquely translate the given set of performance objectives into the required performance weights; a designer's subjective assessment plays a significant role in this process. For instance in this case, a relatively large q, compared to r, implies emphasizing "nearness" to the prescribed speed over "efficiency" in terms of expenditure of fuel. For numerical purposes here, we choose $q = r = 1$, which weighs these performance measures equally. As we shall see shortly in the case considered, q/r is the key factor in determining the control gain.

Once the performance weights q and r have been selected, the solution to the optimization problem can be used as a control law of the form $\delta\theta = k(v_d - v)$ with k as the control gain, and given by

$$k = \frac{1}{k_s}\left(\sqrt{1 + \frac{q}{r}k_s^2} - 1\right)$$ **(EQ 8.16)**

The numeric value of k, given the numeric values of $q = r = 1$ and $k_s = 45.2$, is found as 0.978. We shall not delve into the details of deriving the form of this solution. The appendix to this chapter discusses the solution. The interested reader may refer to a number of references on this topic[190, 394]. Fig 8.5 depicts the architecture of this controller where the complete control law (including a feedforward component) is given by

$$\theta = \theta_0 + \delta\theta$$
$$\delta\theta = k_s^{-1}\delta v_d + k(v_d - v)$$ **(EQ 8.17)**

where $\delta v_d = v_d - v_0$ and k_s^{-1} is the inverse of the steady-state gain of the plant.

FIGURE 8.5   Closed-Loop Cruise Control System

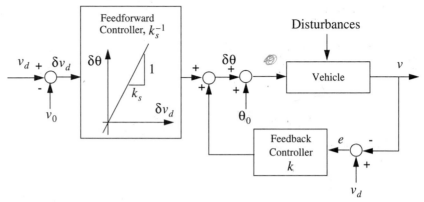

The resulting feedback system takes the form

$$\frac{\tau}{1+kk_s}\delta\dot{v} = -\delta v + \delta v_d - \frac{1}{1+kk_s}\left(\frac{1}{k_v}F_g\right) \qquad \textbf{(EQ 8.18)}$$

which indicates that the plant will reach steady state with no error if no disturbances are present and with a time constant reduced by a factor of $1 + kk_s$, which for the numeric values of $k_s = 45.2$ and $k = 0.978$ is 36.26. Thus the predicted closed-loop time constant is $\tau/(1 + kk_s) = 0.07\,\text{sec}$. The effect of the road grade induced disturbance, F_g/k_v, is reduced by this same factor as well. (Compare the above equation with the open-loop model, Equation 8.14.)

   Figs 8.6 and 8.7 illustrate the response pattern of the closed-loop control system with the nonlinear vehicle model in place.[1] In the first of these figures the vehicle is expected to reach a 45 m/s speed (approx. 100 mph) from a starting speed of 40 m/s (approx. 90 mph) on a flat road, while in the second case the vehicle speed is expected to be maintained at 40 m/s while riding up a steep road of 10 percent grade. We note that the controller performs acceptably in both cases. However, since the throttle position is at most 1 radian or 60 degrees, the performance objectives, particularly in case of response speed, are not completely achieved.

---

[1] We point out that as it is common practice, an *integral* controller has been augmented with the controller in the simulation models. See the Simulink model in the appendix to this chapter.

FIGURE 8.6   Vehicle Response to a Step Reference of 45m/sec on a Flat Road (Initial Velocity 40 m/s)

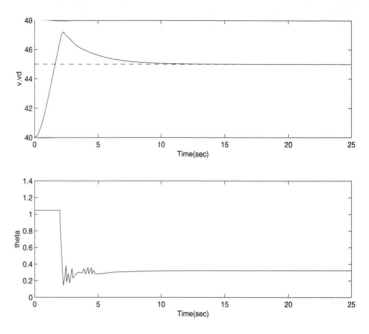

FIGURE 8.7   Response of the Vehicle to 10 Precent Positive Road Grade

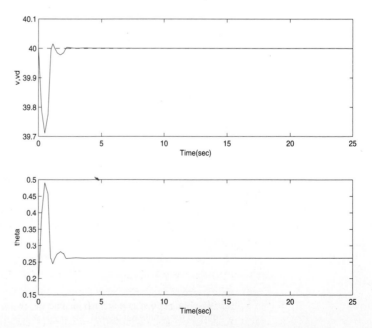

8.3.5 Alternative Formulation of the Automotive Cruise Control Problem

Let us consider the closed-loop system model and the associated controller architecture in Equations 8.17 and 8.18. We note that one can view the resulting controller as a proportional + feedforward control scheme where k affects the closed-loop time constant and the steady-state accuracy of the system. In view of this fact we can simply choose k based on steady-state accuracy and/or response speed requirements. For instance we may consider a steady-state accuracy improvement of 1percent (of the road grade effect F_g/k_v), which effectively means that

$$1 + kk_s > 100 \qquad\qquad \textbf{(EQ 8.19)}$$

and require that the response speed be improved by a factor of 10, in which case we have:

$$1 + kk_s > 10 \qquad\qquad \textbf{(EQ 8.20)}$$

The combined effect of these inequalities is that $k > 99/k_s$, which for the numeric value of $k_s = 45.2$ effectively means that $k > 2.2$, which is larger than that predicted earlier. We note that the 1 rad or 60 degree limit on the throttle position makes this gain not necessarily achievable, however, as it is evident from the vehicle step response in Fig 8.6.

8.4 Semiformal Aspects of the Design Process

While the approach discussed in the previous section is apparently systematic, selection of the performance weights (q and r) requires some understanding of the role they play in shaping the resulting control strategy. In the relatively simple system considered here, this process is fairly transparent; the closed-loop time constant is inversely proportional to k, and k is roughly proportional to q/r. Generally, however, it is not clear how one must choose the so-called performance weights to achieve the desired external performance objectives; for instance while certain qualitative factors are almost always evident(large q implies large k and improved response speed), it is not clear how large or small q and r must be to meet the specific external performance specifications. In this sense the formal nature of the design process is rather misleading in that one cannot entirely rely on formal means to accomplish the desired objective of selecting a controller.

This fact has long been recognized within control engineering and indeed one of the most popular control design techniques, based on the so-called *Proportional plus Integral plus Derivative*(PID) control, is often used to design controllers in a semiformal manner. Quite often the control engineer simply "tunes" the parameters of PID controllers on the basis of understandingthe behavior of the system without relying on any formal strategy. At best, time response of the given plant is used to select control parameters via the so called Ziegler-Nichols method[190].

The use of informal means in control engineering and the role if plays in conjunction with formal design techniques is illustrated in Fig 8.8 where the area marked by dashed lines shows the true boundary of the control design process while the inner circle with solid boundaries represents the limited, formal aspect of this process. As the figure indicates, informal translation of external performance specifications and selection of model structure are the starting points for the application of formal control design techniques. For instance, in a combustion process, performance specifications stated in terms of fuel efficiency and emission characteristics must be restated in terms of control theoretic objectives such as limits on the deviation of one or more plant outputs. Likewise in a machine control design problem, requirements on the quality of machined parts must be translated into control design objectives such as bandwidth of the mechanical positioning system and associated tracking errors. Moreover in these situations choice of plant model structure and subsequent selection of model parameters and associated uncertainty bounds are based on a great deal of subjective assessment concerning the behavior of the given system and its usage and are in turn affected by the external performance objectives.

In brief, the role of informal means in control engineering is a significant aspect of the control design process. Moreover, as we shall see shortly, the role of fuzzy logic in control engineering is to facilitate this aspect of the design process.

FIGURE 8.8   An Outline of the Control Design Process

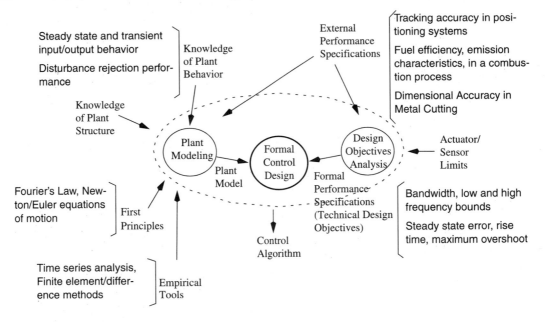

8.5 Fuzzy Logic Control

Broadly stated, fuzzy logic control attempts to come to terms with the informal nature of the control design process. In its most basic form, the so-called Mamdani architecture to be discussed shortly, one may view fuzzy logic control as directly translating external performance specifications and observations of plant behavior into a rule-based linguistic control strategy. This architecture forms the backbone of the great majority of fuzzy logic control systems reported in the literature in the past 20 some years.

An alternative architecture, originating in the work of Takagi and Sugeno[576] and also to be discussed shortly, uses a combination of linguistic rules and linear functions to form a fuzzy logic control strategy. This architecture has gained support because of its efficiency and will most likely be the architecture of choice for many future applications of fuzzy logic control.

In subsequent sections we will discuss these various approaches and illustrate their application via examples.

8.6 Mamdani Architecture for Fuzzy Control

The basic assumption underlying the approach to fuzzy logic control proposed by E.H. Mamdani in 1974[400] is that in the absence of an explicit plant model and/or clear statement of control design objectives, informal knowledge of the operation of the given plant can be codified in terms of *if-then,* or condition-action, rules and form the basis for a linguistic control strategy.

The basic paradigm for fuzzy logic control that has emerged following Mamdani's original work is a linguistic or rule-based control strategy of the form

if OA_1 is — and OA_2 is — and... then CA_1 is — and CA_2 *is* —...

if OA_1 is — and OA_2 is — and... then CA_1 is — and CA_2 *is* —...

. . .

which maps the *observable attributes*(OA_1, OA_2,...,) of the given physical system into its *controllable attributes*(CA_1, CA_2,...,). The controller structure in Fig 8.9, relates this architecture to that of a conventional feedback control system, where appropriately,

$$Output \quad \leftrightarrow \quad Observable\ Attribute$$
$$Input \quad \leftrightarrow \quad Controllable\ Attribute$$

FIGURE 8.9   Architecture for Fuzzy Control

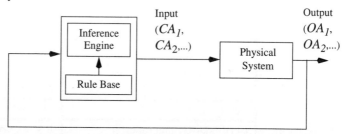

In particular each OA_i, $i = 1, 2, ...,$ is either a directly measurable variable and/or the difference between any such variable and its associated reference value. For instance in the context of an automotive cruise control problem, either speed, v, or its deviation from some desired value, v_d, that is, $v_d - v$, may be used within the context of a control strategy. We should point out that any filtered value of an otherwise observable attribute (such as its rate of change) may also be used within the premise of a rule:

$$OA_1 \equiv e = v_d - v \qquad \text{(EQ 8.21)}$$

$$OA_2 \equiv \frac{de}{dt} \qquad \text{(EQ 8.22)}$$

where e would be the speed error. Alternatively in the case of a discrete time implementation of fuzzy control systems, $OA_2 \equiv e(t) - e(t - T)$ where T is the so-called *sampling period* of the system. It is quite common in fuzzy control to refer to error and its rate of change within the antecedent clause of a given rule set.

Tables 8.3 and 8.4 describe two of the many applications of fuzzy logic control based on the above paradigm. In each case the control strategy takes the form of condition-action rules that relate situations that may occur during the operation of the plant to the desired action

TABLE 8.3   Aircraft Landing Control (Larkin [558])

Observable Attributes	Rate-of-Aircraft-Descent
	Airspeed
	Glide-Slope-Deviation
Controllable Attributes	Engine-Speed-Change
	Elevator-Angle-Change
Condition→Action Rules	If Rate-of-Aircraft-Descent is PM [a] and Airspeed is NB and Glide-Slope-Deviation is PB then Engine-Speed-Change is PM and Elevator-Angle-Change is IC.

[a] PM, NB, etc., are short for Positive Medium, Negative Big, and so on.

TABLE 8.4   Gas Metal Arc Welding (Langari and Tomizuka [352])

Observable Attributes	Workpiece-Temperature
	Workpiece-Temperature-Change
	Arc-Current
Controllable Attributes	Electrode-wire-Feedrate-Change
Condition→Action Rules	If Workpiece-Temperature is *High* and Workpiece-Temperature-Change is *Small* and Arc-Current is *Moderate* then Electrode-wire-Feedrate-Change is *Small*.

8.6.1 Design Issues in Fuzzy Control

While the rule-based approach suggested above offers a rather flexible means of representing a control algorithm, it is not apparent how one must devise such a rule set. To this end, Sugeno and Takagi[576] have suggested the following means as the basis for derivation of the rule set:

- Interrogation of human operator
- Observation of human operator in action
- Fuzzy model of process

The first two of the above methods are applicable in situations in which prior knowledge of the operation of the given physical system is available, such as the case described by Larkin(Table 8.3) where interviews with pilots formed the basis for derivation of the control rules.

Generally, however, one may rely on an *intuitive model* of the behavior of the given physical system and in particular on understanding its response patterns given changes in its inputs. Such an approach is attractive in situations where one has a generic understanding of the behavior of the given physical system. The case in point here would be the automotive cruise control system we considered earlier where almost universally one has a basic understanding of the behavior of given plant(vehicle) even if one's knowledge of the interrelationships involved may not be exact or precise. In the next section we present a generic fuzzy controller that is applicable to the automotive cruise control problem as well as similar problems.

8.7 Design of a Generic Mamdani Type Fuzzy Controller

Consider the vehicle cruise control problem discussed earlier in this chapter. Generally speaking the vehicle behaves such that a positive throttle position drives the velocity upwards and vice versa. Moreover, the response of the vehicle is roughly proportional to the changes in the throttle position. In this sense the vehicle control problem can be viewed as a generic case of controlling the behavior of a physical system whose behavior is largely described in terms of a series of step response graphs as, for instance, is shown in Fig 8.10, where $u_1, u_2, u_3, \ldots$ are the step input values applied to the plant(always starting from the same nominal state, say $y = 0$) and $y_1, y_2, y_3, \ldots,$ are the corresponding steady-state values of the plant output. Note that we generally observe certain limits on the magnitude of the input to the plant, that is

$$|u_i| < \bar{u} \qquad\qquad \textbf{(EQ 8.23)}$$

This limit, $\bar{u}$, is subsequently referred to in the control design process.

The response graphs shown in Fig 8.10 can be summarized in terms of a pair of parameters: *steady state gain*, k_s (average ratio of steady-state output to input) and *time constant*, τ (roughly the average value of the time it take for the plant to reach 2/3 of its steady-state value.) These notions of gain and time constant are borrowed from conventional linear systems theory and are meant to capture the essential aspects of the behavior of the simple generic plant considered here.

From a qualitative standpoint the generic plant described in Fig 8.10 behaves in such a manner as to respond to any stepwise changes in its input (within limits stated above) so as to reach steady state within, say four, time constants as would be expected if the system roughly behaved as one would expect from the response graphs shown in the figure.[2]

FIGURE 8.10   Response Pattern of a Simple Generic Plant

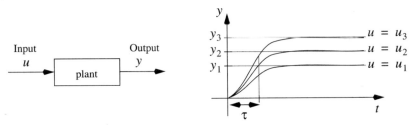

We thus proceed to develop a control strategy that, as shown in Fig 8.11, maps the error, $e = y_d - y$ into the control action, u. At the heart of this control scheme is a fuzzy logic control algorithm that operates in discrete time steps of period T and maps the *normalized* values of error, $e_n(t)$, and change in error, $ce_n(t)$, defined respectively as

$$e_n(t) = n_e e(t)$$
$$ce_n(t) = n_{ce}(e(t) - e(t-T))$$

(EQ 8.24)

where n_e and n_{ce} are the corresponding *normalization factors*, into changes in the control action, $\delta u_n(t)$ via rules of the form   ↳ Tuning Parameters

If $e_n(t)$ is P and $ce_n(t)$ is N then $\delta u_n(t)$ is Z.

Here P, N and Z are short for Positive, Negative, and Zero, defined as fuzzy sets over the normalized domains of definition of the relevant variables as shown in Fig 8.12. Note that the fuzzy sets defined over the domain of definition of δu_n are fuzzy singletons for simplicity.

In reference to Fig 8.11, the computed value of δu is subsequently *denormalized* and integrated (added to the past value of u to form the present value of u) as follows:

$$u(t) = u(t-T) + de_{\delta u}\delta u_n(t)$$

(EQ 8.25)

where $de_{\delta u}$ is the corresponding *denormalization factor*.

---

[2] A simple first-order system would reach within 2 percent of its steady-state output within four time constants.

FIGURE 8.11   Architecture of the Generic Fuzzy Control System

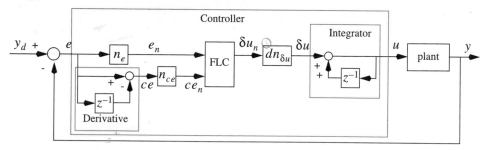

8.7.1 Derivation of the Rules

The rules effectively reflect a typical scenario for operation of the system. For instance as depicted in Fig 8.12 we assume that at time $t = 0$, the system starts at some nominal output value (say $y = 0$) and is expected to reach some desired value y_d. On the normalized scale depicted in the figure, y_d reflects a one-unit difference from the normalized value of y and results in the initial error value of one unit. This starting state [marked as ⓪ in the figure] corresponds to the error being positive (P) and change in error likewise being positive (P) it evidently makes sense to suggest that δu_n be positive as well.

FIGURE 8.12   Control Rules for a Simple Generic Fuzzy Controller

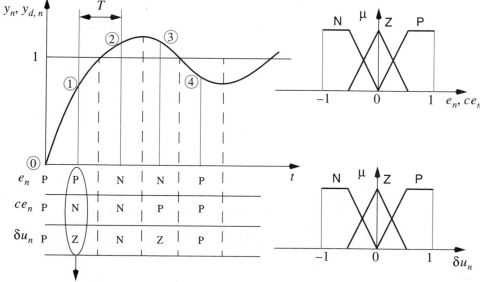

if $e_n(t)$ is P and $ce_n(t)$ is N then $\delta u_n(t)$ is Z.

Assuming now that the plant behaves in a predictable manner, the output will rise, and provided that the sampling time is chosen to be of the same order of magnitude or somewhat smaller than the time constant of the system, the output will reach a level marked as ① in the figure where the error is still positive (P) but change in error is negative (N), One may suggest that the change in input, that is δu_n, should be Zero (Z). Continuing with this approach one gradually determines the majority of the rules necessary to complete the rule set. These are listed in Fig 8.12. The remaining rules are determined by inspection and the entire rule set is tabulated in Table 8.5.

TABLE 8.5   Rules for the Generic Fuzzy Controller

Observable Attributes	error, e_n
	Change-in-error, ce_n
Controllable Attributes	Change-in-input, δu_n
Condition→Action Rules	
I: Starting up, change the input in response to the setpoint change	If e_n is P and ce_n is P then δu_n is P
	If e_n is N and ce_n is N then δu_n is N
II: Plant is not responding; adjust input.	If e_n is P and ce_n is Z then δu_n is P
	If e_n is N and ce_n is Z then δu_n is N
III: Plant is responding normally, keep input the same	If e_n is P and ce_n is N then δu_n is Z
	If e_n is N and ce_n is P then δu_n is Z
IV: Reached equilibrium	If e_n is Z and ce_n is Z then δu_n is Z
V: Error is nil but changing, take action	If e_n is Z and ce_n is N then δu_n is N
	If e_n is Z and ce_n is P then δu_n is P.

8.7.2 ⭐Determining the Normalization Factors

The aforementioned normalization factors and denormalization factors are determined largely based on an assessment of the operating range of the system and the steady-tate gain mentioned earlier. The largest value of e, which is closely related to the largest difference between y and y_d, determines the normalization factors for e and ce:

$$n_e = n_{ce} = \frac{1}{\max|y_d - y|} \qquad \text{(EQ 8.26)}$$

Likewise the sampling time, T, is chosen based on the time constant τ of the system and it makes sense to set this value to about half of the estimated time constant of the system.

Selection of the denormalization factor for δu can be made according to the limit $\bar{u}$ discussed earlier. We can use the following rule of thumb in this situation:

$$de_{\delta u} = 1/\bar{u} \qquad \text{(EQ 8.27)}$$

although a smaller value will likely be best.

8.7.3   Evaluation of the Generic Mamdani Controller

The Mamdani controller devised above is tested on a first-order linear plant and its output is shown in Fig 8.13 for $k_s = 1$ and the time constants is 1 second. The normalization factors are 1 for both e and, ce and the denormalization factor is similarly 1 for δu. The sampling period, $T = 0.5$ sec .The Simulink model used to produce these results is given in the appendix to this chapter.

FIGURE 8.13   Response for $k_s = 1$

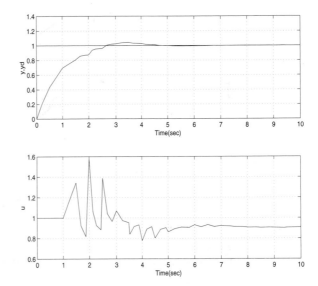

8.7.4   Extending the Generic Mamdani Controller

The Mamdani controller described earlier can be extended by refining the fuzzy sets defined over the relevant domains of definition and likewise refining the rule set as shown in Table 8.6. The explanatory remarks suggest the rationale for the particular form of the rules in each

TABLE 8.6   Rules for the Generic Fuzzy Controller

Observable Attributes	error, e_n
	Change-in-error, ce_n
Controllable Attributes	Change-in-input, δu_n
Condition→Action Rules	

TABLE 8.6　　　Rules for the Generic Fuzzy Controller

I: Starting up, change the input in response to the setpoint change	If e_n is LP and ce_n is LP then δu_n is LP
	If e_n is SP and ce_n is SP then δu_n is SP
	If e_n is SN and ce_n is SN then δu_n is SN
	If e_n is LN and ce_n is LN then δu_n is LN
II: Error not changing, change input accordingly	If e_n is LP and ce_n is Z then δu_n is LP
	If e_n is SP and ce_n is Z then δu_n is SP
	If e_n is SN and ce_n is Z then δu_n is SN
	If e_n is LN and ce_n is Z then δu_n is LN
III: Moving along; maintain input	If e_n is LP and ce_n is SN then δu_n is Z
	If e_n is SP and ce_n is SN then δu_n is Z
	If e_n is SN and ce_n is SP then δu_n is Z
	If e_n is LN and ce_n is SP then δu_n is Z
IV: Getting worse, reverse input somewhat	If e_n is LP and ce_n is SP then δu_n is LP
	If e_n is SP and ce_n is LP then δu_n is LP
	If e_n is SN and ce_n is LN then δu_n is LN
	If e_n is LN and ce_n is SN then δu_n is LN
V: Error changing too fast, adjust input somewhat	If e_n is LP and ce_n is LN then δu_n is SN
	If e_n is SP and ce_n is LN then δu_n is SN
	If e_n is SN and ce_n is LP then δu_n is SP
	If e_n is LN and ce_n is LP then δu_n is SP
VI: Reached equilibrium	If e_n is Z and ce_n is Z then δu_n is Z
VII: Error is nil but changing, take action	If e_n is Z and ce_n is LN then δu_n is SN
	If e_n is Z and ce_n is LP then δu_n is SP
VIII: Error is nil and changing insignificantly, no action-- wait and see	If e_n is Z and ce_n is SN then δu_n is Z
	If e_n is Z and ce_n is SP then δu_n is Z.

category. Alternatively one may view the rules as describing an appropriate control action given certain expected behavior on the part of the plant as illustrated in Fig 8.14.

FIGURE 8.14   Control Rules for a Generic Fuzzy Controller

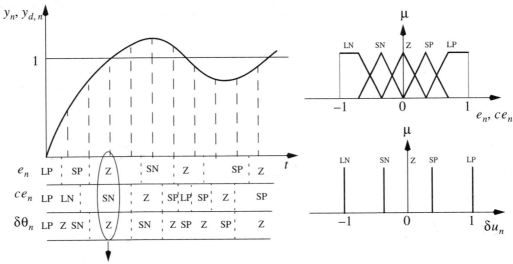

if e_n is Z and ce_n is SN then δu_n is Z

8.7.4.1 Application to Vehicle Cruise Control

The generic Mamdani controller described above is directly applicable to the automotive cruise control problem considered earlier with a simple adjustment of the normalization and denormalization factors as follows:

$$n_e = n_{ce} = 0.2, \ de_{\delta u} = 0.1 \qquad \textbf{(EQ 8.28)}$$

and $T = 0.25$ sec. The response of the system is illustrated in Figs 8.15 and 8.16. The vehicle is expected to start at 40 m/s and reach 45 m/s in the first case, and in the second, a road grade shift of 10 percent must be overcome by the vehicle.

FIGURE 8.15   Mamdani Controller Step Response Test with the Nonlinear Vehicle Model

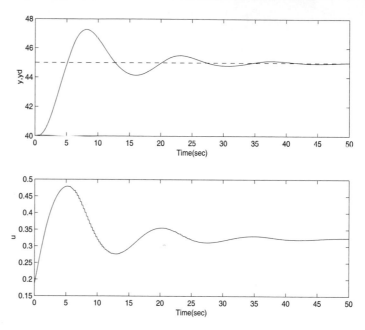

FIGURE 8.16   Disturbance Rejection Using the Mamdani Model

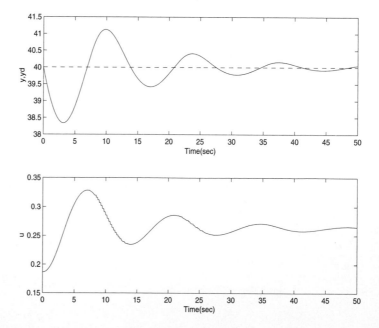

8.8 Additional Design Examples

We look at another design example that is meant to illustrate how fuzzy logic is used in practice to design a simple but yet effective control strategy even in situations in which nonlinear behavior dominates the response of the system. Consider the heat exchange process shown in Fig 8.17. The fluid whose temperature must be regulated flows through the system and in the process exchanges heat with the working fluid. The temperature of the working fluid can be indirectly regulated via two valves, let us say hot and cold, to produce a fluid mixed at the "right" temperature.

FIGURE 8.17   A Heat Exchange Process

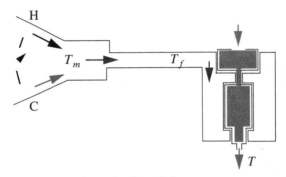

For the purpose of simulating the behavior of the system, we can view the plant model as follows.

- The heat exchanger is modeled via the basic heat transfer equation:

$$\frac{dT}{dt} = \alpha(T_f - T) \tag{EQ 8.29}$$

where θ is the heat transfer coefficient which in general depends on a number factors such as the flow rate Q_f and temperature gradient, $T_f - T$:

$$\alpha = \alpha(Q_f; T_f - T) \tag{EQ 8.30}$$

- The mixing stage is described in terms of a relative simple continuity of an incompressible fluid in an adiabatic mixing chamber (no additional losses to the environment are assumed)

$$\frac{dT_f}{dt} = aT_f + bT_m \tag{EQ 8.31}$$

$$Q_m T_m = Q_c T_c + Q_h T_h \tag{EQ 8.32}$$

$$Q_m = Q_c + Q_h \tag{EQ 8.33}$$

$$Q_f = Q_m \qquad \text{(EQ 8.34)}$$

The key fact is that, based on Equation 8.30, the plant is nonlinear and the nonlinearity is attributed to *nonuniform* heat transfer characteristics manifested in the dependence of α on Q_f and the temperature gradient, $T_f - T$; the system responds more effectively to larger temperature gradients. Beyond this, not much is assumed known about the behavior of the system; for instance, it is not precisely or even formally known how α depends on Q_f.

The conventional approach with respect to the given problem would be to design a linear control law, say PID, based on a simplified plant model derived from empirical observations of plant behavior and to later tune the controller parameters as necessary. Given the potentially dominant nonlinearity present in the plant, however, it is not evident if the controller would function well across a sufficiently broad operating range. Alternatively we design a simple fuzzy controller that, while not radically different from a conventional PID controller, deals with the nonlinearity in the plant. The rule set for this algorithm, depicted in Fig 8.18, shows the relatively intuitive approach to control design based on the operational view of the plant.

FIGURE 8.18   Rule Set for the Fuzzy Logic Controller(Developed with TILShel from Togai Infralogic)

```
IF (Error IS N) AND (Error_Rate IS N) THEN
                      Th_valve_change=Z
                      Tc_valve_change=Z
END
IF (Error IS N) AND (Error_Rate IS Z) THEN
                      Th_valve_change=N
                      Tc_valve_change=P
END
IF (Error IS Z) AND (Error_Rate IS N) THEN
                      Th_valve_change=P
                      Tc_valve_change=N
END
                        .
                        .
                        .

IF (Error IS P) AND (Error_Rate IS P) THEN
                    Th_valve_change=Z
                    Tc_valve_change=Z
END
```

We do state the fact that the algorithm is not, once developed, radically different from a conventional one and in fact, the initial design of the controller is such that it resolves into a linear law as evident from the obvious symmetry in the definition of the membership functions of the relevant fuzzy sets in Fig 8.19.[3] The initial performance is apparently not satisfactory however in this configuration(Fig 8.20.)

[3.] This fact is discussed extensively in Langari[587].

FIGURE 8.19   Initial Membership Functions Defined Over the Normalized Domain of Definition of Process Error.

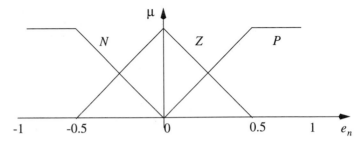

FIGURE 8.20   Results with Original Membership Functions

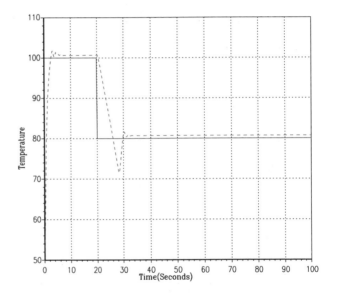

Revising the membership functions so as to emphasize the plant nonlinearity (Fig 8.21) results in the response pattern depicted in Fig 8.22. In the revised form, the nonlinearity is interpreted in view of the rule set as follows. The plant effectively exhibits a higher gain and faster response for larger errors. Thus to make the system respond more uniformly, we deemphasize what is meant by the terms in the vocabulary used to define the rule set. In particular, we redefine *positive*, *negative*, and *zero* over the domain of the definition of process error so as to shift their focus away from the origin. The result would be a broadening of what is meant by *zero* error as opposed to *positive* (*negative*) error, thus effectively normalizing the controller's response characteristics.

FIGURE 8.21   Initial and Revised Membership Functions Defined over the Normalized Domain of
                        Definition of Process Error

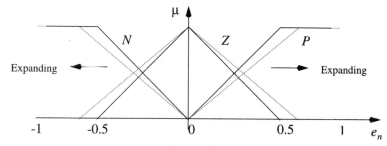

We should point out, however, that in this example what is accomplished is effectively a
nonlinear PI algorithm. Indeed it can be rigorously proven that single-layer fuzzy controllers do resolve into nonlinear PI/PD controllers[349]. The fact remains, however, that the
design process followed above is quite transparent and would be subject to interpretation
whereas the design process that would lead to a nonlinear PI algorithm(perhaps via gain
scheduling) may not be.

FIGURE 8.22   Results with Revised Membership Functions

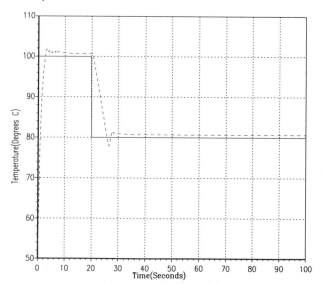

8.9 The Sugeno-Takagi Architecture

The natural extension of the notion of fuzzy logic control is by means of the so called Takagi-Sugeno (alternatively Takagi-Sugeno-Kang, TSK) architecture [576] whereby the rule set defines regions of action of conventional differential algebraic control laws. In view of the discussion in Section we may view the TSK architecture as emphasizing the *outer layer* in Fig 8.8 while relying to some degree on the formal structure(linear state feedback).

In this paradigm the rule set is made up of r rules of the form

$$\text{Rule } i: \text{If } x_1(t) \text{ is } M_{i1}, x_2(t) \text{ is } M_{i2}, \dots, \text{ and } x_n(t) \text{ is } M_{in},$$
$$\text{then } y = a_{i0} + a_{i1}x_1(t) + a_{i2}x_2(t) + \dots + a_{in}x_n(t). \qquad \textbf{(EQ 8.35)}$$

where $x_1, x_2, \dots, x_n$ are the antecedent variables and y is the consequent variable. Moreover, $M_{i1}, M_{i2}, \dots, M_{in}$ are fuzzy sets defined over the respective domains of definitions of $x_1, x_2, \dots, x_n$ while $a_{i0}, a_{i1}, \dots, a_{in}$ are constant coefficients that characterize the linear relationship defined by the ith rule in the rule set, $i = 1, 2, \dots, r$.

From a control engineering standpoint the application of the TSK-based scheme would be in constructing a *nonlinear control* law by augmenting piecewise linear relationships of the form that appears in the consequent of the rule in Equation 8.35. A simple instance of this idea would be a nonlinear PI type controller for a single-input, single-output plant where the antecedent variables would be in error and change in error and the consequent variable would be the change in input to the plan:

$$\text{Rule } i: \text{If } e_n(t) \text{ is } M_{i1}, ce_n(t) \text{ is } M_{i2},$$
$$\text{then } \delta u_n(t) = k_{i0} + k_{i1}e_n(t) + k_{i2}ce_n.(t) \qquad \textbf{(EQ 8.36)}$$

where e_n and ce_n are the normalized error and change in error and δu_n is the normalized change in input.

We may proceed to develop a simple nonlinear PI control scheme based on this idea for the heat exchanger discussed in the previous section. Fig 8.23 illustrates the conceptual idea for controller design where δu_n as a function of e_n is depicted. Evidently the controller gain is smaller for small error and larger for large error. This is based on the observation that the plant exhibits a higher gain for larger error than it does for smaller error.

The net result here is that only three rules completely specify the control algorithm:

$$\text{If } e_n(t) \text{ is } N, \text{ and } ce_n(t) \text{ is } A,$$
$$\text{then } \delta u_n(t) = k_{10} + k_{11}e_n(t) + k_{12}ce_n(t). \qquad \textbf{(EQ 8.37)}$$

$$\text{If } e_n(t) \text{ is } Z, \text{ and } ce_n(t) \text{ is } A,$$
$$\text{then } \delta u_n(t) = k_{20} + k_{21}e_n(t) + k_{22}ce_n(t). \qquad \textbf{(EQ 8.38)}$$

$$\text{If } e_n(t) \text{ is } P \text{, and } ce_n(t) \text{ is } A,$$
$$\text{then } \delta u_n(t) = k_{30} + k_{31}e_n(t) + k_{32}ce_n(t). \tag{EQ 8.39}$$

Note that M_{11} is *negative*(N), M_{21} is *zero*(Z), and M_{31} is *positive*(P) and M_{i2} for $i = 1, 2, 3$ is *any*(A, a fuzzy set with membership function equal to 1 across the entire domain of definition of ce_n).

The response of the system is shown in Fig 8.24. The normalization factors used are $n_e = n_{ce} = 0.03$ and $de_{\delta u} = 0.1$ since we are operating the system from 25 to 50 degree changes. The relatively limited control input makes it sensible to set the denormalization factor for δu to $0.1 (1/10\text{th}$ of the range of u). The particular values of k_{ij} are listed in Table 8.7. Note that k_{i0}'s represent the intersections of the piecewise linear seg-

TABLE 8.7   Values of the Control Gains for the TSK Control of the Heat Exchanger

i	k_{i0}	k_{i1}	k_{i2}
1	-0.25	0.5	1
2	0	1	1
3	0.25	0.5	1

ments with the vertical axis.

FIGURE 8.23   Conceptual Design of the TSK Controller for the Heat Exchanger

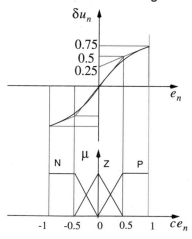

FIGURE 8.24   Response of the Heat Exchanger to the TSK Controller

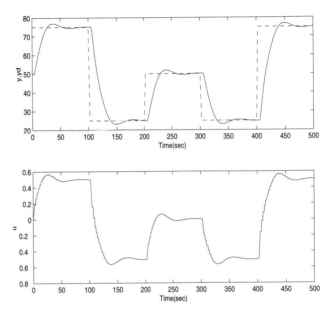

8.9.1 TSK Model-Based Control

A more systematic use of the TSK architecture can be based on a model that captures the behavior of the plant as follows[578]:

$$\text{Rule } i: \text{If } x(k) \text{ is } M_{i1}, x(k-1) \text{ is } M_{i2}, \dots, \text{ and } x(k-n+1) \text{ is } M_{in},$$
$$\text{then } \underline{x}_i(k+1) = A_i \underline{x}(k) + B_i u(k). \qquad \textbf{(EQ 8.40)}$$

where

$$\underline{x}(k) = \left[x(k) \; x(k-1) \; \dots \; x(k-n+1) \right]^T \qquad \textbf{(EQ 8.41)}$$

$i = 1, 2, \dots, r$ and r is the number of rules, $\underline{x}_i(k+1)$ is the output of the ith rule, A_i, and B_i are the state and input matrices of the system as correlated with the ith rule. In addition, $M_{i1}, M_{i2}, \dots$, are linguistic terms, such as *high, low, slightly,* defined in terms of fuzzy sets over the domains of definition of the corresponding variables. The rules in effect suggest what form the plant model takes (in terms of A_i, and B_i) depending on the region of operation and in particular depending on th value of x at the kth instant, that is, $x(k)$ and its past values ($x(k-1), x(k-2), \dots, x(k-n+1)$). The matrix A_i takes the form

$$A_i = \begin{bmatrix} a_{i1} & a_{i2} & \cdots & a_{in-1} & a_{in} \\ 1 & 0 & \cdots & 0 & 0 \\ 0 & 1 & \cdots & 0 & 0 \\ \cdots & \cdots & \cdots & 0 & 0 \\ 0 & 0 & \cdots & 1 & 0 \end{bmatrix} \qquad \textbf{(EQ 8.42)}$$

Generally one may consider $x(k)$ to be the effective output of the system at the kth instant. Moreover,

$$B_i = \begin{bmatrix} 1 & 0 & \cdots & 0 & 0 \end{bmatrix}^T \qquad \textbf{(EQ 8.43)}$$

creating the so-called *control canonical* form.

The behavior of the system is described by

$$\underline{x}(k+1) = \frac{\sum_{i=1}^{r} \mu_i(k)(A_i \underline{x}(k) + B_i u(k))}{\sum_i^r \mu_i(k)} \qquad \textbf{(EQ 8.44)}$$

where

$$\mu_i(k) = \prod_{j=1}^{n} \mu_{M_{ij}}(x(k-j+1)) \qquad \textbf{(EQ 8.45)}$$

is the *truth value* of the ith rule in the rule set.

Next we consider how this framework may be applicable to control problems.

8.9.2 Design Methodology

The TSK model-based architecture is effectively meant to support the construction of piecewise (linear) models of nonlinear systems.[4] In particular let us consider a nonlinear plant of the form

$$\underline{x}(k+1) = \underline{f}(\underline{x}(k)) + \underline{g}(\underline{x}(k))u(k) \qquad \textbf{(EQ 8.46)}$$

where $\underline{x}(k)$ is defined as in Equation 8.41 and $\underline{f}(.)$ and $\underline{g}(.)$ are nonlinear vector functions. We consider the case where $\underline{f}(.)$ takes the form

$$\underline{f}(\underline{x}(k)) = \begin{bmatrix} f(\underline{x}(k)) & x(k) & \cdots & x(k-n+2) \end{bmatrix}^T \qquad \textbf{(EQ 8.47)}$$

where $f(\underline{x}(k))$ is a nonlinear scalar function of the state vector $\underline{x}$ and

---

[4.] The formulation in this part is based on the work of Kazuo Tanaka[580].

$$g(\underline{x}) = \begin{bmatrix} 1 & 0 & \dots & 0 & 0 \end{bmatrix}^{T} \tag{EQ 8.48}$$

With this in mind we can clarify the rationale for the TSK model as follows. If $f(.)$ were linear, it could take the form

$$f(\underline{x}(k)) = a_1 x(k) + a_2 x(k-1) + \dots + a_n x(k-n+1) \tag{EQ 8.49}$$

and therefore Equation 8.46 could take the form

$$\underline{x}(k+1) = A\underline{x}(k) + Bu(k) \tag{EQ 8.50}$$

with A and B similarly defined as in Equations 8.42 and 8.43, making it clear that the TSK approach generalizes on the idea of a linear system by extending the linearized model to nonlinear systems. Precisely how this is done is described next.

Consider a nonlinear system as described above, where, for instance, $f(\underline{x})$ as a function of the jth element of the state vector, x_j (in effect $x(k-j+1)$ based on the above notation), is depicted as in Fig 8.25. The regions of operation are defined by the changes in the slope of f which play the role of an indicator for the TSK model.

FIGURE 8.25   Approximation of a Nonlinear Function via the TSK Method

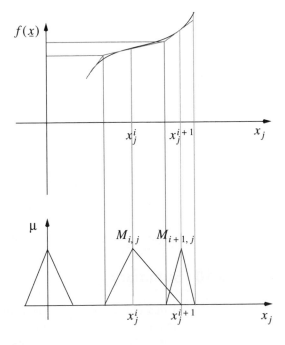

The function f can be viewed as being represented in terms of

$$f^j(\underline{x}) \cong f(\underline{x}^i) + \sum_j \frac{\partial f}{\partial x_j}\bigg|_{x=\underline{x}^i} (x_j - x_j^i) \qquad \text{(EQ 8.51)}$$

where $\underline{x}^j$ is the nominal value of $\underline{x}$ in ith operating region. The TSK approach is to associate with the ith operating region a set of n fuzzy sets, M_{i1}, M_{i2}, ..., M_{in} centered at x_j^i, $j = 1, 2, ..., n$ and of characterizing the system in terms of a Taylor series expansion as above but using linguistic rules of the form of Equation 8.40. Likewise in the situation above we can extract these values of slope of f at the nominal values of $\underline{x}$ for each operating region and construct the model given by Equation 8.40 as follows:

$$a_{ij} = \frac{\partial f}{\partial x_j}\bigg|_{x=\underline{x}^i} \qquad \text{(EQ 8.52)}$$

In other words the coefficients of the TSK model are partial derivatives of the primary nonlinear term of the nonlinear system model of the plant with respect to each of the states and evaluated at each representative or nominal value of the operating region.

8.9.3   Control Design

This model can be used as the basis for a control strategy (as, for example, described by Tanaka). The idea is to take a cue from the linear systems theory wherein linear state feedback can be used to stabilize systems of the form given earlier.

    The control law takes the form

$$u_i(k) = F_i \underline{x}(k) \qquad \text{(EQ 8.53)}$$

where F_i is an n-vector of feedback gains. The resulting system is now given by

$$\underline{x}(k+1) = \frac{\displaystyle\sum_j \sum_l \mu_i(k)\mu_j(k)\{A_i + B_i F_j\}\underline{x}(k)}{\displaystyle\sum_j \sum_l \mu_i(k)\mu_j(k)}. \qquad \text{(EQ 8.54)}$$

Effectively each subsystem is stabilized via a linear state feedback vector.

8.10   Summary

In this chapter we considered the rationale for fuzzy control in view of the historical ingrained approach to control engineering. We also discussed the manner in which fuzzy logic can be introduced into this framework either at the direct control level or at the level of coordination, or supervisory control. We also summarized what may be the similarities and differences between fuzzy logic control and conventional techniques and discussed the aspects of design and evaluation of performance of fuzzy control systems.

8.11 Appendix

8.11.1 Solution to Optimal Control Design Problem in Section 8.7

We formulate the plant model as

$$\delta\dot{v} = \left(-\frac{1}{\tau}\right)\delta v + \frac{k_s}{\tau}\delta\theta \tag{EQ 8.55}$$

ignoring the disturbance effects. This is the standard form for optimal control design with $\dot{x} = Ax + bu$ $A = -1/\tau$ and $b = k_s/\tau$. The control gain is found as[32]

$$k = r^{-1}b^T P = \frac{k_s p}{r\tau} \tag{EQ 8.56}$$

where p is in turn the solution of the algebraic Riccati equation

$$A^T P + PA + Pbr^{-1}b^T P + Q = 0$$
$$-2\frac{p}{\tau} - p\frac{k_s}{\tau}\frac{1}{r}\frac{k_s}{\tau}p + q = 0 \tag{EQ 8.57}$$

which results in

$$p = \frac{r\tau}{k_s^2}\left(\sqrt{1 + \frac{q}{r}k_s^2} - 1\right). \tag{EQ 8.58}$$

Thus

$$k = \frac{1}{k_s}\left(\sqrt{1 + \frac{q}{r}k_s^2} - 1\right) \tag{EQ 8.59}$$

8.11.2 Vehicle Model and Control System in Matlab/Simulink

The nonlinear vehicle model given in Equation 8.1 is captured in terms of a Simulink model as shown in Figs 8.26 and 8.27. The definitions of algebraic expressions and specific values of the related constants are given in the Matlab script file shown in Fig 8.28. Note that the vehicle engine model shown in Fig 8.27 incorporates a first-order *dynamic* model of the vehicle engine as given by

$$\tau_e\frac{dF_e}{dt} = -F_e + F_{e1}(\theta) \tag{EQ 8.60}$$

where

$$F_{e1}(\theta) = F_i + \gamma\sqrt{\theta} \tag{EQ 8.61}$$

The time constant τ_e is different in acceleration and deceleration but in general if it is small, then the combined effect of the above equations is still the same as Equations 8.5.

FIGURE 8.26   Vehicle Model in Simulink

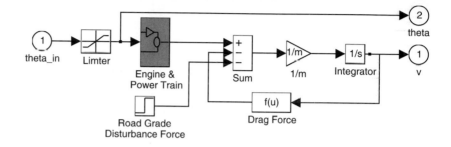

FIGURE 8.27   Vehicle Engine Model in Simulink

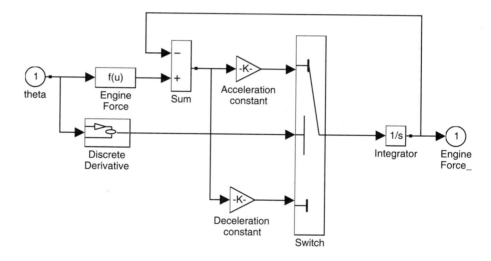

FIGURE 8.28 Simulink Data for the Vehicle Model.

```
% Data definitions for the vehicle cruise control, R. Langari 3/1/97
% VdataM.m
% The vehicle model is assumed to be given by
%    m*dv/dt=Fe-Fd-Fg
% with the engine model given by
%    te_p*dFe/dt=-Fe+Fe1(theta)     in acceleration and
%    te_n*dFe/dt=-Fe+Fe1(theta)     in deceleration
%
%    theta ............... Throttle position
%    v ................... Vehicle velocity
%    Fe .................. Engine force
%    Fe1 ................. Fi+gamma*sqrt(theta)-gamma1*theta-gamma3*theta^3
%    Fd ................. Drag force, Fd(v)=alpha*v^2
%    Fg ................. Road grade disturbance force(gravity)
%    Fi ................. Engine idle force
%    gamma,gamma1,gamma3 .. Engine force coefficients
%    alpha ............... Drag coefficeint
%    v0,theta0,Fe0 ........ Init/oper veloc., throt. pos, engine force
%    vd .................. Desired vehicle velocity
%    thetamax ............. Maximum throttle position
%    te_p/te_n ............ accel/decel time constants of the engine
%
% Units are all SI.
%
m=1000;
gamma=12500; gamma1=0; gamma3=0;
alpha=4;
Fi=1000;
thetamax=pi/3;
Fg=0;
v0=40;
vd=45;
%
% Find theta0 and Fe0 corresponding to v0. Let gamma1=gamma3=0.
%
theta0=((1/gamma)*(alpha*v0^2-Fi))^2;
Fe0=Fi+gamma*sqrt(theta0);
te_n=0.1; te_p=1;
```

FIGURE 8.29   Optimal Vehicle Control System in Simulink

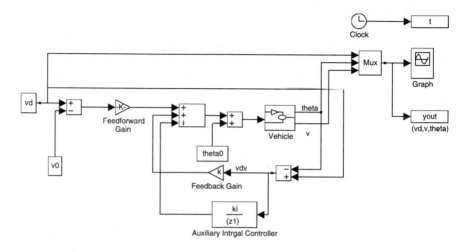

FIGURE 8.30   Simulink Data for the LQ Optimal Vehicle Control System

```
% The following are for LQ vehicle cruise control, R. Langari, 3/17/97
% VdataC.m
%
% Sampling time
%
T=0.25;
%
% Other parameters used in the LQ controller
%
q=1;
r=1;
ktheta=gamma/(2*sqrt(theta0));
kv=2*alpha*v0;
ks=ktheta/kv;
tau=m/kv;
k=(1/ks)*(sqrt(1+(q/r)*ks^2)-1);
ki=0.1;
```

8.11.3 The Simulink Model of the Generic Mamdani Fuzzy Controller

The generic fuzzy controller based on the Mamdani architecture is depicted in Fig 8.31. Note that the eNorm and ceNorm, and likewise, uDenorm are the normalization and denormalization factors used in connection with the controller. Individual rules are constructed using Simulink function lookup blocks as shown in Fig 8.32, which, for instance, implements the rule

$$\text{If } e \text{ is P and } ce \text{ is P then } \delta u \text{ is P.} \qquad \textbf{(EQ 8.62)}$$

Note that the consequent of the rule is implemented in terms of a fuzzy singleton.

This generic controller is tested with a simple first-order plant as shown in Fig 8.33. The response graphs were shown in Fig 8.13. The data for this simulation are defined in Fig 8.34.

FIGURE 8.31   Simulink Model of the Simple Generic Mamdani Controller

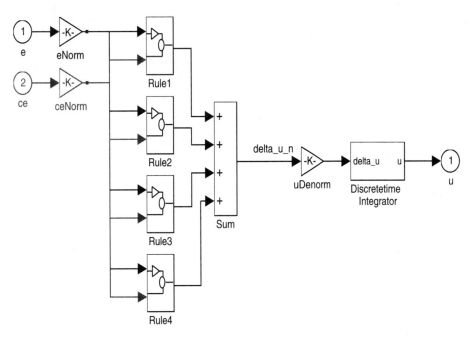

FIGURE 8.32   SImulink Model of a Single Rule

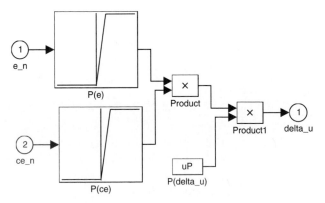

FIGURE 8.33   Mamdani Test Model in Simulink

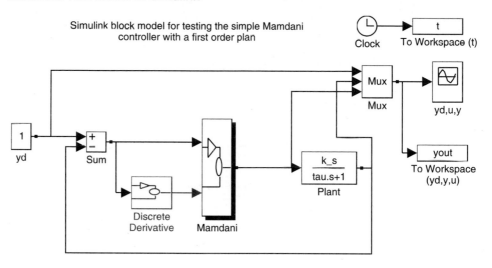

Simulink block model for testing the simple Mamdani
controller with a first order plan

FIGURE 8.34   Simulink Data for the Test Case

```
% Data definitions for the generic Mamdani controller
% R. Langari 3/1/97
%
% Normalized range of error and the related fuzzy sets
e_n=[-1 -0.25 0 0.25 1];
Z=[0 0 1 0 0];
N=[1 1 0 0 0];
P=[0 0 0 1 1];
%
% Term set definition used in plotting the membership
functions
%
V=[N,Z,P];
%
% change-in-error is defined the same as error
%
ce_n=e_n;
% Fuzzy singletons are used for the output of the fuzzy
controller
%
uN=-1;
uZ=0;
uP=1;
% Normalization factor for the error, change-in-error and
the denormalization
% factor for the controller output
%
eNorm=1;
ceNorm=1;
uDenorm=1;
```

```
%
% Initial condition used by the integrator in the Mamdani
controller.
% We set this to the initial/operating value of the throttle
position.
%
u0=0;
%
% Sampling time of the controller. May be over-ridden by
the the user.
%
T=0.5;
%
% Plant test data, gain 1, and time constant of 1 second.
This data
% is for a first order plant used to test the controller. The
response
% with k=1 is very smooth but slow. With k=2 the re-
sponse is oscillatory
% and with k in-between(let us say 1.5) the response is
reasonable.
%
k_s=1.1;
tau=1;
yd=1;
```

8.11.4   Mamdani Vehicle Control System Simulink Model

Figure 8.35 shows a Mamdani fuzzy vehicle control system and a simulation model implemented in Simulink.

FIGURE 8.35   Vehicle Control System in Simulink

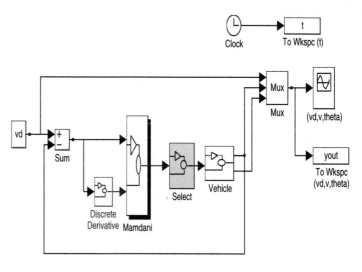

8.11.5   TSK Control System for the Heat Exchanger

Figure 8.36 shows a simulation model and a TSK fuzzy controller for a heat exchanger system.

FIGURE 8.36   TSK Heat Exchanger Control System in Simulink

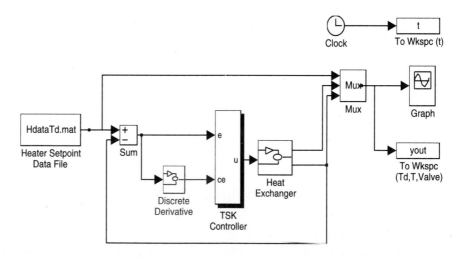

Exercises

8.1   Consider an elevator positioning system and answer the following questions.

(a) What are the input, output, and reference variables? State them in physically meaningful terms.

(b) What are the sensors? actuators? List them.

(c) What would be reasonable performance objectives? List them briefly.

(d) State two sample fuzzy rules that describe your control strategy? What is the logic behind this strategy? Be brief.

8.2   Consider the heat exchanger control system studied in this chapter. Assuming that $a = -1$, $b = 1$, and that the total flow through the valve is 10 GPM, use the generic controller described in this chapter to devise a fuzzy logic controller that would regulate the temperature in the heat exchanger. You can assume that

$$\alpha = 0.1 + 0.01(T_f - T)^2 \qquad \text{(EQ 8.63)}$$

and that the single value produced by the Mamdani controller should be used to determine the hot and cold flows as follows. Assume that the input is between -1 and 1. Scale it to be between 0 and 1 and use it to decide the total flow into hot and cold streams. Assume hot flow temperature to be 100 degrees and cold flow to be at 0 degrees. The starting temperature should be about 50 degrees and the desired temperature can be either 25 or 75 degrees.

8.3   Assume that the simplified dynamics of an aircraft is given by

$$M\dot{v} = -(\alpha v + \beta v^3) + k_e u \qquad \text{(EQ 8.64)}$$

where u is the throttle position, v is the airspeed; α, β, and M are the drag coefficients and mass of the aircraft, respectively. Design a controller that allows the aircraft to follow a desired reference velocity. Specifically

(a) Linearize the plant around the nominal velocity of 100 m/s. Assume M to be 800 Kg, $\alpha = 100\text{N/(m/sec)}$, $\beta = 10^{-2}\text{N/(m/sec)\textasciicircum3}$ and k_e to be 4000 Newtons per unit of control input, u.

(b) Assume the linearized model is given by

$$\tau\delta\dot{v} = -\delta v + k\delta u \qquad \text{(EQ 8.65)}$$

Design a controller that results in steady-state accuracy of 1 percent for operation in a small range around the operating point such that k is at most 10 percent different from its nominal value of 10. You may also assume that $\tau = 2$ sec.

(c) Design a fuzzy controller via the Mamdani approach that accomplishes the same task as in (b). Compare the response of the two different controllers.

(d) Let us assume that the parameter α varies by as much as 10 percent during the aircraft flight. How will the two controllers designed above respond?

8.4   Consider the simple process plant

$$\dot{x} = a(x, u)x + b(x, u)u \qquad \text{(EQ 8.66)}$$

Assume a to be constant and $b(x, u) \equiv b(u)$ to vary as shown below. Design a TSK fuzzy control

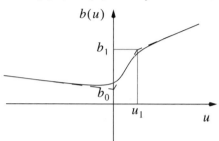

algorithm that enables the system to respond to the desired profile given below:

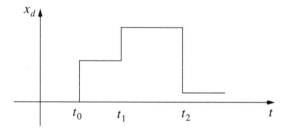

You may assume $a(x, u)$ to be a constant value of -1 and that $b(x, u)$ starts at $b_0 = 1$ with a slope of 2 and reaches $b_1 = 3$ at $u_1 = 1$ and then changes its slope to 1 afterwards. For $u < 0$, b decreases with a slope of 1. Let t_0, t_1, t_2 be 10, 20, and 30 seconds, respectively. Step changes can be 1 and 2 and back to 0.25 units, respectively.

8.5   Suppose a mechanical system is affected by a friction force shown below. The object of mass 1 Kg is

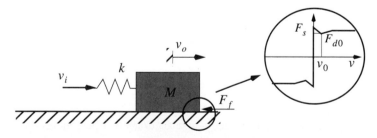

driven by a spring of stiffness 10 N/m and the input velocity, v_i is constant (let us say 1 m/sec.) The static friction threshold F_s is 0.25 N while the minimum dry friction, F_{d0}, is 0.2 N, which occurs at $v_0 = 0.1$ m/s, and F_d is 0.22 N. The objective is to design a fuzzy logic controller that eliminates the effect of friction and produces a smooth velocity profile. *Where would you place the controller and how would you apply the input to the system?*

8.6   Consider the classic moon-landing problem depicted below. The equations of motion for the system are given by $\dot{h} = v$ and $\dot{v} = u$, respectively.

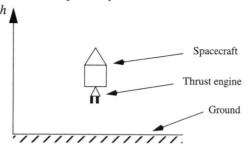

A linguistic vehicle model, based on the above equations, is given in rule-based form in Tables 8.9 and 8.8 (note that u is offset by gravity.)

TABLE 8.8   Height at Time $t + T$, given $h(t)$ and $v(t)$

		$v(t)$		
	$h(t+T)$	**Z**	**NS**	**NL**
	Very High	Very high	High	Medium high
	High	High	Medium High	Medium Low
	Medium High	Medium High	Medium	Ground
$h(t)$	**Medium**	Medium	Medium Low	Hard impact
	Medium Low	Medium Low	Low	Hard impact
	Low	Low	Ground	Hard impact
	Ground	Ground	Hard impact	Hard impact

TABLE 8.9   Velocity at Time $t + T$, given $v(t)$ and $u(t)$

		$u(t)$		
	$v(t+T)$	**Z**	**PS**	**PL**
	Z	Z	DNA[a]	DNA
$v(t)$	**NS**	NS	Z	DNA
	NL	NL	NS	Z

[a] DNA: Does Not Apply.

Determine the control algorithm that lands the vehicle in as short a time as possible starting from *very high* altitude and *large* velocity.

9   HIERARCHICAL INTELLIGENT CONTROL

9.1 Introduction

This chapter discusses the application of fuzzy logic in constructing hierarchically structured intelligent control systems. To this end we first provide an overview of the emerging area of intelligent control and explain how multilayered, hierarchical information processing captures essential features of intelligent behavior, namely *goal directedness* and *information abstraction,* and thus constitutes a potentially viable paradigm for intelligent control.

In this context we will suggest how fuzzy logic, by bridging the gap between symbolic and subsymbolic or numeric computing, can play a potentially significant role in facilitating seamless transition across multiple levels of hierarchical control systems and further how this paradigm may be used to formulate control policies that incorporate a task-level view of the given control engineering problem.

9.2 Intelligent Control

We suggested above that fuzzy logic can serve as a tool in developing intelligent control systems. In this context we note that intelligence connotes

- ability to *plan* via decomposition of a complex task into manageable subtasks,
- *robustness* to variations in the environment, and
- ability to *learn* from experience and/or adapt to new (unforeseen) situations.

In other words, intelligent control implies a level of autonomy that transcends conventional control techniques where following a prespecified trajectory and/or limited disturbance rejection is the essential concern. We can suggest that intelligent control addresses *task-level* control where aspects of planning and execution of a given task(including trajectory generation, execution, and diagnostic and corrective actions) are integral components of the control problem.

9.2.1 Architecture of Hierarchical Control

Precisely how one accomplishes the objectives of intelligent control as stated above and whether there are systematic means of doing so remain points of contention. Arguably heuristics play a strong role in accomplishing these objectives. Likewise, appropriate architecture and associated knowledge structures play a strong role in achieving the objectives of intelligent control. Indeed the complexity of the task and/or the environment requires that some mechanism for *information abstraction* exist in any intelligent control system. Fig 9.1 depicts a hierarchical architecture that embodies this notion that has appeared in the literature in different forms.

FIGURE 9.1   Hierarchical Architecture

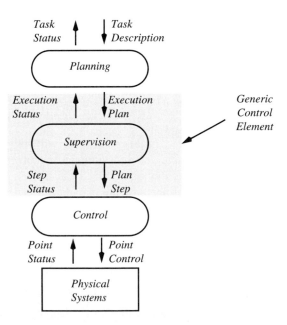

This architecture is based on the ideas that a number of researchers including Saridis[523], Fu[192], and, in recent years, Albus[9] have suggested and is meant to incorporate the following features:

- Information abstraction
- Balancing precision with complexity
- Multiple time scale operations

The general idea underlying hierarchical fuzzy systems is similar to other hierarchical systems in AI such as planning in a hierarchy of abstraction [519, 518], hierarchical planning [150], intelligent systems [9], and hierarchical control [1].

In particular, it is assumed that the higher levels in the hierarchy, that is planning and supervision, deal with a more abstract view of the control problem and do so in less precise terms. Moreover the action taking place at the higher levels affects the behavior of the system over a longer time span whereas the lower levels in the hierarchy operate on a faster time scale, as it were.

9.2.2 Implementation Issues

Precisely how one implements each level within a hierarchical framework and how one might facilitate the interaction among various levels require further elaboration. In this connection, a number of alternative information processing paradigms present themselves as candidates:

- Symbolic vs. numerical representation of information
- Rule-based vs. algebraic information processing
- Data-driven vs. procedural computation

The higher levels in the hierarchy deal with the more abstract view of the problem and thus must represent information in *granular* or *symbolic* form while the lower levels deal more efficiently with numeric information. Likewise the higher levels rely more on heuristics while the lower levels are based largely on differential algebraic formulations. Alternatively the higher levels are based on general-purpose computing mechanisms (logical inference) and are data driven while the lowest levels rely on proceduralized algorithms that specifically address a limited facet of the problem.

9.2.3 The Role of Fuzzy Logic

While it is not a priori evident how one addresses these issues in a universal manner, it does appear that fuzzy set theory can provide a functional basis for addressing some of these issues. In particular fuzzy sets can provide a semantically sound means of representation of granularized information. Moreover the use of fuzzy if-then rules in place of classical logic helps provide the tolerance for imprecision needed in situations such as mentioned above. In addition, there is a close connection between fuzzy logic and neuro-computing and that can be used in conjunction with complex nonlinear mappings that may appear in sensor processing in intelligent systems. Generally speaking, however, fuzzy logic may be best used to implement high-level or supervisory functions in a hierarchically structured intelligent control system. In this mode, one may see a similarity between the use of fuzzy logic and that of classical logic in *hybrid*(combined discrete-event and differential-algebraic-based) systems in control engineering. We will discuss this point briefly below.

9.2.3.1   Connection with Hybrid Systems

The ideas mentioned above are closely connected with the notions of *hierarchical control*, *hybrid systems*, and *large-scale systems*. These notions have long appeared in the literature of control theory[254, 326, 277] but have not received widespread response from the mainstream of the control engineering community. Specifically while the idea that dividing the control task into multiple levels of coordination or supervision and execution has been widely promoted, the issue of how one may design the higher levels within the hierarchy has remained the major stumbling block in connection with the design of such systems.

Likewise, with the exception of a limited class of systems, little has been accomplished in terms of addressing the analytical issues involved in hybrid and/or hierarchical systems. It is nonetheless worth pointing out that fuzzy logic can play a natural role in such systems. For instance, as depicted in Fig 9.2, fuzzy logic control can appear in the higher levels and play the role of supervisory control or coordinator. The advantage that fuzzy logic offers in this setting is twofold. First, one may argue that it can facilitate the synthesis of supervisory control strategies due in part to its ability to better represent the semantics of linguistic terms and constructs often used in supervisory control strategies.

Second, because of the fact that fuzzy logic-based control strategies can be viewed as nonlinear control strategies, analytical study of the resulting hierarchical system may be facilitated. For instance, the existing framework of nonlinear singular perturbation theory may be used in this context[626]. This latter point is indeed pressed in Fig 9.2 where fuzzy logic control is shown to be integrated with conventional control strategies in a hierarchically structured (multitime-scale) system.

FIGURE 9.2 Architecture for Hierarchical Control

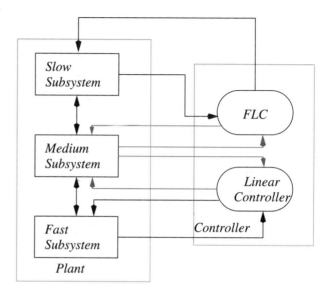

9.2.4 Other Hierarchical Architectures

We should mention that there are also connections between the ideas suggested above and some commonly used and some emerging ideas in control engineering. Among the existing approaches, we can relate the notion of intelligent control in its most rudimentary form to the idea of *gain scheduling* control, *mode fusion* control and the like. We will make appropriate connections to these ideas in the context of presenting some instances of application of fuzzy logic control in hierarchically structured systems.

9.3 Fuzzy Logic in Hierarchical Control

As suggested above, the potential use of fuzzy logic in intelligent hierarchical control is rather broad and multifaceted. The state of the art in using fuzzy logic in this setting is, however, rather limited. This is largely due to the fact that fuzzy logic is a relative newcomer as it were to control engineering; however, it is expected that its application in this field will continue to grow. With this in mind, in this section we will discuss several ways in which fuzzy logic has been used in the context of, albeit limited, hierarchical structures and outline its potential usage in these settings. In the subsequent section we will explore several concrete applications where the ideas discussed in this section are applied in experimental settings.

9.3.1 Fuzzy Logic-based Gain Scheduling and Autotuning

The notion of gain scheduling has long appeared in the control literature. In particular, in applications such as flight control, gain scheduling has always been a standard approach. The idea of gain-scheduled control is schematically depicted in Fig 9.3 where the gain scheduling mechanism driven by a variation of an external parameter(dynamic air density in aircraft control, for instance) varies the parameters of an otherwise fixed (linear)controller as the plant moves through a wide operating range.

FIGURE 9.3   Gain-Scheduled Control

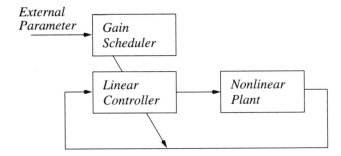

This approach is generally implemented via linear interpolation in conventional control engineering. Fuzzy logic-based gain scheduling essentially amounts to using a fuzzy logic-based strategy within the gain scheduling mechanism. Specifically, instead of simple linear interpolation, one utilizes a set of *if-then* rules to characterize the operating points of the system and to associate a set of controller parameters with those points. This strategy in effect provides a linguistically meaningful basis for interpreting the notion of an operating point. In this connection we should point out that the concept of the Takagi-Sugeno fuzzy control in effect amounts to a form of gain scheduling control.

Specifically in this approach a piecewise linear model of the plant is formed via a combination of if-then rules and linearized plant models and subsequently used to devise a gain scheduling control strategy that may take the form

$$\text{Rule } j : \text{if } \theta(k) \text{ is } \tilde{A}_j \text{ then } u_j(k) = K_j \underline{x}(k) \qquad \textbf{(EQ 9.1)}$$

where θ is the external parameter on which scheduling takes place and

$$\underline{x}(k) = \left[x_1(k) \; x_2(k) \; \dots \; x_n(k) \right]^T \qquad \textbf{(EQ 9.2)}$$

is the *state vector* used in the feedback control strategy (assuming that the control strategy is fundamentally a state feedback one.) Further $j = 1, 2, ..., r$ and r is the number of rules, and $u_j(k)$ is the output of the jth rule at discrete time k. In addition, A_j, s are linguistic terms such as *high*, *low*, *slightly*, defined in terms of fuzzy sets over the domain of the definition of θ. The control law takes the form

$$u(k) = \frac{\sum_j \mu_{\tilde{A}_j}(k) u_j(k)}{\sum_j \mu_{\tilde{A}_j}(k)} \qquad \text{(EQ 9.3)}$$

This is illustrated in Fig 9.4 where a nonlinear control action is devised as a function of θ.

9.3.1.1   Autotuning via Fuzzy Logic

A variant of the fuzzy logic-based gain scheduling control is autotuning or self-tuning of conventional linear control strategies such as PID control. In effect, instead of using an external parameter to adjust the controller, an autotuning controller uses any one of the signals already present within the control system, such as *error* or in certain cases the *reference* or *setpoint* or *disturbance*. In other words the only distinction between an autotuning fuzzy controller and a gain-scheduled fuzzy controller is the source of information used in the antecedent of the rules used in the control strategy. We will look more closely at an example of an application of an autotuning PID controller in the next section.

9.3.2   Fuzzy Logic-based Mode Fusion of Low-level Controllers

A *mode fusion* fuzzy system uses fuzzy rules in a higher level to fuse decisions recommended by multiple lower-level modules, as shown in Fig 9.5. The higher-level mode classification module determines the weight for each decision unit's output based on the current situation. For example, each decision unit can be a low-level controller for a specific goal, while the high-level classification unit dynamically determines the priority of these goals. These priorities are expressed as numerical weights, which are used to aggregate the recommendations generated by the low-level controllers.

FIGURE 9.4   Possible Controller Behavior

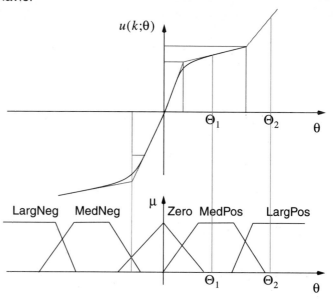

FIGURE 9.5   The Architecture of Mode Fusion Fuzzy System

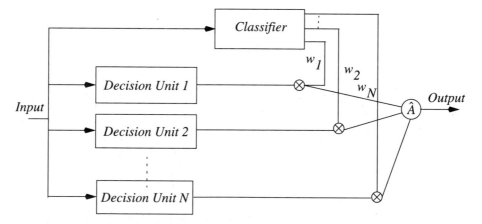

A typical industrial application of the mode fusion hierarchical fuzzy system is a hierarchical controller for a recuperative turboshaft engine developed by P. Bonissone and K. H. Chiang at General Electric [64], discussed at length below. A fuzzy rule based module is used to replace a conventional mode selector at the supervisory level such that control commands of six low-level PID controllers can be smoothly fused when the engine moves from one mode into another. The fuzzy logic-based hierarchical controller improves both the responsiveness and the fuel consumption of the conventional mode switching control scheme.

9.3.2.1   Hierarchical Task Decomposition

A *hierarchical task decomposition* fuzzy system generates high-level tasks to be performed based on external inputs and system goals. These high-level tasks are then decomposed into even more detailed suggestions. The lowest level subtasks are executed by the system. This architecture is certainly similar to that of *task expansion* in hierarchical planning [150]. Both architectures use knowledge about preferred ways to perform a task (often referred to as task decomposition *methods*) under various circumstances. They differ, however, in that the fuzzy logic architecture allows easy fusion of multiple task decomposition methods when the current situation is on the borderline between two regions suitable for two different methods. A notable example of hierarchical fuzzy control system is the helicopter controller developed by M. Sugeno [560, 563].

9.4   Case Studies

In this section we discuss some application areas of the use of fuzzy logic in intelligent hierarchical control. These application areas are by no means comprehensive and the interested reader is referred to the end of the chapter for further references.

9.4.1   Hierarchical Controller for GE's Recuperative Turboshaft Engine

A particular illustrative example of the use of fuzzy logic in hierarchically structured control systems that incorporate fuzzy logic involves a recuperative turboshaft engine developed by Pierro Bonissone and K. Chiang at General Electric [64] and schematically shown in Fig 9.6. This system, used to power a land vehicle, consists of a compressor that supplies pressurized air into the combustion chamber. The high pressure/temperature exhaust gas drives both the high pressure turbine, which supplies the power to the compressor, and a secondary power turbine, which supplies the power needed to drive the vehicle.

FIGURE 9.6   Architecture of the GE Turboshaft Engine

(From P. P. Bonissone and K. H. Chiang, *"Fuzzy Logic Hierarchical Controller for a Recuperative Engine: From Mode Selection to Mode Melding,"* in J. Yen, R. Langari, and L. A. Zadeh (eds.), *Industrial Applications of Fuzzy Control and Intelligent Systems.* © 1995 IEEE.)

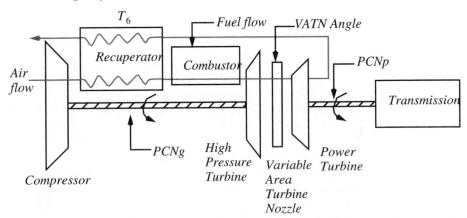

The main concern in this problem is to assure that not only the performance objectives(largely in terms of the power made available to drive the vehicle) are met but also that the vehicle is operated efficiently and in ways that prevents short-term or long-term damage and/or failure of the engine components. In particular, one must be concerned with the potential problems listed in Table 9.1.

TABLE 9.1   Control Issues in a Recuperative Turboshaft Engine

As fuel flow increases:	- Potential for compressor stall
	- Possible stress-related failure of mechanical components
	- Possible overspending of high pressure spool and/or power turbine
As fuel flow decreases:	- Possibility of flameout
	- Potential for loss of power to cooling system increases

9.4.1.1   Control architecture

In order to address the issues listed in Table 9.1, the control strategy already in place had incorporated a *two-stage* approach consisting of a supervisory controller as well as a number of(up to 10) lower-level controllers as is schematically shown in Fig 9.7. Note that the following notation is used in the figure:

- *PLA*   Power level angle(from operator)
- *Fuel-flow*   Used in coarse adjustment control action
- *VATN-Angle*   *Variable Area Turbine Nozzle* for fine adjustment control action
- T_6 Temperature in the recuperator
- N_g Compressor core speed
- N_p Power turbine speed

FIGURE 9.7   Control Architecture for the GE Turboshaft Engine(Adapted from Bonissone and Chiang[64])

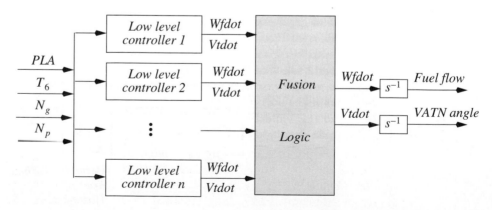

The specific functions of the low-level controllers are listed in Table 9.2. As listed

TABLE 9.2   Low Level Controllers

Operating Mode	Controller	Description
Nominal	Ng-T6 governor[a]	Quasi steady-state operation; responds to operator commands.
Engine protection	Ng bottomer and Ng topper Np-T6 bottomer and Np-T6 topper T6 limiter	Maintains core speed, power turbine speed, and recuperator inlet temperature[b] within limits.
Acceleration	Ndot governor	Maintains core acceleration within limits.
VATN saturation	Ng governor Np bottomer and Np topper	Regulates core speed, power turbine speeds within limits.

a. Core speed and T6 (recuperator inlet temperature) reference values are computed as a function of PLA, corrected core, and power turbine speeds.

b. Recuperator inlet temperature is maintained above the minimum values set by limits on high-pressure turbine outlet temperature and the combustor outlet temperature.

in the middle column of this table, the function of each controller is to address a limited objective (to respond to the driver power-level request or to prevent rapid acceleration of the compressor core, etc.) and to propose the incremental fuel flow (Wfdot) and the change in the variable area nozzle angle (Vtdot.) The mode fusion logic prioritizes the low-level controllers and selects the "right" fuel flow and variable area turbine nozzle angle among those proposed by the low-level controllers and based on the priority associated with each controller.

9.4.1.2 Fuzzy Logic-based Mode Selection

While the control strategy discussed above does deal with the issues of concern in the control of the turboshaft engine as stated in Table 9.1, the use of "hard logic" in the mode selection strategy results in abrupt transitions across operating modes. For this reason Bonissone and Chiang proposed a fuzzy logic-based mode selection strategy. This mode selection strategy amounts to a weighting scheme for various controllers or in effect their proposed action. Likewise, this scheme implicitly prioritizes these controllers in terms of their significance and their contribution to the overall mission of the system (Table 9.3). This contrasts with the approach that has been commonly used

TABLE 9.3   Fuzzy Logic-based Mode Selection

(From P. P. Bonissone and K. H. Chiang, "Fuzzy Logic Hierarchical Controller for a Recuperative Engine: From Mode Selection to Mode Melding," in J. Yen, R. Langari, and L. A. Zadeh (eds.), *Industrial Applications of Fuzzy Control and Intelligent Systems.* © 1995 IEEE.)

T6	PCNp	Ndot	Ng-T6 governor	Np-T6 topper	Np-T6 bottomer	T6 Limiter	Ndot governor
normal	normal	pos high	0.40				0.60
		pos low	1.00				
		zero	1.00				
		neg	1.00				
	low	pos high	0.26		0.14		0.60
		pos low	0.40		0.60		
		zero	0.30		0.70		
		neg	0.20		0.80		
	high	pos high	0.13	0.27			0.60
		pos low	0.33	0.67			
		zero	0.33	067			
		neg	min	min			
high	normal	pos high	0.06			0.34	0.60
		pos low	0.27			0.73	
		zero	0.65			0.35	
		neg	min			min	
	low	pos high	0.06		0.11	0.40	0.60
		pos low	0.10		0.20	0.70	
		zero	0.16		0.67	0.17	
		neg	0.16		0.67	0.17	
	high	pos high	0.09	0.19		0.12	0.60
		pos low	0.23	0.47		0.30	
		zero	0.26	0.48		0.26	
		neg	min	min		min	

wherein the low-level controllers simply propose an action (in terms of Wfdot and Vtdot) and the mode selection logic chooses or rejects the proposed action depending on the priority given to the controller (for instance, compressor stall prevention may have the highest priority while engine overacceleration may take a lower priority.)

9.4.2 Autotuning a PID Controller

Industrial systems such as heating furnaces and gas generation equipment often exhibit load variations that undermine the performance of conventional (PID) controllers. In many instances the performance of the system depends on both the load and load change direction. Hence it is desirable to provide some means of uniform "optimal" performance over a wide operating range. To this end, one can develop a "self-tuning" strategy that incorporates knowledge of operation of the plant whereby

- A parameter table is used to handle load and direction of load changes and
- A switch-over function smooths the transition of control parameters.

An instance of application of this idea is illustrated in Fig 9.8 where it is applied to the problem of control of temperature inside the pipe of an evaporator. As is evident in Figs 9.9 and 9.10, there is a similarity between the resulting control architecture and the general architecture for hierarchical control depicted in Fig 9.2.

FIGURE 9.8   Logical Basis of an Autotuning Controller.

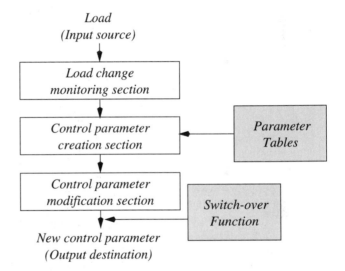

The specific strategy used in the autotuning scheme involves a cascade of two control loops where the primary loop is used to regulate temperate inside the evaporator and the secondary one is used to control the flow rate of steam to the evaporator. Since the characteristics of the evaporator change greatly with the input flow rate, the autotuning logic is used to monitor this quantity and to vary the P and I gains of the primary controller. The resulting strategy tested under the conditions of varying the stepoint from 50 percent to 70 percent while the flow rate changes from 40 percent to 70 percent shows effective performance of the resulting scheme as depicted in Fig 9.11 where the solid line represents the performance of the system with the autotuning scheme.

FIGURE 9.9   Architecture of an Autotuning PID Control System
(From K. Nukuia, M. Arakawa, M. Komido and T. Taniguchi, "An Auto Tuning Method for Obtaining Optimal Control Parameters by Following Changes in Process Characteristics," ISA Paper #91-0459. © ISA, all rights reservered.)

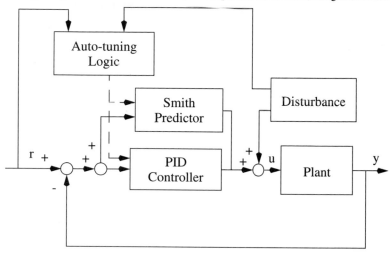

FIGURE 9.10   Schematic of the Autotuning System
(From K. Nukuia, M. Arakawa, M. Komido and T. Taniguchi, "An Auto Tuning Method for Obtaining Optimal Control Parameters by Following Changes in Process Characteristics," ISA Paper #91-0459. © ISA, all rights reservered.

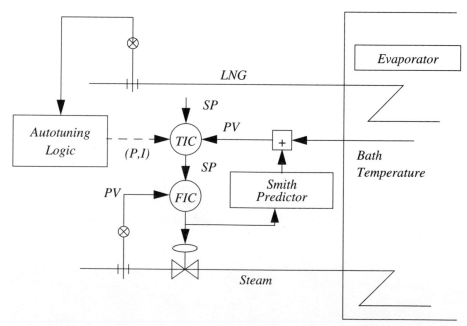

FIGURE 9.11   Performance Results of an Autotuning Controller

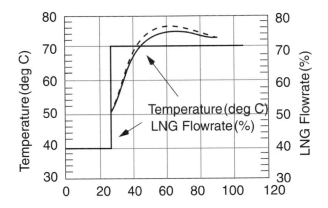

9.4.3 Fuzzy Logic-based Parameter Optimizer in a Hierarchical Process Control System

In process control one of the key issues is variation of operating parameters. On the other hand, using linear control techniques such as PID and/or PID plus feedforward or lead, lag compensators are commonplace in the processing industry. In situations such as these fuzzy logic can augment existing control techniques as discussed earlier through on-line optimization(tuning) of controller parameters[520]. Consider, for instance, a distillation process for a benzene, toluene mixture(Fig 9.12.) In this process and in a 50% toluene, 50% benzene mixture, vapor starts to evolve at 92.5 °C. The resulting vapor is about 20% richer in the volatile component, benzene, while the liquid is richer in Toluene. Moreover, warm vapor rising in the column and coming in contact with cooler liquid, running down the column, loses temperature but becomes richer in benzene while the liquid becomes richer in toluene. In this setting, conventional linear controllers are used to perform feedforward control on the top and bottom flows and fuzzy logic appears within a supervisory mechanism as shown in Fig 9.13.

FIGURE 9.12   Temperature/Concentration vs. Tray Number in a Distillation Column

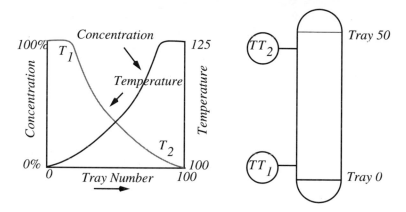

9.4.4 Control Strategy

The control strategy relies on the fact that there does exist a reasonably effective linear model of the behavior of the system. The parameters associated with this linearized model, however, vary with the variation in the in-flow rate of the raw material. For instance a possible ad hoc empirical model of the system may be given by

$$
\begin{bmatrix} T_2 \\ T_1 \end{bmatrix} = \begin{bmatrix} k_{11}\dfrac{e^{-T_{11}s}}{1+8s} & k_{12}\dfrac{e^{-T_{12}s}}{1+11s} \\ k_{21}\dfrac{e^{-T_{21}s}}{1+\tau_{21}s} & k_{22}\dfrac{e^{-\tau_{22}s}}{1+\tau_{22}s} \end{bmatrix} \begin{bmatrix} m_2 \\ m_1 \end{bmatrix} + \begin{bmatrix} k_{1d}\dfrac{e^{-T_{1d}s}}{1+\tau_{1d}s} \\ k_{2d}\dfrac{e^{-T_{2d}s}}{1+\tau_{2d}s} \end{bmatrix} d \qquad \textbf{(EQ 9.4)}
$$

where T_1 and T_2 are the temperatures in the top and bottom trays, m_1 and m_2 are the respective input mass flow rates, and d is the disturbance input.

An effective means of constructing a control strategy is to use two relatively simple single-loop controllers at the top and bottom levels of the column and augment these with two additional compensators, namely a feedforward disturbance compensator that makes use of the feedflow information to aid the top-level controller and a decoupling compensator that performs a similar function with respect to the bottom-level controller. The resulting scheme is further augmented by a fuzzy logic-based supervisory controller that adjusts the parameters of the feedforward and decoupling compensators in response to variations in the feed flow rate. The rule set for the controller is shown in Table 9.4, where depending

TABLE 9.4   Rule Set for the Distillation Column

		Increase in feed flow/decoupling feedforward signal			
		Zero	**Small**	**Medium**	**Large**
	Zero	Zero	Zero	Zero	Zero
Temperature overshoot	**Small**	Zero	Small	Very Small	Very very small
	Medium	Zero	Large	Medium	Small
	Large	Zero	Very very large	Very large	Large

on the temperature overshoot and previous state of the feedforward gain, changes are made to this latter quantity to reduce the overshoot. The response of the system is depicted in Fig 9.14 which clearly shows the improved performance of the system with the addition of the fuzzy supervisory logic.

FIGURE 9.13   Overview of the Fuzzy Logic-based Distillation Column Control System

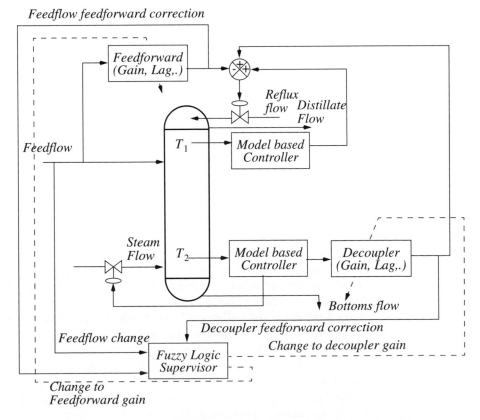

FIGURE 9.14   Response of the Distillation Column Control System

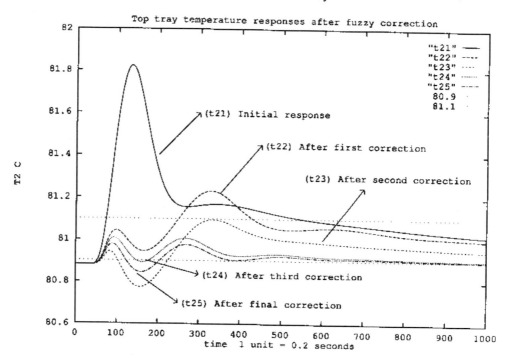

9.5 Summary

The discussion in this chapter centered on the use of fuzzy logic in hierarchically structured control systems. As pointed out, there exists a significant potential for the use of
fuzzy logic in this setting, mainly to address the need to incorporate heuristic knowledge
involved in operating a given plant but also to help facilitate the transition from symbolic
to numeric computation in a multiple time-scale system. The present state of the art, however, makes limited use of the potential of fuzzy logic, primarily in such areas as gain
scheduling control and/or autotuning conventional control strategies such as PID control.
This does, nonetheless, point to the fact that fuzzy logic control is indeed expanding
beyond the boundaries of single layer, Mamdani style control algorithms. This limited
usage also points to a significant issue not addressed in this chapter and that is the overall
stability of the resulting control system. In previous chapters we explored the use of analytical techniques in fuzzy logic control but such techniques are largely limited to the simpler class of fuzzy logic controllers. Analysis of hierarchically structured systems, with or
without fuzzy logic, present special challenges that must indeed be addressed before the
use of this paradigm becomes widely accepted.

Bibliographical and Historical Notes

There is both great interest and a great deal of activity in the area of hierarchical fuzzy control systems not touched on in this chapter. References to some, for instance the works of Raju and colleagues[490] and Gegov and Frank[205], but by no means all of such works in this area, are given in the reference section of this book. Likewise there are various approaches to autotuning PID-type controllers as well as other types of adaptive fuzzy control that in a sense can be viewed as hierarchical type control strategies. Notable among these is the work of L. X. Wang[619].

Exercises

9.1   Consider the problem of controlling a simple first order plant given by

$$\tau \dot{y} = -y + k(u)u + d \qquad \text{(EQ 9.5)}$$

where $k(u)$ represents a nonlinear gain and d is a disturbance. We can use a simple PI controller to deal with this problem, but in this exercise we intend to explore the use of fuzzy logic to tune the PI controller. Suppose $k(u)$ is given by

$$k(u) = \begin{cases} \bar{k} & \dfrac{du}{dt} \geq 0 \\[2mm] \underline{k} & \dfrac{du}{dt} < 0 \end{cases} \qquad \text{(EQ 9.6)}$$

In other words, the system has a different gain depending on the direction of change of u. Let τ be 10 sec and let $\bar{k}$ and $\underline{k}$ be 1.2 and 0.8, respectively. First design a PI controller that performs a reasonable job of tracking step references and rejecting disturbances. Test your controller under both positive shift and negative shifts in reference. Next design an autotuning strategy that improves the response in a way that would generally produce a reasonably symmetric response pattern.

9.2   In the previous problem let us assume that a delay of 2 seconds is added to the controller. In effect the nominal plant transfer function is given by

$$\hat{g}(s) = \frac{ke^{-2s}}{\tau s + 1} \qquad \text{(EQ 9.7)}$$

What would be the consequence of using the control strategy derived in the previous exercise?

10 ANALYTICAL ISSUES IN FUZZY LOGIC CONTROL

10.1 Introduction

Since its emergence in the mid-1970s,[1] fuzzy logic control has been the subject of sometimes heated debate within academic circles. The key issue in this debate is whether one can analyze fuzzy control systems in terms of the established traditions of control theory. As it was pointed out in Chapter 9, however, fuzzy logic control takes a fundamentally nonanalytical view of the control design process; the control algorithm reflects prior knowledge of operation of the given physical system in terms of *if-then* rules, rather than emerging as the outcome of a formal modeling and design process. Nonetheless one may need to ensure that a given fuzzy control system holds certain analytical properties, such as *stability*, and/or characterizes the functional, structural role of the controller in relation to the plant. This chapter explores these issues at some length. Specifically we will consider both past and present work in analytical study of fuzzy control systems, describe trends in this area, and discuss various ways in which important issues such as stability of these systems have been addressed.

We start with a brief historical overview of the treatment of analytical issues in fuzzy logic control.

[1.] Tracing fuzzy logic control to E.H. Mamdani's seminal paper published in 1974 [400].

10.2 Historical Overview and Mamdani's Approach

As mentioned earlier, the motivation for analytical study of fuzzy control systems is to formally characterize certain properties of such systems such as *stability* as well as the *structural* relationship between the plant and the controller. To this end, we note that E. H. Mamdani who is credited with the first application of fuzzy logic control initiated an approach to the analysis of fuzzy control systems that remains of value even to date[315]. Mamdani's approach, based on the classical notion of *describing functions*[681], attempts to characterize a fuzzy control system in terms of a *time-invariant* nonlinear element(a multilevel relay) acting on the plant(approximated as a linear dynamic system) and to further determine the likelihood of sustained oscillations(instability) in the resulting feedback loop(Fig 10.1.)

FIGURE 10.1   A Schematic Diagram Depicting Mamdani's Approach

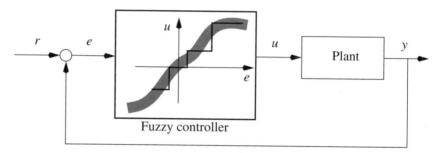

It is worth noting that this approach constitutes an approximate analysis method and as such is not a full-fledged analytical approach. Nonetheless, the notion of describing functions had received a great deal of attention in the decade preceding Mamdani's work and represented an acceptable tool for analysis of nonlinear control systems Moreover, in addition to its emphasis on the issue of stability of fuzzy control systems, Mamdani's work paved the way for further investigation of the role of fuzzy control in facilitating the synthesis of nonlinear control algorithms — an issue that has remained of interest since.

10.2.1 The Describing Function Approach

The key idea in the describing function approach is to view a given nonlinear control system as driven by a single memory-less, possibly time-varying nonlinear function[$\sigma(.)$ in Fig 10.2] in conjunction with a linear dynamic system[$\hat{g}(s)$ in the same figure.] Generally, though not always, the nonlinear element is the controller(incorporating actuator saturation, deadzone, and the like) while the linear element constitutes the plant represented as a strictly proper transfer function. (A strictly proper transfer function has a numerator with a lower order than its denominator and can be thought of as a *low-pass filter*. This fact plays a strong role in the describing function framework.)

In this manner one can study the behavior of the resulting closed-loop system and determine *likelihood* of sustained oscillations(effectively instability.) Fig 10.2 illustrates this idea that sustained oscillations appear as synchronized oscillations of the input, output, and error signals, u, y, e, even when the reference or setpoint, r, is zero ($r \equiv 0$).

FIGURE 10.2   Sustained Oscillations in a Nonlinear System

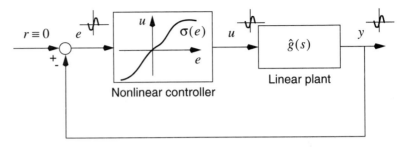

The describing function method relies on approximate frequency-domain analysis of the closed-loop system involved. Accordingly, one characterizes the given nonlinear element, $\sigma(.)$ (say the controller in Fig 10.2) in terms of a complex valued "gain":

$$DF(e;\omega) = \gamma_1 + j\gamma_2 \qquad \textbf{(EQ 10.1)}$$

where γ_1 and γ_2 depend on e and ω, amplitude and frequency of the sinusoidal input to that element respectively. We refer to $DF(.)$ as the sinusoidal input *describing function* associated with the given nonlinear element.

Sustained oscillations are likely to appear in the system if there exists a nonzero e such that

$$e = r - y = r - \hat{g}(j\omega)DF(e;\omega)e = -\hat{g}(j\omega)DF(e;\omega)e \qquad \textbf{(EQ 10.2)}$$

where $(r \equiv 0)$. In other words these oscillations appear if the equation

$$(1 + \hat{g}(j\omega)DFe;\omega)e = 0 \qquad \textbf{(EQ 10.3)}$$

has a nontrivial solution for e, which in turn implies that for sustained oscillations to exist, one must have

$$1 + \hat{g}(j\omega)DF(e;\omega) = 0 \qquad \textbf{(EQ 10.4)}$$

For a memory-less nonlinearity, $DF(e;\omega)$ does not depend on ω and thus is effectively a measure of the "gain" of the given nonlinear element:

$$\hat{g}(j\omega) = -\frac{1}{DF(e)} \qquad \textbf{(EQ 10.5)}$$

which in turn implies that the resulting system would exhibit sustained oscillations if, as shown in Fig 10.3, the point at which the graph of $\hat{g}(j\omega)$ (in the complex plane) crosses the real line coincides with the point on the negative real line that has a magnitude of $(1/DF(e))$.

FIGURE 10.3   Stability Condition Based on the Describing Function Approach

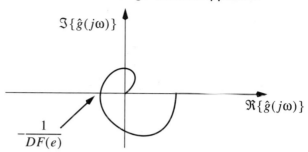

10.2.2 Application to Fuzzy Control Systems

The idea of describing function was applied by E.H. Mamdani to the study of fuzzy control systems[315]. Mamdani's approach was to first show that a fuzzy logic control algorithm with the so called *min, max* operators and *mean of max* defuzzification leads to a multilevel relay and thus enables one to characterize such an algorithm in terms of the following describing function:

$$DF(e) = \frac{4}{\pi e^2} \left\{ \sum_{i=1}^{N-1} U_i [\sqrt{e^2 - E_i^2} - \sqrt{e^2 - E_{i+1}^2}] + U_N \sqrt{e^2 - E_N^2} \right\} \text{(EQ 10.6)}$$

where E_i, U_i are the *center values* of the fuzzy sets defined over the domains of definition of error, e, and input, u, respectively, and $i = 1, 2, ..., N$, with N as the number of rules in the rule set. This is illustrated in Fig 10.4, where on the right the function $DF(e)$ is schematically depicted as a function of e. Note that in this graph, $E_1 \ll E_2$ and $E_2 \approx E_3$ lead to a closer blending of the second and third lobes of the describing function.

FIGURE 10.4   Multilevel Relay Representation of a Fuzzy Logic Control Algorithm and Its Describing Function(Reprinted from *Fuzzy Sets and Systems*, 1(1), W. J. M. Kickert and E. H. Mamdani, "Analysis of a Fuzzy Logic Controller," pp. 29–44, © 1978, with permission of Elsevier Science.)

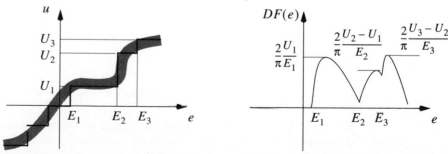

As an example of application of this approach consider the system given by [315],

$$\hat{g}(s) = K\frac{e^{-Ts}}{s+a} \qquad \textbf{(EQ 10.7)}$$

representing a first-order process type plant with delay

$$\tau\dot{y} = -y + k_s u(t-T) \qquad \textbf{(EQ 10.8)}$$

with $\tau = 1/a$ as time constant and $k_s = K/a$ as the steady-state gain of the plant. Typically $T \ll \tau$ while the numerical values of $a = 0.05$, $K = 0.05$, and $T = 12$ sec are used in the following discussion, which results in $\tau_s = 20$ sec and $k_s = 1$.

The fuzzy control scheme to be evaluated has three rules with the corresponding values of E_i and U_i given respectively by 1, 9, 10 and 4.4, 30.8, 35.2 based on Mamdani and Kickert's work [315].

Evaluating the above transfer function at $s = j\omega$ we have

$$|\hat{g}(j\omega)| = \frac{K}{\sqrt{a^2 + \omega^2}}, \quad \angle\hat{g}(j\omega) = -\omega T - \tan^{-1}\frac{\omega}{a} \qquad \textbf{(EQ 10.9)}$$

as magnitude and phase angles associated with $\hat{g}(j\omega)$. The values of ω where $\angle\hat{g}(j\omega)$ is 180°, i.e., where $\hat{g}(j\omega)$ crosses the negative real axis as, for instance, in Fig 10.3, are given by

$$-\frac{\omega}{a} = \tan T\omega \qquad \textbf{(EQ 10.10)}$$

For $T \ll \tau = 1/a$, this condition can be approximated by

$$\omega = \left(k + \frac{1}{2}\right)\frac{\pi}{T}, \quad k = 0, 1, 2, \dots \qquad \textbf{(EQ 10.11)}$$

For numerical values of $a = 0.05$, $K = 0.05$, and $T = 12$ sec, the above condition suggests that oscillations occur at frequencies as low as $\omega = \pi/24$, and the corresponding values of $|\hat{g}(j\omega)|$ and $DF(e)$ that lead to oscillations would be 0.36 and 2.78, respectively. The graph of $DF(e)$ for values of E_i and U_i given earlier is depicted in Fig 10.5.

FIGURE 10.5   Graph of the Describing Function of the Fuzzy Controller

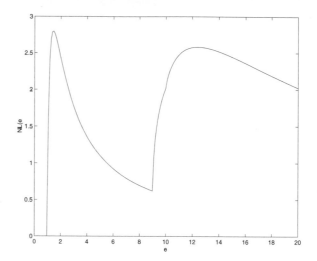

The presence of sustained oscillations(instability) in the system is depicted in Fig 10.6.

FIGURE 10.6   Sustained Oscillations Generated with Fuzzy Control Scheme

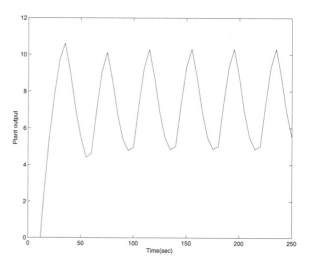

10.3 Reflections on Mamdani's Approach

The approach developed by Mamdani is relatively simple and efficient. Moreover, it does suggest a connection between fuzzy control and nonlinear control theory(equivalence of the fuzzy logic controller with a multilevel relay) that remains of interest to date. This approach suffers from two drawbacks, however:

1. The assumption that the plant is linear is overly restrictive and largely untenable; indeed, one often-stated rationale for using fuzzy logic control is plant nonlinearity

2. The use of the describing function method constitutes an approximate analysis that may or may not yield a viable answer.

Both of these issues were later addressed to varying degrees by researchers who worked through a number of possible alternatives to Mamdani's approach. In particular, several researchers developed and extended the framework developed by Mamdani by introducing more powerful tools from nonlinear systems theory. For instance K. Rao and D. Majumdar[575] applied the concept of *absolute stability* to the study of fuzzy control systems and concluded that instability in a fuzzy control system can be due to high effective gain of the nonlinear feedback action that results from a fuzzy logic controller. This is illustrated in Fig 10.7 where the defuzzified control action(obtained via the *centroid of area* defuzzification rule) lies in the shaded zone in the first and third quadrants of the coordinate frame defined by the e and u axes. The defuzzified control action is thus bounded by the dashed line in the figure, whose slope, k_{max}, acts as the maximum controller gain and places a bound on the excursion of the plant transfer function into the left half- plane as shown in Fig 10.7. Specifically, for the resulting closed-loop system to be stable, the graph of the plant transfer function in the complex plane must not extend to the left of the line defined by $\Re\{\hat{g}(j\omega)\} = -1/k_{max}$. In the extreme case, that is for sufficiently large values of k_{max}, the aforementioned graph must lie entirely in the first quadrant of the complex plane, severely restricting the class of systems that satisfy the proposed stability condition.

FIGURE 10.7   Application of the Idea of Absolute Stability to Fuzzy Control Systems

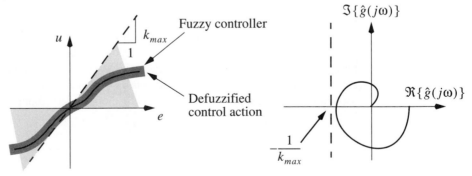

In addition to the work of Rao and Majumdar, other works dealing with the use of nonlinear control techniques have since appeared in the literature (some of which are discussed in the Bibliography and Historical Notes section of this chapter.) Among these works the use of Lyapunov's direct method appears most promising and as such is discussed at length below.

10.4 Analysis of Fuzzy Control Systems via Lyapunov's Direct Method

The approach described in the previous section offers insights into the role of Fuzzy Logic Control (FLC) algorithms as nonlinear controllers and leads to, as shown, sufficient conditions for the stability of the resulting control system. A more general study of FLCs can be taken up, however, via the so-called via Lyapunov stability theory. The advantage of this approach as we shall see shortly is as follows:

- It offers a rather intuitive approach to interpretation of stability as perturbation from an equilibrium state.

- It uses energy-like functions that are potentially less dependent on the specific structure of the plant model.

The idea underlying this approach is based on one of the most fundamental techniques in control theory, namely Lyapunov's direct method [706, 690]. Before we consider the application of this method to fuzzy control systems, however, we consider some basic definitions and in particular that of stability in the sense of Lyapunov. Note that this definition addresses the issue of stability of finite-dimensional autonomous dynamic systems of the form $\dot{x} = f(x)$ where x is an n-dimensional vector that represents the state of the system. Specifically, the definition refers to perturbations of the system from its equilibrium state, x^*, where $f(x^*) = 0$. Note that x^* is often 0, but in cases where it is not one can choose the coordinate system so as to shift x^* to 0.

DEFINITION 17   An equilibrium point, x^*, of a given dynamic system

$$\dot{x} = f(x) \qquad \text{(EQ 10.12)}$$

is stable if and only if given any $\varepsilon > 0$ there exists some $\delta > 0$ such that given any initial state x_0 with $\|x_0 - x^*\| < \delta$, we have, for all $t \geq 0$, $\|x(t) - x^*\| < \varepsilon$..

This definition effectively suggests that, as shown in Fig 10.8, no matter how small ε may be, one can find a δ such that trajectories starting within δ distance of the equilibrium state, x^*, remain within ε of x^*. With this definition in mind we consider the basic concept used in Lyapunov's direct method, namely the notion of a Lyapunov function candidate.

FIGURE 10.8   Lyapunov Stability Shown Schematically for a Two Dimensional System

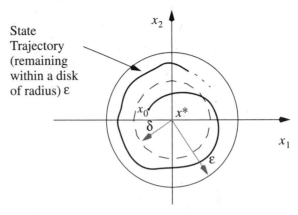

DEFINITION 18   A Lyapunov function candidate, $V(x)$, is a scalar, continuously differentiable function of the state x that satisfies

-   $V(x) \geq 0$, with $V(x) = 0 \Leftrightarrow x = 0$,

-   $V(x) \rightarrow \infty$ as $\|x\| \rightarrow \infty$.

   This function is essentially an *energy-like* function whose behavior along the trajectories of the system determines the stability of the given closed-loop system. Specifically as shown in Fig 10.9, stability of the given system effectively amounts to V consistently decreasing as the state x changes with time. Intuitively $V(x)$ is positive for any value of x and becomes unbounded with x, and is 0 if and only if x is 0. Showing that there exist some $V(x)$, which consistently decrease along the state trajectories, effectively amounts to showing that these trajectories remain bounded. This idea is more formally stated below as a theorem.

THEOREM 9   The equilibrium point, $x^* = 0$, of a dynamic system, $\dot{x} = f(x)$, is stable if a Lyapunov function, $V(x)$, exists such that $\dot{V}(x) \leq 0$ along the trajectories of the system.

A natural extension of the notion of stability is that of *asymptotic* stability.

DEFINITION 19   A given dynamic system is asymptotically stable if it is stable in the sense of Lyapunov and moreover if as $t \rightarrow \infty$, $\|x(t)\| \rightarrow 0$.

   In terms of the application of Lyapunov's method (Theorem 9) asymptotic stability is guaranteed if the inequality affecting $\dot{V}(x)$ is made strict, that is if $\dot{V}(x) < 0$ along the trajectories of the system.

FIGURE 10.9   Schematic of a Lyapunov Function Candidate

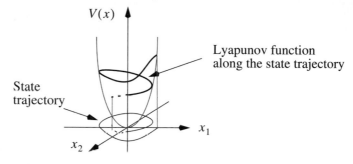

A simple example illustrates the idea of stability in the sense of Lyapunov. Consider the pendulum shown in Fig 10.10. Clearly with friction present, the pendulum stabilizes itself asymptotically at the equilibrium point(the vertical position.) We shall investigate this problem via Lyapunov's stability theory. The model of the system derived via Newton-Euler equations is given by

$$\dot{\theta} = \omega$$

$$I\dot{\omega} = -b\omega - \frac{1}{2}(mgl\sin\theta) \qquad\qquad \textbf{(EQ 10.13)}$$

where θ and $\omega = \dot{\theta}$ are the angular position and velocity of the pendulum, respectively. We define the state of the system as

$$x = \begin{bmatrix} \theta \\ \omega \end{bmatrix} \qquad\qquad \textbf{(EQ 10.14)}$$

FIGURE 10.10   Simple Pendulum

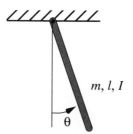

Let us define the Lyapunov function candidate

$$V(x) = \frac{1}{2}I\omega^2 + mgl(1 - \cos\theta) \qquad\qquad \textbf{(EQ 10.15)}$$

as the total mechanical energy of the system(sum of kinetic and potential energies.) We have(using $\dot{\theta}$ in place of ω)

$$\dot{V}(x) = I\dot{\theta}\ddot{\theta} + \frac{1}{2}mgl\dot{\theta}\sin\theta$$

$$= \dot{\theta}\left(-b\dot{\theta} - \frac{1}{2}mgl\sin\theta\right) + \frac{1}{2}mgl\dot{\theta}\sin\theta \qquad \textbf{(EQ 10.16)}$$

$$= -b\dot{\theta}^2$$

which is clearly negative along the trajectories of the system thus showing that the system is stable.

10.4.1   Discrete Time Systems

The notion of Lyapunov stability can be extended to the case of discrete time systems, described in terms of *difference* equations of the form $x_{k+1} = f(x_k)$.

DEFINITION 20   The equilibrium point, $x^* = 0$, of a given discrete time dynamic system $x_{k+1} = f(x_k)$ is stable if and only if given any $\varepsilon > 0$ there exists some $\delta > 0$ such that given any initial state x_0 such that $\|x_0 - x^*\| < \delta$, we have, for all $k \geq 0$, $\|x_k - x^*\| < \varepsilon$.

FIGURE 10.11   Lyapunov Stability for Discrete Time Systems

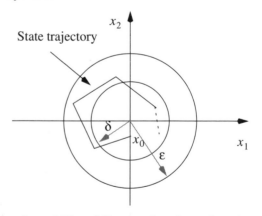

We will next present a theorem concerning the stability of discrete-time dynamic systems.

THEOREM 10   The equilibrium point, $x^* = 0$, of a dynamic system, $x_{k+1} = f(x_k)$, is stable if a Lyapunov function, $V(x)$, exists such that $V(x_{k+1}) - V(x_k) \leq 0$ along the trajectories of the system.

As an example of application of the above theorems, consider a simple first-order difference equation:

$$x_{k+1} = \alpha(x_k)x_k \qquad x \in \mathfrak{R} \qquad -\underline{\alpha} \leq \alpha(x_k) \leq \bar{\alpha} \qquad \underline{\alpha}, \bar{\alpha} > 0 \textbf{(EQ 10.17)}$$

In order to investigate the stability of the equilibrium point, $x^* = 0$, of this system, let us define $V(x) = |x|$. Then

$$V(x_{k+1}) = |x_{k+1}| = |\alpha(x_k)x_k| = |\alpha(x_k)||x_k| \le \max(\underline{\alpha}, \bar{\alpha})|x_k| \qquad \textbf{(EQ 10.18)}$$

Thus for the system to be stable in the sense of Lyapunov, we must ensure that $\max(\underline{\alpha}, \bar{\alpha}) < 1$.

10.4.2 Application to Fuzzy Control Systems

In this section we consider the application of Lyapunov stability theory to fuzzy control systems. The intent here is twofold:

- The plant model should not be treated as linear; moreover, no unnecessary assumptions concerning the plant model should be introduced.

- The fuzzy logic-based control algorithm should not be approximated in a manner that would violate the spirit in which it is originally developed and stated.

For this purpose and following the guidelines stated earlier, we assume the plant model to take the general form

$$x_{k+1} = f(x_k) + g(x_k)u_k \qquad \textbf{(EQ 10.19)}$$

where

$$|f(x)| \le F|x| \qquad |g(x)| \le G \qquad \textbf{(EQ 10.20)}$$

with F and G positive constants. We further let the control law take the simple form

$$\text{if } e_k \text{ is } \tilde{A}_j \text{ then } u_k \text{ is } \tilde{B}_j$$

where $\{\tilde{A}_j\}$ and $\{\tilde{B}_j\}$ are collections of fuzzy sets defined over the domains of the definition of e and u, respectively. We shall further assume that $\{\tilde{A}_j\}$ and $\{\tilde{B}_j\}$ form nearly ideal fuzzy partitions of their respective domains.

Assumption 1   For each e in its domain of definition,

$$\sum_j \mu_{A_j}(e) \cong 1 \qquad \textbf{(EQ 10.21)}$$

(A similar assumption holds for $\{\tilde{B}_j\}$.)

The controller takes the form (via the center average method)

$$u_k = \frac{\sum \mu_j U_j}{\sum \mu_j}, \qquad \textbf{(EQ 10.22)}$$

where μ_j is short for $\mu_{A_j}(e_k)$ and this summation is over all applicable rules.

Development of the proof procedure requires certain additional notation to be defined. The first of these establishes a correlation between the input-output fuzzy sets

DEFINITION 21   Let E_j and U_j be the center values of the $\tilde{A}_j$ and $\tilde{B}_j$, respectively. Then

$$U_j = K_j E_j \qquad j = 1, 2, \ldots \tag{EQ 10.23}$$

Clearly K_j is bounded for all j and thus we define

$$\bar{K} = \max_j K_j \tag{EQ 10.24}$$

as shown in Fig 10.12

FIGURE 10.12   Correlation Between the Input, Output Fuzzy Sets

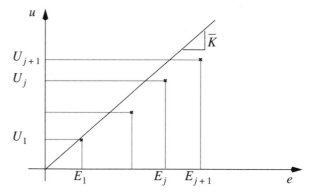

We shall further assume that $\tilde{B}_j$'s are defined in terms of the so-called LR parametrization and thus we have

$$E_j = e_k - R_j^{-1}(\mu_j)$$
$$E_{j+1} = e_k + L_{j+1}^{-1}(\mu_{j+1}) \tag{EQ 10.25}$$

The derivation of sufficient conditions for stability of the resulting closed-loop system requires expansion of Equation 10.22 with the help of the above notation. We thus have (after some algebraic manipulations)

$$u_k = K_j e_k + \frac{K_{j+1} - K_j}{\mu_j + \mu_{j+1}} e_k + \frac{\mu_{j+1} K_{j+1} L_{j+1}^{-1}(\mu_{j+1}) - \mu_j K_j R_j^{-1}(\mu_j)}{\mu_j + \mu_{j+1}} \tag{EQ 10.26}$$

Now recall Assumption 1 and let us define $\delta\mu_j$ as follows:

$$\mu_j + \mu_{j+1} \approx 1 + \delta\mu_j \tag{EQ 10.27}$$

and further define $\Delta E_j = E_{j+1} - E_j$ and $\Delta K_j = K_{j+1} - K_j$ and consider the geometry of the shaded zones in Fig 10.13 where

$$\mu_{j+1}K_{j+1}L_{j+1}^{-1}(\mu_{j+1}) - \mu_j K_j R_j^{-1}(\mu_j) \approx \delta\mu_j(1 + \Delta K_j / K_j)\Delta E_j K_j \qquad \textbf{(EQ 10.28)}$$

essentially indicating that as e_k moves between E_j and E_{j+1}, the areas of the two triangles remain nearly equal. (This is effectively a restatement of Assumption 1.)

FIGURE 10.13   Equivalence of Overlapping Fuzzy Sets

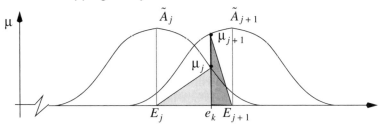

We next define $\Delta E_j = \delta E_j e_k$ and $\delta K_j = \Delta K_j / K_j$, and with the aid of Equation 10.28 rewrite Equation 10.26 as

$$\begin{aligned} u_k &\approx K_j e_k + (1 - \delta\mu_j)\Delta K_j e_k + (1 - \delta\mu_j)(1 + \delta K_j)\delta\mu_j\Delta E_j K_j \\ &= [(1 + (1 - \delta\mu_j)\delta K_j + (1 - \delta\mu_j)(1 + \delta K_j)\delta\mu_j\delta E_j)K_j]e_k \end{aligned} \qquad \textbf{(EQ 10.29)}$$

Thus the effective gain of the fuzzy controller is

$$\overline{K}_j = (1 + (1 - \delta\mu_j)\delta K_j + (1 - \delta\mu_j)(1 + \delta K_j)\delta\mu_j\delta E_j)K_j \qquad \textbf{(EQ 10.30)}$$

where $\overline{K}_j < \overline{K}$ and $j = 1, 2, \dots$.

Now the stability requirement is derived as follows:

$$\begin{aligned} x_{k+1} &\leq |f(x_k)| + |g(x_k)||u_k| \\ &= F|x_k| + G|\overline{K}e_k| \\ &= (F + G\overline{K})|x_k| \end{aligned} \qquad \textbf{(EQ 10.31)}$$

Thus for the resulting fuzzy control system to be stable, the effective control gain must be limited by

$$\overline{K} \leq (1 - F)/G \qquad \textbf{(EQ 10.32)}$$

This implies that for a given fuzzy control system to maintain stability one has to potentially compromise on the extent to which the system may be "pushed." This fact is not entirely unexpected however since it implies that in fuzzy control as in conventional control theory, one must limit the performance in order to be able to guarantee stable behavior.

10.5 A Linguistic Approach to the Analysis of Fuzzy Control Systems

In this approach, which originates in the work of M. Braae and D. Rutherford [79], the plant is assumed to take the form of a nonlinear differential equation:

$$\dot{x} = f(x, u) \qquad y = h(x, u) \qquad \textbf{(EQ 10.33)}$$

which can be stated in linguistic form as

$$\dot{x}''(t) = F(x''(t), u''(t)) \qquad y''(t) = H(x''(t), u''(t)) \qquad \textbf{(EQ 10.34)}$$

where x'' and u'' are linguistic equivalents of x, u, denoting the internal state and input to the plant, respectively. Tables 10.1 and 10.2 list the values $\dot{x}''(t)$ and $y''(t)$ as functions of $x''(t)$ and $u''(t)$, respectively, for a typical situation. Furthermore Table 10.3 lists the

TABLE 10.1 Table of $\dot{x}''$ as a New Function of x'' and u'' [a] (Reprinted from *Automatica*, 15(5), M. Braae and D. A. Rutherford, "Theoretical and Linguistic Aspects of the Fuzzy Logic Controller," © 1979, with permission from Elsevier Science.)

				u''				
	$\dot{x}''$	NB	NM	NS	ZE	PS	PM	PB
	NB	ZE	PB	NS	PB	NM	NM	NB
	NM	NS	ZE	NS	PB	ZE	NS	NB
	NS	NS	NB	ZE	PB	PS	ZE	NB
x''	**ZE**	NM	NB	NM	ZE	NB	PS	ZE
	PS	NB	NB	NM	PS	PB	ZE	PB
	PM	NM	NM	NB	PS	PM	NS	PB
	PB	NS	NB	NB	PS	PB	NB	PB

a. Stable , unstable █, oscillatory .

TABLE 10.2 Table of y'' as a Function of x'' (Reprinted from *Automatica*, 15(5), M. Braae and D. A. Rutherford, "Theoretical and Linguistic Aspects of the Fuzzy Logic Controller," © 1979, with permission from Elsevier Science.)

x''	NB	NM	NS	ZE	PS	PM	PB
y''	NB	NM	NS	ZE	PS	PM	PB

TABLE 10.3   Table of $x''(t + T)$ as a Function of x'' and $\dot{x}''$ (Reprinted from *Automatica*, 15(5), M. Braae and D. A. Rutherford, "Theoretical and Linguistic Aspects of the Fuzzy Logic Controller," © 1979, with permission from Elsevier Science.)

		$\dot{x}''$						
	$x''(t + T)$	**NB**	**NM**	**NS**	**ZE**	**PS**	**PM**	**PB**
	NB	NB	NB	NB	NB	NM	NS	ZE
	NM	NB	NB	NB	NM	NS	ZE	PS
	NS	NB	NM	NM	NS	ZE	PS	PM
x''	**ZE**	NS	NS	NS	ZE	PS	PM	PB
	PS	NS	NS	ZE	PS	PM	PB	PB
	PM	ZE	ZE	PS	PM	PB	PB	PB
	PB	PS	PS	PM	PB	PB	PB	PB

"next state values, i.e., $x''(t + T)$ as a function of x'' and $\dot{x}''$. We illustrate the use of these tables for some typical cases.

For instance, starting with x'' and u'' as Negative Medium(NM) and Positive Small(PS), respectively, the state rate of change, $\dot{x}''$, is Zero(ZE) from Table 10.1. This implies (Table 10.3) that the state remains where it presently is, that is Negative Medium(NM) and thus the plant is stable if started at the state considered. In this situation the plant output, as given by Table 10.2, is Negative Medium(NM.) On the other hand, let us start with x'' and u'' as Negative Small(NS) and Positive Small(PS) respectively. Then from Table 10.1 the state rate of change, $\dot{x}''$, is Positive Small(PS) and from the Table 10.3 the state will move to zero(ZE) where with the same input, u'' , i.e., Negative Small(PS), the state rate of change, $\dot{x}''$, will be Negative Big(NB) from Table 10.1. This implies that from Table 10.3 the state will move back to Negative Small (NS) where the above process reinitiates itself and we will have an *oscillatory behavior*. This behavior or one that is out-right unstable(see Exercise 10.2) calls for a closed-loop control strategy.

10.5.1 Analysis of Behavior of the Controller

The closed loop system incorporates a controller that in principle takes the form

$$\dot{q}'' = F_c(q'', r'', y'')$$
$$u'' = H_c(q'')$$

(EQ 10.35)

where q'' represents the controller's *internal state*. In most cases of interest the controller does not incorporate any internal dynamics, however. In other words, based on the value of the reference, r'' , and plant output, y'' , the control action, u'' , is determined in linguistic form by two tables such as Tables 10.4 and 10.5. For instance given the plant output, y'', as Negative Small(NS) and reference, r'', as Negative Medium(NM) the controller's internal state, q'', is given from Table 10.4 as Negative Big(NB) and subsequently, the control action, u'', given by Table 10.5, is Negative Big(NB), which, based on a similar analysis as was done earlier, leads to determination of the plant's state rate of change from Table 10.1 and subsequently to the determination of the next state, etc.

We note that the design of the controller is not addressed via this approach and remains heuristic. It is worth pointing out, however, that in most cases one makes use of the knowledge of the plant behavior and by effectively inverting this model develops the control strategy.

TABLE 10.4   Table of q'' as a Function of y'' and r'' (Reprinted from *Automatica*, 15(5), M. Braae and D. A. Rutherford, "Theoretical and Linguistic Aspects of the Fuzzy Logic Controller," © 1979, with permission from Elsevier Science.)

						y''		
	q''	NB	NM	NS	ZE	PS	PM	PB
	NB	NB	NS	NM	PS	NM	NS	NS
	NM	NM	PS	NB	NS	NM	NS	NS
	NS	ZE	ZE	NS	NS	NS	NS	NS
r''	**ZE**	ZE	ZE	PS	ZE	NS	NM	NS
	PS	ZE	ZE	ZE	PM	ZE	ZE	NB
	PS	ZE	ZE	ZE	PM	ZE	ZE	NB
	PB	ZE	ZE	ZE	PM	PB	ZE	NB

TABLE 10.5   Table of u'' as a Function of q'' (Reprinted from *Automatica*, 15(5), M. Braae and D. A. Rutherford, "Theoretical and Linguistic Aspects of the Fuzzy Logic Controller," © 1979, with permission from Elsevier Science.)

q''	NB	NM	NS	ZE	PS	PM	PB
u''	NB	NM	NS	ZE	PS	PM	PB

10.5.2 Closed-loop System

The resulting closed-loop system is given by

$$\dot{x}'' = F_{cl}(x'', r'')$$
$$y'' = H_{cl}(x'')$$

(EQ 10.36)

and is illustrated in Tables 10.6 and 10.7 Table 10.3 determines the next state value based on values of x'' and $\dot{x}''$. We note that for any . initial state we can verify that the path that

TABLE 10.6   Table of $\dot{x}''$ as a Function of x'' and r'' (Reprinted from *Automatica*, 15(5), M. Braae and D. A. Rutherford, "Theoretical and Linguistic Aspects of the Fuzzy Logic Controller," © 1979, with permission from Elsevier Science.)

						x''		
	$\dot{x}''$	**NB**	**NM**	**NS**	**ZE**	**PS**	**PM**	**PB**
	NB	ZE	NS	NB	NB	NB	NB	NB
	NM	PB	ZE	NS	NM	NB	NB	NB
	NS	PB	PB	PS	NM	NM	NB	NB
r''	**ZE**	PB	PB	PB	ZE	NM	NM	NB
	PS	PB	PB	PB	PS	ZE	NS	NB
	PM	PB	PB	PB	PS	PS	PS	NS
	PB	PB	PB	PB	PS	PB	PS	NS

TABLE 10.7   Table of y'' as a Function of x'' (Reprinted from *Automatica*, 15(5), M. Braae and D. A. Rutherford, "Theoretical and Linguistic Aspects of the Fuzzy Logic Controller," © 1979, with permission from Elsevier Science.)

x''	NB	NM	NS	ZE	PS	PM	PB
y''	NB	NM	NS	ZE	PS	PM	PB

the plant state takes converges toward the desired reference. For instance starting with x'' and r'' as Negative Small(NS) and Negative Medium(NM) respectively, the state rate of change, $\dot{x}''$, is Negative Small(NS) from the Table 10.6. This implies that, from Table 10.3, the state will move to Negative Medium(NM) in one step where with the same r'', i.e. Negative Medium(NM), the state rate of change, $\dot{x}''$, is zero(ZE) from the Table 10.6. This implies that state will stay at this value and thus stable behavior is achieved.

10.5.3 Reflections on Linguistic Analysis

The approach discussed above is interesting from the standpoint of its use of linguistic, rather than numerical, representation of the behavior of the plant and of the resulting closed-loop system. It is limited, however, from an analytical standpoint due to its exclusive reliance on enumeration and search as a means of assuring that the resulting system is stable. From this perspective one may argue whether the approach can be considered analytical at all. There are, however, reasons to believe that linguistic approaches may indeed lead to a more viable framework for analysis of stability of fuzzy control systems. In particular the connection between the above approach and the ideas discussed in the section on discrete-event dynamic systems suggests that such approaches may be potentially applicable to the study of fuzzy control systems.

10.6 Parameter Plane Theory of Stability

This approach is based on the work of Daley and Gill [145], which is in turn based on an approach initially developed by Siljak for application in uncertain dynamic systems [618]. The general idea of this approach is described below.

10.6.1 Basic Parameter Plane Theory

Consider a polynomial of order m:

$$F(p) = \sum_{k=0}^{m} a_k p^k \qquad \text{(EQ 10.37)}$$

where $a_k = a_k(\alpha, \beta)$ with α and β as two parameter groups. This polynomial may be the *characteristic polynomial* of a given closed-loop linear dynamic system and the intent of the method is to determine the range of α and β such that the resulting system remains stable (have all its characteristic roots in the left half -plane as it were.) Let

$$p = \sigma + j\omega \qquad \text{(EQ 10.38)}$$

Then $F(p) = 0$ translates into

$$\begin{cases} R(\sigma, \omega; \alpha, \beta) = 0 \\ I(\sigma, \omega; \alpha, \beta) = 0 \end{cases} \qquad \text{(EQ 10.39)}$$

where R and I are real and imaginary parts of F. Now Equation 10.39 can be solved for α and β provided that the Jacobian,

$$J = \begin{vmatrix} \dfrac{\partial R}{\partial \alpha} & \dfrac{\partial R}{\partial \beta} \\ \dfrac{\partial I}{\partial \alpha} & \dfrac{\partial I}{\partial \beta} \end{vmatrix} \qquad \text{(EQ 10.40)}$$

exists and is nonzero.

10.6.2 Application to Fuzzy Logic Control Systems

Since fuzzy logic controllers are generally nonlinear, in order to make use of the above method, we make use of the notion of describing function once again. In particular, consider the fuzzy logic controller

$$u = FLC(y, \dot{y}) \qquad \text{(EQ 10.41)}$$

whose describing function is in general given by

$$u = N_1 y + \frac{N_2}{\Omega} \dot{y} \qquad \text{(EQ 10.42)}$$

and thus the parameter groups can be specified as

$$\begin{cases} \alpha = \alpha(N_1, N_2) \\ \beta = \beta(N_1, N_2) \end{cases} \qquad \text{(EQ 10.43)}$$

We illustrate the use of the method in terms of an example as follows.

10.6.3 Application to Spacecraft Altitude Control

In this example from S. Daley and K. Gill's paper [145], we consider the equations of a flexible spacecraft as

$$\begin{cases} \ddot{\theta} = K_1 u \\ \ddot{\alpha} + K_3 \alpha = K_2 u \\ y = \theta + K_2 \alpha \end{cases} \qquad \text{(EQ 10.44)}$$

with the $u = FLC(y, \dot{y})$. The characteristic polynomial of the system is given by

$$F(z) = \sum_{k=0}^{m} a_k z^k \qquad z = \sigma + j\omega \qquad \text{(EQ 10.45)}$$

where $a_k = b_k \alpha + c_k \beta + d_k$ and b_k, c_k, and d_k, are functions of K_1, K_2, and K_3. In this case $\alpha = N_1$ and $\beta = N_2/\Omega$.

Now $F(z) = 0$ resolves into

$$\begin{cases} R \equiv B_1 \alpha + C_1 \beta + D_1 = 0 \\ I \equiv B_2 \alpha + C_2 \beta + D_2 = 0 \end{cases} \qquad \text{(EQ 10.46)}$$

where

$$B_1 = \sum_{k=0}^{m} b_k X_k \qquad B_2 = \sum_{k=0}^{m} b_k Y_k$$

$$C_1 = \sum_{k=0}^{m} c_k X_k \qquad C_2 = \sum_{k=0}^{m} c_k Y_k \qquad \text{(EQ 10.47)}$$

$$D_1 = \sum_{k=0}^{m} d_k X_k \qquad D_2 = \sum_{k=0}^{m} d_k Y_k$$

Now expanding $z^k = X_k + jY_k$ where X_k and Y_k are recursively defined as

$$\begin{cases} X_{k+1} - 2\sigma X_k + (\sigma^2 + \omega^2) X_{k-1} = 0 \\ Y_{k+1} - 2\sigma Y_k + (\sigma^2 + \omega^2) Y_{k-1} = 0 \end{cases} \qquad \text{(EQ 10.48)}$$

with $X_0 = 1$, $X_1 = \sigma$, $Y_0 = 0$, and $Y_1 = \omega$ then

$$\begin{cases} \alpha = \dfrac{C_1 D_2 - C_2 D_1}{B_1 C_2 - B_2 C_1} \\[2mm] \beta = \dfrac{B_2 D_1 - B_1 D_2}{B_1 C_2 - B_2 C_1} \end{cases}$$

(EQ 10.49)

Note that for $B_1 C_2 - B_2 C_1$

$$\begin{cases} \alpha B_1(\sigma) + \beta C_1(\sigma) + D_1(\sigma) = 0 \\ \sigma = \pm 1 \end{cases}$$

(EQ 10.50)

Fig 10.14 depicts the regions of the stable operation of the system. In particular, regions where 1, 2, 3, or 5 unstable roots exist are marked in the figure.

FIGURE 10.14   Parameter Plane (Reprinted from *Computers in Industry*, 7, S. Daley and K. F. Gill, "A Study of Fuzzy Logic Controller Robustness Using the Parameter Plane," pp. 511–522, © 1986 with permission from Elsevier Science.)

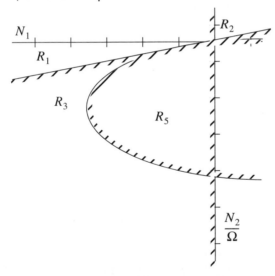

10.6.4 Reflections on the Parameter Plane Approach

The approach described above is in effect an extension of the idea developed by Mamdani based on the use of describing functions. As such it does suffer from similar limitations and in particular by assuming that plant model is linear and approximating the fuzzy controller by a describing function, it in effect constrains the method to a limited set of applications of fuzzy logic control. On the other hand it is relatively systematic and can be used to quickly determine whether a given fuzzy logic control system is potentially unstable in situations where one may not be able to perform a full-fledged analysis (via the Lyapunov approach for instance.)

10.7 Takagi-Sugeno-Kang Model of Stability Analysis

This section is based on the work of K. Tanaka et al. [579] which draws largely on linear systems theory for analytical study of fuzzy control systems. This approach which focuses on the stability of the Takagi-Sugeno model was initially studied by K. Tanaka, and Michio Sugeno at Tokyo Institute of Technology. This work was based on the idea that since the TSK model is fundamentally a *piecewise linear* model of a given dynamic system, linear system theoretic techniques could potentially be applied to the study of fuzzy control systems. Before we discuss the approach, we present a brief overview of the model.

10.7.1 Takagi-Sugeno-Kang Model Architecture

We consider the plant model to be of the form

If $x_1(t)$ is M_{i1}, and $x_2(t)$ is M_{i2}, and ... , and $x_n(t)$ is M_{in},
$$\text{then } \dot{x}(t) = A_i x(t) + B_i u(t) \qquad \text{(EQ 10.51)}$$

where $x(t) = \left[x_1(t) \ x_2(t) \ ... \ x_n(t) \right]^T$ and M_{ij}, $i = 1, 2, ..., m$, $j = 1, 2, ..., n$ are fuzzy sets defined over the domain of definition of the relevant variable. The output of the system represented above takes the form

$$\dot{x}(t) = \sum_{i=1}^{m} \mu_i(t)\{A_i x(t) + B_i u(t)\} / \sum_{i=1}^{m} \mu_i(t), \qquad \text{(EQ 10.52)}$$

where

$$\mu_i(t) = \prod_{j=1}^{n} \mu_{M_{ij}}(x_j(t)) \qquad \text{(EQ 10.53)}$$

denotes the truth value of the ith rule. The model constitutes a *blending* of multiple linear systems with (A_i, B_i) pairs of system matrices.

10.7.2 Stability of the TSK model

Formal study of the TSK model is largely due to the effort of Tanaka and Sugeno in the late 80s and early 90s. Initial effort in this direction focused on deriving conditions that if satisfied would establish the stability of a given TSK system. In this vein the basic theorem applicable to the TSK model is as follows:

THEOREM 11 The equilibrium point of the system (Equations 10.51 and 10.52) is asymptotically stable if there is a *common* positive definite matrix, P, such that $A_i^T P + P A_i < 0$ for $i = 1, 2, ..., m$.

A major issue of concern here is the difficulty of finding a common Lyapunov function that satisfies the given theorem. Thus, Tanaka and co-workers have developed alternative approaches that in principle eliminate the need for finding such a common Lyapunov function. One approach that appears promising is based on the work of P. Khargonekar, et al.[313] on uncertain linear dynamic systems. This approach, described below, essentially relies on the fact that the set of pairs (A_i, B_i) for $i = 1, 2, \ldots$ are not necessarily radically different and one may indeed find a common denominator as it were among these and express each individual pair as a deviation of the common element from its nominal value. We shall explain this idea and its usage in analytical study of the TSK model below.

10.7.3 Stability Conditions for Uncertain Linear Systems

Consider the uncertain system model:

$$\dot{x}(t) = \{A_0 + DF(t)E\}x(t) \qquad \text{(EQ 10.54)}$$

where D and E are known real matrices and $F(t) \in \overline{F} = \{F(t) : \|F(t)\| \le 1\}$. The following theorem is a statement of the main results of the approach due to Khargonekar, et al.[313].

THEOREM 12   The uncertain system (Equation 10.54) is stable if and only if
(i) A_0 is stable and
(ii) $\|E(sI - A_0)^{-1}D\|_\infty < 1$.
Corollaries of Theorem 12 are as follows:
(C1) - Consider the set of matrices $\underline{A} = \{A = A_0 + DEF: F \text{ is real}, \|F\| \le 1\}$, then there exists a positive definite matrix P such that $A^TP + PA < 0$ for all A in $\underline{A}$.
(C2) - Assume that

$$H = \begin{bmatrix} A & -DD^T \\ E^TE & -A^T \end{bmatrix} \qquad \text{(EQ 10.55)}$$

Then all real parts of the eigenvalues of H are nonzero, i.e. $\Re\{\lambda_i(H) \ne 0\}$ for all i.
We shall make use of these notions shortly. Before we do so however we shall look at a generalization of the above results due to Tanaka et al.[579].

10.7.4 Stability of Generalized Class of Nonlinear Systems

Consider the uncertain system given by

$$\dot{x}(t) = (A_0 + D\Delta(t)E)x(t) \qquad \text{(EQ 10.56)}$$

Assume that $\Delta(t) = \text{diag}\{g_1(t), g_2(t), \ldots, g_s(t)\}$ where $\alpha_i \le g_i(t) \le \beta_i$, $i = 1, 2, \ldots, s$. Let

$$M = \text{diag}\left\{\frac{\beta_1 + \alpha_1}{2}, \frac{\beta_2 + \alpha_2}{2}, \ldots \frac{\beta_s + \alpha_s}{2}\right\}$$

$$\qquad \text{(EQ 10.57)}$$

$$N = \text{diag}\left\{\frac{\beta_1 - \alpha_1}{2}, \frac{\beta_2 - \alpha_2}{2}, \ldots \frac{\beta_s - \alpha_s}{2}\right\}$$

These represent a *structure* for the uncertainty that will come into play below. The system given above can be stated as

$$\dot{x}(t) = (\bar{A} + \bar{D}\bar{\Delta}(t)\bar{E})x(t) \qquad \textbf{(EQ 10.58)}$$

where $\bar{A} = A_0 + DME$ and $\bar{D} = DN$, $\bar{E} = E$, and $\bar{\Delta}(t) = N^{-1}(\Delta(t) - M)$. The main result given in [579] is as follows.

THEOREM 13   The uncertain system (10.56) is stable if and only if

$$\left\| DN(sI - A_0 - DME)^{-1}E \right\|_\infty < 1 \qquad \textbf{(EQ 10.59)}$$

We shall make use of this theorem in the study of fuzzy control systems based on the TSK model.

10.7.5 Stability of Fuzzy Logic Control Systems

Consider the TSK model referred to above and further consider the control law

If $x_1(t)$ is M_{i1}, and $x_2(t)$ is M_{i2}, and ... , and $x_n(t)$ is M_{in},
$$\text{then } u(t) = -K_i x(t) \qquad \textbf{(EQ 10.60)}$$

where in effect the control law *parallels* the model in that it captures the "necessary" control action in terms of a structure that is derived from the plant model. The single-valued control action representing a *blending* of the rules is given by

$$u(t) = -\sum_{i=1}^{m} \mu_i(t)K_i x(t) \qquad \textbf{(EQ 10.61)}$$

Note that the same weights as the fuzzy system model are used here. The resulting closed loop system is thus given by

$$\dot{x}(t) = \sum_{i=1}^{m}\sum_{j=1}^{m} \mu_i(t)\mu_j(t)\{A_i - B_i K_i\}x(t) \qquad \textbf{(EQ 10.62)}$$

For the purpose of making use of Theorem 13 we rewrite the above successively as

$$\dot{x}(t) = Gx(t) + \sum_{i=1}^{m}\sum_{j=1}^{m} \mu_i(t)\mu_j(t)\Delta G_{ij}x(t)$$
$$= Gx(t) + \sum_{i=1}^{m} \mu_i(t)\mu_j(t)\Delta G_{ii}x(t) + \sum_{i<j} \mu_i(t)\mu_j(t)\Delta T_{ij}x(t) \qquad \textbf{(EQ 10.63)}$$

where

$$G = \frac{1}{r}\sum_{i=1}^{m} A_i - B_i K_i \qquad \textbf{(EQ 10.64)}$$

and $\Delta G_{ij} = A_i - B_i K_i - G$, and $\Delta T_{ij} = \Delta G_{ij} + \Delta G_{ji}$. Now we let

$$D = \begin{bmatrix} \Delta G_{11} & \Delta T_{12} & \cdots & & \Delta T_{1m} \\ & \Delta G_{22} & \Delta T_{23} & \cdots & \Delta T_{2m} \\ & & & \Delta G_{m-1,m-1} & \Delta_{m-1,r} \\ & & & & \Delta G_{mm} \end{bmatrix} \qquad \textbf{(EQ 10.65)}$$

$$E = \begin{bmatrix} I_n & I_n & \cdots & I_n \end{bmatrix} \qquad \textbf{(EQ 10.66)}$$

$$D(t) = \text{block-diag} \begin{bmatrix} L_{11}(t) & L_{12}(t) & \cdots & & L_{1m}(t) \\ L_{21}(t) & L_{22}(t) & \cdots & & L_{2m}(t) \\ & & & & \cdots \\ & & & L_{m-1,m-1}(t) & L_{m-1,m}(t) \\ & & & & L_{mm}(t) \end{bmatrix} \qquad \textbf{(EQ 10.67)}$$

where $L_{ij}(t) = \mu_i(t)\mu_j(t)$. We then obtain

$$\dot{x}(t) = [G + D\Delta(t)E]x(t) \qquad \textbf{(EQ 10.68)}$$

which now takes the form that is suitable for the application of Theorem 13. This step leads to the following theorem that captures the essential result developed in [579].

THEOREM 14   The TSK fuzzy control system given by (Equations 10.49 and 10.62) is stable if and only if
  (*i*) $G + DME$ is stable, and
  (*ii*) $\|DN(sI - G - DME)^{-1}E\|_\infty \le 1$
where $M = N = \text{block-diag}\left[\dfrac{d_{ij}}{2}I_n\right]$, and $d_{ij} = \max_{x(t)}\mu_i(t)\mu_j(t)$.

We can further state a corollary to the above theory as follows

COROLLARY 1   The TSK model described by (Equations 10.49 and 10.62) is stable if the real parts of eigenvalues of

$$H = \begin{bmatrix} G + DME & -DNN^T D^T \\ E^T E & -(G + DME) \end{bmatrix} \qquad \textbf{(EQ 10.69)}$$

are nonzero.

10.7.6 Reflections

The approach discussed above is clearly of interest and can be applied to many instances in which fuzzy control has been used. On the other hand, the use of the linear system theoretic framework albeit with modifications made to address the needs of fuzzy logic control systems remains questionable. One can justify the use of the linear system's theoretic approach by suggesting that the use of fuzzy logic is no more than to blend multiple linear models and thus FLCs are akin to gain scheduling controllers, in which case however, the approach constitutes a viable framework for the study of fuzzy control systems.

10.8 Summary

A great deal of work has been focused on a connection between fuzzy control and nonlinear control theory. In this chapter we considered the application of the analytical aspects of fuzzy control in terms of the methodologies developed in nonlinear control theory. Much needs to be done in this connection, however, and one may suggest that the field is still in its infancy. Nonetheless the results so far obtained are suggestive of the fact that fuzzy control can be based on a solid analytical foundation and as such one may not be skeptical of this paradigm as some have suggested in the past; stability and robustness of fuzzy control systems can be established, albeit with some effort, along the lines developed by control theorists.

Bibliographical and Historical Notes

There has been a great deal of work on analysis of fuzzy control systems beyond what is described in this chapter. For instance J. Kiszka et al. [321] developed an approach based on an extension of the idea of Lyapunov stability, which they termed *energetistic* approach. Y. Chen [122] developed an approach based on the notion of *cell-to-cell mapping* originally developed by C. S. Hsu[261]. S. Farinwata and G. Vachtsevanos[186, 608] likewise developed an approach that is also based on the concept of the cell-to-cell, suggesting that other alternative approaches may be potentially used to address the problem of stability of fuzzy control systems.

There have also been works on fuzzy relations models of fuzzy control systems not discussed in this chapter. This approach along with the linguistic analysis discussed earlier can potentially offer some significant contributions to the stability of fuzzy control systems. References to works by R. Tong [597], E. Czogala and W. Pedrycz[142], A. Cumani[141] among others are provided in the reference section of the book.

Debate on whether fuzzy logic control systems have an analytical basis has been in industry as well. For instance the feedback section of the *IEEE Control Systems Magazine*, August 1994, contains some responses on the part of fuzzy logic community.

Exercises

10.1 Consider the automotive cruise control system in Chapter 9. Determine, using the describing function technique, whether sustained oscillations do occur in the system.

10.2 Using Table 10.8 show that if given the system considered in Section 10.5 and we start with x'' and u'' as Negative Medium(NB) and Positive Small(PS), respectively, the state will diverge away, leading to unstable response.

10.3 Consider the engine model given by

$$\dot{x} = f(x)[u - g(x)]$$

(EQ 10.70)

TABLE 10.8 Table of $\dot{x}''$ as a Function of x'' and u'' (Reprinted from *Automatica*, 15(5), M. Braae and D. A. Rutherford, "Theoretical and Linguistic Aspects of the Fuzzy Logic Controller," © 1979, with permission from Elsevier Science.)

	$\dot{x}''$	ST	VS	SL	SC	FC	FA	VF
	MN	ZE	ZE	NS	NS	NM	NB	NB
	VL	PS	PS	PS	NS	NM	NM	NB
	LO	PS	PS	PS	PS	NS	NM	NM
u''	BA	PS	PM	PM	PS	PS	NS	NM
	AA	PS	PM	PB	PM	PS	ZE	NS
	HI	PS	PB	PB	PB	PM	PS	NS
	MX	PS	PB	PB	PB	PB	PS	ZE

TABLE 10.9 Table of y'' as a Function of x'' (Reprinted from *Automatica*, 15(5), M. Braae and D. A. Rutherford, "Theoretical and Linguistic Aspects of the Fuzzy Logic Controller," © 1979, with permission from Elsevier Science.)

x''	NB	NM	NS	ZE	PS	PM	PB
y''	NB	NM	NS	ZE	PS	PM	PB

whose linguistic interpretation is given by Tables 10.8 and 10.9. Devise a control strategy that stabilizes the system and show, via linguistic analysis, that the resulting system is indeed stable for any initial state and reference value.

10.3   Consider the problem of studying for exams. Table 10.10 describes the situation on a weekly basis.

TABLE 10.10   Future Grade Table (Reprinted from *Automatica*, 15(5), M. Braae and D. A. Rutherford, "Theoretical and Linguistic Aspects of the Fuzzy Logic Controller," © 1979, with permission from Elsevier Science.)

	Low	Medium Low	Medium	Medium High	High
Low	Low	Low	Medium Low	Medium	Medium High
Medium	Medium Low	Medium	Medium	Medium	Medium High
High	Medium Low	Medium	Medium High	Medium High	High

So, for instance, if the present grade is Medium Low and most of the most study is Medium, then the future grade is Medium.

a. What are the stable equilibrium points of your grade for any given level of study?

b. Suppose you study hard (High amount) for eight weeks followed by slacking off (study Low amount) for one week. What are the possible sustainable grade(s)? What if you study Hard for two weeks and a Medium amount for one week? What if you study hard all the time? What if you study a Low amount all the time?

c. Show why the above answers are different.

PART III

FUZZY LOGIC IN INTELLIGENT INFORMATION SYSTEMS

11 FUZZY LOGIC AND ARTIFICIAL INTELLIGENCE

We mentioned in Chapter 1 that fuzzy logic can be viewed from three perspectives: machine intelligence, control, and information technology. To support the machine intelligence perspective, we have pointed out many times that fuzzy logic can capture and reason about human knowledge, which is imprecise in nature. However, fuzzy logic is not the only approach for machine intelligence. Research activities aiming to enhance the level of machine intelligence have been recognized as a field in computer science known as "Artificial Intelligence" (AI).

A hypothesis that influences much of AI research is the *physical symbol system hypothesis* proposed by A. Newell and H. A. Simon in 1976 [447]. The assumption states that a system that manipulates and processes a collection of symbol structures (which is called a *physical symbol system*) has the necessary and sufficient means for general intelligent action. This assumption has a profound impact on AI research. On the positive side, this assumption encouraged the development of symbol-based knowledge representation and problem solving techniques. Frame-based representation, which became the precursor of object-oriented programming, is an example of these techniques. On the negative side, the belief that "symbol structures" are all we need to develop intelligent systems de-emphasizes the role of quantitative approaches. This belief thus became an obstacle for the AI community to take fuzzy logic seriously. As neural networks became popular in the mid-80s, the AI community became more receptive to alternative approaches to machine intelligence that do not purely rely on symbol manipulation. Since then, the core technology of artificial intelligence has been gradually broadened to include not only neural networks, but also fuzzy logic and evolutionary computing.

Fuzzy logic and AI have one common objective: to develop intelligent systems that can perform tasks and solve problems that require human intelligence. Fuzzy logic has one additional goal that is not shared by AI — to develop cost-effective approximate solutions by exploiting the tolerance for imprecision. As we discussed in Chapter 1, this is one of the major motivations for developing various fuzzy logic techniques. Much of the early work in artificial intelligence focuses on the capabilities of intelligent systems rather than cost-effective approximation to achieve intelligent behaviors. Nevertheless, cost has become increasingly important in the 1990s due to an increasing demand for the AI systems to consider resource constraints such as deadlines (i.e., constraint on computational time) and network bandwidth (i.e., constraint on communication resources).

In this chapter, we will first present a brief historic view of AI, emphasizing the major milestones and its connection to fuzzy logic. We chose to focus on the role of fuzzy logic in three areas in AI: expert systems, intelligent agents, and mobile robot navigation. Finally, we briefly mention the connection between fuzzy logic and a young subfield in AI—emotional intelligent agents (also referred to as social agents).

11.1 Artificial Intelligence

Early AI research focused on various techniques for addressing important issues raised by efforts in developing various kinds of intelligent programs. Attempts to develop programs that can understand natural languages uncovered the problem of representing meaning of words, sentences, and contexts. This motivated the development of various *knowledge representation and reasoning* schemes such as frame-based representations (also known as semantic nets), which are the precursor of object-oriented paradigm.

Efforts to develop intelligent programs for playing chess and other types of games uncovered the *combinatorial explosion* problem — the number of possible situations that need to be analyzed by these programs increases exponentially as the "think-ahead" steps of the program increase. This led to various *search strategies* that aim to focus the analysis on more interesting situations. One of these search techniques, α-β pruning, has been implemented in special-purpose VLSI chip, which was used to build the famous chess machine — IBM's Deep Blue (a successor of a chess machine called Deep Thought developed at Carnegie Mellon University). In a 1997 "man versus machine chess match" that caught the attention of the entire world, Deep Blue challenged and beat the world chess champion Garry Kasparov in a six-games chess match (won 2, lost 1, tied 3). However, the AI community, and the public in general, realized that this result does not suggest that computers are more intelligent than human beings. Human intelligence is still far superior to machine intelligence. We just cannot do what God does.

An important idea adopted early by AI is that an intelligent system for performing a specific task (e.g., diagnosis) in a problem domain (e.g., lung diseases) can often benefit from knowledge about the problem domain. This idea led to the development of many *expert systems*, which are systems that are capable of performing a specific task that requires human expertise. Real-world examples of these systems include Microsoft's WordWizard, which offers intelligent recommendations regarding desktop publishing to users of Microsoft Word, and Digital Equipment Corporations XCON, which configures computer systems based on customer orders. One of the main challenges in developing expert systems is the fact that human experts often rely on judgmental knowledge (e.g., rules of thumb) that they learned from their experience. Even though the experts know that judgmental knowledge is not as certain as first principles, physical laws, or provable theories, they find if useful in drawing plausible conclusions. The principle of expert system technology is to capture human expert knowledge in a way such that a computer can (1) reason about this knowledge, and (2) arrive at a conclusion similar to that of the human experts. A key element in this reasoning process is a scheme for representing uncertainty in the knowledge, for making inferences using uncertain knowledge, and for combining multiple inference results. Such a scheme is often referred to as a scheme for "reasoning under uncertainty," or for "uncertain management." Fuzzy logic offers one of the foundations for developing schemes for reasoning under uncertainty in expert systems.

Approaches for dealing with uncertainty in AI fall into four categories: (1) Bayesian probabilistic approaches, (2) approaches based on the Dempster-Shafer (DS) theory, (3) fuzzy logic approaches, and (4) heuristic approaches. Bayesian probabilistic approaches use probability to represent the degree of certainty and develop their probabilistic inference scheme based on Bayes' theorem. The D-S approaches use an interval probability to represent the degree of certainty, as we mentioned in Chapter 7. Instead of using Bayes' theorem, the DS approaches use the Dempster rule of combination to aggregate information collected from independent sources. The fuzzy logic approaches usually use fuzzy rules to describe knowledge that is uncertain. The degree of certainty is represented by a possibility measure. The most well-known heuristic approach for the management of uncertainty is the Certainty Factor of MYCIN, which is one of the pioneer expert systems [536].

Our discussion in this chapter will focus on fuzzy logic-based approaches for uncertainty management in AI.

11.2 Fuzzy Logic in Frame-based Representation

A frame-based knowledge representation scheme uses a frame, sometimes referred to as a concept or a class, to represent a collection of objects. The typical properties of a class are described using slots, which are also referred to as attributes. Typically, a slot has a name and a value. Even though frame-based representation has become one of the most useful knowledge representation techniques in AI, it can not represent and reason about imprecise attribute values. Suppose we wish to represent the concept of healthy family. Most of the properties of such a class are vague: having **infrequent** clinical visits, having **low cholesterol** meals, having **regular** exercises, etc.

Using fuzzy logic, a frame-based system can be extended to represent these vague properties of a class. To do this the slot of a frame can take a linguistic term as its value, and the meaning of the linguistic term is defined by a memberships function. For example, to represent "infrequent clinical visits" as a common property of healthy family we can use a slot called `clinical visits` whose value can be in linguistic terms such as frequent, infrequent, very seldom, more or less seldom. The meaning of these linguistic terms are defined by their memberships functions. Therefore, such a fuzzy logic-based extension to frame-based systems uses a dual representation to represent imprecise properties of a class: a symbolic representation using linguistic terms, and a quantitative representation using membership functions. In fact, a slot in such a system becomes a linguistic variable. It is obvious that the same kind of extension can also be made to object-oriented systems [26, 206].

Since linguistic terms can be used to represent imprecise properties of a class, an instance can be partially in a class. Therefore, the class membership relationship between a class and an instance becomes a matter of degree. For example, a particular family may be partially a healthy family and partially a rich family. Partial class membership complicates an important issue in frame-based systems: the handling of multiple conflicting inheritance. In a conventional frame-based system, one way to deal with multiple conflicting inheritance is to establish a priority among superclasses of a class. These priorities determine the order in which the parent classes are searched for inheriting values to the instance. When the class-instance relationship becomes a matter of degree, these membership degrees should also affect the selection of the actual source of inheritance along multiple inheritance paths. Several research ideas for this problem have been proposed [206, 81].

Generalizing frame-based systems using fuzzy logic also has significant impact on the reasoning about class subsumption and class membership. In early frame-based systems and object-oriented systems, the subsumption relationship between classes (i.e., IS-A links) and the membership relationship became an instance in these classes (i.e., instance-of links) are all asserted by the developer of a knowledge base. Due to a problem about the meaning of IS-A links in these early frame-based systems [81], a family of more principled frame-based systems, called term subsumption systems, were later developed [82]. In a term subsumption system we can describe the defining characteristics of a class. Based on these class definitions, these systems can infer the subsumption relationship between classes. They can also infer the classes to which an instance should belong. Fuzzy logic can be used. Fuzzy logic has been used to generalize the syntax and the semantics of term subsumption languages so that (1) imprecise defining characteristics can be used to define a class (i.e., a fuzzy class), and (2) subsumption between fuzzy classes can be inferred from the semantics of their definitions [665].

11.3 Fuzzy Logic in Expert Systems

FIGURE 11.1   The Basic Architecture of Expert Systems

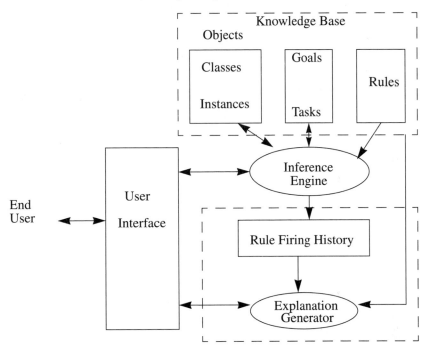

The basic architecture of an expert system is shown in Fig 11.1. It consists of four major components: (1) a knowledge base, (2) an inference engine, (3) an explanation facility, and (4) a user interface module. The knowledge base stores relevant knowledge obtained from human experts, which is often represented in the form of rules. The knowledge base also stores factual information about the system's specific problem domain. The inference engine uses rules and this factual information to infer additional information or to construct pieces of the solution. In either case, the execution of rules modifies objects in the knowledge base. The knowledge base often contains descriptions about goals/subgoals or tasks/subtasks that reflect the human expert's high-level problem solving strategy. The structure of goals and tasks can be used to organize rules into groups based on their functionality. An expert system not only needs to perform a task as competent as a human expert, it also needs to be able to explain the rationale of its conclusions. Hence, an expert system often has an explanation component, which generates explanations by tracing the rules fired by the inference engine.

The user interface component of an expert system interacts with the user to collect input data (e.g., symptoms of a patient for a medical diagnosis system), for displaying conclusions/solutions, and for providing explanations upon a user's request.

As was mentioned in Section 11.3, expert systems often rely on human experts' judgmental knowledge (i.e., heuristics) to make decisions. This judgmental knowledge is usually uncertain. Hence, an expert system needs to be able to reason about how much *certainty* is associated with a particular conclusion drawn by a piece of heuristic knowledge within a particular context. Various fuzzy logic-based approaches have been developed for achieving this kind of reasoning under uncertainty. They differ in the types of uncertain knowledge they emphasize.

There are at least four kinds of uncertainty in expert systems. First, the situation for applying a piece of knowledge may not have a well-defined sharp boundary. For instance, the symptom that "blood pressure is high," which is often used in medical diagnosis, does not have a clear sharp boundary. Second, the knowledge that associates observations to hypotheses can be uncertain. Third, the nature of the conclusion can be imprecise. Typically, this occurs when the conclusion is about something that can be quantified (e.g., severity of a disease, a predicted stock market index, etc.) Fourth, the data to which the knowledge is applied may be uncertain.

Even though many techniques have been developed for dealing with various types of uncertainties [530], fuzzy logic is uniquely suitable for handling the first and the third kind of uncertainty, because both of them are imprecise by nature. Most fuzzy logic-based expert systems use fuzzy if-then rules, discussed in Chapter 5, to associate vague conditions with a conclusion, which can be precise or imprecise. These rules are typically in the following form:

- IF $(A_1(o_1)$ is $v_1)$ AND $(A_2 (o_2)$ is $v_2)$ ......$(A_k (o_k)$ is $v_k)$
  THEN it is likely (τ) that $A_{k+1} (o_{k+1})$ is v_{k+1}

where A_i, o_i, and v_i denote attributes, objects, and values, respectively, while t denotes the truth value of the implication rule. The value v_i can be a fuzzy set. The truth value τ can be a number [0, 1] or a fuzzy subset of [0, 1]. For example, a medical diagnostic system for checking liver function has the following rule [584]:

- IF GOT is Medium AND
  GPT is Medium AND
  Previous GOT is Very High AND
  GOT > GPT
  THEN it is *somewhat likely* that liver function is abnormal.

GOT and GPT in the rule are two blood test results for checking liver function profiles. All attributes in this system refer to one identical object—the patient being diagnosed; therefore, the object is not explicitly represented in the rule. Terms such as Medium and Very High in the rule are fuzzy sets to describe imprecise conditions. The truth qualifier somewhat likely is a fuzzy set that characterizes the uncertainty of the rule.

The foundation of fuzzy rules in expert systems is fuzzy implication rules introduced in Chapter 6. Hence, the inference of these rules is based on approximate reasoning using fuzzy implication functions as described in Section 6.3.

In addition to fuzzy rules, fuzzy logic has also been used to develop other reasoning schemes for diagnostic expert systems. An important system in this category is a medical expert system called CADIAG developed by K. P. Adlassnig [6]. CADIAG (Computer-Assisted DIAGnosis) uses two measures to describe the relationships between symptoms and diseases: the frequency of occurrence and the strength of confirmation. These measures are derived from a large patient database. For instance, CADIAG's knowledge base for diagnosing rheumatic diseases was constructed from about 3500 patient records in a rheumatological hospital [5]. The knowledge base includes 189 rheumatological diseases, 905 symptoms, findings, test results, and about 20,000 symptom-disease relationships (i.e., fuzzy rules).

CADIAG's inference scheme is based on composition of fuzzy relations [4]. Let R_{SC} and R_{fC} denote fuzzy relations recording strength of confirmation and frequency of occurrences between a set of symptoms and a set of diseases. These relations are used to confirm, exclude, and rank diagnoses. For instance, a patient's confirmation degrees of diseases are computed from the composition of fuzzy relations as follows:

$$R_{CD} = R_S \circ R_{SC}$$ (EQ 11.1)

where R_S *and* R_{CD} denote one-dimensional fuzzy relations of the patient's symptoms and confirmed diagnoses respectively. The rational of the composition is that if the patient has a symptom s_i to α degree, and the symptom has a strength of confirmation β for disease d_j, then the patient is likely to have disease d_j to the degree min $\{\alpha, \beta\}$. Similarly, if a symptom s_j always appear for a disease d_j but we do not observe the symptom in a patient, we can conclude that the patient does not have the disease d_j. This leads to the following composition for computing the degree that a disease should be excluded (i.e., rejected):

$$R_{ED} = (1 - R_S) \circ R_{DC}$$ (EQ 11.2)

where R_{ED} denotes a one-dimensional fuzzy relation of excluded diagnoses for a given patient. CADIAG's inference scheme described above can also be justified using fuzzy implications and generalized modus ponens. We will leave this as an exercise.

CADIAG's inference scheme and generalized modus tollens for ranking diagnoses is a weighted sum of R_{CD} and R_{ED}. Even though it involves an ad hoc assignment of weights, it is an effective heuristic approach for employing a large amount of complex diagnostic knowledge into an expert system.

An example of CADIG's rule is given below:

- IF the patient shows low back pain,
  a limitation of motion of the lumbar spine,
  a diminished chest expansion, and
  the patient is male, and
  is between 20 and 40 years old
  THEN the diagnosis may be ankylosing spondylitis
  WITH the frequency of occurrence being very often (0.9) and
      the strength of confirmation being strong (0.8).

CADIAG has been extensively tested using more than 300 real patient cases. For instance, CADIAG-2/ RHEUMA, a second generation of CADIAG systems for diagnosing rheumatic diseases, was able to correctly diagnose 81% of the test cases as either confirmed (68.2%) diseases or diagnostic hypotheses (12.8%) [7].

OMRON uses fuzzy logic in a health management expert system that helps a large corporation by providing a personal health diagnosis and health management plan for its employees [266]. The health management system uses 500 fuzzy if-then rules to capture medical knowledge of human experts regarding health diagnosis, life cycle evaluation, diet, physical fitness, and major diseases. The system provides personal health diagnoses and assists personal health management planning for employees in an organization. The use of fuzzy logic enables the system to process patients' inputs that are described using vague linguistic terms and to implement doctors' linguistic and complex decision-making processes. In addition, it enables the expert system to represent medical knowledge in a smaller number of rules than a conventional expert system. Fuzzy sets represent abstractions of clinical data (e.g., numerical data are low, normal, or high) as well as degree of severity of diseases. The system is able to support more than 10,000 individuals in an organization.

We have introduced the application of fuzzy logic in medical diagnostic expert systems. Medical diagnosis falls into a category of expert systems known as *heuristic classification systems*. A heuristic classification system classifies the descriptions of a given problem into one of preenumerated categories using heuristic knowledge obtained from human experts. Other examples of heuristic classification expert systems include systems for financial loan approval (i.e., classifying loan application data into classes such as approved, not-approved, insufficient data, etc.), systems for engine troubleshooting (classifying engine symptoms into one or multiple likely faults). Fuzzy rules can be used to develop these expert systems in a way similar to that of medical diagnostic systems.

11.3.1 Fuzzy Logic in Synthesis Type Expert Systems

The second category of expert systems is of a design and synthesis type. These systems construct (synthesize) a solution that satisfies certain constraints and achieves certain desired goals. A typical example of expert systems in this category is computer configuration systems. Others include scheduling systems and engineering design support systems.

The primary role of fuzzy logic in design and synthesis type expert systems is to represent imprecise constraints and imprecise objectives. Alternative designs/solutions generated by the system can thus be evaluated by calculating the degrees all fuzzy goals are satisfied. We will illustrate these using an expert system shell for design and configuration tasks (called KONWERK) and its application to the passenger cabin layout of an AIRBUS A340 aircraft [222, 328].

KONWERK's knowledge base includes a fuzzy frame-based representation of domain objects, fuzzy design constraints, and fuzzy design goals. It also includes a set of design optimization methods such as depth-first search, branch and bound, fuzzy multicriteria decision making, etc. For instance, KONWERK could use "branch and bound" optimization to generate a set of alternative designs that satisfy all constraints. These designs can then be evaluated and ranked based on fuzzy multicriteria decision making. We describe below an expert system implemented using KONWERK for the passenger cabin layout of an AIRBUS A340 aircraft.

A cabin layout defines the number, the type, the arrangement, and the position of cabin interior components (e.g., passenger seats, lavatories, galleys, and cabin attendant seats) in the passenger cabin of an aircraft. A good cabin layout needs to satisfy user requirements (e.g., the distance between two rows of seats), government rules regarding aviation safety (e.g., space requirements around emergency exit doors), and technical constraints (e.g., mounting restrictions for a galley). In addition, a good cabin layout needs to consider optimality criteria such as number of seats, comfort level, cost, delivery time (i.e., the time from ordering a cabin interior component to delivering it).

Some of the user requirements or optimality criteria for cabin layout can be imprecise. We give some examples below:

- The distance between two rows of seats (i.e., called seat pitch) should be **comfortable** for the passenger.

- The delivery time for required cabin interior components should be **fast**.

The meaning of linguistic terms "comfortable" and "fast" used above can be characterized by membership functions. Fig 11.2 shows the membership function of fast delivery time. These fuzzy goals not only can be used to evaluate and rank a set of feasible solutions generated by KONWERK's search and optimization module, they can also be used to relax crisp goals (e.g., delivery time should be less than 11 months) if the search optimization module fails to find feasible solutions using the crisp goals.

FIGURE 11.2   The Membership Function of Fast Delivery Time

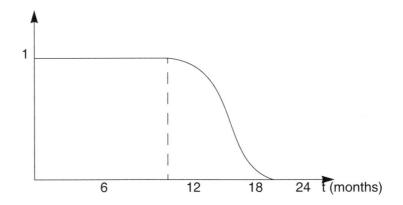

11.3.2 Fuzzy Expert Systems Shells

Expert system shells are software tools that facilitate the development of expert systems. They provide an inference engine, tools for creating and editing the knowledge base, and tools for tracing and debugging rules. Some shells also provide a facility for explanation generation and user interface construction. A fuzzy expert system shell provides additional support in the inference engine for fuzzy reasoning. We describe a few representative fuzzy expert system shells below.

P. Bonissone and his colleague at GE Corporate Research and Development developed a software tool for expert systems called PRIMO that integrated reasoning under uncertainty with default reasoning in AI [15]. PRIMO, like its predecessor RUM [66], uses a combination of fuzzy logic and interval logic to represent and reason about uncertainty. This approach has been successfully demonstrated in two defense-related applications: the Pilot's Associate (an expert system assisting the pilot of a fighter aircraft) and the Submarine Operational Automation System (an expert system that automates the navigation of submarines). One of PRIMO's features is that it allows the expert system designer to choose the fuzzy conjunction/disjunction operators from a large set of built-in operators [63]. By doing this, PRIMO can support a variety of uncertainty reasoning schemes (including probabilistic reasoning) within a uniform framework.

FRIL (Fuzzy Relational Inference Language) is a commercially available fuzzy expert system shell developed by J. Baldwin [21]. It supports possibilistic reasoning and an interval probabilistic reasoning scheme, called mass assignment theory. Rules in FRIL are similar to rules in logic programming (i.e., horn clauses in PROLOG). The inference engine of FRIL extends unification in logic programming to allow partial matching. FRIL has been used to develop fuzzy expert systems as well as fuzzy logic controllers.

FEShell developed at LIFE supports both fuzzy rules and fuzzy frame/objects [583]. System Z-II, developed by K. Leung and W. Lam, integrates possibilistic reasoning and MYCIN's certainty factor [372]. L. Godo, R. Lopez de Mantaras, and their colleagues developed a fuzzy expert system shell called MILORD [210].

11.4   Intelligent Agents

A unified framework for AI systems called *intelligent agents* has been evolved in the 90s. Agent is a software system that interacts with an external environment by performing appropriate actions to the environment based on information received from the environment. An intelligent agent is an agent that demonstrates certain intelligent behaviors in its actions. S. Russel and P. Norvig classified agents into four classes [517]: (1) simple reflex agents, (2) agents that keep track of the world, (3) goal-based agents, and (4) utility-based agents. We briefly summarize these four types of agents below.

A *simple reflex agent* uses condition-action rules to map sensor inputs directly to action. *Agents that keep track of the world* extend the simple reflex agent with an explicit description of the external environment, which is sometimes referred to as a state description or a world model.

The agent keeps track of changes in the environment by updating the internal state using three kinds of information: (1) sensor input of the agent, (2) information about how the world evolves independently of the agent, and (3) information about how the agent's own actions affect the world. The motivation for this extension to the simple reflex agent is that the agent's sensor input often does not provide complete information about the environment. In fact, even if an agent attempts to keep track of the world, it may not be completely certain about the environment due to dynamic changes to the environment. Fuzzy logic and other schemes for reasoning under uncertainty (e.g., probabilistic reasoning) can contribute to this important issue in intelligent agents, as we will see later.

A *goal-based agent* is an agent that extends the second type of agents with an explicit representation of its goals. The goals of the first two types of agents are implicitly incorporated in the condition-action rules. Explicitly representing these goals enables an agent to adapt to the environment by dynamically planning its course of action for accomplishing the goals. Even though these agents improve the adaptability of the first two types of agents, they are still limited in dealing with conflicting goals. That is due to the fact that the notion of achieving a goal in AI is typically a black-and-white matter. A goal-based agent that encounters conflicting goals has to either abandon some of its goals or to modify the goals [640]. It cannot, however, explore an effective trade-off among these goals by considering the possibility of partially satisfying them.

In order to do this, we need the fourth type of agent — the *utility-based agent*. A utility-based agent extends the black-and-white notion of goal accomplishment into a continuum called the utility measure. Such an agent can then evaluate the utility of alternative actions or strategies. In the case that the outcome of an action is uncertain, the agent needs to consider the likelihood of possible outcomes as well as the utility (including cost) of these outcomes. An agent is said to be *rational* if it always chooses a decision/action that maximizes its overall expected utility. Fig 11.3 shows a basic architecture of a utility-based agent. The rectangles in the figure depict knowledge and information the agent should have, while the circles depict reasoning and information processing components of the agent.

FIGURE 11.3   The Architecture of a Utility-based Agent

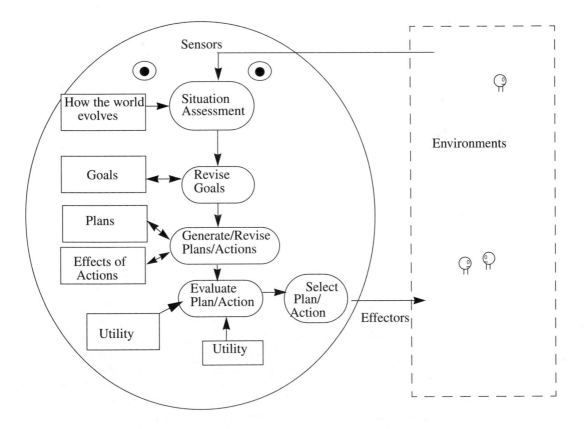

The environment may also include other agents, as shown in Fig 11.3. A set of agents that communicate with each other and collaborate in some capacity forms a *multi-agent system*.

Major techniques that support multi-agent systems include

- Standard languages for agents to communicate with each other. KIF is a standard knowledge representation language. KQML describes the syntax of messages and protocols of communication between agents.

- Organizational structure of agents can facilitate collaboration and reduce communication overhead (e.g., federation system with facilitators, broker agents).

- Problem solving architectures have been designed specifically for multiple agents to share information and to collaborate in accomplishing common goals.

11.5 Fuzzy Logic in Intelligent Agents

What roles can fuzzy logic play in intelligent agents? While the answer to this question is still evolving, we can identify at least four major areas that fuzzy logic can contribute to intelligent agents:

- Fuzzy rules can be used to implement simple reflex agents, especially if possible actions are in a continuous domain (e.g., turning direction of a mobile robot).

- Fuzzy logic can serve as a bridge between goal-based agents and utility-based agents. Fuzzy logic can be used to describe fuzzy goals that can be partially satisfied. This enables an agent to maximize the overall accomplishment by partially satisfying these goals that are conflicting.

- Fuzzy logic can be used to deal with the uncertainty about the environment, especially if the uncertainty is imprecise in nature.

- Fuzzy logic can be used in an agent for situation assessment to combine information from multiple sources, especially if the information is quantitative in nature.

- Fuzzy logic can be used to model emotional intelligence in agents.

11.5.1 Fuzzy Rules in Simple Reflex Agents

As we have discussed in Chapter 5, fuzzy mapping rules approximate a functional mapping. It is thus natural to use these rules to describe the knowledge for simple reflex agents to map sensor data to action output. Compared to condition-action rules in AI, fuzzy rules can often generate a smooth mapping using fewer rules (or, equivalently, a smoother mapping using the same number of rules). This feature is important if the possible actions are in a continuous domain. For example, a reflex agent for a mobile robot's navigation control needs to map inputs from sensors (e.g., sonar) to control the traveling speed and the turning direction. Since it is desirable for the mobile robot to maintain a smooth ride, fuzzy logic-based reflex agents are useful for mobile robots navigation control. We will discuss the application of fuzzy logic in mobile robot in Section 11.6.

11.5.2 Fuzzy Logic Bridges Goal-based Agents and Utility-based Agents

A goal-based agent uses symbols to express its goals, while a utility-based agent uses a utility function to describe the desirability of various possible outcomes that partially accomplish the goals. The symbolic representation and the quantitative representation of goals are complementary — the former enables goal-driven symbolic reasoning such as planning (i.e., generating a course of action that accomplishes a goal for the environment the agent is in), the latter supports a principled trade-off analysis about conflicting goals and likely effects of actions. A bridge between the two representations is needed if an agent needs to possess capabilities for both goal-driven symbolic reasoning and utility-based trade-off analysis. Fuzzy logic can build such a bridge, because it is linked to symbolic reasoning through its linguistic expression and to quantitative analysis through its membership function representation. The essence of the bridge is the notion of *fuzzy goal*.

A *fuzzy goal* is a goal that can be partially accomplished. The degree that the goal is accomplished in various situations can be specified by a mapping, in theory, from all possible situations to the degree the goal is accomplished. In practice, however, its more efficient to describe the mapping as a function from a set of properties of the situation to the degree of goal achievement (i.e., a number in [0,1]). We will call this function an *achievement function* of a goal. For example, the Sendai subway controller mentioned in Chapter 1 has several goals (i.e., control objectives):

* The ride should be comfortable.

* The fuel economy should be high.

* The precision of the train stopping location (relative to the marked stopping positions in the train station) should be high

* The train should be on schedule.

These goals are imprecise in nature and can be satisfied to different degrees. Therefore, they are fuzzy goals. For instance, the stopping precision goal can be represented quantitatively as a mapping from d to [0, 1], where d is the distance between the train's stopping position to its target position marked at the station. Fig 11.4 shows the achievement function of the stopping accuracy fuzzy goal.

FIGURE 11.4   The Achievement Function of the Stopping Accuracy Fuzzy Goal

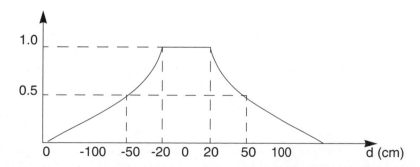

Once we explicitly represent a fuzzy goal in an intelligent agent, we can use it to evaluate alternative actions. Different actions lead to different possible future situations, which may accomplish the agent's goal to different degrees. Hence, it is important not only to estimate the degree each goal is likely to be accomplished by a candidate action, but also to combine these estimations for different goals to an overall measure about the expected utility of the candidate action. Such a measure serves as the basis for the agent to select its actions from the candidates. We use the Sendai subway controller to illustrate this idea.

FIGURE 11.5   The Intelligent Agent for Sendai Subway Control

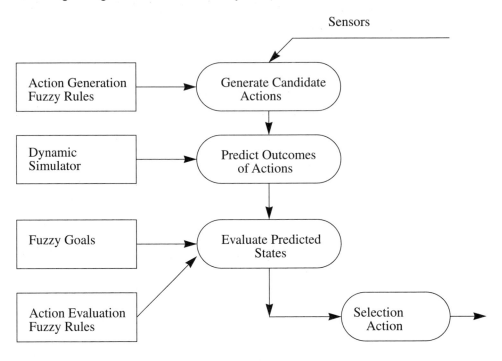

Fig 11.5 shows the architecture of the intelligent agent in the Sendai Subway controller. To be fair, we should point out that the concept of intelligent agents was not yet born when Dr. Seiji Yasunobu and other engineers in Hitachi designed the Sendai Subway controller. However, the system is an excellent example of utility-based agents. The architecture consists of four major components: (1) a generator that proposes candidate decision, (2) a prediction module that predicts the impact of the decision on future system states, (3) an evaluator that evaluates the predicted future states based on the fuzzy goals, and (4) a selection module that selects the best decision based on the evaluation result. Two different sets of fuzzy rules are used for generating candidate actions and for evaluating future states of those actions. From an intelligent agent viewpoint, a novel aspect of the architecture is its use of a dynamical system model for the prediction purpose.

The architecture is related to predictive control. However, it differs from predictive control in that goals are explicitly represented and used for evaluating alternatives. Other applications of the architecture include a road tunnel ventilation control system [196] and a large-scale fuzzy financial support system [663].

11.5.3   Reasoning about Uncertainty of the Environment

An intelligent agent often does not have complete information about the environment due to several reasons. First, its sensing capability is limited. For example, a mobile robot with a sonar sensor can identify obstacles, but not their types. Second, certain information about the environment cannot be directly observed, but rather needs to be inferred. For instance, an agent that monitors a machining tool in a manufacturing environment may not be able to directly observe the wearing condition of the tool but can infer it from rotation speed, vibrations, cutting time, etc. Third, the environment may change dynamically. For example, an obstacle on the path of a mobile robot may be introduced by another mobile robot.

Due to incomplete information and/or the dynamic nature of the external environment, an intelligent agent often needs to reason about the environment under uncertainty. Reasoning of this kind can be best illustrated by two kinds of intelligent agents: diagnostic agents and situation assessment agents.

A *diagnostic agent* identifies the causes of a set of symptoms and anomaly. The environment for these agents is the "system" to be diagnosed (e.g., a patient for a medical diagnosis agent). Diagnosis thus involves reasoning about the unknown state of the environment. Diagnostic expert systems that we discussed earlier can be viewed as this type of agent.

A *situation assessment agent* needs to infer the high-level situation of the environment from various lower-level observations. For instance, an incident detection agent monitors the traffic data on a transportation network for detecting the occurrence of traffic incidents. Such an agent obviously reasons about the state of the dynamic environment, which is a transportation network.

We will use intelligent analysis for military reconnaissance missions to illustrate the application of fuzzy logic in reasoning about the uncertainty of the environment.

11.5.4   Case Study: Intelligent Analysis of Military Reconnaissance Missions

The prototype system described in this case study was developed by J. R. Surdu, D. Ragsdale and B. Cox. They extended the previous approach at the United States Military Academy by including vegetation, weather data information, and operational knowledge in the reasoning process [488].

The *reconnaissance mission* is one of the most challenging missions that a small military unit is required to perform. Many of the intelligence analyses performed in support of this type of mission are accomplished under a great deal of uncertainty. Additionally, the manner in which the various underlying aspects of the intelligence analysis problem interact is very complex. Also contributing to the difficulty of planning a reconnaissance mission is the great risk to the well-being of the personnel who make up the recon team. The risk involved is attributable to the fact that in order to successfully accomplish a reconnaissance mission, a team must normally occupy a position that is close to an enemy force. The reconnaissance planner therefore has two conflicting goals to consider when he selects positions for a reconnaissance mission. On one hand, he must select potential positions that are close enough for the team to gather all required information. On the other hand, he must select a position that does not expose the recon team to undue risk. These opposing goals can be seen clearly in the doctrinal fundamentals of reconnaissance missions as articulated in a related Army Training Manual [7]. The goals of a reconnaissance mission are (1) gain all required information, (2) avoid detection.

This trade-off between potential for information gain versus risk of being detected is the most important consideration that the operational planner must take into account when selecting potential positions from which to conduct a recon. If the planner can successfully balance these two competing goals, there is a far greater likelihood that a particular mission will be successful.

From an intelligent agent's perspective, the action of an agent assisting the reconnaissance mission is to generate an assessment about the suitability of locations for a reconnaissance mission. The two goals described above are agent's goals. Since these two goals are conflicting and require trade-off analysis, utility-based agents are more suitable for this problem than are goal-based agents.

Conventional rule-based or statistical approaches are possible solutions to this problem. However, a fuzzy logic approach provides an alternative technology that alleviates many of the difficulties conventional approaches have in handling uncertainty, imprecision, and ambiguity. Furthermore, fuzzy logic is a viable means to perform reasoning using a combination of spatial information (vegetation, roads, etc.) and subjective operational knowledge in locating potential observer positions. Finally, fuzzy logic methodologies provide a "communication bridge" between domain experts and the system through fuzzy linguistic variables.

11.5.4.1 Architecture

The system includes three modules. The first two modules model risk and information gain associated with the recon locations. Each location is assigned a rating describing the risk and information gain potential of that location. The final module evaluates these ratings in combination with the commander's willingness to accept risk. Furthermore, the analyst bounds the space of possible locations by defining a minimum and maximum radial distance from the enemy location, which results in a doughnut-shaped area (annulus).

From the viewpoint of a utility-based agent, the output of the first two modules can be viewed as utility measures of a location from the viewpoint of information gain and risk, respectively. The third module combines these two utility measures into an overall utility evaluation result.

FIGURE 11.6   Graphical Output of Reconnaissance Analysis

(From D. J. Ragsdale et al., "A Fuzzy Logic Approach for Intelligence Analysis of Actual and Simulated Military Reconnaissance Mission," *IEEE Int. Conf. on Systems, Man, and Cybernetics, pp. 2590-2595,* © 1997 IEEE)

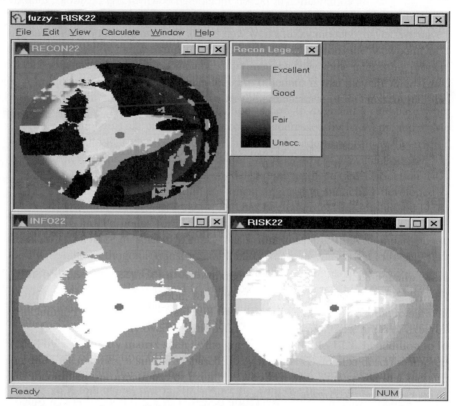

After the user inputs all the necessary information, the analysis process is initiated. After the analysis is complete, the results can be overlaid on any of the base maps supported by the system. The assessment result is displayed in color, where green locations are the most favorable and the black areas are the least favorable. An example output is illustrated in Fig 11.6. This figure contains the results of the three modules. The top image is the overall assessment and the bottom two are the results from the information gain and risk modules, respectively. The information gain image uses dark shades of gray in areas that are not conducive to acquiring the needed information. The risk module uses lighter shades of gray to indicate areas of high risk.

The information gain module and the risk assessment module each consist of submodules that produce intermediate analysis result. Fig 11.8 shows the architecture of the system. Each module is shown as a rectangle with inputs connected from the left and output produced to the right. We describe each one of the three high-level modules below in further details.

Risk Assessment Module

Risk is a measurement that describes the likelihood of loss-of-life occurring during the execution of a mission. The level of acceptable risk is dictated in the overall mission statement. A combination of factors including known enemy positions, their detection capabilities, and the geographical characteristics (surface contour, vegetation, etc.) of the area all play a role in the evaluation of risk. Risk is modeled by considering the threat associated with a position and how visible a position is relative to the enemy. Hence, the risk assessment module consists of two submodules: one of assessing the observer's visibility (from the enemy perspective) and the other for assessing the danger of the location considering enemy threats as well as friendly deception capabilities. Positions that are unobservable by known enemy locations are assumed to be less risky. The objective of this module is to assign low ratings to close locations that are safe and that allow minimal observation by the enemy. Areas that are known enemy positions or that are considered highly visible are assigned high-risk ratings. Environmental factors such as fog or precipitation, canopy cover provided by vegetated areas, terrain obscuration, and darkness all lower the risk of a specific location. Conversely, enemy positions with higher security and more elaborate detection devices will increase the overall risk. However, some of these factors also reduce the visibility of the target, which in turn reduces the information gain potential of the location.

Information Gain Module

The degree to which a specific location supports information gathering is a complex function that takes into account factors such as distance to the target, terrain masking (concealment due to terrain features), weather and astronomical data, vegetation, target illumination, and friendly observation capability. The assessment of the information gain potential of a given position is determined through two submodules: one for assessing the target visibility from the location, the other for assessing the target observability.

The visibility of the target is determined by considering the weather, the terrain and vegetation masks, the target and the ambient illumination, and the obscuring effects of the weather. The observation potential of a given location is a function of the distance to the target and the equipment that the recon team will employ to enhance their ability to observe the target. The weather obscuration is determined by the amount of fog and/or precipitation that is expected at the time of the mission.

Mission Assessment Module

In order to assess the suitability of each location that surrounds a particular reconnaissance target, a commander must take into account the competing goals of minimizing risk while maximizing the potential for information gain. His assessment of how to reconcile these two goals is dependent on the amount of risk that he is willing to accept.   For example, if the information that will be acquired during the mission is highly critical, then the commander would be more likely to accept a higher degree of risk. However, if there might be serious negative ramifications associated with a compromised mission, then he would be less likely to accept a high level of risk. Consequently, the overall assessment of a location is a function of the risk assessment, the information assessment, and the commander's risk acceptance threshold. Fig 11.7 shows the surface of this functional mapping for a particular risk acceptance threshold.

FIGURE 11.7   Inference Surface for Mission Assessment

(From D. J. Ragsdale et al., "A Fuzzy Logic Approach for Intelligence Analysis of Actual and Simulated Military Reconnaissance Mission," *IEEE Int. Conf. on Systems, Man, and Cybernetics, pp. 2590-2595,* © 1997 IEEE)

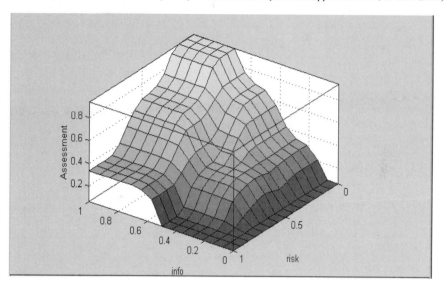

Design of Fuzzy Rule-based Models

Each of the modules discussed above is implemented using a set of fuzzy mapping rules. The inference surface shown in Fig 11.7 was obtained from a MATLAB implementation using the Mamdani model. The design of the rule base was performed by first identifying the necessary intermediate linguistic variables. Then the input variables necessary to infer these intermediate values were identified. This process produced inference chains which are represented in the hierarchical structure shown in Fig 11.8. Each box in the figure represents a fuzzy rule-based model. During the design process, the number of input variables to each model was minimized in order to reduce the number of required rules. After the hierarchical decomposition process, the information gain module was represented using 4 fuzzy models, the risk module using 5 fuzzy models, and the mission assessment module using a single fuzzy model. To represent these 10 fuzzy models, a total of 731 rules were encoded.

11.5.4.2   Results

The majority of data used in this case study were digitized representations of two geographical features: elevation and vegetation data. The elevation data contained a series of elevation postings describing the height of the terrain in meters above sea level. The vegetation feature data describe height and type of vegetation at each location. The digital feature data covered an area in the state of New Mexico. The size of the area covered was approximately 25 kilometers square with an estimated resolution of 30 meters (between elevation postings). Both features are represented in a raster format (Arc Info's Arc/Grid), where each pixel covers a 30 meter square area in both the *x* and *y* directions.

FIGURE 11.8   Inference Hierarchy

(From D. Ragsdale et al., "A Fuzzy Logic Approach for Intelligence Analysis of Actual and Simulated Military Reconnaissance Mission," *IEEE Int. Conf. on Systems, Man, and Cybernetics, pp. 2590-2595,* © 1997 IEEE)

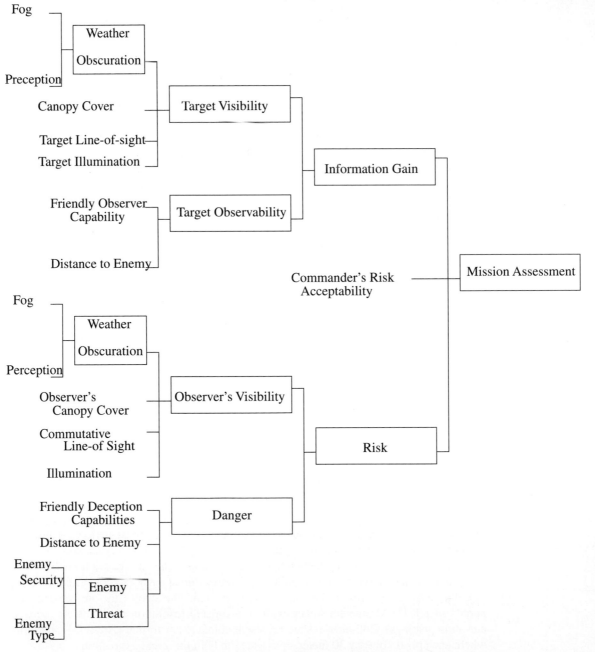

Fig 11.6 shows the result produced by a prototype of the system implemented using Fuzzy Logic Toolbox in MATLAB. The result is consistent with the doctrinal principles for the reconnaissance mission. For example, all tree-line areas in the vicinity of the target were assessed to be very suitable recon locations. This adheres to conventional wisdom for conducting a reconnaissance in that these locations provide maximum observation of the target (i.e., high information gain) while minimizing the chances of being detected by the enemy (i.e., low risk). In addition, all densely vegetated areas, regardless of their proximity to the target, were assessed to be unsuitable. This result is intuitive since dense vegetation normally minimizes the area that can be observed from a given location (i.e., low information gain) and as a result, these locations are generally unsuitable for recon locations.

The system could be extended to perform additional intelligence analyses and planning tasks. For example, the risk assessment module could be used in conjunction with additional user inputs to provide an assessment of the suitability of projected movement routes.

11.6 Fuzzy Logic in Mobile Robot Navigation

The planning and control of a mobile robot in a dynamic environment has been an area of great interest to many AI researchers. Because the sensor data and the control action of mobile robot navigation are typically in the continuous domain, fuzzy logic was found to be effective for designing intelligent agents for this application. In this section, we will first describe the problem of mobile robot navigation from an intelligent agent prospective. We will then briefly introduce three major approaches for mobile robot navigation: (1) model-based approaches, (2) sensor-based approaches, and (3) hybrid approaches. We will then focus our discussion on the application of fuzzy logic in a sensor-based approach — the behavior architecture.

11.6.1 An Intelligent Agent Perspective to Mobile Robot Navigation

The basic problem of autonomous mobile robot path planning and control is to navigate safely to one or several target locations.[1] To help us in discussing this problem from an intelligent agent perspective, we describe below the major concepts of intelligent agents within this context.

- *Environment*: The environment of a mobile robot navigation control agent is the world in which the robot is in. At least two types of information in this environment are important to the agent: (1) anything that can be an obstacle, and (2) the location of the goal. Depending on the application, the environment can be static or dynamic. For example, the terrain on Mars does not change much from day to day; therefore, the environment for Mars Rover (a mobile robot navigating on Mars for collecting scientific data) is mostly static. However, the environment of an unmanned vehicle

---

[1.] The problem can be further complicated by other considerations such as deadlines for reaching those locations, safety considerations of paths, reactiveness to emergent situations and uncertainty about the environment.

for a battle field is highly dynamic because it changes rapidly from minute to minute. The environment of a mobile robot in a manufacturing facility or in an office building (e.g., for postal mail delivery) is somewhat static and somewhat dynamic though the location of fixtures (e.g., assembly lines, machining equipments, parts storage cabinets, office doors, hallways, elevators) in these cases are usually fixed. The environment may have a dynamic component — the locations and motions of other agents (e.g., other mobile robots and human workers). As we shall see later, a dynamic environment and the characteristics of changes to the environment highly influence the agent design.

- *Sensors*: The sensor of a mobile robot ranges from low-cost sonar (range) sensors to costly video cameras. The low-cost sensors provide limited information about the environment, while more costly sensors typically provide more information about the environment.

- *States*: The state information of a mobile robot agent ranges from the distance to the closest obstacle to a complete 3D map of the environment (including the robot's location). A more detailed state description about the environment reduces the agent's uncertainty about the environment, yet it is usually more costly to update the state.

- *Goals*: The two basic goals are (1) to reach the given destination, and (2) to avoid hitting obstacles. Other goals may be added for more advanced mobile robot (e.g., a mail delivery robot that picks up litter and drops it in garbage collectors).

- *Utility*: The goal of obstacle avoidance is obviously more important than reaching the target location (unless the robot is a suicidal one for terrorist activities). Therefore, the utility information is usually represented qualitatively using a priority scheme rather than quantitatively.

11.6.2 Approaches to Mobile Robot Navigation

Existing approaches to mobile robot navigation can be classified in three categories: model-based approaches, sensor-based approaches, and hybrid approaches. *Model-based approaches* use a model of the environment to generate a path for the robot to follow. Techniques for model-based path generation include road mapping[391,357], cell decomposition[357, 158], and potential fields[357, 455]. All of these methods are able to find a path from an initial point to a goal point using a model of the environment. Some of these techniques (e.g., road mapping) can be used to find the shortest path between a starting location and a target location, if such a path exists. However, these methods rely on an accurate model of the environment to generate a safe path (i.e., a path that does not go through or near any obstacles). Since it is usually difficult to obtain an accurate model of a dynamic environment, model-based approaches are primarily used for robot path planning in controlled environments.

   Sensor-based approaches to mobile robot navigation generate control commands based on sensor data. A promising architecture for sensor-based approaches is the *behavioral architecture*, which consists of multiple behaviors, each one of which reacts to sensor input based on a particular concern of the navigation controller[13, 85]. Examples of typical behaviors include goal-attraction, wall-following, and obstacle-avoidance. The main advantage of sensor-based approaches is that the robot can navigate safely in a dynamic environment, because it can easily react to obstacles detected by sensors in real time. The major limitation of purely sensor-based approaches is that the robot may not reach the goal, even if a path to the goal exists.

Model-based approaches and sensor-based approaches can be combined into *hybrid approaches* to mobile robot navigation[465]. A hybrid approach usually uses a model-based planner to generate a path from an incomplete model of the environment. The path is used by a sensor-based controller to navigate the robot such that it follows the path while avoiding obstacles unknown to the model. This approach is thus able to explore the trade-off between the optimality of model-based approaches and the adaptability of sensor-based approaches. A suitable scheme for fusing the control commands thus is critical for a hybrid approach to be successful.

From an intelligent agent perspective, these three approaches differ on their assumption about the environment and the source of their information. A model-based agent assumes the environment is static (or mostly static) and relies primarily on the agent's world model (i.e., map of the environment) as the information source. The sensor input plays a very minor role, if any, in a model-based mobile agent. A sensor-based agent, on the other hand, relies heavily on the sensor data. This type of agent typically has very little information in its model about the environment. An assumption underlying this type of agent is that the environment is highly dynamic. A hybrid agent draws information from its internal world model as well as from the sensor inputs. It considers the two types of information sources equally important. The assumption underlying this agent design is that the world is partially dynamic, and partially static. We summarize these differences in Table 11.1.

TABLE 11.1   A Comparison of Major Approaches to Mobile Robot Agents

	Model-based Agents	Sensor-based Agents	Hybrid Agents
Assumption about the Environment	Static	Dynamic	Partially Static, Partially Dynamic
Sources of Information	Internal World Model (Map)	Sensor Data	Partial Map Sensor Data
Adaptability	Low	High	Medium to High
Optimality of Path	High	Low	Medium

11.6.3   The Behavior Architecture

One of the major sensor-based approaches to mobile robot control is *the behavior architecture*. The term *behavior* comes from biology and refers to the reaction of an agent to a given situation. Therefore, a behavior in a mobile robot navigation system usually represents a concern of the robot, such as *follow the path* or *avoid obstacles*. Each behavior associates a situation detected by the sensors (e.g., an obstacle within 5 feet in front of the robot is detected) to an appropriate action (e.g., turn left or right 10 degrees). A command arbitrator either fuses control decisions recommended by multiple behaviors or selects the control decision generated by the behavior with the highest priority.

From an intelligent agent perspective, we can view a behavior architecture as a two-layer mutli-agent system. Each behavior is a simple reflex agent at a lower level, and the recommended actions of these behaviors are combined by a command fusion agent at a higher level.

Even though goals are not explicitly represented, they are implicit in the behaviors. For example, the wall-following behavior describes a strategy for two implicit goals: (1) to avoid hitting a wall and (2) to avoid being trapped and not knowing how to reach the target location (e.g., if the target location is on the other side of the wall). The relative utility of the robot's goals is implicit in the scheme for command fusion. For example, the obstacle-avoidance behavior is usually given the highest priority in command fusion because the negative utility of hitting an obstacle is much larger than the positive utility of the other goals.

11.6.4   A Fuzzy Logic-based Behavioral Architecture

Fuzzy logic has been used to extend the behavioral approach to mobile robot navigation using the notion of *fuzzy behaviors*, which associate a *soft condition* (e.g., a close obstacle in front of the robot is detected) to an appropriate fuzzy action (e.g., turn left or right slightly). These fuzzy behaviors are typically implemented using fuzzy if-then rules. Their control decisions can also be blended using fuzzy logic.

The American National Conference on Artificial Intelligence (AAAI) has held a robot competition since 1992. One of the robots that has won awards from the competition is *Flaky* designed by E. Ruspini and other AI researchers from SRI. Flaky integrates fuzzy logic with high-level symbolic planning techniques in AI [516]. Flaky uses a representation scheme called *control structure* to explicitly associate goals, goal-achieving behaviors, and fuzzy conditions under which the goal is desired. This extends the fundamental concepts of goals, actions, and conditions in AI planning to allow them to be elastic in nature.

The main benefits that can be offered by a fuzzy logic-based mobile robot navigation system include (1) simplicity, (2) understandability, (3) extensibility, and (4) reduced hardware cost. It should be pointed out, thought, that the degree of these benefits offered by different fuzzy logic approaches may vary. Typically, a fuzzy logic-based navigation controller only needs a small number of rules. The simplicity of the system is due to the interpolative capability of a fuzzy system. A fuzzy logic approach is usually easier to comprehend because it explicitly describes the control strategies using linguistic terms. The simplicity and the understandability together make a fuzzy logic-based navigation system easier to scale up for dealing with more complex navigation problems. Finally, a fuzzy logic mobile robot controller can potentially use low-cost sensors because of its tolerance of sensor noise. This could help in reducing the overall cost in implementing the robot.

One issue encountered in applying fuzzy logic to mobile robot navigation is the problem of using the Center of Area (COA) defuzzification technique [473]. A modification to the COA technique, called Center of Largest Area (CLA) defuzzification, has been proposed as a solution to the problem [473, 667]. We will discuss this in Section 11.6.4.

Fig 11.9 shows an example of a situation in which the path-following behavior suggests the robot turn left, but the robot must go straight a little longer to avoid the obstacle on the left (i.e., obstacle A). Using this example, we will describe and demonstrate each step of the algorithm in the following subsections.

FIGURE 11.9   An Example of Hybrid Approach to Mobile Robot Navigation

(From J. Yen and N. Pfluger. "A Fuzzy Logic Based Extension to Payton and Rosenblatt's Command Fusion Method for Mobile Robot Navigation," *IEEE Trans. on Systems, Man, and Cybernetics,* Vol. 25, No. 6, pp. 971-978, June 1995. © 1995 IEEE.)

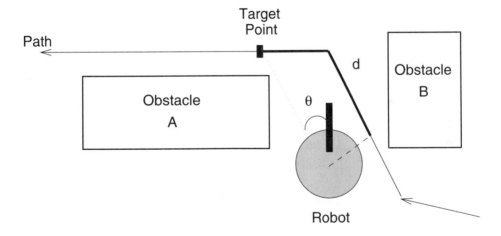

11.6.4.1 Fuzzy Behavior for Goal Following

A *fuzzy rule-based behavior* uses a set of fuzzy mapping rules to describe a mapping from sensor data to control actions of a mobile robot. The action recommended by a fuzzy logic-based behavior is usually imprecise. A common fuzzy rule-based behavior is for the robot to follow/track a goal, an intermediate goal, or an imperfect path generated from an incomplete and uncertain model of the environment (in a hybrid approach). Typically, this type of behavior computes a target turning angle based on the robot's current location, current heading, and the location of the goal (or an intermediate goal on a path toward the final goal location). It then uses fuzzy rules to map the target angle to a more general desired turning direction, which gives the robot more flexibility in avoiding obstacles while following the path. Fig 11.10 shows two fuzzy rules of a goal-following behavior.

FIGURE 11.10   Two Fuzzy Rules for Goal Following
(From J. Yen and N. Pfluger. "A Fuzzy Logic Based Extension to Payton and Rosenblatt's Command Fusion Method for Mobile Robot Navigation," *IEEE Trans. on Systems, Man, and Cybernetics,* Vol. 25, No. 6, pp. 971-978, June 1995. © 1995 IEEE.)

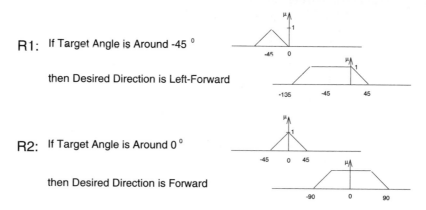

R1: If Target Angle is Around -45 °

then Desired Direction is Left-Forward

R2: If Target Angle is Around 0 °

then Desired Direction is Forward

The goal-following behavior's fuzzy inference module combines desired directions recommended by all goals following fuzzy rules using a weighted sum. This process is illustrated in Fig 11.11 for a target angle of - 30 degrees using rules R1 and R2 in Fig 11.10. The antecedent membership functions (i.e., `Around 0 degrees`, `Around -45 degrees`, etc.) are designed such that the sum of their membership values for an angle is exactly 1. We chose weighted sum fuzzy composition instead of other fuzzy reasoning methods (e.g., max-min) for the goal-following behavior because the behavior performs, in effect, a linear interpolation between rules with adjacent antecedent membership functions.

FIGURE 11.11   An Example of Computing Desired Direction
(From J. Yen and N. Pfluger. "A Fuzzy Logic Based Extension to Payton and Rosenblatt's Command Fusion Method for Mobile Robot Navigation," *IEEE Trans. on Systems, Man, and Cybernetics,* Vol. 25, No. 6, pp. 971-978, June 1995. © 1995 IEEE.)

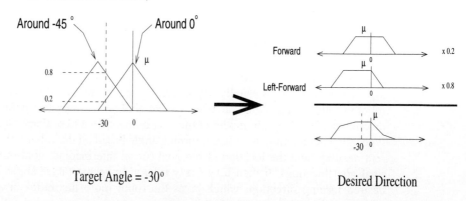

Target Angle = -30°

Desired Direction

11.6.4.2 Fuzzy Behaviors for Obstacle Avoidance

A fuzzy rule-based obstacle-avoidance behavior uses sonar sensor data to generate a fuzzy set that represents the disallowed directions of travel (i.e., directions that lead into or near any obstacles in the short term). Two fuzzy mapping rules for obstacle avoidance are shown in Fig 11.12. The behavior operates by first comparing each sensor input, which measures the distance of the closest obstacle detected by the sensor, to a fuzzy set NEAR associated with the sensor. Based on the result of the comparison, the behavior determines the degree to which the general direction of the sensor is considered disallowed. In general, this type of fuzzy-rule based obstacle avoidance behavior consists of n rules, where n is the number of sonar sensors. In this example, we assume that the robot has seven sonar sensors that are distributed as shown in Fig 11.13. The membership functions of disallowed turning direction associated with it are designed such that (1) it partially overlaps those of neighboring sensors, and (2) it has a major influence on the sensor's direction. For example, the membership function of disallowed direction of rule R3 is not symmetric with respect to - 45 degrees. More precisely, the -45 degree sensor has less influence in the forward direction than in the left direction. This is due to the presence of a - 20 degree sensor towards the front, while there is no sensor between - 45 degrees and - 90 degrees.

FIGURE 11.12   Fuzzy Rules Used for Obstacle Avoidance Behavior

(From J. Yen and N. Pfluger. "A Fuzzy Logic Based Extension to Payton and Rosenblatt's Command Fusion Method for Mobile Robot Navigation," *IEEE Trans. on Systems, Man, and Cybernetics,* Vol. 25, No. 6, pp. 971-978, June 1995. © 1995 IEEE.)

R3: If -45° sensor distance to nearest obstacle is *Near*

then Disallowed Direction is

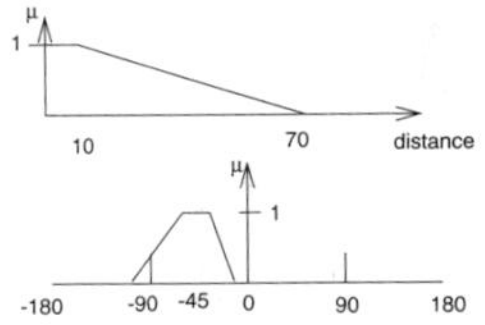

R4: If 0° sensor distance to nearest obstacle is *Near*

then Disallowed Direction is

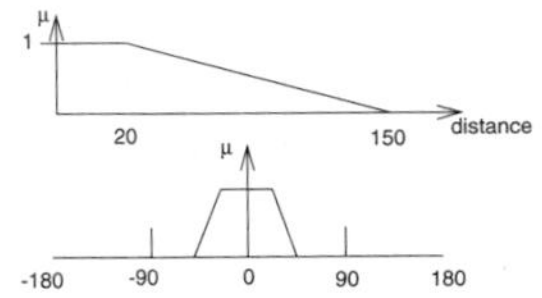

Once all the fuzzy rules associated with the obstacle-avoidance behavior have been fired, their fuzzy conclusions are combined using the *MAX* operator. Fig 11.14 shows an example of this combination with sensor inputs based on the situation in Fig 11.9. The behavior's fuzzy inference module uses the *MAX* operator instead of other t-conorm operators (i.e., other union operators in fuzzy set theory) because it is consistent with the intuition that the degree a travel direction is disallowed should be determined by the sensor source that has the strongest opinion about it.

FIGURE 11.13 Sensor Directions of an Exemplar Mobile Robot

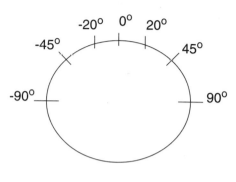

It is worthwhile to point out that the membership functions NEAR for different sensors are different, as illustrated in Fig 11.12. This is because an obstacle in the robot's traveling direction poses more of a threat than an obstacle that is on the side. From an obstacle-avoidance viewpoint, an obstacle of distance d, detected by the front sensor, is thus considered "closer" to the robot than an obstacle of distance d detected by the left side sensor. In fuzzy logic, the meaning of a linguistic term is always associated with the context in which the term is used. A term could therefore have different meanings in different contexts. This flexibility in fuzzy logic makes it possible to capture the desired meanings of the term NEAR for various sensor directions by associating with them the appropriate membership functions.

FIGURE 11.14   An Example of Sensor Fusion and the Resultant Fuzzy Set

(From J. Yen and N. Pfluger. "A Fuzzy Logic Based Extension to Payton and Rosenblatt's Command Fusion Method for Mobile Robot Navigation," *IEEE Trans. on Systems, Man, and Cybernetics,* Vol. 25, No. 6, pp. 971-978, June 1995. © 1995 IEEE.)

Sensor Number	Sensor Angle	Distance Returned	Rule Firing Strength
1	-90	200	0.0
2	-45	15	1.0
3	-20	200	0.0
4	0	200	0.0
5	20	50	0.6
6	45	30	0.8
7	90	200	0.0

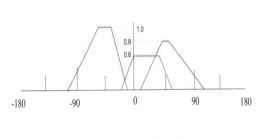

Disallowed Direction

Even though our examples so far have focused on fuzzy rule-based behavior for controlling the turning direction similar strategies can be applied to the design of fuzzy rule-based behavior for controlling the speed of robot. We will leave this as an exercise.

11.6.4.3 Fuzzy Logic-based Command Fusion

Once we use fuzzy rule-based behaviors to generate imprecise control action recommendations, it is natural to use fuzzy logic to combine these action recommendations to obtain a final (typical precise) action. A fuzzy logic-based approach to command fusion involves two steps: (1) fuse fuzzy recommendations generated from multiple behaviors, and (2) convert the combined fuzzy action into a precise one so that it can be executed. The second step can be viewed as a defuzzification task. However, as we will show below, it can present problems for conventional defuzzification techniques. We illustrate these two steps using the output of fuzzy logic behaviors described earlier.

The goal-following behavior generates a fuzzy conclusion about desired turning direction, while the obstacle-avoidance behavior generates a fuzzy conclusion about disallowed turning direction. These two fuzzy conclusions should be combined to produce desired directions that are allowed. This can easily be achieved using a fuzzy conjunction operator, since the final turning angle should be both desired from the path-following viewpoint and *not disallowed* from the obstacle-avoidance consideration.

$$\mu_{Turing-Direction}(x) = \mu_{Desired\ AND\ (NotDisallowed)}(x) \qquad \textbf{(EQ 11.3)}$$

$$= \mu_{Desired}(x) \otimes (1 - \mu_{Disallowed}(x))$$

If we choose the min operator for fuzzy conjunction, we have

$$\mu_{Turing-Direction}(x)= \min\{\mu_{Desired}(x), (1-\mu_{Disallowed}(x))\} \quad \text{(EQ 11.4)}$$

Returning to our example situation in Fig 11.9, the command fusion step under the situation is illustrated in Fig 11.15. Even though the target angle in the example is approximately -30 degrees, most of the combined fuzzy command for turning direction is around 0 degrees, which is the correct direction for the robot to take given the presence of obstacle A. The next section will describe how to convert a combined fuzzy command into a crisp one.

11.6.4.4   The Center of Largest Area (CLA) Defuzzification

Defuzzification is the process of converting a fuzzy command into a crisp command, (e.g., turn 4.3 degrees to the right). As we discussed in Chapter 5, the two major methods of defuzzification are (1) the Mean of Maximum (MOM) method and (2) the Centriod method. The MOM defuzzification method computes the average of those values with the highest membership degree in the fuzzy command. The centroid method computes the center of gravity of the entire fuzzy command.

The major drawback of the MOM method is that it does not use all of the information conveyed by the fuzzy command and thus has difficulty in generating commands that turn the robot smoothly over time.

FIGURE 11.15   An Example of Command Fusion

(From J. Yen and N. Pfluger. "A Fuzzy Logic Based Extension to Payton and Rosenblatt's Command Fusion Method for Mobile Robot Navigation," *IEEE Trans. on Systems, Man, and Cybernetics,* Vol. 25, No. 6, pp. 971-978, June 1995. © 1995 IEEE.)

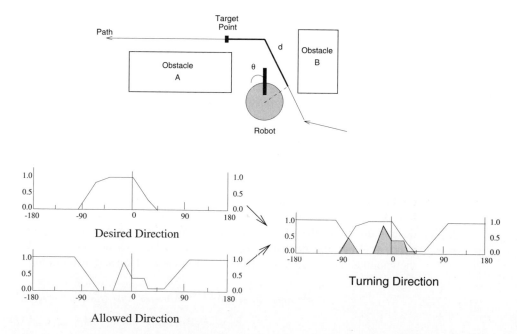

The centroid defuzzification method also has problems when applied to mobile robot control. Let us consider the situation shown in Fig 11.16. The combined fuzzy command for the mobile robot's turning direction in this situation has a twin-peak membership function. Applying the centroid defuzzification technique to this twin-peak fuzzy set yields a bad command that will bring the robot even closer to the obstacle. In general, the centroid method could create problems when it is applied to applications that involve prohibitive information, for it does not ensure that the defuzzified decision avoids those regions that are prohibited.

To alleviate this difficulty, a defuzzification method called Centroid of Largest Area (CLA)was developed. The technique partitions a multiple-peak fuzzy command into several disjoint fuzzy subsets, each corresponding to a feasible fuzzy command [473]. The fuzzy subset with the largest area is then selected and defuzzified using the COA method. This new defuzzification technique is illustrated in Fig 11.16. The twin-peak fuzzy control command is partitioned into two fuzzy subsets, one corresponding to each feasible fuzzy command. The fuzzy subset on the right is then chosen because its area is larger than the left one. Applying COA defuzzification to the right fuzzy subset results in a turn of 42 degrees to the right. Consequently, the robot will go around the obstacle while maintaining its proximity to the path at the same time.

FIGURE 11.16 **An Example of Defuzzification**
(From J. Yen and N. Pfluger. "A Fuzzy Logic Based Extension to Payton and Rosenblatt's Command Fusion Method for Mobile Robot Navigation," *IEEE Trans. on Systems, Man, and Cybernetics,* Vol. 25, No. 6, pp. 971-978, June 1995. © 1995 IEEE.)

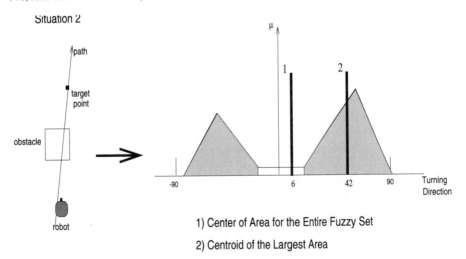

1) Center of Area for the Entire Fuzzy Set

2) Centroid of the Largest Area

11.6.4.5   Benefits

A fuzzy logic-based behavior approach to mobile robot navigation offers three important benefits: simplicity, extensibility, and understandability.

1. *Simplicity:* The fuzzy logic-based navigation agent only needs a small number of rules. In our exemplar implementation, the goal-following behavior uses seven rules, and the obstacle avoidance behavior uses one rule for each sonar sensor. The simplicity of the system is due to its modular design and the interpolative reasoning capability of a fuzzy system.

2. *Extensibility:* The approach is easily extensible. For instance, one can easily modify the agent using different numbers of sensors. This is because the desired interaction between sensors is easily achieved by (1) designing the membership functions of the sensors' disallowed directions such that they partially overlap, and (2) choosing the appropriate fuzzy inference scheme (i.e., the max-min inference).

3. *Understandability:* The knowledge of each behavior is easy to comprehend because it is captured in linguistic form by fuzzy rules.

11.7   Fuzzy Logic in Emotional Intelligent Agents

Do emotions play a role in human intelligence? If so, can emotions play a useful role in intelligent agents? While complete answers to these questions may not be clear yet, it is not premature to say that fuzzy logic has a role in addressing the latter question. Ancient philosophers like Plato tried to identify the role of emotions in the human thinking process. He and others identified the emotional process to be a process that is quite separate from our rational thinking. Descartes' famous quote, " I think therefore I am," divided the mind from the body and thus projected Plato's words in a more profound theory. Descartes later published a book, *Passions of the Soul* [153], to address the particular role of emotions in human life. His book seems to emphasize the fact that emotions are nothing but a physiological process that subjects urges and needs of the body to the mind[153]. Moreover, his definition seemed to separate the body from the mind. Emotions were defined as urges or needs of the body, where they originate. While, rational thinking was regarded as a product of the mind's thinking process.

Decades later, A. Demasio not only came up with entirely different answers to the questions above, but he also refuted Descartes' theory of the mind and the body that had dominated the field of philosophy of the mind for more than 300 years. He brought out some neurological evidence that suggested that emotions have a great impact on the human decision-making process. Through his work with patients suffering a myriad of disorders, including brain damage, problems with memory, language, and reason, Demasio was led to believe that both cognition and emotion play a role in human intelligence. A patient named Elliot was referred to Demasio after others had failed. Elliot had been diagnosed with brain tumor. The tumor grew in his brain and put enormous pressure on both frontal lobes. After what the surgeons called successful surgery, in which the tumor and the damaged tissue were removed, Elliot had experienced a radical personalty change. He changed from a successful businessman with a happy and stable family life to a person who could not hold a job, follow a schedule, motivate himself enough to get dressed in the morning, and was incapable of making the most basic decisions. For instance, he would spend an entire afternoon at work deciding whether to classify a group of data by date or place. After counseling Elliot, Demasio said, " I never saw a tringe of emotion in my many hours of conversation with him: no sadness, no impatience, no frustration with my incessant and repetitious questioning" [146]. Elliot was given tests to examine his IQ, memory, decision-making and many more. These tests indicated that his IQ was above average. However, Demasio argued that real-world decisions are far more complex than the tests that were given. In the real-world we have to act under uncertainty. Demasio hypothesized that emotions play a role in a human being's efficient decision-making process, which is often referred to as "gut feeling." He argues that the "high reasoning" advocated by Plato, Descartes and Kant, who employ formal logic and pure reasoning, rejecting emotions and feelings from their formula, is flawed. Pure reason is what Elliot was bound to and that did not get him very far.

Long before Demasio, Herb Simon, writing on the foundations of cognition, emphasized that a general theory of thinking and problem solving must incorporate the influences of emotion [539]. Furthermore, Marvin Minskey, one of the pioneer AI researchers, also acknowledged the role of emotions by his famous statement, "the question is not whether intelligent machines can have any emotions, but whether machines can be intelligent without any emotions" [415]. Even though these ideas stimulated AI researchers in the 60s, 70s, and 80s to model emotional intelligence, cognitive reasoning and knowledge representation were still the dominant areas in the AI research field at that time. It was not until the mid 90s that AI researchers became interested in incorporating emotions into robots and intelligent agents. Much of the research results generated from the 1960s through the 1980s was summarized by R. Pfeifer [472].

Research on emotional modeling then raised another set of questions. How can we model emotions? How can we evaluate these models? Why does an intelligent agent need emotions? When does an agent need to use emotional intelligence and when not to? What applications can benefit from emotional intelligent agents?

We will not attempt to answer all these questions. In fact, most of them are still left unanswered. However, in the following subsections, we will illustrate the potential contributions of fuzzy logic in this area by describing a fuzzy logic-based model of the emotional process. We will also mention potential applications of intelligent agents with both High IQ and high EQ (Emotional Quotient). Readers who are interested in emotional modeling are strongly recommended to refer to Rosalind Picard's book *Affective Computing* [474], which summarizes computational models of emotions.

11.7.1 Reactive versus Deliberate Emotional Agents

Most of the emotional agents that were implemented and cited in Picard's book tend to be more reactive than deliberate, i.e., they tend to treat an agent's response to emotions as a reflex action (if this emotion occurs, then the agents does this action, or if this emotion occurs with this intensity, then that agent does this action) [396]. Furthermore, the emotion generation process itself is entirely determined by external stimulus (i.e., if this event occurs, then this emotion will be generated to this intensity). Do we think about our emotions before acting on them? Can we control our emotions? As pointed out by Psychologists such as Erickson and more recently by Howard Gardner and Daniel Golemman [202, 217], the answer to these questions is 'YES.' Yes, we can control and perceive our emotions. Such a capability is fundamental to emotional intelligence. Emotional intelligence is our ability to understand our emotions, control our emotions, and understand the emotions of others. Our behavior can be influenced by understanding our emotions as well as the emotions of others. Moreover, the generation of emotion is affected by our value system and our self-esteem. An attempt to incorporate this kind of insight in a model of an emotional intelligence agent will require a representation of internal states (e.g., self-esteem, self-perception, self-control, self-awareness, etc.) that affect an agent's capability to sense emotions, to act on them, and know how these internal states evolve.

11.7.2 Toward a Computational Model of Emotion

We will first describe a nonfuzzy computational model of emotion, and then outline an extension of this model using fuzzy logic. Emotional agents contain a computational model of emotions to simulate their emotional process. Our discussion below is based on a psychology model of emotions developed by I. Roseman [507]. However, the idea of incorporating fuzzy logic can be applied to all the other models as well. Roseman's model is shown in Table 11.2. Emotions are generated according to events. These events are evaluated according to the following five categories:

(1) *Situational State*: An event is categorized as motive-consistent or motive-inconsistent. A motive-consistent event is an event that can lead to a goal, whereas, a motive-inconsistent event is an event that can threaten a goal.

(2) *Probability*: This category describes the likelihood of the event to occur.

(3) *Agency*: Describes who caused the event.

(4) *Motivational State*: Describes whether the event is motivated by obtaining a reward or by avoiding punishment.

(5) *Power*: This categorizes the self-perception that is, does a person perceive him/her self as being strong or weak with respect to the event.

TABLE 11.2   Roseman's Model
(From J. Roseman, "Cognitive Detereminants of Emotions: A Structural Theory," in P. Shaver (Ed.) *Review of Personality and Social Psychology,* Vol. 5, p. 31. © 1984 by Sage Publications. Reprinted by permission of Sage Publications, Inc.)

Circumstance-Cased	Motive-consistent		Motive-inconsistent		Self
	Appetitive	Aversive	Appetitive	Averasive	
Unknown	Surprise				
Uncertain	Hope		Fear		Week
Certain	Joy	Relief	Sadness	Distress, Disgust	Week
Uncertain	Hope		Frustration		Strong

TABLE 11.2   Roseman's Model (Continued)

(From J. Roseman, "Cognitive Detereminants of Emotions: A Structural Theory," in P. Shaver (Ed.) *Review of Personality and Social Psychology,* Vol. 5, p. 31. © 1984 by Sage Publications. Reprinted by permission of Sage Publications, Inc.)

Circumstance-Cased	Motive-consistent		Motive-inconsistent		Self
	Appetitive	Aversive	Appetitive	Averasive	
Certain	Joy	Relief	Frustration		Strong
Other-Caused					
Uncertain	Liking		Dislike		Week
Certain	Liking		Dislike		Week
Uncertain	Liking		Anger		Strong
Certain	Linking		Anger		Strong
Self-Caused					
Uncertain	Pride		Shame, Guilt		Week
Certain	Pride		Shame, Guilt		Weak
Uncertain	Pride		Regret		Strong
Certain	Pride		Regret		Strong

We all have experienced emotions with different intensity levels (e.g., fear of not being popular, fear of losing a job, fear of losing a spouse, fear of death). If we introduce intensity into emotions, then a model of emotional agent needs to map an emotion with a certain intensity level to a certain behavior. Thus, one direction for using fuzzy logic in emotional intelligent agents is to describe such a mapping using fuzzy rules. For example, if we have the following rule:

- IF Anger is between 10 and 20
  THEN Action is Aggressive.

It is clear that the mapping of an emotional intensity to a behavior is not clear-cut. What if the agent's intensity level of anger is 9.5? Does that make him behave aggressively? Fuzzy logic offers an alternative way of mapping behaviors from emotional intensity. Instead of a clear-cut interval mapping, we will define different intensity levels using the fuzzy sets defined in Chapter 3. Moreover, we will use a fuzzy mapping rule-based system to map the different emotional intensity levels to a behavior. Thus the rule above will be reformulated into

- IF Anger is *High*
  THEN Action is Aggressive.

We can also use hedges in descriptions such as VERY sad, SOMEWHAT sad. For example, we may have a rule such as

- IF the agent's expectation was HIGH
    AND its reward was LOW
  THEN the agent is VERY disappointed.

- IF the agent's self-esteem is HIGH
    AND the agent is VERY disappointed
  THEN the agent's mood is SLIGHTLY reduced.

Another usage of fuzzy logic in modeling emotion is to generalize the emotion-generation rules in Table 11.2 such that motive-consistent and motive-inconsistent become a matter of degree. An event may be precisely categorized as a motive-consistent or a motive-inconsistent event. However, it is not surprising to categorize an event as 70 percent consistent with a motive and 30 percent inconsistent with another motive or two. Thus, if the event can be categorized in more than one category (categories being motive-consistent or motive-inconsistent), then we need to introduce the degree to which an event belongs to a certain category. This kind of uncertainty could be modeled easily by fuzzy logic.

11.7.3 Case Study: A Fuzzy Logic-based Model of Emotion

In this section we will present one aspect of fuzzy logic. We will present an example of a model that was developed by M. Seif El-Nasr at the Center for Fuzzy Logic, Robotics, and Intelligent Systems at Texas A&M University [529]. This model uses fuzzy logic to manipulate emotional intensities. We present a model that uses the ideas discussed above to simulate an emotional intelligent agent. In the first subsection, we will outline the key concepts of the model. In the following subsection, we describe a prototype system called PETEEI that implements part of the model.

11.7.3.1 The Architecture of a Model of Emotion

The architecture of the fuzzy logic-based model of emotion includes five types of internal states of an agent and three processes, which generate emotions and behaviors using the internal states. The internal states are defined as follows:

- *Expectations:* This internal state is crucial to the emotional reasoning process. Expectations are known to change the intensity of various emotions. For example, if a motive-consistent event happens, but this event was 70 percent expected, then the intensity of the joy emotion will not be as intense as if the event were 100 percent expected. Thus, the intensity level can be formalized as a function of an expectation level as well as a function of motive-consistency.

- *Attention*: The perception of an event is one of the most important issues in modeling behaviors. If an event is not perceived, will it affect the agent's emotional state?

- *Values and Ethnic Standards:* Some emotions are more related to ethnic standards than others. Thus, the emotional generation process had to consider the ethnic standards.

- *Goals*: Goals are predefined in the hierarchical structure and are domain specific.

- *Emotion:* An emotional state consists of emotions with different levels. Thus, the agent could be feeling anger, joy, sadness, and shame at the same time. The different intensities of these emotions will then shape the agent's behavior.

- *Motivational States*: Motivational states are states like hunger, thirst, fatigue, and pain. These motivational states affect different emotions. In fact, they might inhibit or empower some emotions [61]. Thus it is important to model these states and to study the effect of each of them on the emotional states that the agent might have.

In addition to these internal states, the model contains three processes that map the internal states listed above into an emotion, and in a later step to an action. These processes are described below:

- *Emotion generation:* The emotional generation process will map an event, a situation, and the agent's internal state into an emotion. Furthermore, emotion can be aggregated and intensified. For example, if the agent was angry before and the user hits the agent or punishes him for something that he did. Moreover, if the agent thinks that he didn't do anything wrong (by evaluating his action with his standards). Then the anger should be enforced.

- *Emotion Decay:* People often experience that the intensity of an emotion decays as time progresses. However, we have a decay constant for negative emotions which is different from the decay constant of positive emotions. However, the negative emotions tend to have a more profound impact on the agent's state, and they decay slower than the positive emotions. Two decay constants are used: one for positive emotions and the other for the negative emotions. This might also depend on the personality of the agent. Some personalities are easier to please than others. Some others find it easier to forget and forgive than others. Even though personality is not yet explicitly represented in our model, it is likely to play a role in future agent's models of emotion.

- *Behavior Generation:* This process is simulated as a fuzzy mapping of emotional intensity and situation to an agent's action. However, this can be further generalized to include the personalty, moods, etc.

For a more detailed description of the model, the reader is referred to [529].

11.7.3.2 PETEEI

PETEEI (PET with Evolving Emotional Intelligence) is a prototype of a fuzzy-logic based emotional intelligent agent. The agent is a virtual pet that simulates a dog. Fuzzy sets were used to represent emotional intensities, while fuzzy mapping rules were used to map these emotions into behaviors according to the intensity level of each emotion and the situation as described above. PETEEI simulated 14 emotions: sad, joy, anger, remorse, admiration, fear, hope, relief, disappointment, gratitude, gratification, pride, shame, and reproach. We implemented four motivational states in PETEEI: hunger, thirst, tiredness, and pain.

FIGURE 11.17   The User Interface of PETEEI

Fig 11.17 shows the graphical interface of PETEEI. The agent is a virtual pet which is still inside the little house. This picture shows the first scene of our prototype. The user can interact with the pet and the environment through the buttons that are shown in the bottom of the figure. The user can (1) open an object, (2) take objects, (3) close an object, (4) touch an object, (5) hit an object, (6) talk to the dog, and (7) look at an object. Everything in the scene is an object (e.g., grass, house, tree, sky, the dog); but not all of them can be taken. The user can initiate a talk with the virtual pet using a talk box. The witch in the picture holds the inventory of the user.

The dog initially has innate fear of humans; thus he will fear and probably hate the user at first. However, through interactions with the user this feeling might change for the worse or the better according to the user's actions. If the user keeps ignoring the dog or hitting him, then the dog will probably hate the user more and more. Currently, the goals of the dog are really primitive. At the very high level we have two goals: survival and entertainment. Subgoals under survival are not(hunger), not(thirst), not(pain), shelter. Under the entertainment goal we have subgoals: not(ignored), play(toys).

One of the fuzzy rules in PETEEI is shown below:

- IF UserAction is "Touch-Dog"
  AND Emotion.LoveUser is None
  AND Emotion.HateUser is Very High
  THEN increase AngerLevel slightly.

This rule denotes that if the user touches the dog and the dog's emotional state toward the user is hate, then the anger emotion will be incremented by one level.

PETEEI also contains processes that check the impact of certain events on the goals of the agent. Thus, an event can be rated according to its impact on the agent's fuzzy goals. To generate an emotion like joy, PETEEI considers the desirability of an event and the level of expectation using rules like the one below:

- IF EventMeasure is Good
  AND Expectation is Low
  THEN increase Joy moderately.

However, if the event had a very high expectation rate, then the Joy level will not be increased as much, as

- IF Event Measure is Good
  AND Expectation is High
  THEN increase Joy slightly.

In conclusion, this model was evaluated by collecting survey answers to specific questions on the behavior and the believability of the agent's reaction. The survey was conducted for people with varying backgrounds ranging from medicine, dentistry, psychology, electrical engineering to computer science. The survey indicated that the believability of the pet is significantly enhanced by the fuzzy logic-based emotional model.

11.8 Applications in the Twenty-first Century

Fuzzy logic-based intelligent agents have found applications in a wide range of emerging areas that aim to improve the level of "intelligence" of the transportation system, communication system, and even financial investment system of our society. The success of these applications will have a major and lasting impact on the daily life of human beings in the twenty-first century. To help us in developing such a vision, we briefly discuss a small sample of these applications.

Fuzzy logic has been used to detect incidents on freeways or urban streets [237, 369]. It has also been used to control vehicles on future intelligent freeways so that drivers can use the traveling time in a more productive way. Fuzzy logic-based intelligent agents have also been used for adjusting critical parameters in ATM network admission control to adapt to changing network traffic situations [116, 477]. Finally, fuzzy logic has been used in financial trading [34, 664].

11.9 Summary

The general theme of this chapter was fuzzy logic and AI. However, most of the chapter focused on fuzzy logic and intelligent agents, because the notion of intelligent agent has evolved into a uniform framework for discussing all aspects of AI with a modern perspective. More, specifically, we have discussed the following concepts.

- The similarities and difference between the goals of AI and the goals of fuzzy logic.
- Fuzzy logic provides an approach for reasoning under uncertainty in expert systems.
- Fuzzy logic can be used to extend frame-based representation.
- Fuzzy logic can be used in heuristic classifications expert systems to capture knowledge regarding heuristic matches.
- Fuzzy logic can be used to describe mapping an agent's internal states, and emotions to behaviors.
- Fuzzy logic can be used in synthesis type expert systems to describe imprecise constraints and fuzzy goals.
- The basics of an intelligent agent.
- Four different roles fuzzy logic can play in intelligent agents.
- A case study of a fuzzy logic-based intelligent agent for a reconnaissance mission.
- The use of fuzzy logic in behavioral approaches to mobile robot navigation.
- A new defuzzification technique for considering prohibitive information that was motivated by mobile robot navigation.
- Fuzzy logic can be used to model emotions and related concepts such that an agent can demonstrate a certain degree of emotional intelligence.

Bibliographical and Historical Notes

Detailed discussions about probabilistic approaches and DS approaches to reasoning under uncertainty in AI can be found in [467,466]. Research results in this area have also been published in the Proceedings of Uncertainty in Artificial Intelligence Conference (UAI) since 1989 and in workshop proceedings under the same name from 1985 to 1988.

A comprehensive coverage of expert systems can be found in [182, 276]. MYCIN was developed by E. H. Shortliffe in the 1970s [536]. J. McDermott developed R1 [25], which was later extended by Digital Equipment Corporation into XCON [406]. Many lessons were learned from DEC's experience in maintaining XCON [406]. W. J. Clancey introduced the term "heuristic classification" to refer to a large class of expert systems whose high-level problem solving involves three steps: data abstraction, heuristic match, and solution refinement [131].

A book edited by T. Terano, K. Asai, and M. Sugeno contains a chapter (Chapter 4) that describes several good applications of fuzzy expert systems [588]. C. V. Negoita wrote one of the earliest books on fuzzy expert systems [441]. The applications of fuzzy logic (including fuzzy arithmetical) in engineering designs have been investigated in [644, 645, 494, 457]

A good collection of papers regarding intelligent agent technologies can be founded in a volume edited by Bradshow [83].

Pin and Watanabe implemented fuzzy behavioral approaches to mobile robot navigation which have also been implemented using special-purpose VLSI fuzzy inference chips, that can process a large number of fuzzy inferences in parallel [475, 476].

Several research efforts have been made to incorporated various models of emotions into intelligent agents. The OZ project at Carnegie Mellon University led by J. Bates [30] has been developing emotional agents using a psychological model by A. Ortony et al. [456]. J. Velasquez at M.I.T. [612] is also building emotional agents using a psychology model by I. Roseman [507]. Both models are reactive in nature.

Exercises

11.1   Give an example of each type of intelligent agent.

11.2   What is the physical symbol system hypothesis? How does it affect the development of artificial intelligence technology?

11.3   In what ways did fuzzy logic contribute to artificial intelligence?

11.4   In what ways did artificial intelligence contribute to fuzzy logic?

11.5   Justify CADIAG's inference scheme using fuzzy implication, generalized modus ponens, and generalized modus tollens.

11.6   Suppose the Sendai Subway controller predicts future states of actions using probabilistic reasoning rather than dynamic simulation. How would you modify the architecture in Figure? How would you modify the way predicted states are evaluated?

11.7   Design fuzzy rule-based behaviors for controlling the traveling speed of a mobile robot with seven sonar sensors as shown in Fig 11.13.

11.8   Implement the two fuzzy logic behaviors for mobile robot navigation described in Sections 11.6.4.1 and 11.6.4.2 using MATLAB Fuzzy Logic Toolbox or an alternative fuzzy logic development environment.

11.9   Implement the fuzzy logic-based command fusion described in Sections 11.6.4.3 and 11.6.4.4 using MATLAB Fuzzy Logic Toolbox or an alternative fuzzy logic development environment.

11.10   Modify the fuzzy logic-based obstacle avoidance behavior for a robot that has 15 sensors in the following directions: 0°, +/- 10°, +/- 20°, +/- 30°, +/- 45°, +/- 60°, +/- 90°.

11.11   What is emotional intelligence? Why is it useful to model an agent with emotional intelligence?

11.12   Table 11.2 shows Roseman's model of emotions. Give some examples from your everyday life showing how the events and emotions encountered in your examples can be explained by Roseman's
model. Additionally, show how fuzzy logic can potentially act as a useful tool in the design and
implementation of this model.

12   FUZZY LOGIC IN DATABASE AND INFORMATION SYSTEMS

Database systems evolved out of generalized file management systems. The first generation of databases consisted of the network and the hierarchical data models. The relational model was not introduced until the early 1970s. However, it has dominated the field since the 1980s. Traditionally, these database models focused on describing precise information. However, this is not always the case in real-world data. Hence, there is a need to represent and retrieve information that is imprecise in nature. Such a need has inspired the development of fuzzy logic-based extensions to database management systems and information retrieval systems.

12.1   Fuzzy Information

Fuzzy information or fuzzy data can appear for different reasons. First, it may be due to the imprecision of real data. For instance, a sensor data may be a distribution, rather than a precise value. Second, fuzzy information can arise from subjective judgments. For example, a database containing information about real estate for family housing may need to describe the quality of public schools, the safety of the neighborhood, the estimated appreciation of housing prices, and so on. Representing this information using precise values would often fail to capture the soft boundaries between qualitative descriptions such as poor, fair, good, excellent, etc. Thus, it is important to find a data model that can represent and manipulate fuzzy information. Third, the information that a user is interested in may not be precise. For example, a college senior may be interested in finding a university that has a good graduate engineering program and low living costs. The meaning of "good" and "low" in the previous sentence is imprecise. Formulating this query using thresholds such as "annual living cost is less than x dollars" will exclude those universities whose annual living cost is slightly above x, but whose graduate engineering school is excellent. In other words, representing an imprecise query using a precise fomalism is likely to miss information that the user wishes to obtain.

Traditional databases were designed primarily for the efficient storage and convenient retrieval of large amounts of precise data. However, the rapid advancement of telecommunication and information technology in the 1990s has introduced a new "cyberspace" in which a user can have direct access to a tremendous amount of information on the World Wide Web. As information service providers, software companies, and related industries (e.g., financial industry, entertainment, media, etc.) explore a wide range of exciting opportunities that were not possible a decade ago, one thing is for sure — the nature of information of interest to a user has become far more imprecise than what was conceived by the designers of traditional databases and information systems.

12.2 Fuzzy Logic in Database Systems

Fuzzy logic has been used to extend database systems in two areas: (1) for storing and updating information that is imprecise by nature, and (2) for processing queries that are imprecise. In this section, we give an overview of the major roles that fuzzy logic plays in extending conventional database systems in these two areas.

12.2.1 Imprecise Information

Imprecise information in a database system can be classified into two types: (1) imprecise attribute values in a tuple, and (2) partial membership of a tuple in a relation. We briefly review techniques for representing these two types of imprecise information below.

12.2.1.1 Two Approaches for Representing Imprecise Attribute Values

An attribute value in a tuple (also referred to as a domain value in the database literature) may be imprecise due to the subjective nature of the attribute (e.g., the rating of movies) or a lack of precise information (e.g., the height of the suspect for a robbery case is around six feet). In the latter case, the attribute itself (i.e., the height of a person) is not always subjective. Hence, its value can be precise if it is measured objectively (e.g., in a hospital). In the former case, however, the value of an attribute is intrinsically imprecise due to its subjectiveness. These two different causes of imprecise attribute values in database systems motivated two approaches for representing fuzzy data:

- similarity-based approaches and
- possibility-based approaches.

A *similarity-based approach* uses linguistic terms (e.g. poor, fair, good, excellent) to describe attribute values. The impreciseness of these terms is characterized by a similarity matrix, which records the degree of similarity between pairs of linguistic terms in a domain. For example, let us consider a database that stores information about job openings and applicants. Suppose the speciality areas of an applicant may include robotics, AI, expert systems, and statistics. A similarity-based fuzzy database system can use a similarity matrix such as the one shown in Table 12.2 to determine the matching degree between a job opening and an applicant, as we will explain later. Similarity-based approaches were first introduced by B. Buckles and E. Petry in the late 1970s [470].

An alternative approach for representing imprecise data is to use a possibility distribution as the value of an attribute. We call such approaches *possibilistic-based approaches*. For instance, the height of a bank robbery suspect, who has been described by a witness as around six feet tall, can be represented by a possibility distribution associated with the linguistic description "around 6 feet." If the police later want to find all cases that may be linked to another murder case whose suspect is known to be 5 feet 11 inches, a possibilistic representation of the suspect's height information would enable the police to retrieve the bank robbery suspect, because the possibility distribution partially matches the height description of the murder suspect.

Based on possibility theory introduced in Chapter 3, two kinds of matching degrees can be calculated for possibilistic-based approaches: the possibility measure and the necessity measure. These two measures represent the minimal degree and the maximal degree that possibilitic data satisfy a fuzzy condition in a query.

12.2.1.2 Partial Membership in a Relation

The second type of imprecise information in a database is the partial membership of a tuple in a relation. Membership in a relation is a black-and-white matter in a conventional database. For instance, a person is either an employee or not an employee of a company. However, there are relations whose membership has a gray area. For example, the relation `Endangerd_species` may include wild animals whose quantity has been significantly decreasing over the years but have not been officially declared as "endangered species." These animals can thus be considered as somewhat endangered, and hence a partial member of the `Endangered_species` relation. Even though membership degree in a relation can be represented internally as an attribute of the relation, its meaning is fundamentally different from other attributes. In a fuzzy object-oriented database, for instance, partial membership of an object in a class directly affects the way the object inherits its properties from its classes. We will discuss this in Section 12.6. A tuple with a partial membership in a relation is often referred to as a "weighted tuple."

DEFINITION 22   **Weighted Tuples:**

A *weighted tuple* t in a relation R is a tuple associated with its membership degree in R, denoted $\mu_R(t)$.

12.2.2 Imprecise Queries

In addition to representing imprecise information, fuzzy logic can also be used to retrieve imprecise information from a database or an information repository such as the World Wide Web. The imprecision of information retrieved can be due to two reasons: (1) The information stored in the system is imprecise; or (2) the query (i.e., question) posted by the user is imprecise in nature. We call the latter an *imprecise query*. A query is imprecise if it includes at least one of the following components:

- imprecise conditions,
- imprecise operators,
- imprecise quantifiers.

We will give an example of each of these below. Consider the following query to a database of the International Revenue Service (the U.S. government agency for collecting, processing, and auditing federal income tax returns):

> Find all tax payers who have been audited in 1997 and
> whose annual family income *is less than 30K*.

The condition "annual family income is less than 30K" may be more naturally represented by an imprecise condition for describing a low-income family:

> Find all tax payers who have been audited in 1997 and
> whose annual family income *is LOW*.

where the meaning of LOW is characterized by a membership function.

A query often involves comparison operators such as "equal to," "greater than," "less than." For instance, the following query uses the equality operator:

> Find all countries whose top three imported goods
> *are identical to* the top three imported goods of the US.

Sometimes users are interested in information that is best described through imprecise operators such as "about equal to," "somewhat greater than," "significantly less than." An example of such a case is given below:

> Find all countries whose top three imported goods
> are *about the same as* the top three exported goods.

where "about the same" is an imprecise operator. An imprecise operator "about the same" can be represented as a fuzzy relation characterized by the following membership function:

$$\mu_{About-the-same}(I_3(c), I_3(\text{US})) = \frac{\|I_3(c) \cap I_3(\text{US})\|}{3} \qquad \textbf{(EQ 12.1)}$$

where $I_3(c)$ denotes the set of the top three imported goods.

A query often involves quantifiers such as "for all" or "there exists." For instance, the former is used in the query below:

> Find companies whose customers are *ALL* from govern-
> ment agencies.

An imprecise quantifier allows us to weaken or strengthen a classical quantifier so that the query can reflect the user's intent more closely. For instance, the query below uses the quantifier "MOSTLY," which is not as rigorous as "ALL."

> Find companies whose customers are *MOSTLY* from
> government agencies.

The quantifier "Mostly," in the example above, can be expressed by a fuzzy set whose universe of discourse is the ratio

$$r = \frac{\|The\ company's\ customers\ from\ government\ agencies\|}{\|All\ customers\ of\ a\ company\|} \qquad \textbf{(EQ 12.2)}$$

and whose membership function is shown in Fig 12.1.

FIGURE 12.1   Membership Function of the Quantifier *MOSTLY*

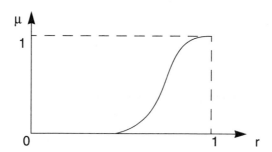

12.2.3 Redundancy and Functional Dependency

Incorporating imprecise information into conventional databases does introduce new issues that need to be resolved. Redundancy and functional dependency are two of the important ones.

A tuple in a relation is redundant if its content is identical to another tuple in the relation. A database management system needs to remove redundant tuples so that it not only avoids wasting storage space but also avoids anomalies (e.g., inconsistency) due to redundancy. The issue of redundancy removal becomes more complicated if a database system allows imprecise information. This is because **the "identity" relationship between two fuzzy tuples is no longer black and white**. Two fuzzy tuples can be partially identical, and yet partially different at the same time. The notion of redundancy thus needs to be extended to deal with this problem.

The concept of functional dependency is important in designing a relational database. More specifically, it plays a critical role in determining the keys of a relation (i.e., attributes that are unique for each tuple in the relation). We illustrate the basic idea of functional dependency using a relation R that has five attributes: A, B, C, D, E. If any two tuples in R with identical values for A and B also have identical values for $C, D,$ and E, we say $C, D,$ and E are functionally dependent on A and B. Under such circumstances, A and B can be used as the primary key of the relation. If these attributes can have imprecise values, the fact that these attribute values can be partially identical complicates the issue of functional dependency. Both issues mentioned above are due to a common fundamental question regarding fuzzy databases:

How similar do two fuzzy tuples need to be for
them to be considered identical?

We will discuss the techniques for addressing these issues in the Sections 12.4.2 and 12.5.

12.3 Fuzzy Relational Data Models

A *data model* is an abstract model of the data stored in a Database Management System (DBMS). This abstraction allows the user of a DBMS to focus on the context of the information, rather than the details of its physical data storage (i.e., file systems). To provide such an abstraction, a data model has two components: (1) a notation for describing data, and (2) a set of operations for manipulating the data. Major data models include relational models, entity-relationship models, and object-oriented models.

In this section and the following section, we will introduce fuzzy relational models and fuzzy object-oriented models. We skip fuzzy network models because they are not as significant as the other two.

TABLE 12.1   PROFILE Relation

Fname	Lname	JobType	Expertise
Bob	McLedon	Academic	AI
Rob	Mucker	Industry	Expert System
Nancy	McCay	Government	Statistics
John	Hunt	Government	Robotics

12.3.1 Relational Data Models

A relational model organizes data into *relations*, which can be viewed as tables with meaningful names. The columns of this table are called *attributes* of the relation, while the rows are called *tuples*. For example, Table 12.1 shows a relation whose attributes are Fname(First Name), Lname (Last Name), JobType, and Expertise. Conceptually, the relation describes the preferred careers and the expertise of job candidates in a job search agency's database. The relation is named PROFILE.

To formally define a relation, we need to first introduce the notion of domains. A *domain* is a set of values. A domain can be discrete (e.g., the set {Academic, Industry, Government }) or continuous (e.g., the real numbers between 0 and 100). Each attribute in a relation has a domain, which specifies all possible values for the attribute. Let $D_1, D_2, ..., D_k$ be the domains of k attributes in a relation R. The Cartesian product of these domains $D_1 \times D_2 \times ... \times D_k$ forms all possible combinations of values of the k attributes. The relation R, hence, is a subset of the Cartesian product $D_1 \times D_2 \times ... \times D_k$. A *tuple* is a member in a relation.

Alternatively, a relation can be viewed as a set of mappings, each of which maps each domain to a value. The main advantage of this formalism is that changing the order of a relation's attributes does not change the relation. In other words, a relation is defined solely by the content of its table, independent of the order of its columns.

In the original relational model, the domains for each attribute have a set of predefined values. Using the similarity matrix approach, the imprecision is implicit in the similarity matrix and so the representation is still the same. Thus, this approach is very close to the conventional relational model. The possibility approach requiring changing the data representation to a representation of fuzzy attribute values. For example, if we have an age attribute in a given database, then originally this attribute will have numeric values. Using possibility distributions, we will introduce terms like *Young, Old, Middle-aged* that are characterized by corresponding possibility distributions. In the next two subsections, a more detailed description of the two approaches will be presented. Furthermore, these two approaches will be discussed through examples to enable the reader to grasp how the fuzzy query operations are handled.

TABLE 12.2   A Similarity Matrix for Expertise Areas

	Robotics	Expert Systems	AI	Statistics
Robotics	1.0	0.6	0.6	0.2
Expert Systems	0.6	1.0	0.9	0.2
AI	0.6	0.9	1.0	0.2
Statistics	0.2	0.2	0.2	1.0

12.3.2 Similarity-based Fuzzy Relations

In a similarity-based fuzzy relation, the impreciseness of attribute values is primarily in their meaning. For example, the values of the attribute EXPERTISE in the relation shown in Table 12.1 often partially overlap. The overlapping degree of these areas of expertise are specified by a similarity matrix such as the one shown in Table 12.2. In addition to imprecision due to semantics, a similarity-based fuzzy relation could also describe imprecision due to incomplete information. For example, we may wish to describe that the age of a suspect is 19 or 20. To do this, an attribute in a tuple is allowed to have a set of values, which describe all possible values for the attribute in the tuple. A similarity-based fuzzy relation involving domains $D_1, D_2, ..., D_k$ is thus a subset of the Cartesian product $2^{D_1} \times 2^{D_2} \times ... \times 2^{D_k}$. Furthermore, as we discussed earlier, the values in domains $D_1, D_2, ..., D_k$ are related by a similarity relations $S_1, S_2, ..., S_k$ that map each pair of values in a domain to the interval [0, 1]. A similarity measure of 1 means that the pair is identical, whereas a similarity measure of 0 means the pair is entirely different.

A *similarity relation S* for a domain D is a fuzzy relation that maps pairs of domain values in D to the unit interval, i.e., $S: D \times D \rightarrow [0, 1]$, with the following three properties for all x, y, z in D:

1. Reflexive: $S_D(x, x) = 1$
2. Symmetric: $S_D(x, y) = S_D(y, x)$
3. Transitive: $S_D(x, z) \geq \max_{y \in D}(min(S_D(x, y), S_D(y, z)))$

The first two properties are intuitive. The usefulness of the third property will become clear when we discuss the issue of redundancy in a fuzzy database in Section 12.4.2.

An example of the similarity relation is given in Table 12.2 in a matrix form. The matrix describes the similarity within domain values of a given attribute. In our example, for the attribute `Expertise`, the domain values are {`Robotics`, `Expert Systems`, `AI`, `Statistics`}. For each pair of expertise areas, the similarity matrix describes one degree to which each expertise area is similar to the others.

A similarity relation cannot be freely constructed. It has to satisfy the three properties listed above. In fact, whenever we specify two different similarity measures $s(x, y)$ and $S(y, z)$, the similarity measure between x and z must be the smaller of $S(x, y)$ and $S(y, z)$. This is formally stated in the theorem below.

THEOREM 15   Suppose S is a similarity relation of domain D, and x, y, and z are elements of D. If $S(x, y) \neq S(y, z)$, then we have

$$S(x, z) = min(S(x, y), S(y, z)) \qquad \text{(EQ 12.3)}$$

Proof:
Let $S(x, y) = a$, $S(y, z) = b$, $S(x, z) = c$. We will first assume that $a < b$ and prove that $c = min(a, b) = a$. We will prove this by showing that $c > a$ and $c < a$ are both contradictory to the basic properties of similarity relations.
Case 1: Assuming $c > a$
Since both b and c are greater than a, we have

$$min(c, b) > a \qquad \text{(EQ 12.4)}$$

Since S is symmetric, $S(y, z) = S(z, y)$. Equation 12.4 can be rewritten as

$$S(x, y) < min(S(x, z), S(z, y)) \qquad \text{(EQ 12.5)}$$

This is contradictory to the transitivity property of similarity relation.

Case 2: Assuming $c > a$
Since $b > a$, we have

$$min(a, b) = a > c \qquad \text{(EQ 12.6)}$$

Equation 12.6 can be written as

$$S(x, z) < min(S(x, y), S(y, z)) \qquad \text{(EQ 12.7)}$$

This is contradictory to the transitivity property of similarity relation. Since cases 1 and 2 both lead to contradiction, c must be equal to
$$c = a = min(a, b)$$
Similarly, we can show that if $b < a$, c must be equal to b
$$c = b = min(a, b)$$
Hence, we have proved that
$$S(x, z) = min(S(x, y), S(y, z))$$
It is easy to see that the similarity relation in Table 12.2 satisfies the constraint imposed by this theorem.

12.3.3   Possibility-based Fuzzy Relations

The possibility-based approach is, as the name implies, based on the possibility distribution theory. The possibility/necessity approach is more general than the similarity approach in the sense that it handles all types of information. As described above, the similarity based approach uses domains with discrete values. Even though an extension was added to it to handle the fuzzy number, the similarity approach still heavily depends on discrete values.

A possibility-based fuzzy relation generalizes a relation by allowing the value of an attribute A to be a possibility distribution $\Pi_{A(t)}$ of the attribute's domain. Let $D_1, D_2, ..., D_k$ be the attributes of a fuzzy relation R. A tuple in a possibility-based fuzzy relation is denoted $(v_1, v_2, ..., v_k)$ where $v_1, v_2, ..., v_k$ are fuzzy subsets of $D_1, D_2, ..., D_k$. The interest of such an approach is its ability to represent, in a unified manner, precise values (singletons), NULL values, as well as fuzzy values. This is further illustrated in Fig 12.2

FIGURE 12.2   Possibility-based Representations

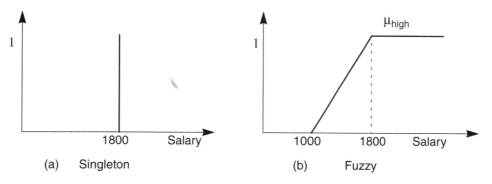

It should be noted that since the data are imprecise, then the result of the query will also be imprecise. In theory, both possibility measures and necessity measures can be used for processing queries in fuzzy databases. In practice, however, the necessity measures are rarely used.

To characterize the impression of the response to a query, we use possibility measure and necessity measure introduced in Chapter 4.

For example, if we are given a possibility distribution of a suspect's age, and we want to find all *young* suspects, we can calculate the possibility degree and necessity degree as follows:

$$Poss(Young|\Pi_x) = \sup_x[\min(\Pi_x, \mu_{young}(x))] \qquad \textbf{(EQ 12.8)}$$

$$Nec(Young|\Pi_x) = \inf_x[\max(\Pi_x, 1-\mu_{young}(x))] \qquad \textbf{(EQ 12.9)}$$

EXAMPLE 10   Let the universe of discourse of a person's age be { 10, 15, 20, 25, 30, 35, 40, 45, 50}, and the age possibility distribution of a suspect (denoted J) be:

$$\Pi_{Age}(J) = 0.2/15 + 0.5/20 + 1/25 + 0.8/30$$

Suppose that the membership function for the linguistic term Young is defined as a discrete fuzzy set as follows:

$$Young = 1/10 + 1/15 + 1/20 + 0.8/25 + 0.4/30 + 0.2/35$$

Using Equation 12.8, we calculate the possibility degree for suspect J to satisfy the condition "young suspect,":

$$Poss(Young|\Pi_{Age}J) = \max\{0.2 \wedge 1, 0.5 \wedge 1, 1 \wedge 0.8, 0.8 \wedge 0.4\}$$
$$= \max\{0.2, 0.5, 0.8, 0.4\}$$
$$= 0.8$$

To calculate the necessity measure, we first calculate the complement of the possibility distribution of a suspect J's age:

$$1 - \Pi_{Age}(J) = 1/10 + 0.8/15 + 0.5/20 + 0/25 + 0.2/30 + 1/35 + 1/40 + 1/45 + 1/50$$

The necessity measure is obtained using Equation 12.9

$$Nec(Young|\Pi_{Age}(J)) = \min\{1 \vee 1, 1 \vee 0.8, 1 \vee 0.5, 0.8 \vee 0, 0.4 \vee 0.2, 0.2 \vee 1, 0 \vee 1, 0 \vee 1, 0 \vee 1\}$$
$$= \min\{1, 1, 1, 0.8, 0.4, 1, 1, 1, 1\}$$
$$= 0.4$$

Therefore, the possibility that suspect J is young is 0.8, while the necessity that he/she is young is 0.4.

12.4 Operations in Fuzzy Relational Data Models

Operations in the relational data model are often expressed in algebraic notation, called *relational algebra*, where queries are expressed by applying specialized operators to relations. This section describes basic operations in relational algebra and their generalizations for handling fuzzy relations. We need to address at least two issues regarding the operations in fuzzy relational algebra. First, we need to determine the membership of a tuple in the new relation produced by an operation. Some of these generalizations (e.g., for union, difference and Cartesian product) are based on fuzzy set theory introduced in Chapters 3 and 4, while others (e.g., selection, projection and join) are defined specifically for managing fuzzy relational databases. Second, we need to remove redundant tuples in the new relation generated by an operation. To do this, a uniqueness criterion needs to be established first to determine how dissimilar two tuples need to be in order for them to be considered "not identical". Using a uniqueness criterion, tuples that are "identical" in a fuzzy relation are collected and the tuple with the highest membership degree in the relation is kept. Others are discarded. We discuss these two issues regarding operations in fuzzy relational models below.

12.4.1 Fuzzy Relational Algebra

(1) *Union.* The union of two relations R and S, denoted $R \cup S$, is the set of tuples that are in at least one of the relations. R and S should have the same number of attributes. The union of two fuzzy relations, as we have seen in Chapter 4, is defined as

$$\mu_{R \cup S}(t) = \mu_R(t) \oplus \mu_S(t) \qquad \text{(EQ 12.10)}$$

where $\approx\oplus$ denotes a fuzzy disjunction operator (e.g., max), and t denotes tuples.

(2) **Set difference.** The difference of relations R and S, denoted $R - S$, is the set of tuples in R but not in S. Like the union operation, R and S need to have the same arity. The difference of two fuzzy relations is defined based on intersection and compliment of fuzzy relations:

$$\mu_{R-S}(t) = \mu_R(t) \otimes (1 - \mu_S(t)) \qquad \text{(EQ 12.11)}$$

where $\otimes$ denotes a fuzzy conjunction operator (e.g., min).

(3) **Cartesian product.** Let R and S be relations of arity k_1 and k_2, respectively. The Cartesian product of R and S, denoted $R \times S$, is the set of all possible $(k_1 + k_2)$ tuples formed by concatenating a tuple from R with a tuple from S. The membership in the Cartesian product of fuzzy relations R and S, as we have seen in Chapter 4, is defined as

$$\mu_{R \times S}(t_{12}) = \mu_R(t_1) \otimes \mu_R(t_2) \qquad \text{(EQ 12.12)}$$

where t_{12} is a tuple formed by concatenating t_1 and t_2.

(4) **Projection.** The projection of a relation R selects a subset of R's attributes (i.e., columns) that the user is interested in. Let $a_1, a_2, ..., a_k, a_{k+1}, ..., a_n$ be attributes of relation R, a projection of R on a_2, a_k and a_1 denoted as $\pi_{a_2, a_k, a_1}(R)$. For example, if we form a projection of the PROFILE relation, given in Table 12.1, on JobType, we will generate the relation shown in Table 12.3.

TABLE 12.3	The Projection of PROFILE Relation on JobType

JobType
Academic
Industry
Government

The projection of a fuzzy relation is typically followed by the redundancy removal step to be discussed in the next section.

(5) **Selection.** The selection of a relation R based on condition C, denoted $\sigma_C(R)$ is the set of tuples in R that satisfy the condition C. The condition C can be described by comparing one or more attribute values of R with constants or other attributes using $=, \neq, \leq, \geq, <, >$. Multiple simple conditions can be combined using logical operators $\wedge (and)$, $\vee (or)$, and $\neg (not)$.

EXAMPLE 11   Returning to our job search example, suppose we want to find the profile information of applicants whose last name is McLedon. We can retrieve the information using selection on Lname = McLedon followed by a projection on all attributes of the PROFILE relation. This query can be expressed using relational algebra as follows:

$$R1 = \sigma_{Lname = McLedon}(PROFILE)$$ **(EQ 12.13)**

The result is shown in Table 12.4

TABLE 12.4   Result of query in Example 11

Fname	Lname	JobType	Expertise
Bob	McLedon	Academic	AI

The selection operation is extended in two ways in fuzzy relational algebra. First, a selection involving a condition about fuzzy attributes can be defined using a threshold, on the degree the condition is satisfied. We will call such a threshold the *acceptance threshold*. We can calculate the degree that a fuzzy equality condition $A_i \approx c$ is satisfied by a fuzzy tuple using a possibility measure (if the value of A_i is a possibility distribution) or a similarity measure (if the value of A_i is represented using terms related by a similarity matrix).

The second extension to selection operation is to allow the result of selection to be a relation with weighted tuples. The weight (i.e., the membership) of a tuple is obviously the degree to which the selection condition is satisfied by the tuple, i.e., $\mu_{\sigma_{A_i \approx c}}(R)^{(t)} = Eq(R \cdot A_i(t), c)$.

EXAMPLE 12   Suppose we wish to find all the people in the PROFILE whose expertise is approximately equal to AI (rather than a crisp equal operation) with the acceptance threshold of 0.8. This query can be expressed using fuzzy relational algebra as

$$R2 = \sigma_{Eq(Expertise,AI) \geq 0.8}(PROFILE)$$

where P is shorthand for PROFILE and Eq denotes a function that returns the degree its arguments are approximately identical. The result of this query is shown in Table 12.5.

TABLE 12.5   Result of Query in Example 12

Fname	Lname	JobType	Expertise
Bob	McLedon	Academic	AI
Rob	Mucker	Industry	Expert Systems

Since the threshold is set to be 0.8, from the similarity matrix in Table 12.2 we know that anyone whose expertise is AI or Expert Systems will be selected.

The second type of fuzzy selection is illustrated using the example below.

EXAMPLE 13   Find job applicants whose expertise is approximately equal to Expert Systems.

This query is analogous to the question that was answered above. The main difference is that this query does not use acceptance threshold. Thus, the relational algebra operation for the query could be formulated as follows:

$$R2 = \sigma_{Eq(P,\ Expertise, ExpertSystems)}(RPROFILE) \qquad \textbf{(EQ 12.14)}$$

where Eq denotes the fuzzy equality operator. The resultant relation for this selection is shown in Table 12.6.

TABLE 12.6   Result of Query in Example 13

μ	Fname	Lname	JobType	Expertise
0.9	Bob	McLedon	Academic	AI
1	Rob	Mucker	Industry	Expert Systems
0.6	John	Hunt	Government	Robotics
0.2	Nancy	McCay	Government	Statistics

Compared to the threshold approach, the main advantage of this approach is that it allows us to rank retrieved tuples based on their membership degree. The main disadvantage is that it may retrieve a much larger list.

In addition to the basic relational algebra operators introduced above, there are several additional operations for convenience. Even though these operations can be expressed in terms of the basic ones, they serve as convenient shorthand. The join operation falls in this category.

(6) Join   The q-join of relations R and S on attributes a_i and b_j, written as $R \bowtie a_i \theta b_j$ S, is the shorthand of $\sigma_{R \bullet a_i \theta S \bullet b_j}(R \times S)$, where θ is an arithmetic comparison operator (=, <, and so on). That is, the q-join R and S is the set of tuples in the Cartesian product $R \times S$ such that the relation θ holds between the attribute a_i in R and the attribute b_j in S. If θ is the equality relation =, the join operation is often called an equi-join.

The natural join of R and S, written $R \bowtie S$, is formed by (a) finding the equi-join on all attributes common (i.e., having identical name) to both relations, and (2) removing the redundant attributes in the resulted join. We illustrate the join operation using the JOB relation in Table 12.7. The JOB relation contains the following information about job openings: employer name, jobtype, expertise required, and location of the opening.

TABLE 12.7   JOB Relation

Employer	JobType	Expertise	Location
Texas A&M	Academic	AI	Texas
SUN Micro-systems	Industry	Expert Systems	California

TABLE 12.7   JOB Relation

Employer	JobType	Expertise	Location
NIH	Government	Statistics	DC
NASA Ames	Government	Robotics	California

EXAMPLE 14   Suppose we wish to find all job applicants in the PROFILE relation whose preferred job type matches the type of the openings in the JOB relation. This can be achieved by preforming a join of the PROFILE relation and the JOB relation on the condition P.JobType = J.JobType. Therefore, the join can be expressed as

$$R1 = \sigma_{P.JobType = J.JobType}(P \times S) \qquad \textbf{(EQ 12.15)}$$

$$P \underset{P.JobType = J.JobType}{\bowtie} J = \prod_{Name, Fname, Lname, P.JobType, Expertise, Location} R1 \qquad \textbf{(EQ 12.16)}$$

where P and J are the shorthand notations for the PROFILE relation and the JOB relation (defined in Tables 12.1, 12.7), respectively. The result of the join is shown in Table 12.8.

TABLE 12.8   Natural Join between the PROFILE relation and the JOB relation

Fname	Lname	JobType	Expertise	Location
Bob	McLedon	Academic	AI	Texas
Rob	Mucker	Industry	Expert Systems	California
Nancy	McCay	Government	Statistics	DC
John	Hunt	Government	Robotics	California

Having introduced the basics of the join operation, we are ready to discuss the *fuzzy join* operation. Generally speaking, the join of two fuzzy relations can be derived from the Cartesian product, the selection, and the projection operation on fuzzy relations defined earlier. For example, the θ-join of fuzzy relations R and S on attributes a_j and b_j is a fuzzy relation whose membership is

$$\mu_{R \underset{a_i \theta b_j}{\bowtie} S}(t_{12}) = min(\mu_R(t_1), \mu_S(t_2), \mu_\theta(R.a_i(t_1), S.b_j(t_1))) \qquad \textbf{(EQ 12.17)}$$

where t is a tuple that concatenated t_1 and t_2. We will leave the derivation of this formula as an exercise. We illustrate fuzzy join below using the job search example.

EXAMPLE 15   Suppose we want to find for each job applicant all the locations that have job openings matching approximately the applicant's profile (considering both JobType and expertise). This can be achieved by performing a fuzzy natural join on the PROFILE relation and the JOB relation followed by a projection on Fname, Lname, and Location. The new fuzzy relation formed by the join can be computed using the following expression:

$$\mu_{P \bowtie J}(t_{ij}) = min(\mu_P(t_i), \mu_J(t_j), Eq(P.Expertise(t_i), J.Expertise(t_j)))$$

where t_i and t_j are tuples such that $P.JobType(t_i) = J.JobType(t_j)$. The fuzzy equality operator Eq is calculated based on the similarity matrix of expertise For example, Nancy Hunt's job type (government) matches two entries in JOB relation. The job opening at NIH (located in DC) matches her expertise (statistics) completely, while the job opening at NASA Ames (located in California) matches her expertise slightly (i.e., 0.2 degree). The relation produced by this fuzzy join is shown in Table 12.9. Notice that the first column of the table is not an attribute of the relation, but rather it records the weight (i.e., membership) of tuples in the relation. To avoid confusion, we separate this column from the rest of the table using double vertical lines.

TABLE 12.9   Fuzzy Join between the PROFILE Relation and the JOB Relation

μ	Fname	Lname	Location
1	Bob	McLedon	Texas
1	Rob	Mucker	California
1	Nancy	McCay	DC
0.2	Nancy	McCay	California
1	John	Hunt	California
0.2	John	Hunt	DC

12.4.2 Uniqueness and Redundancy

As we have mentioned earlier, an important issue in fuzzy relational databases is how to determine whether two tuples are identical (i.e., redundant). Redundant tuples not only waste space but also post a danger for introducing inconsistency into the database (e.g., inconsistency due to updating only one of the set of identical tuples).

In conventional databases, two tuples are identical if all of their attribute values are identical. This notion of redundancy needs to be extended in fuzzy databases, because the equality test between imprecise attribute values is no longer a black-and-white one; it has gray areas. To define redundancy in fuzzy database we first formally define the equality measure for fuzzy tuples below.

DEFINITION 23   Let t_1 and t_2 be two tuples in a relation with attributes $a_1, a_2, \dots a_k$. The degree of equality between two fuzzy tuples t_1 and t_2, denoted $Equal(t_1, t_2)$, is the fuzzy conjunction of the fuzzy equality measure between corresponding attribute values, i.e.,

$$Equal(t_1, t_2) = \overset{k}{\underset{i=1}{\otimes}}\ Eq(a_i(t_1), a_i(t_2)) \qquad \text{(EQ 12.18)}$$

where $Eq(a_i(t_1), a_i(t_2))$ denotes a similarity measure or a possibility measure between ith attribute values, whichever is applicable to the attribute a_i.

Since the decision of whether to remove a tuple (due to its redundancy) is a black-and-white decision, we need to choose a threshold for the redundancy test, which we call a *redundancy threshold*.

DEFINITION 24   A fuzzy tuple t_1 is *redundant* in relation R if there exists another tuple t_2 such that
$$Equal(t_1, t_2) \geq \alpha$$
where α is a redundancy threshold.

We illustrate the concept of redundancy threshold using the following example. In the example, we use the min operator as the conjunction operator in Equation 12.18.

EXAMPLE 16   Consider the fuzzy relation shown in Table 12.10, and the similarity matrix shown in Table 12.11. Suppose the redundancy threshold is 0.8. We can calculate the equality measure between the first two tuple in Table 12.10:
$$Equal(t_1, t_2) = min\{1, 1, 0.95\} = 0.95$$
Because the equality measure of t_1 and t_2 exceeds the threshold 0.8, they are considered redundant.

TABLE 12.10   An Example of Redundant Tuples in a Fuzzy Relation

A	B	C
a	c	f
a	c	g
b	d	e
a	d	g
a	c	h

TABLE 12.11   Similarity Matrix of attribute C

	e	f	g	h
e	1	0.3	0.3	0.3
f	0.3	1	0.95	0.9
g	0.3	0.95	1	0.9
h	0.3	0.9	0.9	1

A very desirable property of the redundancy test for fuzzy tuples is transitivity. That is, if t_1, t_2 are redundant, and t_2, t_3 are redundant, then we would like to see that t_1 and t_3 are also redundant. Unfortunately, this property is not always established in fuzzy databases. If the equality of an attribute value is computed using possibility measure, the transitivity is not guaranteed to hold. We will leave this as an exercise. If the equality of attribute values is determined by a similarity matrix, then the transitivity of tuple equality (i.e., tuple redundancy) follows from the transitivity of the similarity relation. The following theorem can then be easily proved.

THEOREM 16   Let R be a fuzzy relation with attributes a_1, a_2, ... , a_k. If all attributes are represented in a similarity-based approach, and the fuzzy conjunction in Equation 12.18 is the min operator, then the tuple equality relationship is transitive for any redundancy threshold α.

We will leave the proof of this theorem as an exercise.

For a fuzzy relation containing weighted tuples, removing redundant tuples involves two issues: (1) identifying redundant tuples, and (2) determining the weight (i.e., membership degree) of the remaining tuples. So far our discussion has been focused on the first issue. The second issue is easily handled by choosing the highest weight among each set of equivalent tuples.

DEFINITION 25   Let R be a fuzzy relation containing weighted tuples t_i. Suppose T is a set of equivalent tuples,

$$T = \left\{ t_i \mid Equal(t_i, t_j) \; t_j \neq t_i, t_j \in T \right\} \qquad \text{(EQ 12.19)}$$

Let R' denotes the fuzzy relation after redundancy removal. R' contains exactly one tuple from T, whose membership is

$$\mu_{R'}(t_i) = \max_{t_j \in T} \mu_R(t_j) . \qquad \text{(EQ 12.20)}$$

Redundancy removal is typically applied to the result of a fuzzy projection and a fuzzy join, because these two operations typically generate a relation with redundant tuples.

12.4.3 Fuzzy SQL

In relational databases, SQL has become a standard query language. This section introduces the basics of SQL and an extension of the language called SQLf for fuzzy relation databases [74, 73, 469].

A query expressed in SQL consists of three major components: a **select** clause, a **from** clause, and a **where** clause. The select clause specifies the attributes to be projected (i.e., retrieved). The from clause lists all relations involved in the query. The where clause describes the selection condition of the query. For example, the following SQL expression retrieves the names and expertise of the job applicants interested in an academic career from the PROFILE relation:

```
SELECT Fname, Lname, Expertise
FROM PROFILE
WHERE JobType = 'Academic'
```

To express a fuzzy query, the syntax of SQL needs to be slightly extended to allow the specification of acceptance threshold. Two types of extensions are made. The first extension specifies one acceptance threshold for the entire query in the select clause. For example, the following SQLf expression retrieves from the PROFILE relation job applicants whose expertise matches "AI" to at least 0.8 degree:

```
Q1: SELECT 0.8 Fname, Lname, Expertise
       FROM PROFILE
       WHERE Expertise= 'AI'
```

The second extension specifies that the query retrieve the best *n* responses (i.e., the top *n* tuples in the resulting fuzzy relation ranked by tuple membership). For instance, we can use the following SQLf query to retrieve the top three job applicants whose expertise matches "Expert Systems" approximately:

```
SELECT 3 Fname, Lname, Expertise
FROM PROFILE
WHERE Expertise = 'Expert Systems'
```

A fuzzy join can be expressed in SQLf without additional changes to the syntax of SQL. For example, finding the locations for each job applicant which have job openings matching the profile of each job applicant to 0.8 degree can be expressed as

```
SELECT 0.8 Fname, Lname, Location
FROM PROFILE, JOB
WHERE PROFILE.JobType = JOB.JobType AND
       PROFILE.Expertise = JOB.Expertise
```

We should point out that the same equality symbol "=" is used in SQLf for precise (nonfuzzy) equality conditions (e.g., the equality of job type) and for imprecise (fuzzy) equality conditions (e.g., the approximate match of expertise). Hence, the meaning of the equality symbol is determined by the attributes involved.

The processing of queries in SQLf is based on transforming them into one or multiple regular SQL queries. For example, the query Q1 (finding job applicants whose expertise matches AI to at least 0.8 degree) can be transformed into the following SQL expression based on the similarity relation between areas of expertise (see Table 12.2):

```
SELECT Fname, Lname, Expertise
FROM PROFILE
WHERE Expertise = 'AI' OR
       Expertise = 'Expert Systems'
```

FIGURE 12.3   Membership Functions for Large Population and Low GNP

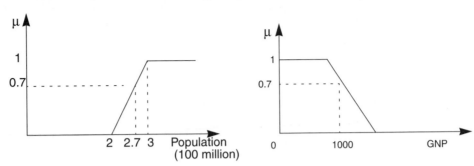

Similar transformations can also be made to imprecise conditions about precise attribute values. Suppose NATIONS is a relation that contains numerical attributes such as POPULATION and GNP (Gross National Product per person), an economic index about the financial strength of a nation. If we wish to find nations that have approximately (0.7) low GNP and large population, we can express this in SQLf as follows:

```
SELECT 0.7 Name
FROM NATIONS
WHERE Population = LARGE AND
      GNP = LOW
```

This query can be transformed into SQL queries by finding the 0.7-cut of LARGE population and 0.7 α-cut of LOW GNP. Based on the membership functions in Fig 12.3, the two α-cuts obtained are

$$LARGE_{0.7}(p) = \{p \mid p \geq 270,000,000\}, \text{ and}$$
$$LOW_{0.7}(g) = \{g \mid (g \leq 1000)\}$$

The SQLf query above can then be transformed into the following SQL query:

```
SELECT Name
FROM NATIONS
WHERE Population ≥ 100,000,000 AND
      GNP ≤ 1000
```

12.5 Design Theory for Fuzzy Relational Databases

12.5.1 Functional Dependencies

A good database design should consider the dependency relationship between attributes based on their meaning in the real world modeled by the database. Analyzing such a dependency relationship could avoid redundancy and anomalies in the database. Let us

consider the job opportunity relation used earlier. Suppose we add the name and the address of the employer into the relation. The scheme of the relation becomes
          `JobEmp(Ename, Eaddr, JobType, Expertise, Location)`.
There are several problems with this scheme. First, the employee address will be duplicated for an employer that has multiple job openings. Second, it is difficult to insert an employee into the relation that does not currently have job openings unless we allow null values. Even if null values are allowed, the design would require a user who wishes to insert a job opening into the relation to first check if the employer has a null-valued job entry. If such entry is found, it needs to be replaced with the new (and real) job opening. Obviously, such a design is not a good one. The cause of the problem is that the name of an employer (`Ename`) determines its address, independent of other attributes in the relation. Hence, it is better to remove the employer address from the job relation and create a new relation for information about employers using the scheme
          `EMP(Ename, Eaddr, Tel, Fax, Email, Webaddr)`
where `Email` and `Webaddr` are E-mail address and home page address on the Web of the employers.

Functional dependency is a restriction on relations about equality or inequality of attribute values.

DEFINITION 26   Let R be a relation with attributes $A_1, A_2, ..., A_n$. Let X and Y be subsets of $\{ A_1, A_2, ..., A_n \}$. We say $X \to Y$, read as "X functionally determines Y" or "Y functionally depends on X" if identical values of X imply identical values of Y i.e.,

$$\forall t_1, t_2 \ (\forall A_i \in X \ \ A_i(t_1) = A_i(t_2)) \to (\forall A_j \in Y \ \ A_j(t_1) = A_j(t_2)) \quad \textbf{(EQ 12.21)}$$

where A_i and A_j denote attributes in X and Y, respectively, and t_1, t_2 denote elements in the Cartesian product of R's attribute domains.

In the definition above, we used logic rather than relational algebra to formally define functional dependency because it is a constraint imposed by the meaning (i.e., semantics) of attributes. This is clearly pointed out by Jeffery D. Ullman in his textbook *Principles of Database and Knowledge-base Systems* [604]:

> "We cannot look at a particular relation r for scheme R and
> deduce what functional dependencies hold for R."

For example, if we have a `JobEmp` relation mentioned earlier, let us assume that each employer in the relation happens to have only one job opening. Hence, tuples with different `Ename` all have different `Expertise`. However, we cannot infer from this that `Expertise` functionally depends on `Ename`. In fact, based on the meaning of these attributes in the real world, we know an employer could have job openings that require different areas of expertise. Therefore, the functional dependency does not hold between Ename and Expertise, even though employer in a `JomEmp` relation may happen to have openings in only one area of expertise.

12.5.1.1   Keys

Functional dependency enables us to identify the smallest set of attributes in a relation, called *keys*, that can uniquely determine a tuple. This is formally stated in the following definition.

DEFINITION 27 Let R be a relation with attributes $A_1, A_2, ..., A_n$, and X be a subset of $\{A_1, A_2, ..., A_n\}$. We say X is a key of relation R if
1) $X \rightarrow A_1, A_2, ..., A_n$,
2) There is no proper subset Y of X ($Y \subset X$) such that
$Y \rightarrow A_1, A_2, ..., A_n$.

For example, if each employer has only one contact address, telephone number, FAX number, e-mail address, and home page on the Web, then Ename in the employer relation is a key.

12.5.2 Functional Dependency in Fuzzy Relations

Like most other concepts in a database, the notion of functional dependency needs to be extended for fuzzy databases. Such a need is related to the generalization "attribute equality" discussed in Section 12.2.3. Because the equality between two imprecise attribute values becomes a matter of degree in a fuzzy database, the logic formula in Equation 12.21 that defines functional dependency needs to be extended. More specifically, we need not only to replace "=" with "Eq" (i.e., fuzzy equality measure), but also to replace logic implication " $\rightarrow$ " with a fuzzy implication. Since the functional dependency should be a black-and-white notion (i.e., the selection of key is a black-and-white matter), we need to choose a fuzzy implication operator that maps degrees of truth in the antecedent and consequent to 0 or 1.

The standard sequence fuzzy implication introduced in Section 6.3.4 satisfies this criterion. This fuzzy implication operator is defined as

$$t(F_1 \rightarrow F_2) = \begin{cases} 1 & t(F_1) \le t(F_2) \\ 0 & t(F_1) > t(F_1) \end{cases}$$

where $t(F)$ denotes the truth value of a logic formula F. Using this fuzzy implication, we can define fuzzy functional dependency.

DEFINITION 28 Let R be a fuzzy relation with attributes $A_1, A_2, ..., A_n$. Let X and Y be subsets of $\{A_1, A_2, ..., A_n\}$. We say "Y functionally depends on X" if the following holds:

$$\forall t_1, t_2 \left(\bigwedge_{A_i \in X} Eq(A_i(t_1), A_i(t_2)) \right) \rightarrow \bigwedge_{A_j \in XY} Eq(A_j(t_1), A_j(t_2)) \quad \textbf{(EQ 12.22)}$$

where A_i and A_j denote attributes in X and Y, respectively, Eq denotes a fuzzy quality predicate based on similarity measure or possibility measure, and $\rightarrow$ denotes standard sequence fuzzy implication.

Functional dependency of fuzzy relations can also be defined using other fuzzy implication functions that map to [0, 1] (e.g., the Godel implication introduced in Chapter 6). Conceptually, these approaches say "X functionally depends on Y" if the degree of truth of the fuzzy implication in Equation 12.22 exceeds a threshold value

$$\forall t_1, t_2 \bigwedge_{A_i \in X} Eq(A_i(t_1), A_i(t_2)) \rightarrow \bigwedge_{A_j \in Y} Eq(A_j(t_1), A_j(t_2))$$

It is significant to note that in these various approaches, the notion of functional dependency defined for fuzzy relations remains Boolean, i.e., it is either fully satisfied or fully unsatisfied, which make it compatible with the notion of a regular integrity constraint.

12.6 Fuzzy Object-Oriented Databases

In the 1980s, the rapid advancement in artificial intelligence technology affected the database community in two important ways — there was a growing interest to (1) capture the "meaning" of data and (2) to augment the functionality of a DBMS with automatic reasoning capabilities. The former interest led to the development of object-oriented databases, whereas the latter interest contributed to the development of deductive databases. Obviously, object-oriented data models combine ideas in Object-Oriented Programming (OOP) and knowledge representation in AI with those in databases; while deductive databases incorporate into databases the techniques of rule-based reasoning and/or PROLOG (an AI programming language based on horn clauses in logic). In this section, we briefly review issues in manipulating imprecise information in object-oriented databases.

In object-oriented data models, data is stored in *objects*. An object is analogous to a tuple in a relational model with one important difference — each object has its own implicit identity. Therefore, two objects with identical attribute values are considered different, even though two tuples with identical contents are considered equivalent (and hence one of them is redundant). An object can be viewed as a tuple with an implicit unique identifier. *Instance variables* in an object, like attributes of tuples, store information about the object.

Objects belong to one or more *classes*. A class plays a role similar to that of a relation in a relational model — both of them group sets of lower-level information units (i.e., tuples or objects) into more abstract units. The main difference between relations and classes is that classes themselves can be further grouped into more abstract classes or specialized into more specific classes, but relations can't. The generalizations (i.e., more abstract classes) of a class is called its superclass, and its specialization (i.e., more specific classes) are called subclasses. Hence, classes and their links to superclass/subclass form a hierarchical structure. This hierarchical structure enables an object-oriented data model to offer an important reasoning capability — to inherit the attribute values of an object from its classes. This capability, which originates in frame-based representation in AI, is called *inheritance*.

In addition to inheriting data, object-oriented data models also inherit procedures for manipulating data. This is achieved by using an important idea in object-oriented programming — separating the desired functionality of a procedure from its various specific implementations (depending on the classes of its argument). A desired functionality is referred to as a *message*, whereas the implementation of the message for objects in a class is specified as a *method* of the class.

Unlike relational data models, object-oriented data models do not have standard operations for retrieving information. They need to be coded in methods. Typical methods include those for retrieving and updating the value of an object's instance variables, for changing the class of an object, for finding all objects in a given class, and for selecting objects in a class that satisfy a given condition.

Imprecise information in an object-oriented database can be classified into three categories: (1) imprecision of the data encapsulated in the object, (2) partial class membership, and (3) fuzzy superclass/subclass relationship.

The first category, where the data encapsulated within an object is imprecise, is quite similar to the approaches discussed in the relational database section. Approaches like similarity measures or possibility distributions can be taken to capture this kind of imprecision.

FIGURE 12.4   Partial Class Membership

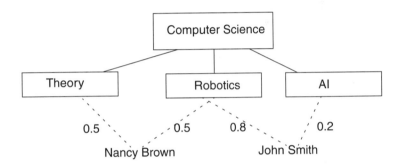

In previous sections, we have mentioned the need to represent partial membership in relational databases. For object-oriented databases, this need corresponds to representing partial membership of an object in a class. Suppose we want to use object-oriented databases to store the areas of faculty in a computer science department. Suppose Professor John Smith conducts research in robotics and AI areas, with more emphasis on robotics, and Professor Nancy Brown conducts research in theory and robotics with an equal emphasis. Suppose that this database is used by the department head to calculate the total number of faculty members in each area. Professors Smith and Brown wish to be counted as a partial member in each of the two areas in which they are involved. A possible representation of this imprecise information is shown in Fig 12.4. Classes are depicted as rectangles in the figure, solid lines indicate superclass/subclass relationships, and dotted lines indicate class membership of objects.

Introducing partial membership, however, does affect the inheritance scheme. A common strategy is to use the membership degree of an object in a class as the priority of the class in searching for a source (i.e., a parent class) for inheriting the object's data and methods. In the example above, the object John Smith inherits information and methods first from the Robotics Class (e.g., having a physical robot lab). The object inherits information and methods from the `AI` class only if they cannot be found in the `Robot-ics` class. Obviously, the object inherits from `Computer Science` class if neither AI nor Robotics has the data or method needed by the object.

The third kind of imprecise information — fuzzy superclass/subclass relationship — is possible only for object-oriented databases. Once we allow a class to have partial members, a class becomes a fuzzy set of objects. A class C_1 is a superclass of another class C_2, also called C_1 subsume C_2 (denoted $C_1 > C_2$), if and only if all objects in C_2 are also in C_1. This can be formally written as

$$C_1 > C_2 \leftrightarrow \forall x \ C_2(x) \rightarrow C_2(x)$$

(EQ 12.23)

Alternatively, we can define class subsumption in terms of the relationship of their extensions (i.e., the entire set of other objects):

$$C_1 > C_2 \leftrightarrow \varepsilon(C_1) \supset \varepsilon(C_2)$$

(EQ 12.24)

where $\varepsilon(C)$ denotes the extension of class C. If C_1 and C_2 becomes fuzzy sets, the formula can be extended using a fuzzy inclusion operator. Conceptually, this means that the degree C_1 is a superclass of C_2 is the degree C_1's extension includes C_2's extension.

DEFINITION 29   Let C_1 and C_2 be two fuzzy classes. The degree that C_1 is a superclass of C_2, denoted by *subsume(C_1, C_2)*, is defined as

$$Subsume(C_1, C_2) = Subset(\varepsilon(C_2), \varepsilon(C_1))$$

(EQ 12.25)

where *Subset* denotes a fuzzy subsethood measure.

Fuzzy subsethood operators introduced in Chapter 3 can thus be used to formalize superclass/subclass relationships. We will illustrate the concept of fuzzy superclass/subclass using our previous example. Suppose the Dean of the College of Engineering wished to establish a new program in computer engineering. Let us assume that most of the faculty in Robotics are considered in the computer engineering area, while some faculty in AI are in computer engineering, and almost none in Theory (except Professor Nancy Brown) is in computer engineering. We may represent this imprecision using fuzzy superclass/subclass relationships, as shown in Fig 12.5.

Once we introduce fuzzy subsumption relationships, we need to be able to calculate the degree an object belongs to any class that can be reached by traversing one or more superclass links. For instance, we should be able to answer questions such as "To what degree does Prof. Nancy Brown belong to computer engineering faculty?" One solution is to multiply the weights of all the links on a path that reaches the "ancestor" class. If there are multiple paths, use the highest membership calculated among all paths. For instance, we can calculate the membership degree that Nancy Brown is a Computer Engineering faculty member as follows:

$$\mu_{Computer\ Engineering}(Nancy\ Brown) = max\{0.5 \times 0.9, 0.5 \times 0.1\} = 0.45$$

Alternatively, we can replace the product with the min operator or other fuzzy conjunction operators. Similarly, max can be replaced by other fuzzy disjunction operators.

FIGURE 12.5   Fuzzy Inheritance

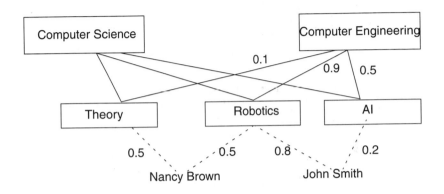

The issue of multiple inheritance is very complicated in the original object-oriented methodologies, because it introduces some ambiguities such as which function of which class should it be called if there happens to be some overriding functions. These issues are moreover reinforced by having partial inheritance. Which superclass should we use? Should we take the super class with the highest membership degree?

12.7 Fuzzy Information Retrieval and the Web Search

Information retrieval deals with the issue of retrieving documents of interest to a user from a large repository of documents. The user typically describes his/her interests using keywords. Documents can be retrieved by matching user's keywords with keywords associated with documents, or by analyzing the frequency that the user's keywords appear in the documents. Keywords can also be weighted so that general and common words carry less weight than words that are novel or not commonly used.

Even though both database systems and information retrieval systems deal with the storage and retrieval of a large amount of information, they differ significantly in the degree the information is structured. Information in a database system is well structured, and the content of a field (e.g., an attribute value in a relation) is relatively simple (e.g., number, single words, short phrases). In contrast, documents in an information retrieval system are usually not well structured, and the content of those documents is typically complex (e.g., paragraphs or even sections of free text). Because of these differences, the advancement of information retrieval technology was not as exciting as that of database technology until mid 1990s. The birth of the World Wide Web (WWW) and the advancement of computer network technology enable every computer user to access a huge amount of rich information on the Web from anywhere in the world. However, finding the needed information on the Web can be time consuming. Hence, information retrieval techniques have been reincarnated in many Web search engines (Web tools that assist the user in retrieving Web documents matching the user's interests) or metasearch engines (Web tools that send a user's query to multiple search engines and integrate their responses before presenting them to the user).

Fuzzy logic can be used to develop novel information retrieval techniques in three ways. First, it can be used to capture the similarity relationship between keywords so that documents that contain keywords similar to (but not identical to) the keywords in the user's query can also be retrieved. RUBRIC is an example of such a fuzzy information retrieval system. Second, it can be used to compute a fuzzy relevance measure between two terms based on the frequency of their co-occurrences in documents. The third usage of fuzzy logic in information retrieval is for representing heuristics in determining the weights of different fields in an HTML document for intelligent Web search. We discuss these three usages below.

RUBRIC, a fuzzy information retrieval system that was developed by R. Tong, V. Askman and J. Cunningham, retrieves documents based on the relevance of the document to the semantics of the user's query [596]. RUBIC used rules to link words to concepts, which are also connected to semantically related concepts by rules. Each rule has a "relevance value" to indicate the strength of the association between its antecedent concepts/words and consequent concept. For example, if both "killing" and "politician" occur in a document, it suggests the document is somewhat related to assassination. This can be expressed as a rule with relevance value 0.5. During the information retrieval, RUBIC views the user's query as a goal. Through goal-driven backward chaining, it determines the degree the document is relevant to the query. Fig 12.6 shows an example inference tree for finding documents related to the concept terrorism. The leaves that are found in a specific document are assigned a relevance value of 1. Otherwise, the term is given a relevance value of 0. These relevance values are propagated upward in the inference tree using rules. Several rules have "auxiliary antecedents", whose appearance in the document modifies (usually strengthens) the relevance value of the rule consequent. For example, a violent action is quite (0.8 degree) relevant to a terrorist event. However, a violent action occurring together with assassination is completely (1.0) relevant to a terrorist event, as shown in the figure.

The relevance of an inferred concept is calculated by RUBIC based on the following equation:

$$v[consequent] = \alpha \times v[primary-antecetent] + (\beta - \alpha) \times v[auxiliary-antecedent] \quad \textbf{(EQ 12.26)}$$

where α and β are relevance values associated with the primary antecedent and the auxiliary antecedent respectively, $v[c]$ denotes the relevance of the current document to the concept c. Relevance values of a concept inferred from multiple rules are combined using fuzzy disjunction.

Using a measure of *recall* (i.e., the ration of the number of relevant documents retrieved to the total number of relevant documents in the database) and a measure of *precision* (i.e., the ratio of the number of relevant documents retrieved to the total number of documents retrieved), RUBIC's performance in retrieving information has been shown to be superior to a comparable approach using nonfuzzy rules.

FIGURE 12.6   Terrorism Example

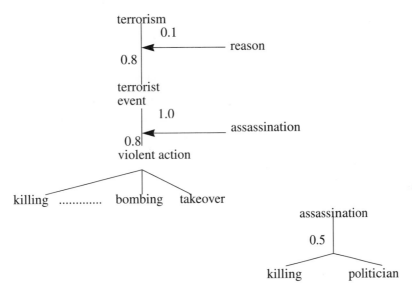

Even though RUBRIC's approach is effective for retrieving documents related to a small set of predefined concepts, it is difficult to scale up the approach for general information retrieval systems. Constructing rules relating all possible keywords and concepts is a formidable task. An alternative approach is to automatically generate a relevance measure between keywords by analyzing all documents in the database of an information retrieval system. S. Miyamoto proposed an approach for generating a fuzzy relevance measure (called *fuzzy pseudothesaurus*) based on the assumption that if two terms occur together frequently in documents, then they are relevant [419, 420]. A fuzzy information retrieval system was implemented using this technique. Outputs of the system are classified into layers based on a relevance threshold dynamically determined by the system [418, 417].

The third usage of fuzzy logic in information retrieval was demonstrated by A. Molinari and G. Pasi in 1996 [426]. They use fuzzy logic to develop a principled approach for assigning weights to different components, which are specified by tags, in an HTML document. The rationale is that a word in the title carries much more weight than the same word appearing in other portions of the document. Therefore, we can sort tags based on their degree of importance. For instance, a sorted list may be Title, Header1, Header2, emphasis, delimiters, etc. Based on the order in the list, a fuzzy weight can be calculated for each tag:

$$w_i = \frac{(n - i + 1)}{\sum_{i = 1., ..., n} i}$$ **(EQ 12.27)**

where n is the total number of tags in the sorted list. The total relevance measure of a document, denoted F, to a query q is a weighted sum of relevance measure of each tag:

$$F(q) = \sum_{i=1}^{n} W_{t_i} \times F_{t_u}(q)$$ **(EQ 12.28)**

where $F_{t_u}(q)$ denotes the degree that the content of tag t_i is relevant to the query q. The aggregation is just the summation of the $F_{tagi} \times w_i$ for all the i values.

12.8 Summary

In this chapter, we discussed the most evident past, present, and future research in the area of fuzzy databases. There are other types of databases (example, rough sets, geographical information systems). However, we will not explore more these database types. We have presented two types: network models and object-oriented models. We chose these two types to give the reader the flavor of the past versus the future. We discussed the past through the network data models and the future through the object oriented databases. As M. Umano stated, "FOOD [Fuzzy Object-Oriented DataBase] will grow up to a new generation of fuzzy database system" [606]. Furthermore, the future of fuzzy information system's research was also depicted in the area of fuzzy information retrieval. A discussion of the most widely used database type in the present day, the relational model, was given. Moreover, we provided a discussion of the different approaches that have been followed to develop a fuzzy relational database. We also looked at how queries are formulated and executed by each of these different approaches.

Bibliographical and Historical Notes

Relational data models were introduced by E. F. Codd in the early 70s [137, 136]. The foundation of fuzzy databases and fuzzy information systems is the fuzzy relation introduced in Chapter 4 and possibility theory introduced in Chapter 3. More specifically, the similarity-based fuzzy relation is based on Zadeh's similarity relation [703]. Obviously, the possibility-based fuzzy relation is based on possibility theory, which was introduced by Zadeh [693], and further developed by D. Dubois, H. Prade and others [176].

As we mentioned earlier in this chapter, similarity-based fuzzy relations were first proposed by B. Buckles and F. Petry in early 80s [96]. About the same time, possibility-based fuzzy relations were introduced by M. Umano [605, 606] and H. Prade [486]. Ruspini developed a possibilistic approach for attributes whose values are organized into a lattice structure [514]. M. Zemankova and Kandel advocated the fuzzy relational model and the use of linguistic quantifiers [714, 713]. Fuzzy logic was also incorporated into relational models through proximity relationships by S. Shenoi [534,533]. Different types of fuzzy relational models were combined by J. Baldwin and others [22,409,513]. The issue of functional dependency in fuzzy relational models is investigated in [99,100,489,535].

Entity-Relationship (ER) data models, which were originally developed by Y. Chen [119], have been extended using fuzzy logic [154, 721]. D. Dubois and H. Prade have discussed the handling of uncertainty in a class hierarchy, which is important for both fuzzy frame-based representation and fuzzy object-oriented models [166]. Several approaches to fuzzy object-oriented data models have been proposed in the 90's [206,230,71,364].

S. Chang and J. Ke proposed a scheme for translating fuzzy queries into relational databases in the late 70s [107, 110]. The fuzzy SQL query language (fSQL) described in this chapter was developed by P. Bosc and his colleague [74, 75]. They have also extended database query languages toward "flexible queries" such that the user can express the degree of importance of various conditions in a query. Based on L. Zadeh's test-score semantics [689], J. Kacprzyk incorporated fuzzy linguistic quantifiers into database queries [289].

An introduction to information retrieval systems can be found in [347, 408]. Fuzzy set-based generalizations to information retrieval systems were developed by C. Negoita, D. Kraft, and others [444, 338]. M. Zemankova proposed a model of an information system with learning capabilities, the FLIP (Fuzzy Intelligent Learning Information Processing) system, based on deductive relational database models [712].

A more detailed treatment of the subject of fuzzy database systems can be found in a text by F. Petry [469].

Exercises

12.1 Give a new example (different from those we have discussed) for each of the following two types of imprecise information: (1) imprecise attribute value, and (2) partial membership in a relation.

12.2 Give a new example for each of the following three components of imprecise queries: (1) imprecise condition, (2) imprecise operator, and (3) imprecise quantifier.

12.3 It has been stated that the three main approaches to fuzzify information within database systems are :
1. Maintain the standard data model and allow for imprecise queries.
2. Retain the standard language (SQL) and extend the data model to handle imprecision.
3. Extend the data model to handle imprecision and allow imprecision for queries.
Elaborate on each of these approaches explaining how each could be realized. Feel free to use examples of your own, if it will explain the point better.

12.4 What is the difference between similarity and possibility approaches for fuzzy databases? What are the advantages and disadvantages of these approaches? Give examples where you would tend to favor one approach over the other.

12.5 Modify the similarity matrix in Table 12.2 such that similarity between AI and robotics is 0.7, and the similarity between AI and statistics is 0.3. (Hint: The new matrix needs to satisfy all properties of a similarity relation.)

12.6 In addition to the modifications to Table 12.2 described in the previous problem, can we also change the similarity measure between Expert Systems and Robotics to 0.5? Explain your answer.

12.7   Table 12.12 below is a relation about student information. The similarity relationship between different areas of expertise is the one used in this chapter (i.e., the similarity relation in Table 12.2).   The
attribute for expected graduation date has a discrete domain D = {F97, S98, F98, S99, F99} where F
and S denote fall semester and spring semester, respectively.

TABLE 12.12   STUDENT Relation

ID	Name	Expertise	GPA	Expected Grad Date
08992	John	Robotics	3.7	[0 0.2 0.5 1 0.8]
09901	Susan	Statistics	3.8	[0.2 0.9 1 0.6 0]
09911	Bob	AI	4	[0 0 0.1 0.5 1]
09099	George	Expert Systems	3.4	[0 0.1 0.3 0.4 1]

Based on the STUDENT relation, construct the following queries using (1) fuzzy relational algebra,
and (2) SQLf.   Then find the response of these queries.

(a) Find the graduate students whose expertise is approximately equal to AI and whose GPA is HIGH,
     where HIGH GPA is characterized by the membership function in Fig 12.7.
(b) Using an acceptance threshold of 0.5, find the graduate students whose expertise is approximately
     equal to Expert Systems and whose expected graduation date is about-F98, where about-F98 is
     characterized by the membership function: $\mu_{about\text{-}F98}(x)$ = [0 0.7 1 0.2 0].

FIGURE 12.7   Membership Function of High GPA

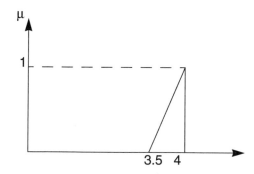

12.8   Using the example discussed in Section 12.3.2, express the following queries using fuzzy relational
algebra and show their results:

(a) Find the job applicants whose expertise matches a job opening in California.
(b) Find the employer who has a job opening in an approximate AI area.

12.9   Calculate the equality measure for all tuples in Table 12.10 using the information described in
Example 16.

12.10   Show that possibility measure is not transitive.

12.11   Prove Theorem 16.

12.12   Prove that the set of equivalent tuples in a fuzzy relation R forms an equivalent partition of R if the tuple quality relation is transitive.

12.13   Derive the formula for fuzzy-Join (Equation 12.17) from the definition of basic operations in fuzzy relational algebra (i.e., fuzzy selection, fuzzy projection, etc.).

13 FUZZY LOGIC IN PATTERN RECOGNITION

13.1 Introduction

Pattern recognition techniques can be classified into two broad categories: *unsupervised* techniques and *supervised* techniques. An unsupervised technique does not use a given set of unclassified data points, whereas a supervised technique uses a dataset with known classifications. These two types of techniques are complementary. For example, unsupervised clustering can be used to produce classification information needed by a supervised pattern recognition technique. In this chapter, we first introduce the basics of unsupervised clustering. The Fuzzy C-Means algorithm (FCM), which is best known unsupervised fuzzy clustering algorithm is then described in detail. Supervised pattern recognition and knowledge-based pattern recognition using fuzzy logic will follow. Finally, we describe hybrid approaches and an application of hybrid fuzzy systems in medical image segmentation.

13.2 Unsupervised Clustering

Unsupervised clustering is motivated by the need to find interesting patterns or groupings in a given set of data. For example, a market analysis company may want to collect data about a group of customers (e.g., through a survey questionnaire or interviews) and analyze these data by finding interesting groupings of customers. The result of such an analysis can be used by a company to develop its marketing strategy or to develop a family of product lines that are targeted toward various customer groupings.

In the area of pattern recognition and image processing, unsupervised clustering is often used to perform the task of "segmenting" the images (i.e., partitioning pixels on an image into regions that correspond to different objects or different faces of objects in the images). This is because image segmentation can be viewed as a kind of data clustering problem where each datum is described by a set of image features (e.g., intensity, color, texture, etc.) of a pixel.

Conventional clustering algorithms find a "hard partition" of a given dataset based on certain criteria that evaluate the goodness of a partition. By "hard partition" we mean that each datum belongs to exactly one cluster of the partition. More formally, we can define the concept "hard partition" as follows.

DEFINITION 30   Let X be a set of data, and x_i be an element of X. A partition $P = \{C_1, C_2, ..., C_l\}$ of X is **"hard"** if and only if

$$i) \forall x_i \in X \quad \exists C_j \in P \qquad \text{such that} \qquad x_i \in C_j$$
$$ii) \forall x_i \in X \quad x_i \in C_j \Rightarrow x_i \notin C_i \quad \text{where } k \neq j, C_k, C_j \in P$$

The first condition in the definition assures that the partition covers all data points in X, the second condition assures that all clusters in the partition are mutually exclusive.

In many real-world clustering problems, however, some data points partially belong to multiple clusters, rather than to a single cluster exclusively. For example, a pixel in a Magnetic Resonance Image (MRI) may correspond to a mixture of two different types of tissues. A particular customer may be a "borderline case" between two groups of customers (e.g., between moderate conservatives and moderate liberals). These observations motivated the development of the "soft clustering" algorithm.

A soft clustering algorithms finds a "soft partition" of a given dataset based on certain criteria. In a soft partition, a datum can partially belong to multiple clusters. We formally define this concept below:

DEFINITION 31   Let X be a set of data, and x_i be an element of X. A partition $P = \{C_1, C_2, ..., C_l\}$ of X is **soft** if and only if the following two conditions hold:

1) $\forall x_i \in X \quad \forall C_j \in P \quad 0 \leq \mu_{c_j}(x_i) \leq 1$:
2) $\forall x_i \in X \quad \exists C_j \in P$ such that $\mu_{c_j}(x_i) > 0$.

where $\mu_{c_j}(x_i)$ denotes the degree to which x_i belongs to cluster C_j .

It is easy to see that the concept of soft partition is similar to the concept of fuzzy partition introduced in Chapter 14 in the context of fuzzy models for function approximation. The two differ in a subtle way, though: A soft clustering algorithm partitions a given dataset, not an input space. Theoretically speaking, a soft partition is not necessarily a fuzzy partition, since the input space can be larger than the dataset. In practice, however, most soft clustering algorithms do generate a soft partition that also forms a fuzzy partition.

A type of soft clustering of special interest is one that ensures the membership degree of a point x in all clusters adding up to one, i.e.,

$$\sum_j \mu_{c_j}(x_i) = 1 \qquad\qquad \forall x_i \in X$$

A soft partition that satisfies this additional condition is called a constrained soft partition. The fuzzy c-means algorithm, which is best known fuzzy clustering algorithm, produces a constrained soft partition.

A constrained soft partition can also be generated by a probabilistic clustering algorithm (e.g., maximum likelihood estimators). Even though both fuzzy c-means and probabilistic clustering produce a partition of similar properties, the clustering criteria underlying these algorithms are very different. While we focus our discussion on fuzzy clustering in this chapter, we should point out that probabilistic clustering has also found successful real-world applications. Fuzzy clustering and probabilistic clustering are two different approaches to the problem of clustering.

The fuzzy c-means algorithm generalizes a hard clustering algorithm called the c-means algorithm, which was introduced in the ISODATA clustering method. The (hard) c-means algorithm aims to identify compact, well-separated clusters. Fig13.1 shows a two-dimensional dataset containing compact well-separated clusters. In contrast, the dataset shown in Fig 13.2 contain clusters that are not compact and well separated. Informally, a *compact* cluster has a "ball-like" shape. The center of the ball is called *the center* or *the prototype* of the cluster. A set of clusters are *well separated* when any two points in a cluster are closer than the shortest distance between two clusters in different clusters. Fig 13.3 shows two clusters that are not well separated because there are points in C_2 (such as the one labeled x) that are closer to a point in C_1 than a point in C_2. We formally define well-separated clusters below.

DEFINITION 32   A partition $P=\{C_1, C_2, ..., C_k\}$ of the dataset X has Compact Separated (CS) clusters if and only if any two points in a cluster are closer than the distance between two points in different cluster, i.e, $\forall x, y \in C_p \ \ d(x, y) < d(z, w)$ where $z \in C_q$, $w \in C_r$, $j \neq k$, and d denotes a distance measure.

Assuming that a dataset contains c compact, well-separated clusters, the goals of hard c-means algorithm is twofold:
(1) To find the centers of these clusters, and
(2) To determine the clusters (i.e., labels) of each point in the dataset.

In fact, the second goal can easily be achieved once we accomplish the first goal, based on the assumption that clusters are compact and well separated. Given cluster centers, a point in the dataset belongs to the cluster whose center is the closest, i.e.,

$$x_i \in C_j \ \text{ if } \ \|x_i - v_j\| < \|x_i - v_k\| \quad k = 1, 2, ..., c, k \neq j \qquad \textbf{(EQ 13.1)}$$

where v_j denotes the center of the cluster C_j.

FIGURE 13.1   An Example of Compact Well Separated Cclusters.

FIGURE 13.2   An Example of Two Clusters that are not Compact and Well Separated

FIGURE 13.3   Two Clusters that are Compact, but not Well Separated

In order to achieve the first goal (i.e., finding the cluster centers), we need to establish a criterion that can be used to search for these cluster centers. One such criteria is the sum of the distance between points in each cluster and their center.

$$J(P, V) = \sum_{j=1}^{c} \sum_{x_i \in C_j} \|x_i - v_j\|^2 \qquad \text{(EQ 13.2)}$$

where V is a vector of cluster centers to be identified. This criterion is useful because a set of true cluster centers will give a minimal J value for a given dataset. Based on these observations, the hard c-means algorithm tries to find the cluster centers V that minimize J. However, J is also a function of partition P, which is determined by the cluster centers V according to Equation 13.1. Therefore, the Hard C-Means algorithm (HCM) searches for the true cluster center by iterating the following two steps:
(1) Calculating the current partition based on the current cluster,
(2) Modifing the current cluster centers using a gradient descent method to minimize the J function.

The cycle terminates when the difference between cluster centers in two cycles is smaller than a threshold. This means that the algorithm has converged to a local minimum of J.

13.3 Fuzzy c-Means Algorithm

The Fuzzy C-Means algorithm (FCM) generalizes the hard c-means algorithm to allow a point to partially belong to multiple clusters. Therefore, it produces a soft partition for a given dataset. In fact, it produces a constrained soft partition. To do this, the objective function J_1 of hard c-means has been extended in two ways:
1) the fuzzy membership degrees in clusters were incorporated into the formula, and
2) an additional parameter m was introduced as a weight exponent in the fuzzy membership.
The extended objective function, denoted J_m, is

$$J_m(P, V) = \sum_{i=1}^{k} \sum_{x_k \in X} (\mu_{C_i}(x_k))^m \|x_k - v_i\|^2 \qquad \text{(EQ 13.3)}$$

where P is a fuzzy partition of the dataset X formed by $C_1, C_2, ..., C_k$. The parameter m is a weight that determines the degree to which partial members of a cluster affect the clustering result.

Like hard c-means, fuzzy c-means also tries to find a good partition by searching for prototypes v_i that minimize the objective function J_m. Unlike hard c-means, however, the fuzzy c-means algorithm also needs to search for membership functions μ_{C_i} that minimize J_m. To accomplish these two objectives, a necessary condition for local minimum of J_m was derived from J_m. This condition, which is formally stated below, serves as the foundation of the fuzzy c-means algorithm.

THEOREM 17   **Fuzzy c-Means Theorem**

A constrained fuzzy partition $\{C_1, C_2, ..., C_k\}$ can be a local minimum of the objective function J_m only if the following conditions are satisfied:

$$\mu_{C_i}(x) = \frac{1}{\sum_{j=1}^{k} \left(\frac{\|x - v_i\|^2}{\|x - v_j\|^2} \right)^{\frac{1}{m-1}}} \qquad 1 \le i \le k, x \in X \qquad \textbf{(EQ 13.4)}$$

$$v_i = \frac{\sum_{x \in X} (\mu_{C_i}(x))^m \times x}{\sum_{x \in X} (\mu_{C_i}(x))^m} \qquad 1 \le i \le k \qquad \textbf{(EQ 13.5)}$$

Based on this theorem, FCM updates the prototypes and the membership function iteratively using Equations 13.4 and 13.5 until a convergence criterion is reached. We describe the algorithm in Fig 13.4.

FIGURE 13.4   The Fuzzy C-Means (FCM) Algorithm

FCM (X, c, m, ε)

$\quad$ *X*: an unlabeled data set
$\quad$ *c*: the number of clusters to form
$\quad$ *m*: the parameter in the objective function
$\quad$ ε: a threshold for the convergence criteria

Initialize prototype $V = \{v_1, v_2, ... , v_c\}$
Repeat

$\quad$ $V^{Previous} \leftarrow V$
Compute membership functions using Equation 13.5.
Update the prototype, v_i in V using Equation 13.4

Until $\sum_{i=1}^{c} \| v_i^{Previous} - v_i \| \le \varepsilon$

We will use the following example to illustrate the FCM algorithm.

EXAMPLE 17   Suppose we are given a dataset of six points, each of which has two features F_1 and F_2. We list the dataset in Table 13.1. Assuming that we want to use FCM to partition the dataset into two clusters (i.e., the parameter $c = 2$), suppose we set the parameter m in FCM at 2, and the initial prototypes to $v_1 = (5, 5)$ $v_2 = (10, 10)$.

TABLE 13.1   A Dataset to Be Partitioned

	f_1	f_2
x_1	2	12
x_2	4	9
x_3	7	13
x_4	11	5
x_5	12	7
x_6	14	4

FIGURE 13.5   An Example of Fuzzy c-Means Algorithm

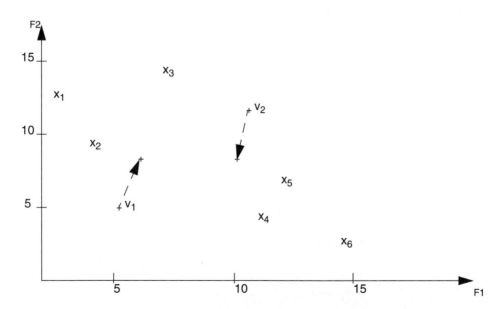

The initial membership functions of the two clusters are calculated using Equation 13.4:

$$\mu_{C_1}(x_1) = \cfrac{1}{\displaystyle\sum_{j=1}^{2}\left(\cfrac{\|x_1 - v_1\|}{\|x_1 - v_j\|}\right)^2} \qquad \text{(EQ 13.6)}$$

$$\|x_1 - v_1\|^\sim = 3^\sim + 7^\sim = 9 + 49 = 58$$

$$\|x_1 - v_2\|^2 = 8^2 + 2^2 = 64 + 4 = 68$$

$$\mu_{C_1}(x_1) = \frac{1}{\dfrac{58}{58} + \dfrac{58}{68}} = \frac{1}{1 + 0.853} = 0.5397$$

Similarly, we obtain the following:

$$\mu_{C_2}(x_1) = \frac{1}{\dfrac{68}{58} + \dfrac{68}{68}} = 0.4603$$

$$\mu_{C_1}(x_2) = \frac{1}{\dfrac{17}{17} + \dfrac{17}{37}} = 0.6852$$

$$\mu_{C_2}(x_2) = \frac{1}{\dfrac{37}{17} + \dfrac{37}{37}} = 0.3148$$

$$\mu_{C_1}(x_3) = \frac{1}{\dfrac{68}{68} + \dfrac{68}{18}} = 0.2093$$

$$\mu_{C_2}(x_3) = \frac{1}{\dfrac{18}{68} + \dfrac{18}{18}} = 0.7907$$

$$\mu_{C_1}(x_4) = \frac{1}{\dfrac{36}{36} + \dfrac{36}{26}} = 0.4194$$

$$\mu_{C_2}(x_4) = \frac{1}{\dfrac{26}{36} + \dfrac{26}{26}} = 0.5806$$

$$\mu_{C_1}(x_5) = \frac{1}{\dfrac{53}{53} + \dfrac{53}{13}} = 0.197$$

$$\mu_{C_2}(x_5) = \frac{1}{\dfrac{13}{53} + \dfrac{13}{13}} = 0.803$$

$$\mu_{C_1}(x_6) = \frac{1}{\dfrac{82}{82} + \dfrac{82}{52}} = 0.3881$$

$$\mu_{C_2}(x_6) = \frac{1}{\dfrac{52}{82} + \dfrac{52}{52}} = 0.6119$$

Therefore, using these initial prototypes of the two clusters, the membership function indicated that x_1 and x_2 are more in the first cluster, while the remaining points in the dataset are more in the second cluster.

The FCM algorithm then updates the prototypes according to Equation 13.5:

$$v_1 = \frac{\sum_{k=1}^{6} (\mu_{C_1}(x_k))^2 \times x_k}{\sum_{k=1}^{6} (\mu_{C_1}(x_k))^2} \qquad \text{(EQ 13.7)}$$

$$= \frac{0.5397^2 \times (2, 12) + 0.6852^2 \times (4, 9) + 0.2093^2 \times (7, 13) + 0.4194^2 \times (11, 5) + 0.197^2 \times (12, 7) + 0.3881^2 \times (14, 4)}{0.5397^2 + 0.6852^2 + 0.2093^2 + 0.4194^2 + 0.197^2 + 0.3881^2}$$

$$= \left(\frac{7.2761}{1.0979}, \frac{10.044}{1.0979} \right)$$

$$= (6.6273, 9.1484)$$

$$v_2 = \frac{\sum_{k=1}^{6} (\mu_{C_2}(x_k))^2 \times x_k}{\sum_{k=1}^{6} (\mu_{C_2}(x_k))^2} \qquad \text{(EQ 13.8)}$$

$$= \frac{0.4603^2 \times (2, 12) + 0.3148^2 \times (4, 9) + 0.7909^2 \times (7, 13) + 0.5806^2 \times (11, 5) + 0.803^2 \times (12, 7) + 0.6119^2 \times (14, 4)}{0.4603^2 + 0.3148^2 + 0.7909^2 + 0.5806^2 + 0.803^2 + 0.6119^2}$$

$$= \left(\frac{22.326}{2.2928}, \frac{19.4629}{2.2928} \right)$$

$$= (9.7374, 8.4887)$$

The updated prototype v_1, as shown in Fig 13.5, is moved closer to the center of the cluster formed by x_1, x_2, and x_3; while the updated prototype v_2 is moved closer to the cluster formed by x_4, x_5, and x_6.

We wish to make a few important points regarding the FCM algorithm:

- FCM is guaranteed to converge for $m > 1$. This important convergence theorem was established in 1980 [40, 49].

- FCM finds a *local minimum* (or saddle point) of the objective function J_m. This is because the FCM theorem (Theorem 17) is derived from the condition that the gradient of the objective function J_m should be 0 at an FCM solution, which is satisfied by all local minima and saddle points.

- The result of applying FCM to a given dataset depends not only on the choice of parameters m and c, but also on the choice of initial prototypes.

13.3.1   Cluster Validity

One of the main issues in clustering is how to evaluate the clustering result of an algorithm. The problem is called *clustering validity* in the literature. More precisely, the problem of clustering validity is to find an objective criterion for determining how good a partition generated by a clustering algorithmis . This kind of criterion is important because it enables us to achieve three objectives:

(1) To compare the output of alternative clustering algorithms for a dataset.

(2) To determine the best number of clusters for a given dataset (e.g. the choice of parameter c for FCM).

(3) To determine whether a given dataset contains any structure (i.e., whether there exists a natural grouping of the dataset).

It is worthwhile to point out that both hard clustering and soft clustering need to address the issue of clustering validity, even though their methods may differ. We will focus our discussion below on validity measures of a fuzzy partition generated by FCM or, more generally, a constrained soft partition of a dataset.

Validity measures of constrained soft partition fall into three categories: (1) membership-based measures, (2) geometry-based measures, and (3) performance-based measures. The membership-based validity measures calculate certain properties of the membership functions in a constrained soft partition. The geometry-based validity measure considers geometric properties of a cluster (e.g, volume) as well as geometric relationships between clusters (e.g, separation) in a soft partition. The performance-based validity measures evaluate a soft partition based on its performance for a predefined goal (e.g., minimum error rate for a classification task). We describe the first two types of clustering validity measures below.

13.3.1.1   Membership-based Validity Measures

Jim Bezdek introduced the first validity measure for soft partition in 1973 — *the partition coefficient* — which is a membership-based measure. The partition coefficient aims to measure the degree of fuzziness of clusters. The rationale is that the fuzzier the clusters are, the worse the partition is. The formula for the validity measure, denoted v_{PC}, is

$$v_{PC} = \frac{1}{n}\sum_{i=1}^{c}\sum_{j=1}^{n}\mu_{C_i}(x_j) \qquad \textbf{(EQ 13.9)}$$

Subsequently, Bezdek also introduced another membership-based validity measure called *partition entropy*, which we denote as v_{PE}

$$v_{PE} = -\frac{1}{n}\sum_{k=1}^{n}\sum_{i=1}^{c}[\mu_{C_i}(x_k)log_a(\mu_{C_i}(x_k))] \qquad \textbf{(EQ 13.10)}$$

where $a \in (1, \infty)$ is the logarithmic base. The entropy measure increases as the fuzziness of a partition increases. Therefore, a cluster with a lower partition entropy is preferred. The two membership-based validity measures are related in the following ways for constrained soft partition:

1. $\nu_{PC} = 1 \Leftrightarrow \nu_{PE} = 0 \Leftrightarrow$ The partition is hard.

2. $\nu_{PC} = \dfrac{1}{c} \Leftrightarrow \nu_{PE} = log_a(c) \Leftrightarrow \mu_{C_i}(x_k) = \dfrac{1}{c}$ $\forall i, k.$

We will leave the proof of these as an exercise. The two situations above correspond to two extreme cases. The first situation is least fuzzy and therefore is most preferred by these measures. The second situation is the fuzziest, and hence the least preferred by the validity measures.

Even though these measures can be used to compare alternative partitions with identical cluster numbers, they suffer from two limitations. First, they have a tendency to be improved (i.e., increasing ν_{PC} and decreasing ν_{PE}) as the number of clusters increase. Hence, they cannot be used to determine the optimal number of clusters. Second, they do not consider geometrical properties such as the degree of separation between clusters. These limitations motivated the development of the second type of cluster validity measures.

13.3.1.2 Geometry-based Validity

X. Xie and G. Beni introduced a validity measure that considers both the compactness of clusters as well as the separation between clusters. Intuitively, the more compact the clusters are and the further the separation between clusters, the more desirable a partition. To achieve this, the Xie-Beni validity index (denoted as ν_{xB}) was defined as

$$\nu_{xB} = \left(\frac{\sum_i \sigma_i}{n} \right) \times \frac{1}{d_{min}^2} \qquad \text{(EQ 13.11)}$$

where σ_i is the variation of cluster C_i defined as

$$\sigma_i = \sum_j \mu_{C_i}(x_j) \| x_j - v_i \|^2 \qquad \text{(EQ 13.12)}$$

n is the cardinality of the dataset and d_{min} is the shortest distance between cluster centers defined as

$$d_{min} = \min_{\substack{i, j \\ i \neq j}} \| v_i - v_j \| \qquad \text{(EQ 13.13)}$$

The first term in Equation 13.12 is a measure of noncompactness, and the second term is a measure of nonseparation. Hence, the product of the two reflects the degree that clusters in a soft partition are not compact and not well separated. Obviously, the lower the cluster index ν_{xB}, the better the soft partition is.

13.3.2 Extensions to Fuzzy c-Means

A major extension to FCM is to generalize the distance measure between data x_i and prototype v_j from $dist(x_i, v_j) = \| x_i - v_j \|^2$ to

$$dist(x_i, v_j) = (x_i - v_j)^T A_j (x_i - v_j) \qquad \text{(EQ 13.14)}$$

where A_j is a symmetric $d \times d$ matrix (d is the dimensionality of x_j and v_j). This enables an extended FCM to adapt to different hyperellipsoidal shapes of different clusters by adjusting the matrix A_j. To illustrate this, let us consider a two-dimensional clustering problem. Suppose data $\overline{x_i} = (x_1^i, x_2^i)$ in cluster C_4 does not form a circular shape, but rather they form an ellipse shape described by the following equation:

$$\left(\frac{x_1^i - v_1^4}{a} \right)^2 + \left(\frac{x_2^i - v_2^4}{b} \right)^2 \leq c \qquad \text{(EQ 13.15)}$$

which can be expressed as

$$(\hat{x}_i - \hat{v}_4)^T A_4 (\hat{x}_i - \hat{v}_4) \qquad \text{(EQ 13.16)}$$

where $\overrightarrow{x_i}$ is the vector representation of a point being clustered

$$\overrightarrow{x_i} = \begin{bmatrix} x_1^i \\ x_2^i \end{bmatrix} \qquad \text{(EQ 13.17)}$$

$\overrightarrow{v_4}$ is the center of cluster C_4,

$$\overrightarrow{v_4} = \begin{bmatrix} v_1^4 \\ v_2^4 \end{bmatrix} \qquad \text{(EQ 13.18)}$$

and A_4 is the matrix below:

$$A_4 = \begin{bmatrix} 1/a^2 & 0 \\ 0 & 1/b^2 \end{bmatrix} \qquad \text{(EQ 13.19)}$$

In general, we would like the matrix (A_j) associated with cluster C_j to be dynamically determined based on the shape of the cluster.

D. Gustafsun and W. Kessel were the first to propose such an extension of the matrix A_j (i.e., $|A_j| = l_j$), they developed a modified FCM that dynamically adjusts $(A_1, A_2, ..., A_c)$ such that these matrices adapt to the different hyperellipsoidal shape of each cluster [228].

There are at least two additional directions in generalizing FCM. J. Bezdek replaced prototypes in FCM's objective function with linear varieties of arbitrary dimensions [50, 132]. R. Krishnapuram and J. Keller developed a possibilistic clustering algorithm by relaxing FCM's sum-to-one constraint of cluster membership [343]. The objective function of FCM was then modified with an additional term to avoid assigning 0 to all cluster memberships. R. Krishnaparam and R. Dave have each combined several directions for extending FCM into a *possibilistic shell clustering algorithm* [344, 149, 341, 342].

13.4 Classifier Design and Supervised Pattern Recognition

We have discussed several techniques and issues regarding fuzzy clustering. These techniques are "unsupervised" in the sense that they do not require points in the dataset to be labelled with their correct classification. In contrast, *supervised pattern recognition* uses data with known classifications, which are also called labeled data, to determine the classification of new data.

Supervised and unsupervised pattern recognition techniques differ in their advantages and limitations. The main benefit of unsupervised pattern recognition techniques is that they do not require training data; however, their computation time is relatively high due to a large number of iterations needed before the algorithm converges. In contrast, a supervised pattern recognition technique is relatively fast because it does not need to iterate; however, it cannot be applied to a problem unless training data or relevant knowledge are available. For image processing, the availability of training data has two major implications. First it implies that one or a set of processed images (for training purposes) is not difficult to obtain. Second, and more importantly, it implies that the general content (e.g., objects, regions, etc.) of the images that we are interested in can be anticipated before processing them. For example, the general content of a brain MRI image is predictable (e.g., white matter, gray matter, CSF, etc.), even though there are many variations in their shape and volume from one image to another. In comparison, the general content of an image collected by an autonomous battlefield vehicle is much more difficult to predict. The image could include everything a soldier can see. Therefore, it is more difficult to generate training images for all possible objects in this type of open-ended images. The use of existing supervised image segmentation techniques in these applications will be limited to recognizing a small set of well-defined objects (e.g., roads) in the images.

13.4.1 Fuzzy K-Nearest Neighborhood

The fuzzy K-nearest neighborhood algorithm is a fuzzy classification technique that generalizes the K-Nearest Neighborhood (K-NN) algorithm. Given a set of classified data, the K-NN algorithm determines the classification of an input based on the class labels of the K closest neighbor in the classified dataset.

The main advantage of K-NN is its computational simplicity. Even though good results have also been obtained with several applications, it has two major problems. First, each of the neighbors is considered equally important in determining the classification of the input data. A far neighbor is given the same weight as a close neighbor of the input. Second, the algorithm only assigns a class to the input data, it does not determine the "strength" of membership in the class. These two limitations motivated the development of the fuzzy K-NN algorithm.

The fuzzy K-NN algorithm assigns to an input x a membership vector $(\mu_{C_1}(x), \mu_{C_2}(x), ..., \mu_{C_l}(x))$ where l is the total number of classes. The class memberships are calculated based on the following formula:

$$\mu_{C_i}(x) = \frac{\displaystyle\sum_{j=1}^{k} \mu_{C_i}(x_j) \frac{1}{\|x - x_j\|^{2/(m-1)}}}{\displaystyle\sum_{j=1}^{k} \frac{1}{\|x - x_j\|^{2/(m-1)}}} \qquad \text{(EQ 13.20)}$$

where $x_1, x_2, ..., x_k$ denotes the k nearest neighbors of x. The K-NN algorithm simply contains two major steps:

1. Find the K-nearest neighbors of the input x.
2. Calculate the class membership of x using Equation 13.20.

The parameter m in Equation 13.20 serves as a function similar to the parameter m in FCM. They determine how heavily the distance is weighted when calculating the class membership. As m increases toward infinity, the term $\|x - x_j\|^{2/(m-1)}$ approaches one regardless of the distance. Consequently, the neighbors x_j are more evenly weighted. As m decreases toward 1, however, the closer neighbors are weighted far more heavily than those further away. This has the effect of reducing the number of neighbors that contribute to the membership value of the input data point. For example, fuzzy 10-NN with $m = 0.01$ behaves very much like fuzzy 1-NN. We will leave this as an exercise.

13.5 Knowledge-based Pattern Recognition

In addition to the use of training data, pattern recognition can also benefit from the use of knowledge. For example, a radiologist often has a large amount of knowledge about characteristics of lesion in a brain MRI image. He or she uses this knowledge to identify lesions from a patient's MRI image. Hence, it can be beneficial to encapsulate this knowledge in a computer system for medical image processing. We will refer to such approaches as *knowledge-based pattern recognition*.

An important scheme for representing human knowledge regarding pattern recognition is fuzzy if-then rules introduced in Chapter 5. Fuzzy rule-based systems have been used to segment an image, to recognize an object in the image, or to analyze relationships between objects in the image. For example, a fuzzy rule for segmenting an image may be

IF the Blue Intensity of the pixel is *Strong* AND the y-coordinate of the pixel is *High* THEN the pixel is labeled *sky*.

Fuzzy rules can also be used for recognizing objects for higher level image processing. For example, the following fuzzy rule can be used to identify central sulci in a 3D brain MRI system:

IF the sulcus is *near the middle* of the brain
  AND is *long*
  AND has *appropriate orientation*
THEN it is central sulcus.

While the structure of fuzzy rules can be acquired from human experts, their parameters are often difficult to determine. This is especially true for a rule-based image segmentation system. The remedy to this problem is to combine rule-based segmentation with supervised and/or unsupervised segmentation. The combined system is called a hybrid pattern recognition system.

13.6 Hybrid Pattern Recognition Systems

Different approaches to pattern recognition have different merits and limitations, some of which are summarized in Table 13.2. Therefore, it is often desirable to develop hybrid pattern recognition systems that use more than one technique to achieve synergism. Obviously, a hybrid system can also be designed by combining a fuzzy approach with a non-fuzzy approach (e.g., a neural network approach or a maximum likelihood approach). Combining fuzzy rules and neural networks gives us neuro-fuzzy pattern recognition system, which we will discuss in Chapter 16.

Hybrid systems for image processing can be classified into two categories: (1) staged hybrid systems and (2) parameter-tuning hybrid systems. A staged hybrid system uses different techniques in different stages of an image processing system. For example, we know unsupervised segmentation is slow, rule-based segmentation is fast. On the other hand, the accuracy of the latter approach depends on the design of rules. To enhance the robustness of rule-based segmentation while improving the speed of an unsupervised segmentation, a two-stage hybrid segmentation system can be used. Rules are first used to label pixels whose classes are clearly identifiable. Subsequently, an unsupervised segmentation technique (e.g., FCM) is used in the second stage to segment the pixels not labelled by rules.We will describe such a hybrid system for brain MRI segmentation in the next section.

TABLE 13.2   A Comparison of Three Classes of Segmentation Techniques

	Processing Speed	Requires Training Data	Requires Knowledge
Unsupervised Segmentation	Slow	No	No
Supervised Segmentation	Fast	Yes	No
Knowledge Segmentation	Fast	Maybe (hybrid)	Yes

The second type of hybrid systems (i.e., parameter-tuning hybrids) adjust the parameters of an image processing system. Such a hybrid system has two components: a parameter-tuning subsystem and an image processing subsystem. The two subsystems can (are likely to) use different techniques. For instance, both supervised and unsupervised techniques can be used to adjust the parameters in a rule-based segmentation system. We will describe an example of this hybrid system in the next section also.

13.7 Applications In Medical Image Segmentation

One of the most important application areas of fuzzy pattern recognition is the segmentation of medical images in general, and the segmentation of Magnetic Resonance Images (MRIs) in particular. Typically three pulse sequences are used in clinical Magnetic Resonance (MR) imaging: T1-weighted, T2-weighted and Proton Density (PD). Each pulse sequence produces images with contrast characteristics that permit specific types of tissue to be visualized. For brain tissue, Cerebrospinal Fluid (CSF) is much brighter than gray and white matter in T2-weighted images and therefore can be used to estimate CSF volume or intracranial volume. Pathologic lesions are seen most clearly as hyperintensities in the PD images, which also show gray matter/white matter contrast. These two pulse sequences often provide sufficient contrast to segment the images into four compartments (i.e., white matter, gray matter, CSF, and abnormal tissues) [135, 221]. T1-weighted images provide excellent atomic detail and good tissue contrast.

Segmentation of an MR image classifies pixels in the image into categories representing anatomical structures. Many different segmentation methods — automatic, semi-automatic or manual — have been developed [51, 193, 221, 248, 319, 378, 460]. Manual tracking of the boundaries between different tissues is the simplest and most widely used segmentation technique, and computer software is available for this purpose [10, 389, 395]. The chief drawbacks of this technique are the tedious and time-consuming nature of the segmentation task, and the lack of consistency among persons preforming it. Thresholding, another basic technique, segments images according to pixel intensity [314, 610, 611]. Because of image noise or insufficient contrast, the technique sometimes fails. When combined with human interaction or other techniques, nevertheless, thresholding can produce useful results [62, 130, 324], but the introduction of operator control also introduces inter-operator variability. Segmentation can sometimes be realized by edge detection, dividing an image into regions with connected boundaries [312, 386, 603, 639]. However, edge detection is sensitive to noise, sometimes resulting in erratic performance. Statistical methods have also been developed, segmenting according to an assumed statistical distribution of the intensities [135, 375, 376, 585]. Spatial information in images, such as contiguity and linkage among groups of pixels, has also been used for segmentation: for example, morphological operators, including erosion, dilation, opening and closing operators, have been used to extract structural information [247, 298].

Image segmentation can also be achieved by clustering techniques, such as the c-means algorithm, neural networks and fuzzy clustering algorithms [235, 459, 646]. FCM and its variants have been used successfully for MR image segmentation [84,235]. Neural networks require that a large amount of training data be available, and learning usually requires a long time. These algorithms segment images according to the distribution of the pixel feature vectors in feature vector space. These approaches can be successful only if the features that are selected for use, such as intensity, gradient, or texture measures, have values that actually differ among the tissues in question; this choice of features is one way in which expert knowledge could be incorporated into the system. Using a threshold prior to the FCM clustering may simplify the problem to a certain extent [379].

The FCM approach cannot easily incorporate certain types of expert knowledge that are useful for the segmentation, such as shapes and locations of structures or lesions. In contrast to clustering methods, rule/knowledge-based systems permit expert knowledge to be explicitly represented in the form of if-then rules [157, 236, 546, 613]. In such a system, however, parameters in the rules often need to be adjusted from image set to image set because of wide variation of intensity distribution at the time of image acquisition. Another class of knowledge-based systems is an anatomical-model-based system; these systems incorporate specific knowledge using means other than rules [16, 155, 287, 290]. However, it is often difficult to develop new anatomical models based on the existing ones [374, 496].

13.7.1   A Case Study: A Two-stage Hybrid Fuzzy System for MRI Segmentation

To illustrate the application of fuzzy pattern recognition techniques discussed in this chapter, we describe a two-stage hybrid fuzzy system for segmenting human brain MR images with lesions. The system was developed by Hao Yong, Gilbert R, Hillman, Thomas A. Kent and Leena M. Ketonen of the University of Texas Medical Branch at Galveston (UTMB) and by Chih-Wei Change and John Yen of Texas A&M University (TAMU).

The system was designed to deal with two specific problems commonly associated with purely rule-based segmentation systems. One is that the rules alone may not be sufficient to segment the entire image successfully, because of noise unexpected variability in the image data; this variability may be caused by instrument instability, movement artifacts, or operator error at the site of image acquisition. Another problem lies in formulating the membership functions that are related by the rules. The medical observer has no obvious way to know the actual intensity distribution of each compartment or structure, or to know how to translate the intensity distribution into membership functions. The problem is complicated by the fact that the observed intensities to tissues are expected to be different when different MR pulse sequences are used; the system must be flexible enough to interpret images as they are acquired at any time.

These two problems are addressed by combining the advantages of rule-based methods and unsupervised methods. The system addresses the first difficulty by augmenting a pure rule-based system with a modified FCM algorithm that classifies those pixels that the rules cannot classify with high confidence. To improve the efficiency of the standard FCM algorithm, the pixels classified by the rules are used to construct the initial prototype for the modified FCM algorithm. The rule-classified pixels are then treated by the modified FCM algorithm as data belonging to a known cluster. The second problem is addressed by automatically identifying the parameters of the membership function for each MR image set. This is achieved by clustering both intensities and the edge values of MR images using the FCM algorithm. The clusters generated by the FCM are used to determine the parameters of the membership functions in rules.

FIGURE 13.6　　The Architecture of a Hybrid Fuzzy Segmentation System with Automatic Identification of Rule Parameters

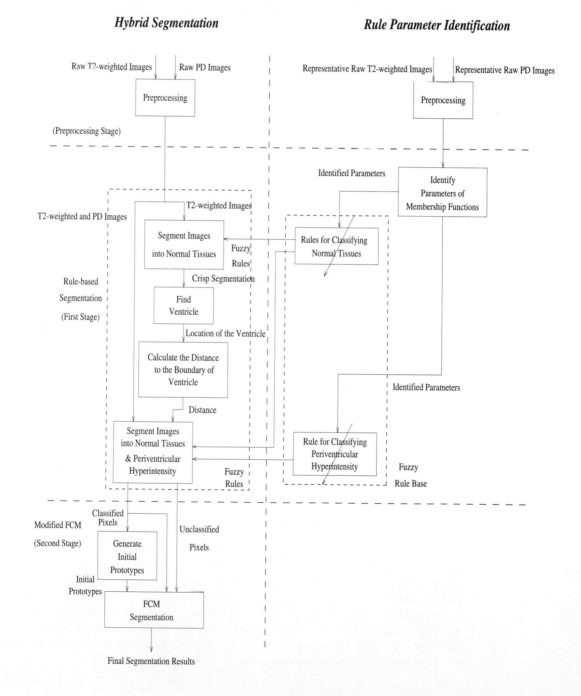

13.7.2 The Rule-based Fuzzy System

For normal images, the system segments the images into gray matter, white matter, and CSF. For abnormal images the system also identifies preventricular hyperintensity regions, which are high intensity areas around the ventricle in PD images. The rule-based segmentation system, shown in Fig 13.6, captures the expert reasoning process used to locate this lesion: First locate the ventricle and then use it as a "landmark" for identifying preiventricular hyperintensity.

The system first identifies the ventricle using a T2 image. This is accomplished by segmenting the image into three normal tissues (i.e., white matter, gray matter, and CSF) using fuzzy rules. The ventricle is always close to the center of the brain and contains CSF. Hence, the system finds the ventricle by identifying large CSF regions that are close to the center of the region of interest (ROI). Once the ventricle is identified, the system calculates the distance of each pixel to the boundary, both inside and outside the boundary, by dilating and eroding the boundary of the ventricles. This distance measure is used to determine how close a pixel is to the ventricle boundary, which is a condition for identifying perinventricular hyperintensity.

Finally, the system uses rules to segment T2 and PD images into four classes (e.g., the three normal tissues plus the periventricular hyperintensity class). The rule that classifies periventricular hyperintensity uses "close to the boundary of the ventricle" as a condition its antecedent. We describe below the design of the fuzzy rules and the fuzzy inference used in our segmentation system.

The rule-based segmentation system uses two different groups of fuzzy rules for two purposes: one group for preliminary segmentation of the T2-weighted images into normal tissues: gray matter, white matter, and CSF; and another group for segmenting the T2-weighted and PD images into four classes with the lesions being the fourth class. The first group of rules is as follows:

> R1:IF *pixel in T2 is Dark* THEN *pixel is White Matter*;
> R2:IF *pixel in T2 is Gray* THEN *pixel is Gray Matter;*
> R3:IF *pixel in T2 is Bright* THEN *pixel is CSF*

where *Dark, Gray* and *Bright* are linguistic descriptions of the intensity of a pixel, as represented by the fuzzy sets whose parameters are generated. There are five rules in the second group; the first two rules are the same as R1 and R2 in the first group. The remaining three rules are

> R4:IF *pixel in T2 is Bright* AND *in PD is Dark-Gray* THEN *pixe*l is CSF
> R5:IF *pixel in T2 is Bright* AND in PD is Very Bright AND *is NOT close to the Ventricle* THEN *pixel is CSF;*
> R6:IF *pixel in T2 is Bright* AND *in PD is Very Bright* AND *is close to the boundary of Ventricle* THEN *pixel is periventricular hyperintensity*

where Dark-Gray, Very Bright, close to the Ventricle, and NOT close to the boundary of the Ventricle are also fuzzy sets. Trapezoidal membership functions are used for all the fuzzy sets describing pixel intensity.

These fuzzy rules capture the essence of the medical expert knowledge relating pixel intensity in T2-weighted and PD images to normal brain tissues and periventricular hyperintensity. The inference of fuzzy rules is based on Zadeh's max-min compositional rule of inference [701]. For each pixel, the rules can compute membership in each of the four tissue classes (i.e., white matter, gray matter, CSF, and periventricular hyperintensity). The highest membership among the four is selected and compared to a threshold value. If the membership is greater than or equal to the threshold, the pixel is assigned to that class. Otherwise, the pixel is labeled as unclassified. If a pixel has equal membership values for two different classes, it is labeled as unclassified so that it will be processed along with other unclassified pixels by the second stage of the fuzzy system. A membership threshold of 0.5 was shown to be adequate for achieving satisfactory segmentation results.

13.7.3 Automatic Parameter Identifier for the Membership Functions

The hybrid system automatically generates the membership function parameters for the fuzzy sets used in the fuzzy rules stated above, except for the close to the Ventricle function, which is fixed because it is not sensitive to intensity variations of the MR images acquired by different clinical MR scanners. The automatic membership function generation technique is adaptive to intensity variations among image sets acquired by different MR scanners, or by the same scanner at different times.

FIGURE 13.7   Parameter Identification of Membership Functions

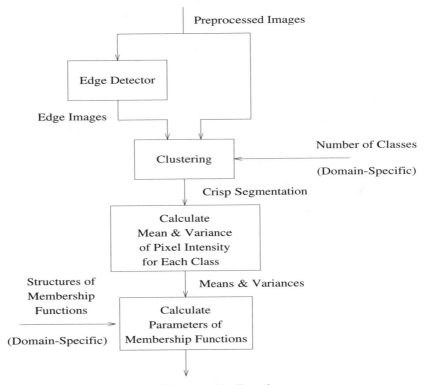

The subsystem for identifying the parameters of membership functions, shown in Fig 13.7, uses FCM unsupervised clustering, based on both pixel intensity and edge strength as determined by the Chen edge operator [193], to create five clusters. These clusters represent the three major tissue compartments and the boundary regions between them. We then use the means and standard deviations of the intensity distributions within the clusters to define the parameters of the three trapezoidal membership functions as shown in Fig 13.8, where μ_i and σ_i denote the mean and the variance of the ith class. For instance, the left slope of the Gray membership function as well as the right slope of the Dark membership function are defined by $\mu_2 - \sigma_2$ and $\mu_3 - \sigma_3$. Similarly, the right slope of the Gray membership function and the left slope of the Bright membership function are defined by $\mu_3 + \sigma_3$ and $\mu_4 + \sigma_4$.

FIGURE 13.8   Membership Function of T2-weighted Images

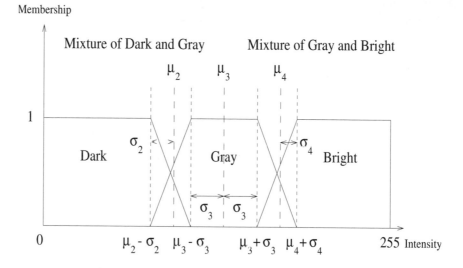

The parameters of membership functions for PD images are identified. PD images are used primarily for identifying periventricular hyperintensities. For this purpose, two fuzzy sets were used to describe intensity of PD images containing lesions: Dark-Gray-3 for all the normal brain tissues and Very Bright for periventricular hyperintensity. However, a PD image of a normal brain actually contains two dominant intensity classes, which we call Dark-Gray-1 and Dark-Gray-2. However, the class of bright, abnormal pixels that we are interested in covers a much smaller area than Dark-Gray-1 and Dark-Gray-2, because the lesions are small, and it is often difficult to form a cluster of pixels in this class, if there are also other small clusters generated by noise. To overcome this difficulty, the parameters of membership function Very Bright are identified indirectly from the mean and variance of its neighboring classes. The PD image pixel intensities and edge values are clustered into three classes (i.e, Dark-Gray-1, a mixture of Dark-Gray-1, and Dark-Gray-2) using the FCM algorithm. We then calculate the left slope parameters of the Very Bright membership function, and the right slope parameters of the Dark-Gray-3 membership function, from $\mu_3 - \sigma_3$ and μ_3, as shown in Fig 13.9.

A novelty of this approach to generate rule parameters is to use the edge values of pixels to segment the image so that the boundary area between two neighboring regions can form a class by itself. The properties of that class can help to define the membership functions for the true tissue classes.

FIGURE 13.9   Membership Function of PD-weighted Images

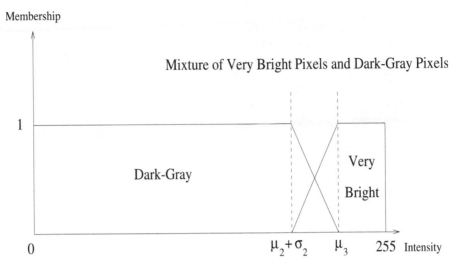

13.7.4 Results

Fig 13.10 shows one set of original T1-weighted, T2-weighted, and PD images with a pathological lesion, identified already by an expert as a periventricular hyperintensity; in such a case the objective of our system is to measure the extent of the lesion. To assist in locating this particular lesion, the system calculates membership values for the fuzzy set "close to the boundary of ventricle." These values are shown as an intensity map in Fig 13.11; the brighter is a point in the image the higher is its membership degree in the fuzzy set. Fig 13.12 demonstrates the segmentation outcome, where white pixels represent the lesion, dark-gray pixels CSF, medium-gray pixels gray matter, and ligh- gray pixels white matter.

The system was applied to a total of 15 image sets (12 normal brain image sets and 3 image sets containing lesions). The result of the segmentation result was verified by two experts.

FIGURE 13.10   One Set of Abnormal Brain Raw Images: (a) T1-weighted Image, (b) T2-weighted
Image and (c) PD Image

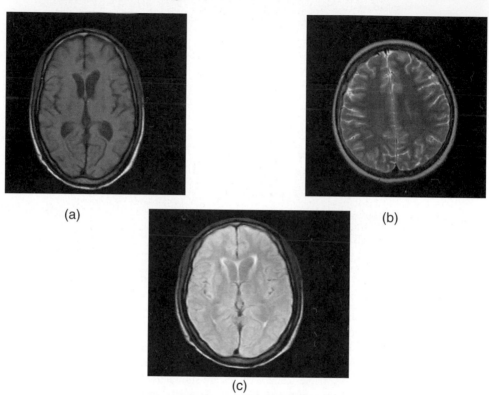

(a)

(b)

(c)

FIGURE 13.11   The Image of the Membership Value of "Close to the Ventricle"

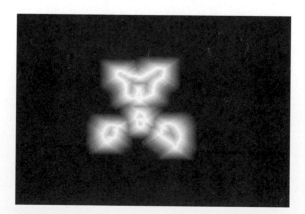

FIGURE 13.12   Segmentation Result of One Abnormal Brain Image Set

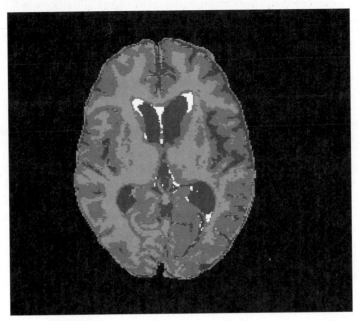

13.8 Summary

In this chapter, we introduced several fuzzy logic-based techniques for pattern recognition. These techniques can be grouped into three categories: unsupervised techniques, supervised techniques, and knowledge-based techniques. These techniques can also be combined to form hybrid systems. Some of the major concepts regarding these techniques are summarized below

- The fuzzy c-means (FCM) algorithm is a fuzzy logic-based extension to the hard c-means clustering algorithm. Both of them are unsupervised techniques.

- The foundation of the FCM algorithm is to optimize an objective function that measures the compactness and the separation of clusters formed.

- Given the number of classes into which we wish to partition a dataset, the FCM algorithm updates the cluster centers (i.e., prototypes) and the membership function of each cluster iteratively until a convergence criterion is reached.

- Criteria for evaluating fuzzy clusters fall into three categories: (1) criteria based on properties of the membership function of each cluster, (2) criteria based on geometric properties of a cluster and geometric relationship between clusters, and (3) criteria based on certain performance objectives (e.g., minimizing classification error).

- The fuzzy K-nearest neighborhood algorithm is a fuzzy classification algorithm (i.e., a supervised pattern recognition technique) that computes an input's membership degrees in a set of classes based on the fuzzy classification K closest neighbors.

- Fuzzy rules can be used to capture human knowledge useful for pattern recognition.

- Multiple approaches to pattern recognition can be integrated sequentially into staged hybrid systems or hierarchically into parameter-tuning hybrid systems.

- Fuzzy pattern recognition techniques can be applied to the segmentation of human brain MR images with abnormal tissue.

Bibliographical and Historical Notes

Zadeh's initial development of the theory of fuzzy sets, as we mentioned in Chapter 1, was motivated in large measure by problems in pattern classification and cluster analysis. It is not surprising, even though it is not well known, that Bellman, Kalaba, and Zadeh wrote the first paper on the application of fuzzy set theory to pattern classification in 1966, shortly after the publication of the first fuzzy set paper. A pioneer work on cluster analysis was Ruspini's concept of a fuzzy partition in 1969 [515]. At about the same time, Azriel Rosenfeld became interested in the use of fuzzy sets in image analysis [509]. One of the best known fuzzy clustering algorithms, the fuzzy ISODATA, and the Fuzzy C-Means (FCM) algorithms, were introduced by J. Dunn in 1973 [181].

The convergence issue of FCM was first addressed by J. Bezdek in 1980 [40]. A correction to a technical flow in the paper was later published by J. Bezdek, M. Windham and others [49].

J. Bezdek's thesis introduced the first cluster validity measure for fuzzy partitions generated by FCM [43]. The measure, called the partition coefficient, is based on properties of the cluster's membership. He also introduced a cluster validity measure called the partition entropy [42]. M. Windham combined the properties of the dataset with the properties of the membership function into a measure called *uniform data function* for evaluating the validity of a fuzzy cluster [642]. E. Backer and A. Jain proposed a performance-based validity criteria called *goal-directed comparison* [18]. X. Xie and G. Beni introduced a geometry-based cluster validity [648].

Determining the optimal number of clusters is an important problem in clustering. I. Gath and and A. Geva developed an approach for assessing the number of clusters using a performance based validity criteria [204]. N. Pal and J. Bezdek reported an empirical evaluation of several different validity criteria [461].

The fuzzy K-nearest neighborhood algorithm was developed by J. Keller [310]. He and his colleague also introduced another supervised pattern recognition technique that (1) uses the histograms obtained from training images to generate conditional fuzzy measures and (2) uses fuzzy integrals to combine fuzzy measures obtained from different image features [309, 568].

L. Hall and his colleagues developed a knowledge-based approach to identify abnormal tissues in brain MR images [374]. A technique for generating fuzzy rules for high-level computer vision using gradient descent method are described in [500]. A comprehensive review of pattern recognition techniques (supervised and unsupervised) for MRI segmentation can be found in [45, 46]. J. Bezdek et al. gave an update to the development of fuzzy models for segmenting medical images [50]. J. Keller et al. reviewed fuzzy rule-based models in computer vision [55]. Fuzzy sets have also been used to develop fuzzy mathematical morphology [55, 56], which have been applied to 3D reconstruction of blood vessels. An edited volume by J. Bezdek and S. K. Pal contains important papers related to fuzzy pattern recognition techniques [44].

Exercises

13.1   Show that a fuzzy K-NN algorithm with an m value approaching 1 behaves like a fuzzy 1-NN algorithm.

13.2   Prove the following relationship between Bezdek's partition coefficient and partition entropy for a constrained soft partition:

(1) $(V_{PC} = 1) \Leftrightarrow (V_{PE} = 0) \Leftrightarrow$ The partition is a hard partition.

(2) $V_{PC} = \dfrac{1}{C} \Leftrightarrow (V_{PE} = \log{}_a(c)) \Leftrightarrow \mu_{C_i}(x_k) = \dfrac{1}{c}$.

13.3   Calculate the prototypes and cluster memberships of the second FCM iteration for Example 17.

13.4   Implement FCM and apply it to a dataset in one of the repositories available on the Web.

13.5   Compare empirically the results of applying FCM and the hard c-means algorithm to the dataset above.

13.6   Suppose you were asked to modify FCM so that each cluster can have more than one prototype.

(a) How would you modify the objective function?
(b) Derive the formula for updating the prototypes and the cluster membership from the multiprototype objective function.

13.7   Develop an algorithm that automatically determines the number of classes. The algorithm should modify FCM by allowing (1) existing classes to be merged and (2) new classes to be added.

13.8   Implement the algorithm in the previous problem and test it using a dataset available on the Web.

PART IV

FUZZY MODEL IDENTIFICATION AND SOFT COMPUTING

14 FUZZY MODEL IDENTIFICATION

A fuzzy model is a nonlinear model that consists of a set of fuzzy if-then mapping rules, which we discussed in Chapter 5. It has a transparent and interpretable model structure and is capable of representing a highly nonlinear functional relation using a reasonable number of fuzzy rules. The problem of *fuzzy model identification* or *fuzzy modeling* is generally referred to as the determination of a fuzzy model for a system or a process by making use of one or both of two kinds of information: linguistic information obtained from human experts, and numerical information obtained from measuring apparatus.

As we have mentioned in previous chapters, one of the most important issues about fuzzy rule-based modeling is the construction of their membership functions. The design of early fuzzy models often involved a manual tuning of their membership functions based on the performance of the model. Even though such a design procedure has led to a large number of successful applications, it was time consuming and difficult to deal with high-dimensional problems. Furthermore, it is subject to criticism for its lack of principles and systematic methodologies. In fact, the construction of suitable membership functions for a fuzzy rule-based model is only part of a more general problem — the problem of *fuzzy model identification*, which includes the following issues:

- Establishing the criteria for evaluating fuzzy models.
- Selecting input variables relevant to the model.
- Selecting the type of fuzzy models.
- Choosing the structure of membership functions.
- Determining the number of fuzzy rules.

- Identifying the parameters of antecedent membership functions.
- Identifying the consequent parameters of rules

These issues can be grouped into three categories: model validation, structure estimation, and parameter estimation.

This chapter presents a stat- of-the-art description of identification techniques for fuzzy models. Although the emphasis is on the essential ideas and tools, many issues that are of theoretical and practical interest are also addressed. We therefore hope that this chapter will be of use for both students and researchers.

14.1 Fuzzy Rule-based Models — A System Identification Perspective

The concept of a mathematical model is fundamental to systems analysis and design which require the representation of systems phenomena as a functional dependence between interacting input and output variables. For example, mathematical models are essential for prediction and control purposes. Conventionally, a mathematical model for a system is constructed by analyzing input-output measurements from the system. These *numerical* measurements are important because they represent the behavior of the system in a quantitative fashion. Very often, there exists another important information source for many engineering systems: knowledge from human experts. This knowledge, known as *linguistic* information, provides qualitative instructions and descriptions about the system and is especially useful when the input-output measurements are difficult to obtain. Fuzzy models are capable of incorporating this kind of information naturally and conveniently, while conventional mathematical models usually fail to do so. Moreover, it is interesting to note that fuzzy models have the same ability to process numerical information as conventional models, if such information is available. In fact, most of this chapter will discuss how to identify fuzzy models from input-output measurements. Being able to deal with linguistic information and numerical information in a systematic and efficient manner is one of the most important advantages of fuzzy models.

The second important property of fuzzy models is their ability to *handle nonlinearity*. It is a well known fact that most engineering systems are nonlinear to some extent. Although it may be possible to represent such systems by a linear model over a restricted operating range (e.g., around the operating point), in general, nonlinear processes can only be adequately characterized by a nonlinear model. Since a mathematical description of a process is often a prerequisite to analysis and controller design, the study of system identification techniques has become an established branch of control theory [384]. However, whereas system identification techniques for linear systems are now well developed and have been widely applied, very few results exist for the identification of nonlinear systems. This may be attributed to the inherent complexity of nonlinear systems and the difficulty of deriving identification algorithms. Fuzzy models provide a valuable addition to the increasing need for nonlinear modeling. By ingeniously constructing fuzzy partitions, a fuzzy model is capable of approximating a highly nonlinear functional relation using a small number of fuzzy rules [576].

Interpretability is another salient feature of fuzzy models. A fuzzy model has a transparent model structure. Each rule in the model acts like a "local model" in the sense that it only covers a local region of the input-output space, and its contribution to the whole output of the model is easily understood. Neural networks are another powerful tool for modeling nonlinear systems [438], but they are essentially a "black box" model and difficult to interpret.

This chapter is organized as follows. Section 14.2 mentions the approximate ability of fuzzy models. Section 14.3 gives a basic framework of fuzzy system identification that consists of three interconnected subproblems, including structure specification, parameter estimation, and model validation. Methods for solving these subproblems are discussed separately in Sections 14.4 through 14.6. The connection between fuzzy model and control is established in Section 14.7. Several numerical examples are provided in Section 14.8.

For notational simplicity, all fuzzy models discussed in this chapter are assumed to be multi-input single-output models. Extension to the multi-input multi-output models, however, is direct. In particular, we want to find a mapping f from an s-dimensional input space $U = U_1 \times U_2 \times \ldots \times U_s \subset R^s$ to a one-dimensional output space V, i.e., $f: U \subset R^s \to V \subset R$.

14.2 Approximation Ability of Fuzzy Models

One of the major applications of fuzzy models is to approximate a desired control or decision function. Mathematically, this amounts to finding a mapping from the input space to the output space. Hence, to a certain extent, the problem of constructing fuzzy models can be viewed as a function approximation problem. A natural question to ask, then, is how well can a fuzzy model approximate an unknown function.

Several researchers have proved that fuzzy models generated by different inferential and defuzzification methods, as well as different membership functions, are universal approximators, which means that the models can uniformly approximate any real continuous function to an arbitrary degree of accuracy in a compact domain [620, 331, 94]. These theoretical analyses not only confirm the approximation ability of fuzzy models, but also lay a mathematically sound foundation for fuzzy model identification.

We should also point out that the universal approximation capability is not unique to fuzzy mode. Many other types of function approximators, such as *feedforward neural networks*, *radial basis function networks*, and *nonparametric regressors* also possess the same (or even better) universal approximation ability. The unique feature of a fuzzy model, which is not shared by other types of function approximators, is its facility for explicit knowledge representation in the form of fuzzy if-then rules, the mechanisms of reasoning in humanly understandable terms, and the capacity of taking linguistic information from human experts and combining it with numerical information.

14.3 Three Subproblems of Fuzzy System Identification

Fuzzy system identification consists of three basic subproblems: structure specification, parameter estimation, and model validation, as shown in Fig 14.1. *Structure specification* involves finding the important input variables from all possible input variables, specifying membership functions, partitioning the input space, and determining the number of fuzzy rules comprising the underlying model. *Parameter estimation* involves the determination of unknown parameters in the model using some optimization method based on both linguistic information obtained from human experts and numerical data obtained from the actual system to be modeled. Structure specification and parameter estimation are interwoven, and neither of them can be independently identified without resort to the other. *Model validation* involves testing the model based on some performance criterion (e.g., accuracy). If the model cannot pass the test, we must modify the model structure and re-estimate the model parameter. It may be necessary to repeat this process many times before a satisfactory model is found. We will discuss these three subproblems, respectively, in great detail in the following sections.

FIGURE 14.1   Fuzzy System Identification Diagram

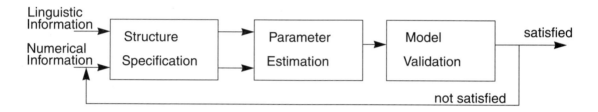

14.4 **Structure Specification**

14.4.1 **Input selection**

The selection of appropriate input variables is a major issue in constructing any model (including fuzzy models). This is because there are usually many factors that can potentially affect the output of the model. However, we only want to include the most relevant factors. Including irrelevant factors not only complicates the task of the model identification but also may lead to misleading interpretations of the model. Therefore, there are variety of practical and economical reasons for reducing the number of input variables in the final model. In addition, variable deletion may be desirable in terms of the statistical properties of the parameter estimates and the construction of the final model [255]. On the other hand, we do not wish to exclude any important relevant factor. Doing so will obviously limit the accuracy of the model being constructed. The decision as to which of these input variables should be included in the final model is not an easy one. Various input selection techniques have been proposed for linear regression modeling in the statistical community, but the criterion employed in each technique seems quite arbitrary and is known to provide different solutions for the same problem [160].

The input selection problem is even more difficult for fuzzy modeling as a fuzzy model is a nonlinear model. Moreover, the identification of a fuzzy model consists of intertwined steps, and the addition or the deletion of an input variable will affect (also be affected by) other identification steps. For example, the number of input variables may affect the number of fuzzy partitions in the input space.

In this section we introduce several methods of selecting input variables for fuzzy models and discuss their merits and drawbacks. Like those used in linear regression modeling, these methods cannot guarantee that the selected input variables are the best. Instead, they only suggest an "acceptable" subset of input variables.

14.4.1.1 Forward Selection Procedure

This procedure starts with one input variable in the model and adds one variable at a time based on a specific performance criteria until the model is satisfactory. The two most commonly used performance criteria are the *Mean Squared Error (MSE)* [78] and the *Unbiasedness Criterion (UC)* [559], which we will discuss in Section 14.6. The input variable considered for inclusion at any step is the one giving the best performance among those eligible for inclusion.

FIGURE 14.2 An Example of Forward Selection Procedure

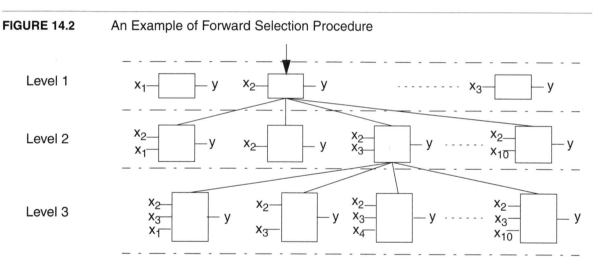

We illustrate the forward selection procedure using the example shown in Fig 14.2. Suppose we have identified ten potential input variables $(x_1, x_2, ...x_{10})$ for a fuzzy modeling problem whose output variable is y. Applying the forward selection procedure to this problem, we first construct fuzzy models using only one input variable. For ease of discussion, we call these models level-one fuzzy models, as shown in the figure. A total of ten fuzzy models are constructed. These fuzzy models can be constructed using any techniques for identifying their fuzzy partitions and their consequent parameters, which we will discuss later. Once these level-one models have been constructed, they are evaluated based on a model validation criteria. Suppose the model with x_2 as input gives the best performance. It is used as a baseline for constructing fuzzy models in the next level (i.e., level two). More specifically, we construct fuzzy models with x_2 and another input variable as inputs (i.e., $x_2 x_1$, $x_2 x_3$, $x_2 x_4$, ..., $x_2 x_{10}$). We also include a fuzzy model containing only the input variable from the previous level (i.e., x_2). This model, however, will have a finer fuzzy partition (i.e., more fuzzy rules). The rationale is that one way to improve the best fuzzy model from the previous level is, instead of introducing new input variables, to partition the original input space into smaller subregions. We then construct all the level two fuzzy models, and evaluate them. The best level two fuzzy model (say the model with $x_2\ x_3$ as inputs) is chosen as the baseline for constructing level three fuzzy models.

FIGURE 14.3   The High-level Algorithm of Forward Selection Procedure

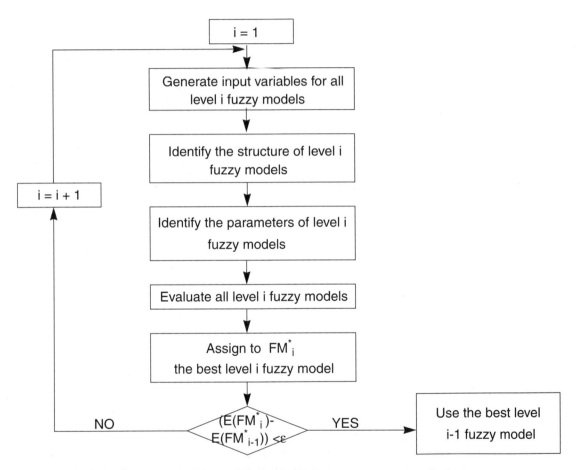

These steps iterate until the performance improvement from one level to the next level is less than a threshold. The high-level algorithm of forward selection procedure is shown in Fig 14.3, where $E(FM_i^*)$ and $E(FM_{i-1}^*)$ denote the evaluation results of the best fuzzy model at level i and level $i-1$, respectively.

This procedure is computationally efficient because it avoids considering all possible combinations of input variables by focusing on improving the best model obtained at every stage. It is one of the most commonly used procedures for input selection in linear regression analysis and also applied by T. Takagi and M. Sugeno to fuzzy modeling [562]. However, a criticism of this procedure is that there is no reason why the subset of k variables which gives the best performance should contain the subset of $(k$-1$)$ variables that gives the best performance for $(k$-1$)$ variables. A. Miller provided an input selection example in linear regression modeling in which the best-fitting subsets of three and two variables selected using this procedure have no variables in common [414]. The same problem can appear in fuzzy modeling. Hence there is no guarantee that forward selection will find the best-fitting subsets of any size except for $k = 1$ and $k = m$. Nevertheless, it suggests an "acceptable" subset of input variables.

14.4.1.2 Backward Elimination Procedure

The forward selection procedure begins with the smallest model, using one variable, and subsequently increases the number of variables in the model until a decision is reached on the model to use. The backward elimination procedure is an attempt to achieve a similar conclusion working from the other direction. That is, it removes variables in turn until the model is satisfactory. The order of removal is determined by using the same performance criterion as that used in the forward selection procedure such as MSE or UC. This basic procedure is as follows. First we build a model using all m input variables, and let MSE_m and UC_m be the corresponding values of mean squared error and regularity criterion of the model. Then we build models using m - 1 variables of the m variables and totally get $C_{m-1}^m = m$ - 1 such models. We check the MSE_{m-1} or UC_{m-1} of these models and choose the model with the smallest MSE_{m-1} or UC_{m-1} as the best model of this stage. This meanwhile determines the variable that should be removed. Similarly, at the following stage, the variable from the remaining m - 1 variables which yields the smallest MSE_{m-2} or UC_{m-2} is deleted. The process continues until there is only one variable left, or until some stopping criterion is satisfied.

This is a satisfactory procedure, especially for users who like to see all the variables in the model in order "not to miss anything" [160]. Like forward selection, backward elimination is widely used by statisticians for selecting input variables for linear regression models. K. Tanaka, M. Sano and H. Watanabe also used this procedure for the input selection of fuzzy models [580].

Backward elimination usually requires more computation than forward selection. If for instance we have 20 variables available and expect to select a subset of less than 5 of them, in forward selection we would only proceed until about 5 or so variables have been included. In backward elimination we start with 20 variables, then 19, 18, until eventually we reach the size of interest.

One computationally efficient backward elimination procedure was recently proposed by S. Chiu [127]. This procedure only needs to build a model which consists of all m input variables and does not reestimate the parameters of the model when a variable is removed, and thus it can find a subset of input variables with acceptable accuracy based on a specific performance (say MSE or UC) very quickly. However, one possible criticism of this procedure is that the coefficients of the remaining variables in the model will change when one variable is removed from the model. Nevertheless, this procedure provides a simple and practically feasible alternative for selecting input variables in building very high dimensional fuzzy models.

Similar to forward selection, backward elimination cannot guarantee the selected subset of input variables to be the best; rather, it only suggests an acceptable subset of input variables. Moreover, we must recognize that once a variable has been eliminated in this procedure it is gone forever. Thus all alternative models using eliminated variables are not available for possible examination.

14.4.1.3 "Best Subset" Procedure

This procedure tries to build a model using a "best subset" of r variables from the m variables available, where r is known *a priori*. The "best subset" here means the subset will lead to a model that has the highest accuracy in terms of a specific performance criterion such as MSE or UC. This is an exhaustive search procedure, and the best subset of input variables can always be found with the given number r. For instance, if we have a modeling problem with 10 candidate input variables and we want to find the most important 3 variables as the inputs of the model, we need to construct $C_3^{10} = 120$ models (each with different combination of 3 input variables) and compute their MSEs or UCs. The model that has the smallest MSE or UC is selected as the best model, and correspondingly the subset of 3 variables comprising the model is taken as the best subset.

Obviously, such a procedure is computationally costly. J. Jang improved the method by changing the procedure of computing MSE or UC [279]. In particular, J. Jang computed the MSEs or UCs of the models just after one epoch of training rather than when the training process is completed, which may take a great number of epochs of training. He then chose the model with the best performance and proceeded for further training to get the final parameters of the model. The improved procedure is based on the assumption that the model with the smallest MSE or UC after one epoch of training has a greater potential of achieving a lower MSE or UC when given more epochs of training. This assumption is not absolutely true, but it is a reasonable heuristic.

The limitation of this procedure is that it needs to know in advance the exact number of variables that are needed to construct the model. Another problem with this procedure is that the number of models constructed increases exponentially as the number of candidate input variables increases. The main benefit of this procedure is that it is the only one that can find the "best subset" of input variables, as its name implies.

14.4.1.4 Other Procedures

An alternative procedure used in selecting input variables for fuzzy models is to fold the input selection problem into the fuzzy partitioning problem [559]. In this case, fuzzy models with increasingly complex partition patterns are generated based on some performance criterion such as UC; input variables that require no partitioning are the ones that can be removed from the model. This approach is novel conceptually, but it is computationally expensive. The search for the optimal partition pattern in the procedure must start by considering all input variables and still requires generating different models to test each partition pattern.

There exist other procedures for input selection, most of which appear in the statistical literature. For a more detailed collection and discussion of these procedures, we refer the reader to [414]. In principle, all these procedures can be applied to the selection of input variables for fuzzy models, but there always exist practical difficulties in applying these procedures and much work needs to be done before they can become a valuable addition to the fuzzy modeling workers' tool kit.

14.4.2   Membership Function Specification

The choice of membership functions affects how well a fuzzy model behaves. We have introduced major types of membership functions in Chapter 3. From the viewpoint of fuzzy model identification, the Gaussian membership functions have some interesting properties.

A unique property of Gaussian membership functions is that they are *factorizable*. To demonstrate this important property, consider the simple case of a Gaussian membership function with unit width in two dimensions:

$$
\begin{aligned}
gaussian(\grave{x};\bar{m}) &= gaussian(\|\grave{x} - \bar{m}\|^2) \\
&= \exp(-\|\grave{x} - \bar{m}\|^2) \\
&= \exp(-(x_1 - m_1)^2 - (x_2 - m_2)^2) \quad \textbf{(EQ 14.1)} \\
&= \exp(-(x_1 - m_1)^2)\exp(-(x_2 - m_2)^2) \\
&= gaussian(x_1;m_1)gaussian(x_2;m_2)
\end{aligned}
$$

Equation 14.1 shows that a two-dimensional Gaussian membership function $gaussian(\grave{x};\bar{m})$ with a center $\bar{m} = [m_1, m_2]^T$ is equivalent to the product of a pair of one-dimensional Gaussian membership functions $gaussian(x_1; m_1)$ and $gaussian(x_2; m_2)$, where x_1 and x_2 are the elements of the vector x, and m_1 and m_2 are the elements of the center $\bar{m}$.

The result of Equation 14.1 may be readily generalized for the case of a multivariate Gaussian membership function that computed the weighted norm, assuming that weighting matrix Λ is a diagonal matrix. To be specific, suppose that we have

$$
\Lambda = diag[\sigma_1, \sigma_2, ..., \sigma_s] \quad \textbf{(EQ 14.2)}
$$

We may then factorize $gaussian(\grave{x};\bar{m}, \Lambda)$ as follows:

$$
\begin{aligned}
gaussian(\grave{x};\bar{m}, \Lambda) &= gaussian(\|\grave{x} - \bar{m}\|_\Lambda^2) \\
&= \prod_{k=1}^{s} \exp\left(-\frac{(x_k - m_k)^2}{\sigma_k^2}\right) \quad \textbf{(EQ 14.3)} \\
&= gaussian(x_1;m_1, \sigma_1)...gaussian(x_s;m_s, \sigma_s)
\end{aligned}
$$

where m_k denotes the kth element of the center $\bar{m}$.

Given such a property, we may synthesize a multidimensional membership function as the product of one-dimensional membership functions. Also, we may obtain the parameters of one dimensional membership functions by first determining the parameters of the multidimensional membership function. More importantly, since the model can be expressed in a compact form (in multidimensional membership function), the "curse of dimensionality" caused by the exponential increase in the number of rules resulting from the exhaustive combination of membership functions of separable variables can be dealt with. Whercas it is believed that the choice of membership functions does affect the performance of the resultant models, there exists no sound principles yet for guiding the choice of membership functions.

14.4.3　Fuzzy Partitioning of Input Space

A fuzzy partition of the input space defines the antecedents of a fuzzy model. The number of fuzzy subregions in a fuzzy partition corresponds to the number of fuzzy rules in the model. Fuzzy partition techniques can be classified into three categorics:

(1) grid partition,
(2) tree partition, and
(3) scatter partition.

We will discuss each one of these techniques below.

One of the most commonly used fuzzy partitioning methods in practice (particularly in control applications) is the *grid partition*. A grid partition, as shown in Fig 14.4(a), is formed by "fuzzy cuts" that divide the input space into several fuzzy slices along each dimension. Each fuzzy slice is specified by a membership function.[1] The intersection of these fuzzy slices from "fuzzy cells" is a grid partition. For instance, each fuzzy cell in the grid partition of a two-input modeling problem is described in the form of "x_1 is A_i and x_2 is B_j." Fig 14.4(a) shows a grid partition of a two-dimensional input space formed by three membership functions (i.e., three "slices") in each dimension.

Obviously, a grid partition can be uniform or nonuniform. The former, illustrated in Fig 14.4(a) is formed by uniformly spaced symmetric membership functions. The latter, illustrated in Fig 14.4(b), is formed by nonuniformly spaced asymmetric membership functions. While a uniform grid partition is easier to construct, a nonuniform partition is more flexible in adapting to the specific nonlinear characteristics of the function being approximated.

An *adaptive* fuzzy grid partition can be obtained if we use learning techniques to construct the partition. Specifically, we may take the uniformly partitioned grids as an initial partition and then use the learning procedure to optimize the parameters of the antecedent membership functions. As the parameters in the antecedent membership functions are adjusted, the grid evolves. Two typical learning procedures used in practice are the gradient descent method suggested by J. Jang and genetic algorithms suggested by C. Karr [278, 299, 300]. We will discuss these learning algorithms in Section 14.5 and in the following chapters.

[1.] Fig 14.4(a) suggests that the membership function used is of trapezoid type. However, the concept of a grid partition can be applied to any type of membership function.

Grid partition is convenient to use, but it may encounter serious "rule explosion" problems when the number of input variables is large. This problem is closely related to the so-called "curse of dimensionality", the well-known problem of exponentially increasing model complexity as the number of input variables increases [478]. For instance, a fuzzy model with 10 inputs and 2 membership functions on each input would result in $2^{10}=1024$ fuzzy if-then rules, which is prohibitively large. Thus, the grid partition method is usually applied to fuzzy models with few input variables. Most fuzzy models used in control applications fall into this category.

Tree partition is the second method used in partitioning an input space. Fig. 14.4(c) gives an example of a tree partition in a two-dimensional input space. A tree partition results from a series of *guillotine cuts*. By a guillotine cut, we mean a cut that is made across the subspace to be partitioned; each of the regions so produced can then be subjected to further independent guillotine cutting. At the beginning of the ith iteration step, the input space is partitioned into i regions. Then another guillotine cut is applied to one of the regions to further partition the entire space into $(i + 1)$ regions. There are various strategies to decide which dimension to cut and where to cut it at each step. Some are based merely on the distribution of training examples; others take the parameter identification methods into consideration. For a brief and clear discussion of these strategies, we refer the reader to C. Sun [564].

FIGURE 14.4   Various Methods for Partitioning the Input Space: (a) Grid Partition (uniform); (b) Grid Partition (Nonuniform); (c) Tree Partition; (d) Scatter Partition

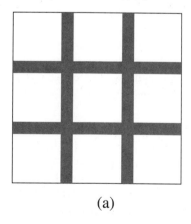

(a)

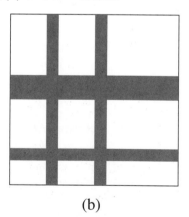

(b)

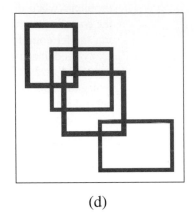

(c)   (d)

Tree partition relieves the problem of rule explosion to a great degree, but it is not easy to use in practice. A number of heuristics are usually needed to find a proper tree structure, and there may be difficulties in designing an optimal tree partition. Due to the variety and arbitrariness of partition strategies, different users may design quite different tree structures, and thus get fuzzy models with quite different performance. Also, errors may accumulate from level to level in a large tree.

The third type of fuzzy partition is the *scatter partition*. An example of such a partition is shown in Fig. 14.4(d). Instead of covering the whole input space, this method tries to find a subset of the input space that characterizes the fuzzy regions of possible occurrence of training examples. Each fuzzy region is associated with one (multidimensional) membership function or a combination of several (one-dimensional) membership functions. It is usually difficult to find such a partition intuitively or manually. Hence, learning techniques have to be adopted to automate the construction of scatter partition. We will introduce these techniques in Section 14.6.

14.4.4 Determining the Number of Fuzzy Rules

When deciding the number of fuzzy rules for a fuzzy model, there are two contradictory concerns. On one hand, it is desired that the model include as many rules as possible so that it minimize errors from training data with sufficient "patches." On the other hand, it is also desired that the model includes as few rules as possible because the generalization capability of the model decreases as the number of fuzzy rules increases. By *generalization capability,* we are referring to the model's ability in approximating testing data, which are not used in training the model.

FIGURE 14.5   An Example of Model Overfitting: (a) A Good Model; (b) An Overfitted Model

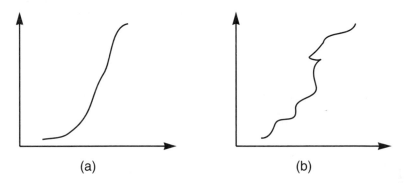

(a)   (b)

Theoretically, the more fuzzy rules, the better the approximate ability of the model [677]. In practice, however, good fit on training data does not always assure good performance on future data, especially if the data have noise and the number of adjustable parameters in the model is large compared to the available training data. A fuzzy model with a large number of rules is liable to encounter the risk of "overfitting," capable of fitting training data completely but incapable of generalizing to untrained data satisfactorily. Fig. 14.5 shows an example of model overfitting.

The trade-off between goodness of fit and simplicity is a fundamental principle underlying various general theories of statistical modeling and inductive inference [8, 14]. Several research efforts have been made in the fuzzy logic community to strike a balance between reducing the fitting error and reducing the model complexity. In particular, *orthogonal transformation* methods have been used for selecting important fuzzy rules from a given rule base [429, 668, 669, 671]. Unlike conventional methods where multiple iterations are usually required to find the "optimal" number of fuzzy rules, orthogonal transformation methods are a noniterative procedure. They start with an oversized rule base and then remove redundant or less important fuzzy rules through a "one-pass" operation. Therefore, orthogonal transformation methods are computationally less expensive compared to the conventional methods, especially when the number of input variables is large. Moreover, they are extremely robust numerically and can be implemented using some well-established numerical computation packages. In this section we introduce how to use *Singular Value Decomposition (SVD)* to select the most important fuzzy rules from a given rule base and construct compact fuzzy models with better generalization ability. In this section we will introduce several methods that determine the number of fuzzy rules by finding a good trade-off between fitting training data and high generalizing ability.

14.4.5 Model Reduction Using Singular Value Decomposition

The Singular Value Decomposition (SVD) of an m-by-n matrix A is a factorization of A into a product of three matrices. That is $A = U\Sigma V^{T}$, where $U \in R^{m \times m}$ and $V \in R^{n \times n}$ are orthogonal matrices, $\Sigma = diag(\sigma_1, \sigma_2, ..., \sigma_p) \in R^{m \times n}(p = min\{m, n\})$

is a diagonal matrix with $\sigma_1 \geq \sigma_2 \geq \ldots \geq \sigma_p \geq 0$. The σ_i are called the *singular values* of A and they are the positive square roots of the eigenvalues of $A^T A$. The columns of U are called the left singular vectors of A (the orthonormal eigenvector of AA^T) while the columns of V are called the right singular vectors of A (the orthonormal eigenvectors of $A^T A$).

SVD is one of the most useful and powerful tools of numerical linear algebra and has found successful applications in various areas such as statistical analysis, image and signal processing, system identification, and linear control. For a survey of SVD, its theory, numerical details, and some interesting applications, we refer the reader to [156].

In order to illustrate the basic principle of using SVD for fuzzy rule selection, we use the fuzzy model with constant consequent constituents as an example. This type of fuzzy model, which is usually referred to as the *zero-order TSK model* in the literature [278, 126], has the following form:

$$R_i\colon If \quad x_1 \text{ is } A_{i1} \text{ and } x_2 \text{ is } A_{i2} \text{ and } \ldots \text{ and } x_m \text{ is } A_{im}$$
$$Then \quad y \text{ is } c_i, i = 1, 2, \ldots, M.$$

(EQ 14.4)

where c_i are the constant constituents. The total output of the model is computed by

$$y = \frac{\displaystyle\sum_{i=1}^{M} w_i c_i}{\displaystyle\sum_{i=1}^{M} w_i}$$

(EQ 14.5)

where w_i is the matching degree (i.e., the firing strength) of the ith rule defined by Equations 14.6 or 14.7 (i.e., similar to Equation 5.42 in Chapter 5).

$$w_i = \min(\mu_{A_{i1}}(a_1), \mu_{A_{i2}}(a_2), \ldots, \mu_{A_{im}}(a_m))$$

(EQ 14.6)

or

$$w_i = \mu_{A_{i1}}(a_1) \times \mu_{A_{i2}}(a_2) \times \ldots \times \mu_{A_{im}}(a_m)$$

(EQ 14.7)

Let us define the *normalized firing strength* of the ith rule as

$$\bar{N}_i \equiv \frac{w_i}{\displaystyle\sum_{i=1}^{M} w_i}$$

(EQ 14.8)

Equation 14.5 can then be rewritten as

$$y = \sum_{i=1}^{M} \bar{N}_i c_i \qquad \text{(EQ 14.9)}$$

This equation can be viewed as a special case of linear regression model [620]

$$y = \sum_{i=1}^{M} p_i \theta_i + e \qquad \text{(EQ 14.10)}$$

with p_i and θ_i given by

$$p_i \equiv \bar{N}_i, \quad \theta_i \equiv c_i \qquad \text{(EQ 14.11)}$$

where p_i are known as the regressors, θ_i are the parameters, and e is an error signal which is assumed to be uncorrelated with the regressors p_i. Given N input-output pairs $\{\dot{x}(k),$ $y(k)\}$, $k = 1, 2, ..., N$, where $\dot{x}(k) = [x_1(k), x_2(k), ..., x_m(k)]^T$, it is convenient to express Equation 14.10 in the matrix form

$$\dot{y} = P\dot{\theta} + \dot{e} \qquad \text{(EQ 14.12)}$$

where $\quad \dot{y} = [y(1), ..., y(N)]^T \in R^N$,

$\qquad P = [\vec{p}_1, ..., \vec{p}_M] \in R^{N \times M}$ with

$\qquad \vec{p}_i = [p_i(1), ..., p_i(N)]^T \in R^N$,

$\qquad \dot{\theta} = [\theta_1, ..., \theta_M]^T \in R^M$, and

$\qquad \dot{e} = [e(1), ..., e(N)]^T \in R^N$.

Note that each column of P corresponds to one of the fuzzy rules in the rule base. We will call P the *firing strength matrix* and $P\dot{\theta}$ the *predictor* of $\dot{y}$ throughout this section for notational simplicity. In building a fuzzy model, the number of available training data is usually larger than the number of fuzzy rules in the rule base. This implies that the row dimension of the matrix P is larger than its column dimension, that is, $N > M$.

The firing strength matrix P may be singular (or nearly singular) because of the existence of *less important* or *redundant* fuzzy rules in the rule base. By *less important* we mean the contribution of the rules to the total output is minor; by *redundant* we mean the contribution of the rules can be replaced by that of other rules. A less important rule can appear in a rule base when the normalized firing strengths (defined by Equation 14.8) of the rule are zeros or near-zeros in the whole input space, while a redundant rule can appear in a rule base when the normalized firing strengths of the rule are identical with or linearly dependent on those of one or more other rules.

Mathematically, the singularity or near-singularity of a matrix is indicated by the existence of zero or near zero singular values in the matrix. Thus, we may determine the less important or redundant fuzzy rules in a rule base by checking the singular values of the firing strength matrix P. Specifically, we may compute the SVD of P by $P = U_P \Sigma_P V_P^T$ where the number of zero or near-zero singular values in Σ_P indicates the number of less important or redundant fuzzy rules in the rule base. Removing these less important or redundant fuzzy rules from the rule base amounts to finding a predictor $P\hat\theta$ in which $\hat\theta$ has at most r nonzero components, where r is the number of important fuzzy rules retained in the rule base after the less important or redundant rules are removed. The position of the nonzero entries determines which columns of P, i.e., which rules in the rule base, are to be used in constructing the model and approximating the observation vector $\hat y$.

In the following we introduce a method that is used to select the r important rules (or equivalently the M-r less important or redundant rules) in the rule base. The method starts by computing the SVD of P, that is

$$P = U_P \Sigma_P V_P^T \qquad\qquad \text{(EQ 14.13)}$$

Partition V_P as

$$V_p = \begin{bmatrix} V_{11} & V_{12} \\ V_{21} & V_{22} \end{bmatrix} \begin{matrix} r \\ M - r \end{matrix} \qquad\qquad \text{(EQ 14.14)}$$

$$\qquad\qquad r \quad M - r$$

Applying a *QR with a column pivoting factorization algorithm* [218] to $[V_{11}^T \ \ V_{21}^T]$ yields

$$Q^T [V_{11}^T \ \ V_{21}^T]\Pi = [R_{11} \ \ R_{12}] \qquad\qquad \text{(EQ 14.15)}$$

$$\qquad\qquad r \quad M - r$$

where $Q \in R^{r \times r}$ is orthogonal, $R_{11} \in R^{r \times r}$ is upper triangular, and $\Pi \in R^{M \times M}$ is a *permutation matrix*. Define

$$\begin{bmatrix} P_r \ P_{M-r} \end{bmatrix} \equiv P\Pi \qquad\qquad \text{(EQ 14.16)}$$

where $P_r \in R^{N \times r}$ comprises the desired columns of P whose original position in P indicates the position of the associated rules in the rule base.

The key of the method is to find the permutation matrix Π and then obtain the desired subset P_r. A permutation matrix is the *identity* with its rows reordered and one of its functions is to interchange the columns of a matrix. Suppose, for instance, we have

$$P = \begin{bmatrix} p_{11} \ p_{12} \ p_{13} \\ p_{21} \ p_{22} \ p_{23} \\ p_{31} \ p_{32} \ p_{33} \end{bmatrix} \qquad\qquad \text{(EQ 14.17)}$$

and we want to interchange its first and the third columns. If we can find a permutation matrix

$$\Pi = \begin{bmatrix} 1 & 0 & 0 \\ 0 & 0 & 1 \\ 0 & 1 & 0 \end{bmatrix} \qquad \textbf{(EQ 14.18)}$$

and apply it to the right side of P, then we have

$$P\Pi = \begin{bmatrix} p_{11} & p_{13} & p_{12} \\ p_{21} & p_{23} & p_{22} \\ p_{31} & p_{33} & p_{32} \end{bmatrix} \qquad \textbf{(EQ 14.19)}$$

Several other rule selection methods where SVD also plays an important role were introduced in J. Yen and L. Wang [671, 668]. One of the methods, known as the *direct-SVD method*, showed the best performance in terms of accuracy and simplicity. This method uses a single SVD step and does not need to form the permutation matrix Π explicitly in extracting the desired subset of P. The key of the method is to compute the SVD of the firing strength matrix P in which the singular values appear in their original order rather than a descending order as done in the above. J. Yen and L. Wang used a *two-sided Jacobi algorithm* to compute SVD for guaranteeing this [218].

The basic principal behind this method can be understood from the following observations. If the jth column of P consists of zero or near-zero entries, then a zero or near-zero singular value will appear on the main diagonal in the jth column of the triangular matrix obtained from the nonordered SVD. If the jth column of P consists of entries which are identical with or linearly dependent on those of one or more other columns, say column l, then a zero singular value will appear on the main diagonal in either the jth or lth column of the triangular matrix. We may illustrate the idea using the matrix P given in Equation 14.17. Suppose that the entries in the second column of P are all zeros, then the singular values on the main diagonal of the triangular matrix look like this

$$\begin{bmatrix} * & & \\ & 0 & \\ & & * \end{bmatrix}$$

where * denotes an arbitrary non-negative real constant. If we suppose that the entries in the second column are identical with those in the third column, then the singular values on the main diagonal of the triangular matrix may look like

$$\begin{bmatrix} * & & \\ & 0 & \\ & & * \end{bmatrix}$$

or

$$\begin{bmatrix} * & & \\ & * & \\ & & 0 \end{bmatrix}$$

In practice, the singular values of the firing strength matrix P are usually not clear-cut. Typically, a number of singular values will have similar small positive numbers. Thus, some criteria or rules of thumb should be established for determining whether these small numbers are significantly different from zero. The information-theoretic criterion Schwarz-Rissanen Information Criterion (SRIC) introduced in the previous section may serve as such a criterion [669]. We will illustrate this in the later examples.

The SVD-based rule selection methods can remove both the less important rules and the redundant rules from a rule base, but they cannot identify which rules are less important and which rules are redundant among the removed ones. In order to distinguish between less important and redundant, we may plot the columns of P which are associated with these removed rules using the two-dimensional graphic as shown in Fig 14.4.

14.5 Parameter Estimation

In this section we will introduce several commonly used methods for estimating fuzzy model parameters. As pointed out in Section 14.3, the parameter estimation and the structure identification of a fuzzy model are two interwoven issues and neither of them can be performed independently of the other. However, as far as most parameter estimation techniques are concerned, the structure of the fuzzy model is given.

The goal of parameter estimation is to find the best values for a set of model parameters. The optimality of parameter values is determined by (1) how well the model fits the training data, and (2) how well the model performs for a given task. The first approach is often used for modeling a plant, a process, or a set of time series data. The second approach is often used for designing a controller (i.e., the direct controller approach to be discussed in Section 14.8. The two approaches also imply different modeling requirements. The former requires training data, while the latter requires clearly specified performance objectives. Finally, the techniques used by the two approaches are often different. Techniques for parameter estimation using training data include backpropagation learning, recursive least square algorithm, the least mean square algorithm, Kalman filters, Singular Value Decomposition (SVD), and the genetic algorithm. Techniques for parameter estimation based on performance include reinforcement learning and genetic algorithms. Even though genetic algorithms can be applied to both types of parameter estimation, it is more commonly used for the latter type. We will discuss the application of the genetic algorithm to fuzzy model identification in Chapter 17. The use of backpropagation learning and reinforcement learning for parameter estimation will be discussed in detail in Chapter 16.

There are two types of parameters in a fuzzy model: (1) parameters of the antecedent membership functions, and (2) parameters in the consequent part of rules. Antecedent parameters are often identified using clustering techniques. Consequent parameters can be estimated using most of the techniques mentioned earlier. We will focus on three of these techniques: SVD-based techniques, the Recursive Least Square (RLS) algorithm, and the Least Mean Square (LMS) algorithm.

14.5.1   Estimation of Antecedent Parameters

Antecedent parameters of a fuzzy model can be estimated independent of consequent parameters using various clustering techniques. The rationale is to find "groupings" of input training data and use the clustering result to determine the parameters of a fuzzy partition (i.e., parameters of antecedent membership functions).

Clustering can be applied either to the input training data or to the input-output training data. All clustering techniques, whether they find hard clustering or soft clustering, can be applied. Examples include fuzzy c-means and fuzzy k-nearest neighborhood, which we introduced in Chapter 13. Another clustering technique proposed for finding a fuzzy partition is the *mountain method* proposed by R. Yager [654].

Clustering usually produces a scattered fuzzy partition of the input space. The parameters of the antecedent membership functions are usually constructed from the mean and the variance of each cluster. For instance, we may use multidimensional Gaussian membership functions

$$\mu_{A_i}(\vec{x}) = \exp\left[\frac{-(\vec{x} - \vec{m}_i)^T(\vec{x} - \vec{m}_i)}{\sigma_i^2}\right] \qquad \textbf{(EQ 14.20)}$$

whose mean $\vec{m}$ are cluster centers.

Once the centers are established, the width σ_i of each cluster can be determined. The goal in setting the widths is to achieve a certain amount of response overlap between each cluster and its neighbors so that they form a smooth and contiguous interpolation over those regions of the input space which they represent. An appropriate width can be determined from the K-nearest neighbor heuristic [166]:

$$\sigma_i = \frac{1}{k}\sum_{l=1}^{k} \|\vec{m}_i - \vec{m}_l\|_2^2 \qquad \textbf{(EQ 14.21)}$$

where $\vec{m}_l$ are the K-nearest neighbors of $\vec{m}_i$. In J. Moody and C. Darken, $k = 2$ was suggested [428]. The K-nearest neighbor heuristic is one of the simplest methods for determining the width σ_i. Several more complicated methods which tend to provide improved performance were discussed in M. Musavi et al. and S. Roberts and L. Tarassenko [434, 505].

As compared with the basic Backpropagation (BP) algorithm (to be discussed in Chapter 16), this approach can converge very quickly [278]. However, a drawback of the algorithm is that it may not provide an "optimal" parameter estimate. The reason is obvious: the K-means clustering and the K-nearest neighbor heuristic determine the antecedent parameters based *only* on input data and do not take into account any information in the output data. Thus, the performance of the model identified using this clustering approach is usually inferior to that of the model identified using the BP algorithm.

14.5.2 Estimation of Consequent Parameters

The parameter estimation algorithms introduced in this section can be applied to all additive fuzzy models. In our discussion, however, we will use the zero-order TSK model to illustrate these techniques, which is rewritten as follows:

$$R_i : If \quad x_1 \text{ is } A_{i1} \text{ and } \quad x_2 \text{ is } A_{i2} \text{ and } \dots \text{ and } x_s \text{ is } A_{is}$$
$$Then \quad y \text{ is } c_i \text{ , } i = 1, 2, \dots, M. \tag{EQ 14.22}$$

The membership functions in rule antecedents are assumed to be of the Gaussian type:

$$\mu_{A_{ij}}(x_j) = \exp\left(-\frac{(x_j - m_{ij})^2}{\sigma_{ij}^2}\right) \tag{EQ 14.23}$$

Algorithms for other types of additive fuzzy models with various membership functions can be derived in a similar manner and are left to the reader as an exercise.

As shown in Equation 14.5, the total output of the zero-order TSK model using product as the fuzzy conjunction operator is computed by

$$y = \frac{\displaystyle\sum_{i=1}^{M} w_i c_i}{\displaystyle\sum_{i=1}^{M} w_i} \equiv \frac{\displaystyle\sum_{i=1}^{M}\prod_{j=1}^{s} \mu_{A_{ij}}(x_j) c_i}{\displaystyle\sum_{i=1}^{M}\prod_{j=1}^{s} u_{A_{ij}}(x_j)} \tag{EQ 14.24}$$

Note that the "product" operator instead of the "min" operator has been used in calculating w_i in Equation 14.24. The parameters to be estimated in the model are m_{ij} and σ_{ij} in the antecedent membership functions and the constant constituents c_i in the consequents.

We can analyze Equation 14.24 from two points of view [620]. First, if we view all the parameters m_{ij}, σ_{ij}, and c_i in the model as free design parameters, then the estimation problem is nonlinear-in-the-parameters. In order to compute these parameters, we must use some nonlinear optimization technique such as the gradient descent algorithm. On the other hand, we can fix the parameters m_{ij}, σ_{ij} in the antecedent membership functions at the beginning of the estimation procedure so that the only free design parameters are the constant constituents c_i in the consequents. This is the approach adopted by S. L. Chiu and L. Wang and R. Langari where the parameters of antecedent membership functions were predetermined using different clustering algorithms [127, 625]. Note that the problem becomes linear in the parameters in this situation, and thus a wide range of linear optimization techniques would be at our disposal for solving c_i. Below we will describe three algorithms based on the two different points of view: (1) SVD-based estimation algorithm, (2) recursive least squre (RLS) estimation algorithm, and (3) least mean square (LMS) estimation algorithm.

14.5.2.1   SVD-based Estimation Algorithm

Equation 14.24 can be viewed as a special form of the linear regression (Equation 14.10) with the matrix expression (Equation 14.12), rewritten as follows:

$$\grave{y} \,=\, P\grave{\theta} + \grave{e} \tag{EQ 14.25}$$

Now we will choose $\grave{\theta}$ in such a way that the following objective function J is minimized:

$$J \,=\, \left\| \grave{y} - P\grave{\theta} \right\|_2^2 \equiv (\grave{y} - P\grave{\theta})^T (\grave{y} - P\grave{\theta}) \tag{EQ 14.26}$$

To carry out the minimization, we differentiate J with respect to $\grave{\theta}$ and equate the result to zero. Thus

$$\frac{\partial J}{\partial \vec{\theta}} \,=\, -2P^T \grave{y} + 2P^T P\grave{\theta} \,=\, 0 \tag{EQ 14.27}$$

from which $\grave{\theta}$ can be solved as

$$\grave{\theta} \,=\, (P^T P)^{-1} P\grave{y} \tag{EQ 14.28}$$

This result is called the *Least Squares (LS) solution* of $\grave{\theta}$.

Let P^+ denote the *pseudoinverse* of P, which is defined by

$$P^+ \,=\, (P^T P)^{-1} P \tag{EQ 14.29}$$

Then Equation 14.28 can be rewritten in a more compact form

$$\grave{\theta} \,=\, P^+ \grave{y} \tag{EQ 14.30}$$

In practice, the most reliable method of computing the pseudoinverse of a matrix is the SVD. Specifically, applying SVD to P yields

$$P \,=\, U\Sigma V^T \equiv \sum_{i=1}^{r} \sigma_i \vec{u}_i \vec{v}_i^T \tag{EQ 14.31}$$

where r is the *rank* of P, which corresponds to the number of nonzero singular values σ_i in Σ, $\vec{u}_i$ and $\vec{v}_i$ are the columns of U and V, respectively. Substituting Equation 14.31 Equation into 14.29 and after some algebraic manipulations, we obtain

$$P^+ \,=\, \sum_{i=1}^{r} \frac{1}{\sigma_i} \vec{v}_i \vec{u}_i^T \tag{EQ 14.32}$$

Correspondingly, the least squares solution of $\grave{\theta}$ in Equation 14.30 becomes

$$\hat{\theta} = \sum_{i=1}^{r} \frac{1}{\sigma_i} \vec{v}_i \vec{u}_i^T \vec{y} \equiv \sum_{i=1}^{r} \frac{(\vec{u}_i \vec{y})}{\sigma_i} \vec{v}_i \qquad \text{(EQ 14.33)}$$

TABLE 14.1   Summary of the SVD-Based Parameter Estimation Algorithm
Step 1: Compute the SVD of P using Equation 14.31 and
get σ_i, $\vec{u}_i$, and $\vec{v}_i$, where $i = 1, 2, ..., r$.
Step 2: Compute the pseudoinverse of P using Equation 14.32.
Step 3: Solve the unknown parameter vector $\hat{\theta}$ from Equation 14.33.

The heart of the algorithm is the SVD, and thus we call it the *SVD-based parameter esti-mation algorithm*. This is a *noniterative* algorithm and capable of obtaining the parameter estimates very quickly and reliably. Many well-established numerical computing packages (e.g., Matlab) are available for performing the computations involved in the algorithm. A summary of the algorithm is presented in Table 14.1.

14.5.2.2 Recursive Least Squares (RLS) Estimation Algorithm

In this subsection we give a recursive algorithm for the basic LS solution in Equation 14.28. The need for a recursive solution arises when fresh experimental data are continu-ously in supply and we wish to improve our parameter estimates by making use of this new information. With a recursive formula, the estimates can be updated step by step with-out repeatedly computing the matrix solution of Equation 14.28. This recursive solution procedure is often referred to as *sequential*, or *on-line estimation*.

$$\begin{bmatrix} y(1) \\ y(2) \\ \dots \\ y(N) \end{bmatrix} = \begin{bmatrix} p_1(1) & p_2(1) & \dots & p_M(1) \\ p_1(2) & p_2(2) & \dots & p_M(2) \\ \dots & \dots & \dots & \dots \\ p_1(N) & p_2(N) & \dots & p_M(N) \end{bmatrix} \begin{bmatrix} \theta_1 \\ \theta_2 \\ \dots \\ \theta_M \end{bmatrix} + \begin{bmatrix} e(1) \\ e(2) \\ \dots \\ e(N) \end{bmatrix} \qquad \text{(EQ 14.34)}$$

Recall that Equation 14.25 actually consists of a set of N equations. Let us express the kth equation as

$$y(k) = \vec{p}(k)\hat{\theta} + e(k) \qquad \text{(EQ 14.35)}$$

where $\vec{p}(k) \equiv [p_1(k), ..., p_M(k)]$ is the kth row of P. Let $\hat{\theta}(k-1) \equiv [\theta_1(k-1), ..., \theta_M(k-1)]^T$ denote the parameter estimate up to the k-1th observation. Then at the kth observation, the parameter estimate is updated as follows according to [219].

$$\hat{\theta}(k) = \hat{\theta}(k-1) + \frac{L(k-1)\vec{p}^T(k)[y(k) - \vec{p}(k)\hat{\theta}(k-1)]}{1 + \vec{p}(k)L(k-1)\vec{p}^T(k)} \qquad \text{(EQ 14.36)}$$

The result simply shows that the new estimate is given by the old estimate plus a correction term. The M-by-M matrix $L(k-1)$ in the correction term can be updated by the recursive formula

$$L(k) = \frac{L(k-1) - L(k-1)\vec{p}^T(k)\vec{p}(k)L(k-1)}{1 + \vec{p}(k)L(k-1)\vec{p}^T(k)}$$ **(EQ 14.37)**

The recursive Equation 14.36 has a very strong intuitive appeal. We notice that the correction term is proportional to the quantity $y(k) - \vec{p}(k)\hat{\theta}(k-1)$, which represents the error of fitting the previous estimate $\hat{\theta}(k-1)$ to the new data $y(k)$ and $\vec{p}(k)$. The vector

$$\frac{L(k-1)\vec{p}^T(k)}{1 + \vec{p}(k)L(k-1)\vec{p}^T(k)}$$ **(EQ 14.38)**

determines how the fitting error is weighted in the correction of $\hat{\theta}(k-1)$.

By starting with an initial estimate $\hat{\theta}(0)$ and the corresponding $L(0)$, we can recursively update $\hat{\theta}$ while new observations are continuously obtained. Table 14.2 summarizes the recursive least squares algorithm.

TABLE 14.2   Summary of the RLS Algorithm

Step 1: Set $\hat{\theta}(0) = \vec{0}$ and $L(0) = \alpha I$ where α is a large positive scalar, and I is
an identity matrix; and set $k = 1$.
Step 2: Update $\hat{\theta}$ using Equation 14.36.
Step 3: Update L using Equation 14.37.
Step 4: If some stopping criterion is satisfied, stop; otherwise set $k = k+1$ and go to
step 2.

The RLS algorithm provides a powerful tool for on-line model construction. Further, by introducing a *forgetting factor into* Equations 14.36 and 14.37, the algorithm can be used to estimate the parameters in a time-varying system [219, 384, 243]. For an application of the algorithm to the identification of time-varying fuzzy systems, we refer the reader to L. Wang and R. Langari [625].

As compared with other recursive algorithms (e.g., the LMS algorithm presented in the following subsection), RLS is generally fast and capable of converging to a globally optimal solution. It is fair to say that RLS is one of the best algorithms for linear parameter estimation problems. Even if an estimation problem is nonlinear and some nonlinear optimization algorithm has to be used to solve the parameters, it is highly advocated that where possible, the positive features of the RLS should be incorporated into the algorithm.

14.5.2.3 Least Mean Square (LMS) Algorithm

The LMS algorithm is an important member of the family of *stochastic approximation-based algorithms* [304]. It was originally formulated by Widrow and Hoff for use in a single-layer linear neural network, known as *Adaline* [636]. We introduce the algorithm in this subsection for two reasons. First, its implementation is simple: It does not require measurements of the pertinent correlation functions, nor does it require complicated matrix computations. In addition, it is the precursor to the *Backpropagation (BP)* algorithm that will be presented in the next chapter.

Let $\hat{y}(k)$ denote the actual output of the model for the kth observation, as computed by

$$\hat{y}(k) = \vec{p}(k)\hat{\vec{\theta}}$$

(EQ 14.39)

We may then define the *error signal*,

$$e(k) = y(k) - \hat{y}(k)$$

(EQ 14.40)

As a performance measure, we introduce the objective function $J(k)$, as defined by

$$J(k) = e^2(k)$$

(EQ 14.41)

which represents the *instantaneous* error squares. Given the *old* value of the parameter vector $\hat{\vec{\theta}}(k-1)$ up to the k-1th observation, the *updated* value of the vector at the kth observation is computed as

$$\hat{\vec{\theta}}(k) = \hat{\vec{\theta}}(k-1) - \eta \frac{\partial J(k)}{\partial \hat{\vec{\theta}}}\bigg|_{\hat{\vec{\theta}} = \hat{\vec{\theta}}(k-1)}$$

(EQ 14.42)

where η is a positive constant called the *learning-rate parameter*; $\dfrac{\partial J(k)}{\partial \hat{\vec{\theta}}}\bigg|_{\hat{\vec{\theta}} = \hat{\vec{\theta}}(k-1)}$

denotes the *gradient* of the objective function with respect to the parameter vector $\hat{\vec{\theta}}$ at observation k-1, which is given by

$$\frac{\partial J(k)}{\partial \hat{\vec{\theta}}}\bigg|_{\hat{\vec{\theta}} = \hat{\vec{\theta}}(k-1)} = -2e(k)\vec{p}^T(k)$$

(EQ 14.43)

Note that the parameter estimate $\hat{\vec{\theta}}(k-1)$ has been substituted for $\hat{\vec{\theta}}$ in Equation 14.38 for computing $\hat{y}(k)$ and then $e(k)$. Substituting Equation 14.42 into Equation 14.41, we have

$$\hat{\vec{\theta}}(k) = \hat{\vec{\theta}}(k-1) + 2\eta e(k)\vec{p}^T(k)$$

(EQ 14.44)

This is the basic formula of the LMS algorithm, which is also referred to as the *delta rule* or the *Widrow-Hoff learning algorithm*.

It was shown that the LMS algorithm is always *convergent in the mean square* (i.e., in $E[e^2(n)]$, where $E[.]$ denotes the *expected value*) if the learning-rate parameter is chosen as follows:

$$1 < \eta < \frac{1}{tr(R_{\hat{p}})} \qquad [243, 637]$$

(EQ 14.45)

where $tr(R_{\hat{p}})$ is the *trace* of the correlation matrix $R_{\hat{p}}$, which is defined by

$$R_{\hat{p}} = E[\hat{p}^T(k)\hat{p}(k)] \equiv \frac{1}{N}\sum_{i=1}^{N}\hat{p}(k)^T\hat{p}(k) \qquad \text{(EQ 14.46)}$$

In practical applications it might not be practical to specify the learning-rate parameter η by first forming $R_{\hat{p}}$; instead, η could be selected by trial and error [231]. The choice of η reflects a trade-off between parameter misadjustment and the speed of adaptation: a small η might give a small misadjustment, but it could accompany a long convergence time; on the other hand, a large η could generate a rapid convergence, but it might have the danger of parameter blowup [637, 243]. The learning-rate parameter η in the basic formula (Equation 14.43) of the LMS algorithm has been assumed to be a constant. There also exist extensive works which discuss the issue of optimization of the learning-rate parameter or methods of varying the learning-rate parameter to improve performance. For example, R. Kwong and E. Johnston presented a variable learning-rate parameter LMS algorithm in which η increases or decreases as the mean square error increases or decreases and thus allows the algorithm to track changes in the system as well as produce small steady state error [345]. Table 14.3 gives a summary of the LMS algorithm.

TABLE 14.3   Summary of the LMS Algorithm
Step 1: Set $\hat{\theta}(0) = \vec{0}$ and $k = 1$.
Step 2: Compute $\hat{y}(k)$ using Equation 14.39 where $\hat{\theta}$ is replaced by $\hat{\theta}(k-1)$;
Calculate $e(k)$ using Equation 14.40.
Step 3: Update $\hat{\theta}$ using Equation 14.44.
Step 4: If some stopping criterion is satisfied, stop; otherwise set $k = k+1$ and go to step 2.

A significant feature of the LMS algorithm is its simplicity. Indeed, it is the simplicity of the LMS algorithm that has made it the *standard* against which other adaptive algorithms are benchmarked in the field of signal processing [243]. Also, this algorithm paves the way for the derivation of a more general parameter estimation algorithm, know as the BP algorithm, which is presented in the next subsection. However, as compared with the RLS algorithm, LMS is generally slow (Essentially, LMS is a first-order gradient-descent algorithm, while RLS is a second-order one [304]). For problems that are linear in the parameters, e.g., the present consequent parameter estimation problem, the RLS algorithm is generally preferred.

14.5.3   Simultaneous Estimation of Antecedent and Consequent Parameters

14.5.3.1   Backpropagation (BP) Algorithm

There is a rapidly growing interest in the fusion of neural networks and fuzzy systems to obtain the advantages of both paradigms while avoiding their individual drawbacks. On the one hand, the theory of fuzzy logic provides a formal framework to abstract the approximate-reasoning characteristics of human decision making and, furthermore, conveys an excellent mode of knowledge representation in the form of if-then rules. However, a common bottleneck in fuzzy logic systems is their dependence on the specification of good rules by human experts. Neural networks, on the other hand, attempt to replicate the learning capabilities possessed by biological species, but it is not always possible to extract and interpret the learned knowledge contained within them. The possibility of the integration of these two paradigms has given rise to a rapidly emerging field of *fuzzy neural systems* that are intended to capture the capabilities and advantages of both neural and fuzzy logic systems. Fuzzy neural systems have become an area of much activity in practice and many interesting problems have been successfully addressed. The detail of applying backpropagation learning in neural networks to identify parameters in fuzzy models will be discussed in Chapter 16.

We should remember that a fuzzy model has its own unique structure that should be taken into consideration in developing the training algorithms. In particular, a fuzzy model comprises an antecedent component and a consequent component. If the parameters in the antecedent component are fixed (temporally!), the estimation problem is linear with respect to the parameters in the consequent component; if, on the other hand, the parameters in the consequent component are fixed (temporally!), the estimation problem is then nonlinear with respect to the parameters in the antecedent component (of course, the estimation problem is always a nonlinear with respect to the antecedent parameters no matter whether the consequent parameters are fixed or not). Thus, it is possible to estimate the parameters in the two components using different types of optimization algorithms (linear or nonlinear). This is exactly the idea proposed J. Jang where an LMS algorithm is used to train the antecedent parameters while an RLS algorithm is used to train the consequent parameters [278]. Jang shows that the combination of RLS with LMS could in general cut down the convergence time substantially. This algorithm may be viewed as a *hybrid algorithm* in the sense that a linear algorithm (RLS) and a nonlinear algorithm (LMS) are incorporated into a single procedure for estimating the parameters. In the following section we will introduce another hybrid type algorithm in which, as in Jang's algorithm, the consequent parameters are computed using the RLS, but the antecedent parameters are computed using a *cluster nearest neighbor method*.

14.5.3.2 Evolutionary Computations

Evolutionary Computation (EC) is a class of computation techniques that are based on Darwin's models of biological evolution. EC includes genetic algorithms, Evolutionary Strategies, and evolutionary programming. These techniques can perform search and optimization in a large and complex space. Since parameter identification can be viewed as a global optimization problem, it is natural to use EC for parameter identification, whether it is for identifying the consequent parameters only or for identifying the antecedent parameters and the consequent parameters simultaneously. We will discuss the use of genetic algorithms for identifying parameters of fuzzy model in Chapter 17.

14.6   Model Validation

Once a model structure has been chosen, the parameter estimation procedure provides us with a particular model within this structure. This model may be the best available one, but the crucial question is whether it is good enough for the intended purpose. Testing whether a given model is appropriate is known as *model validation* [384].

This section discusses three techniques for model validation:

(1) cross-validation,
(2) residual analysis, and
(3) information-theoretic criteria.

14.6.1   Cross-validation

It is not so surprising that a model will perform well when evaluated by its performance for the data set to which it was adjusted. The real test is whether it will be capable of also describing fresh data sets from the process. A suggestive way of comparing two different models is to evaluate their performance when applied to a dataset (known as *validation or testing a dataset*) to which neither of them was adjusted. We would then favor that model that shows the better performance. Such a procedure, known as cross-validation, is a standard tool for parameter estimation and model validation in statistics [567].

An attractive feature of cross-validation is its pragmatic character: The comparison among candidate models needs neither any probabilistic arguments nor any assumptions about the real process. Its disadvantage is that a fresh data set has to be saved for the validation and therefore we cannot use all our information to build the models. In the situations where only limited training samples are available, this procedure cannot be used.

The basic idea of the cross-validation method is as follows: Divide the available training data into an estimation subset (denoted as N_E), and a validation subset (denoted as N_V). The estimation subset N_E is used for estimation of the model (i.e., training the model which consists of a certain number of fuzzy rules) and the validation subset N_V is used for evaluation of the performance of the model. The motivation here is to validate the model on a dataset different from the one used for parameter estimation. In this way, we may use the training set to assess the performance of various candidate model structures that contain different number of fuzzy rules, and thereby choose the "best" one. The particular model with the best-performing number of fuzzy rules is then trained on the full training set.

Cross-validation does not necessarily use one subset exclusively for one purpose (estimation or validation); it allows all the training data to be used for both purposes. For instance, we can partition the full training set into p subsets; compute an estimate from all the subsets but one; and validate the estimate from the left-out subset. Then, we perform the estimation-validation with a different left-out. We can repeat this p times.

There are various formulas for expressing cross-validation [188]. The most commonly used performance measure for model validation is the *Mean Squared Error (MSE)*, as defined by

$$MSE = \frac{1}{n_v} \sum_{k=1}^{n_v} [y(k) - \hat{y}(k)]^2 \qquad \text{(EQ 14.47)}$$

where $y(k)$ and $\hat{y}(k)$ are the desired (target) and the actual outputs of the model at time k, respectively, and n_v is the number of samples in the validation data set.

Another performance measure, which is used by T. Takagi and M. Sugeno in their pioneer work in fuzzy model identification, is the *Unbiasedness Criterion (UC)* [559, 562]. As shown in Figure 14.6, this model validation scheme divides the data set into two subsets: dataset A and dataset B. Each data set is used to construct a fuzzy model. We call these two fuzzy models FM_A and FM_B, respectively. We then compute the difference between the output of the two fuzzy models for dataset A:

$$\sum_{i=1}^{n_A} (FM_A(YA_i) - FM_B(YA_i))^2 \qquad \text{(EQ 14.48)}$$

where n_A is the size of data set A, and YA_i denotes the input of the ith data in dataset A. Similarly, we compute the difference between the output of the two fuzzy models for dataset B.

FIGURE 14.6   Calculating the Unbiased Criteria

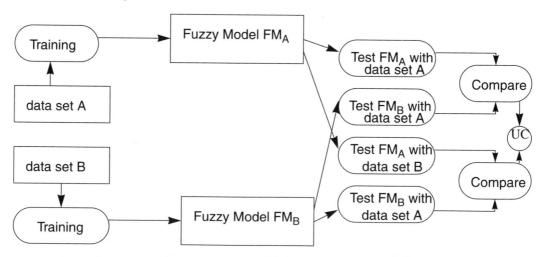

Combining these two error calculations gives us the Unbiased Criterion (UC):

$$UC = \sqrt{\sum_{i=1}^{n_A} (FM_A(YA_i) - FM_B(YA_i))^2 + \sum_{i=1}^{n_B} (FM_A(YB_i) - FM_B(YB_i))^2} \quad \textbf{(EQ 14.49)}$$

The rationale of the criterion is that a good model should not be sensitive to the choice of training data. Hence, it should have low UC value.

Cross-validation essentially tries to find the model structure by checking two subsets of training data and does not take into account the complexity of the resulting model. If enough different model structures are considered, cross-validation may often find a model that has a low error on the two subsets of training data but will not generalize well to new untrained data [28]. The information-theoretic criteria introduced below will take into account the structure complexity explicitly when building a fuzzy model and thus avoid the problem encountered in applying cross-validation.

14.6.2 Residual analysis

(From J. Yen and L. Wang, "Application of Statistical Information Criteria for Optimal Fuzzy Model Construction," *IEEE Trans. Fuzzy Systems*, Vol. 6, No. 3, pp. 362-392 © 1998 IEEE)

This method is based on the analysis of the residuals $e(k) \equiv y(k) - \hat{y}(k)$ of the identified model. In general, if the identified model is not adequate (for example, some needed variables are left out), then the residuals will be correlated and hence will not behave the way the random error component is supposed to. In other words, the model will not capture all of the pattern component in the data set, and part of the pattern component will remain in the residuals.

The traditional residual analysis method used in linear systems identification is to test whether the autocorrelation function of the residuals is an impulse and whether the cross-correlation function between the residuals and the input is zero [78]. The generalization of this method to the nonlinear case is considered by S. Billings and W. Voon where several higher-order correlation functions are introduced to detect the presence of unmodeled linear and nonlinear terms in the residuals [53].

Alternatively, I. Leontaritis and S. Billings propose using a *chi-squared statistical test method* in the validation of nonlinear models [371]. This method is a generalization of the same test method used in the validation of linear models [384]and has been successfully applied to the validation of nonlinear autoregressive moving average models, radial basis function networks [117], and fuzzy basis function models [620]. Here we introduce the basic principle of this method.

This method starts by defining a *d*-dimensional vector-valued function

$$\Omega(k) = [w(k), w(k-1), ..., w(k-d+1)]^T \quad \textbf{(EQ 14.50)}$$

where $w(k)$ is some chosen function of the residuals and the past observations. Let

$$\Gamma^T \Gamma = \frac{1}{N} \sum_{k=1}^{N} \Omega(k) \Omega^T(k) \quad \textbf{(EQ 14.51)}$$

The chi-squared statistic is then calculated according to

$$\zeta = N \upsilon^T (\Gamma^T \Gamma)^{-1} \upsilon \qquad \text{(EQ 14.52)}$$

where

$$\upsilon = \frac{1}{N} \sum_{k=1}^{N} \Omega(k) e(k) / \sigma_e \qquad \text{(EQ 14.53)}$$

and σ_e^2 is the variance of the residual $e(k)$. Under the null hypothesis that the data are generated by the model, the statistic ζ is asymptotically chi-squared distributed with d degrees of freedom. If the values of ζ for several different choices of $w(k)$ are within the 95% acceptance region, that is

$$\zeta < \chi_d^2(\alpha) \qquad \text{(EQ 14.54)}$$

the model is regarded as adequate, where $\chi_d^2(\alpha)$ is the critical value of the chi-squared distribution with d degree of freedom for the given significance α (0.05). The choices of $w(k)$ should generally include some nonlinear functions, say, $w(k) = e^2(k-1)y^2(k-1)$.

Note that the statistic ζ has been formed based on the dataset used to identify the model. When there are sufficient training samples, a fresh dataset should be saved and then used to construct the statistic ζ. This is exactly the idea used in cross-validation.

14.6.3 Information-theoretic Criteria

Various information-theoretic criteria for model-complexity selection have been proposed in the statistical community. For a more detailed discussions of these criteria, we refer the reader to A. Grasa [220]. Although these criteria do indeed differ from each other in their exact details, they share a common form of composition, as described in M. Priestley and S. Haykin [482, 242]:

$$\begin{pmatrix} Model-complexity \\ criterion \end{pmatrix} = \begin{pmatrix} model \\ accuracy \end{pmatrix} + \begin{pmatrix} model-complexity \\ penalty \end{pmatrix} \qquad \text{(EQ 14.55)}$$

The basic philosophy of these information-theoretic criteria is best described by advice from Professor Lennart Ljung: "If I am going to accept a more complex model (according to my own complexity measure) it has to prove to be significantly better!"[384].

The difference between the various criteria lies in the definition of the model-complexity penalty. The structure of these criteria reflects a balance between the ideas of goodness of fit and complexity; that is, the selected model yields a balance between two information sources: the sum of squares of the residuals and the number of parameters of the model.

In principle, all these criteria can be applied to the structure identification of fuzzy models. However, unlike statistical modeling where a single model is used to describe the global behavior of a system, fuzzy modeling is essentially a *multimodel* approach in which individual rules (where each rule acts like a "local model") are combined to describe the global behavior of the system. The complexity of a fuzzy model is determined not only by the number of parameters in the antecedents and consequents of the rules but also by the number of fuzzy rules in the model. Thus, we need to revise these criteria in determining the structure of a fuzzy model. In particular, the penalty given for model complexity should be no longer only limited to the number of parameters in the model. it should take into account the number of fuzzy rules in the model as well. Based on these considerations, several modified information-theoretic criteria are suggested in J.Yen and L. Wang [669]. One typical criterion, known as *Schwarz-Rissanen Information Criterion (SRIC)*, which is based on the one derived independently by Schwarz andssanen is given below [526, 504]:

$$SRIC(m_a, m_c, m_r) = \log(\hat{\sigma}_\varepsilon^2) + \frac{\log(N)c(m_a, m_c, m_r)}{N} \qquad \textbf{(EQ 14.56)}$$

where $\hat{\sigma}_\varepsilon^2$ is the estimated variance of model residuals, N is the number of training data, m_a is the number of antecedent parameters, m_c is the number of consequent parameters, m_r is the number of fuzzy rules constituting the model, and $c(.)$ is a complexity function defined by

$$s(m_a, m_c, m_r) = m_a + m_c + \eta m_r \qquad \textbf{(EQ 14.57)}$$

The number of antecedent parameters m_a depends on the dimensionality of the antecedent membership functions. Most fuzzy models combine one-dimensional membership functions using fuzzy conjunction (i.e., the *AND* operator) in the rule antecedents. For these models, the number of parameters associated with an antecedent variable x_i is the product of the number of membership functions for the variable (denoted as k_i) and the number of parameters for each membership function of x_i (denoted by α_i). The total number of antecedent parameters is thus

$$m_a = \sum_{i=1}^{p} (k_i \times \alpha_i) \qquad \textbf{(EQ 14.58)}$$

where p is the number of antecedent (input) variables. Some fuzzy models use p-dimensional membership functions in the antecedents of rules, where p, as before, denotes the number of antecedent variables. For fuzzy models of this kind, each antecedent membership function corresponds to a fuzzy rule. Suppose that all membership functions have the same number of parameters (denoted as α). Then we have

$$m_a = m_r \times \alpha \qquad \textbf{(EQ 14.59)}$$

Depending on the type of fuzzy models, the consequent parameter count m_c may or may not be related to the rule number m_r. For TSK type models, the number of consequent parameters is proportional to the number of rules. More precisely, if the consequent part of a rule in the model consists of a linear regression equation, then we have

$$m_c = q \times m_r \times (p + 1) \qquad \textbf{(EQ 14.60)}$$

where q is the number of consequent (output) variables. For Mamdani type models, the number of consequent parameters is not directly related to the number of fuzzy rules; rather, it depends on the number of membership functions for each consequent variable y_i (denoted as l_i), the number of parameters needed for each membership function of y_i (denoted as β_i), and the total number of consequent variables (i.e., q). More precisely, we have

$$m_c = \sum_{i=1}^{q} (l_i \times \beta_i) \qquad \text{(EQ 14.61)}$$

The parameter η in Equation 14.57 represents the relative cost for each fuzzy rule. A larger value for η will lead to fewer rules being used and thereby a more "parsimonious" model. There exists no simple rule for choosing η, however. The best value for η in a practical modeling problem may depend on the number of available training data N, and even the type of fuzzy model being used. Although it is theoretically possible to regard η as a parameter of the procedure and solve it using some optimization algorithm, it is computationally too costly to be useful in practice. It was shown that the resulting fuzzy model and its accuracy were fairly insensitive to the value chosen for η in a particular range and choosing a value for η in the interval of 2 to 5 often resulted in a good parsimonious model [669].

SRIC (also other criterion) may be used in two ways. First, it can be used to compare fuzzy models with a different number of fuzzy rules and pick the model with the minimum SRIC value as the best one. Second, it can be used to find the most important rules in a given rule base. Several recent researches have shown that if the less important fuzzy rules are removed from the rule base, the performance of the model constructed using the retaining rules can be high [268, 510, 668].

14.6.4   Model Interpretability

An important feature of a fuzzy model is its *interpretability*: Each rule in the model acts like a "local model" in the sense it only covers a local region of the input-output space, and its contribution to the total output of the model is easily understood. However, this advantage may get lost for some fuzzy models (typically the TSK model) if they are not adequately identified. The problem appears mainly in the parameter estimation stage where the parameters of the model are obtained by minimizing an objective function defined based on the errors between the desired (global) output and the actual (global) output of the model. If we are lucky enough to catch the "optimal" parameters of the model, then the resulting fuzzy rules can describe not only the correct global behavior but also the correct local behavior. Unfortunately, the nonlinearity of the parameter estimation problem may often make us trap a "suboptimal" solution. In this situation the fuzzy rules may show a very erratic local behavior, while they together generate a correct global behavior.

This important issue has not been addressed until recently (Yen and Gillespie [666] and Babuska et al. [17]) and much work remains to be done.

14.7   A Nonlinear Plant Modeling Example

To illustrate the application of the techniques introduced in this chapter, we consider the second-order nonlinear plant

$$y(k) = f(y(k-1), y(k-2)) + u(k) \qquad \text{(EQ 14.62)}$$

where

$$f(y(k-1), y(k-2)) = \frac{y(k-1)y(k-2)[y(k-1)-0.5]}{1 + y^2(k-1) + y^2(k-2)} \qquad \text{(EQ 14.63)}$$

We will use the following techniques for constructing a fuzzy rule-based model from a set of training data for this nonlinear plant: (1) The fuzzy rule-based model is a zero-order TSK model (Equation 14.4): (2) Gaussian membership functions are used to partition the input space: (3) the K-means clustering algorithm and the nearest-neighbor heuristic (Section 14.5.1) are used to identify the antecedent parameters for a scattered partition of the input space: (4) the consequent parameters are identified using the SVD-based technique introduced in Section 14.5.2.1: (5) the number of rules is determined using the SVD-based model reduction technique described in Section 14.4.5.

The nonlinear component f in this plant, which is usually called the "unforced system" in the control literature[437], has an equilibrium state (0,0) in the state space. This implies that while in equilibrium without an input, the output of the plant is the sequence {0}. Fig 14.7 shows the trajectory of the unforced system in the state space. We want to approximate the nonlinear component f using the zero-order TSK fuzzy model (Equation 14.4). For this purpose, 1200 simulated data points were generated from the plant model in Equation 14.62. The first 1000 data points were obtained by assuming a random input signal $u(k)$ uniformly distributed in the interval [-1.5, 1.5], while the last 200 data points were obtained by using a sinusoid input signal $u(k) = \sin(2\pi k/25)$. The 1200 simulated data points are shown in Fig 14.8.

FIGURE 14.7   Trajectory of the Unforced System (From J. Yen and L. Wang [669]. Reprinted with permission of IEEE.)

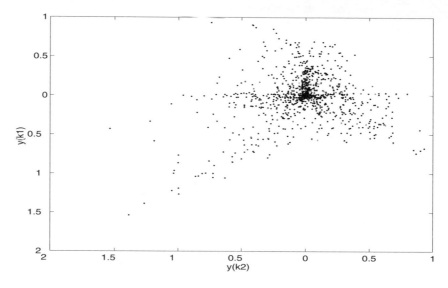

FIGURE 14.8   Output of Plant Model (From J. Yen and L. Wang [669]. Reprinted with permission of IEEE.)

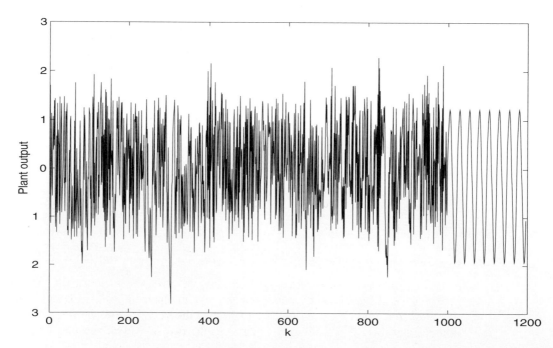

We used the first 1000 data points to build a fuzzy model. The performance of the resulting model was tested using the remaining 200 data points. We chose $y(k-1)$ and $y(k-2)$ as the input variables and arbitrarily set the number of fuzzy rules to 25. The Gaussian functions were used to express the membership functions of $y(k-1)$ and $y(k-2)$ with the parameters shown in Table 14.4. These 25 rules were labeled by the numbers 1, 2, ..., 25 which indicated the position of the rules in the rule base as well as the associated combinations of membership functions. For example, the number "1" indicates the first rule in the rule base which is associated with the membership function combination $\{A_{11}(x_1),$ $A_{12}(x_2)\}$ (where $x_1 \equiv y(k-1)$, $x_2 \equiv y(k-2)$), while the number "25" indicates the 25th rule in the rule base which is associated with the membership function combination $\{A_{25,1}(x_1), A_{25,2}(x_2)\}$.

Initially, only 20 two-dimensional Gaussian membership functions were considered, each of which can be seen as the product of two one-dimensional membership functions for the input variables $y(k-1)$ and $y(k-2)$. The centers c_{ij} and the widths σ_{ij} of the 20 Gaussian membership functions were determined using the K-means clustering algorithm and the nearest-neighbor heuristic, respectively. We then introduced 5 additional Gaussian membership functions with artificially selected parameters in order to emulate the effect caused by the redundant and less important fuzzy rules.

TABLE 14.4   Parameters of Gaussian Membership Functions

i	c_{i1}	c_{i2}	σ_{i1}	σ_{i2}
1	**0.0930**	**-0.3630**	**0.7095**	**0.7095**
2	**0.0933**	**-0.3632**	**0.7095**	**0.7095**
3	1.3828	-0.6617	0.6271	0.6271
4	-1.0414	1.5397	0.7969	0.7969
5	**-1.8130**	**-1.6470**	**1.3205**	**1.3205**
6	**-1.8125**	**-1.6469**	**1.3205**	**1.3205**
7	0.7776	-1.1555	0.7800	0.7800
8	0.1898	1.0142	0.6141	0.6141
9	-0.4052	0.2798	0.8099	0.8099
10	**-0.6613**	**-0.4846**	**0.0100**	**0.0100**
11	-0.6613	-0.4846	0.7051	0.7051
12	0.9529	-0.3965	0.6313	0.6313
13	0.7860	0.7723	0.6177	0.6177
14	0.4329	0.1910	0.6652	0.6652
15	**1.2940**	**1.0740**	**0.6474**	**0.6474**
16	**1.2942**	**1.0738**	**0.6474**	**0.6474**
17	0.6801	1.4083	0.6370	0.6370
18	1.2656	0.2698	0.7156	0.7156
19	-0.3846	1.1827	0.6772	0.6772

Each row of Table 14.4 is associated with one of the fuzzy rules in the rule base. The first two rows give the nearly same parameters of membership functions, and correspondingly the first two rules in the rule base will have the nearly the same firing strengths. This implies that there exists redundancy between the two rules and removing either of them will not affect the performance of the model significantly. The same observation holds for rules 5 and 6 as well as rules 15 and 16, which correspond to rows 5 and 6 as well as the rows 15 and 16, respectively. The widths of the Gaussian functions in the rows 10 and 20 have a small value of 0.01, and thus the resultant membership functions will have a low grade (Geometrically, the membership functions will become very "narrow" [547].) This implies that rules 10 and 20 have a low firing strength and removal of the two rules will not sacrifice the accuracy of the model greatly.

For each of the 1000 input data points $\{y(k\text{-}1), y(k\text{-}2)\}$, k = 1, 2, ..., 1000 to the model, we computed the normalized firing strengths of the 25 fuzzy rules using Equation 14.8 to form a 1000-by-25 firing strength matrix P. Fig 14.9 shows the singular values of the matrix in both descending and original order. It can be seen that there exist 5 zero or near-zero singular values among the 25 singular values. This indicates that 5 rules in the rule base are redundant or less important. Based on the singular values, we determine to retain 20 rules and eliminate 5 rules in the rule base. Using the SVD-QR with column pivoting algorithm, the position of the retained 20 rules in the rule base were identified as $\{25, 4, 7, 19, 3, 24, 8, 23, 14, 13, 21, 18, 17, 22, 12, 9, 11, 2, 16, 5\}$, and the position of the removed 5 rule base were identified as $\{10, 15, 20, 6, 1\}$. Clearly, the SVD-QR with column pivoting algorithm has successfully detected and eliminated the two less important rules (rules 10 and 20) and the three redundant rules (rules 15, 6, and 1) in the rule base.

We computed the consequent parameters of the original 25-rule model and the reduced 20-rules model using the SVD-based parameter estimation method introduced in Section 14.5.2.1. Table 14.5 shows the Mean SquaredEerrors (MSEs) of the two models in both the training stage and the testing stage. Compared to the original model, the reduced model gives larger MSE in the training state, but a smaller MSE in the testing stage. This implies that removal of those redundant or less important rules from the rule base can result in a fuzzy model with better generalizing ability. This finding is consistent with the results reported in several previous studies [268, 669].

TABLE 14.5   MSEs of Two Fuzzy Models for Training and Testing Data

Model	MSE (training)	MSE (testing)
Original Model (25-rules)	2.3092e-4	4.0717e-4
Reduced model (20-rules)	6.8341e-4	2.3836e-4

FIGURE 14.9   Singular Values of the 1000-by-25 Firing Strength Matrix

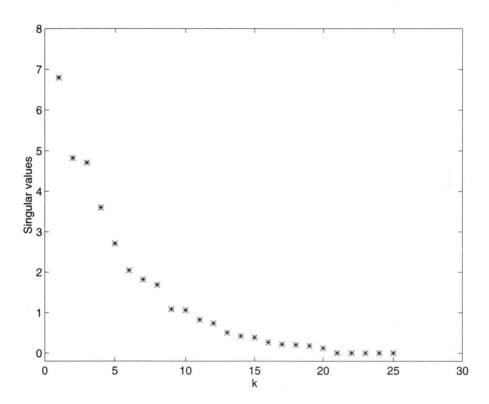

14.8 Fuzzy Model-based Control

The application of fuzzy logic for the solution of control problems has been the focus of numerous studies [558, 673]. The motivation is often that the system knowledge and dynamic models are uncertain, and as an alternative to traditional modeling and control design, fuzzy logic appears to provide a suitable representation of such knowledge and models [701]. Whereas various fuzzy controllers have been designed for numerous applications, they roughly fall into two categories: *direct* fuzzy controllers and *indirect* fuzzy controllers [239, 619].

Fig 14.10 shows the basic architecture of a direct fuzzy control system where the parameters of the controller are manipulated directly without recourse to physical system identification. The controller may be constructed using any of the models introduced in the previous subsections. For example, [560] designed a direct fuzzy controller using the TSK model and successfully applied it to the control of a model car. Since this type of fuzzy controller does not consider any information from the physical system, it is usually called a *model-free* fuzzy controller. Most of existing fuzzy controllers belong to this category.

FIGURE 14.10   Architecture for Direct Fuzzy Control (From J. Yen and L. Wang [670]. Reprinted with permission from IOP.)

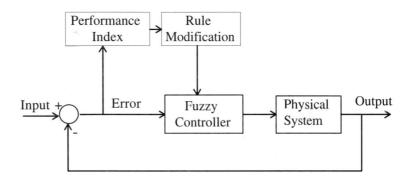

The indirect fuzzy controller has only recently been developed. This type of fuzzy controller, which is usually called *model-based* fuzzy controller, can be distinguished by the fact that a separate fuzzy model is constructed of the physical system, then a design procedure is used to calculate the control signal [239]. Fig 14.11 shows the architecture of an indirect fuzzy control system. The separation of physical system modeling and controller design makes it possible to analyze the performance of this type of system, thus yielding convergence and stability theories [285, 619]. The training data for the model is directly available, unlike the direct fuzzy controller which has to try and infer the control error which caused the physical system output error.

FIGURE 14.11   Architecture for Indirect Fuzzy Control (From J. Yen and L. Wang [670]. Reprinted with permission from IOP.)

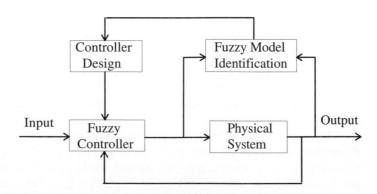

In principle, both types of fuzzy controllers can be applied to any kind of physical systems. This is especially true for the direct fuzzy controller. For the indirect fuzzy controller, however, the following type of physical systems is usually considered [622]:

$$y(k) = F(y(k-1), y(k-2), ..., y(k-p)) + \sum_{i=0}^{q-1} \beta_i u(k-i-d) \qquad \text{(EQ 14.64)}$$

where p and q are two nonnegative integers; d is the time-delay of the system; $F(.)$ is a nonlinear function supposed to be unknown; β_i ($i = 0, 1, ... , q$-1) are real parameters which are also unknown. This type of model, which corresponds to *Model-II* in [438], is known to be particularly suited for the control problem. Concretely, we may model $F(.)$ using one of the fuzzy models (or other granule-based models) introduced in the previous subsections, say, the TSK model, and then synthesize the controller using some conventional design techniques [219, 544] based on the resulting fuzzy model.

14.9 Summary

In this chapter, we discussed the following important issues and techniques regarding fuzzy model identification

- Fuzzy models have been shown to be able to approximate any real continuous functions to an arbitrary degree of accuracy.

- Three major subproblems in fuzzy model identification are (1) structure specification, (2) parameter estimation, and (3) model validation.

- A fuzzy model identification approach typically iterates between these three steps until a satisfactory model is constructed.

- Structure specification of a fuzzy model involves four tasks: (1) the selection of input variables, (2) the selection of membership functions, (3) choosing the structure of the fuzzy partition of the model's input space, and (4) determining the number of fuzzy rules.

- There are three techniques for selecting input variables: forward selection, backward elimination, and "best subset" procedure.

- Techniques for fuzzy partition can be classified into three categories: grid partition, tree partition, and scatter partition.

- Determining the optimal number of fuzzy rules is an area that requires further research. We described several approaches that attempt to reduce the number of fuzzy rules by assessing their degree of importance using Singular Value Decomposition (SVD).

- Parameter estimation problems, in general, involve the optimization of antecedent membership parameters and the consequent parameters.

- Antecedent parameters can be identified using clustering techniques.

- Consequent parameters can be identified using the Recursive Least Square (RLS) algorithm and the Least Mean Square (LMS) algorithm.

- Parameter estimation can also be achieved using neural networks and genetic algorithms, which we will discuss in Chapters 16 and 17, respectively.

- We introduced three criteria for evaluating a fuzzy models: cross-validation, residual analysis, and information-theoretic criteria.

- We illustrated some of the fuzzy model identification techniques using a nonlinear plant modeling problem.

- There are two approaches for applying fuzzy model identification techniques to the design of fuzzy controllers: (1) identifying fuzzy controllers directly, and (2) identifying a model of the plant (i.e., the physical system being controlled) and constructing a fuzzy controller from the plant model.

Bibliographical and Historical Notes

T. Takagi and M. Sugeno's pioneer work in fuzzy model identification in the mid 80s established much of the ground work in this area [562]. They proposed solutions to all three subproblems in fuzzy model identification. They used a forward selection procedure for selecting relevant input variables, a linear regression technique for estimating consequent parameters of their models (later referred to as Takagi-Sugeno model or TSK model), and an unbiased criterion (UC) for evaluating their models.

In early 90s, there were a significant amount of interest in proving the approximation capability of various types of fuzzy models. Using the famous *Stone-Weirstrass theorem* from real analysis, L. Wang and J. Mendel proved that a fuzzy model constructed using product inference, height defuzzification, and Gaussian membership functions is a universal approximator [620]. Similar results were proved by B. Kosko for a fuzzy model that uses product inference, centroid defuzzification, and any type of membership function [331], by X. Zeng and M. Singh for a fuzzy model that uses product or min inference, height defuzzification, and trapezoid membership functions [715], and by J. Buckey for a TSK-type model that uses product or min inference and any type of antecedent membership function [94]. Further, H. Ying, X. Zeng and M. Singh established the approximation error bounds for various fuzzy models [677, 715].

Empirical evaluation of different membership functions can be useful in guiding the choice of membership functions. A. Lotfi and A. Tsoi compared the performance of a fuzzy model constructed using triangular, bell-shaped, and Gaussian membership functions. Their result suggested that triangular membership function is inferior to bell-shaped and Gaussian membership functions [387]. In S. Mitaim and B. Kosko, triangular, trapezoidal, bell-shaped, and Gaussian membership functions are compared with other membership functions, especially with the *sinc* function *sin(x)/x*, based on how closely the resultant fuzzy models approximate the real systems. They found that the *sinc* function performed best or nearly best in most cases [416]. More extensive empirical investigation is needed in this area before a general conclusion can be made.

In addition to techniques for determining the number of fuzzy rules introduced in Section 14.4.4, there are other methods for reducing the number of fuzzy rules in a given rule base. One approach is to fuse (rather than eliminate) rules based on some *similarity measure*. This is the work done by B. Song et al., C. Sun, B. Babuska et al., C. Chao, Y.

Chen and C. Teng, C. Lin and C. Lee, and C. Juang and C. Lin [547, 564, 17, 117, 380, 288]. Yager and D. Filev proposed an *entropy criterion* to find a simple structure of fuzzy model by minimizing the rate of interaction between fuzzy rules [656]. The number of fuzzy rules in this approach was determined using an *unbiasedness criterion* suggested in M. Sugeno and K. Kang [559] (see also [562]). In H. Berenji and P. Khedkar [36], a *pruning and merging* strategy was proposed to eliminate redundant fuzzy rules in a given rule base. In J. Yen and L. Wang [669], several modified *statistical information criteria* were suggested to determine the optimal number of fuzzy rules comprising the underlying model. *Genetic algorithms* based methods have also been used for extracting fuzzy rules for control and classification problems [258, 268, 299, 367, 589]. Another equivalently important problem with rule base simplification is how to *augment* a rule base when the rules are not complete. This problem has been well addressed by L. Wang and J. Mendel and T. Sudkamp and R. Hammell [620, 552]. Interested reader may follow the given references for further details.

The SVD-based method for fuzzy model reduction discussed in Section 14.4.5 is based on a *subset selection* algorithm for linear regression modeling developed by G. Golub, V. Klema, and G. Stewart in [214] and has been used by P. Kanjilal and D. Banerjee in selecting hidden nodes for a feedforward neural networks [297], and by G. Mouzouris and J. Mendel in selecting fuzzy basis functions [429]. A comparison of this method with other orthogonal transformation based methods was given in [671].

The literature in using the backpropagation algorithm to learn fuzzy models is numerous. Books by M. Brown and C. Harris, Jang, C. Sun and Mizutani, C. Lin and C. Lee, as well as Wang and Mendel give a clear and detailed introduction to this subject [89, 564, 382, 620].

Exercises

14.1   We discussed three methods of selecting input variables: forward selection, backward elimination, and "best subset" procedure. What are the pros and cons of the three methods?

14.2   We discussed four methods of partitioning input space: static grid partition, adaptive grid partition, tree partition, and scatter partition. What are the pros and cons of the four methods?

14.3   In building a fuzzy model, why do we need to consider the trade-off between data fitness and model complexity? How can we achieve such a trade-off? Could you give two real-life examples where some kind of trade-off is needed.

14.4   If the only unknown parameters in a fuzzy model are the consequent parameters, then the parameter estimation problem is a linear optimization problem. In this case we can solve the consequent parameters using either the SVD-based estimation algorithm or the RLS algorithm. What are the pros and cons of the two algorithms?

TABLE 14.6   A Data Set for Fuzzy Model Identification Problems

	Group A					Group B				
No.	.r1	.r2	.r2	.r3	y	No.	.r1	.r2	.r3	.r4	y
1	1.40	1.80	3.00	3.80	3.70	26	2.00	2.06	2.25	2.37	2.52
2	4.28	4.96	3.02	4.39	1.31	27	2.71	4.13	4.38	3.21	1.58
3	1.18	4.29	1.60	3.80	3.35	28	1.78	1.11	3.13	1.80	4.71
4	1.96	1.90	1.71	1.59	2.70	29	3.61	2.27	2.27	3.61	1.87
5	1.85	1.43	4.15	3.30	3.52	30	2.24	3.74	4.25	3.26	1.79
6	3.66	1.60	3.44	3.33	2.46	31	1.81	3.18	3.31	2.07	2.20
7	3.64	2.14	1.64	2.64	1.95	32	4.85	4.66	4.11	3.74	1.30
8	4.51	1.52	4.53	2.54	2.51	33	3.41	3.88	1.27	2.21	1.48
9	3.77	1.45	2.50	1.86	2.70	34	1.38	2.55	2.07	4.42	3.14
10	4.84	4.32	2.75	1.70	1.33	35	2.46	2.12	1.11	4.44	2.22
11	1.05	2.55	3.03	2.02	4.63	36	2.66	4.42	1.71	1.23	1.56
12	4.51	1.37	3.97	1.70	2.80	37	4.44	4.71	1.53	2.08	1.32
13	1.84	4.43	4.20	1.38	1.97	38	3.11	1.06	2.91	2.80	4.08
14	1.67	2.81	2.23	4.51	2.47	39	4.47	3.66	1.23	3.62	1.42
15	2.03	1.88	1.41	1.10	2.66	40	1.35	1.76	3.00	3.82	3.91
16	3.62	1.95	4.93	1.58	2.08	41	1.24	1.41	1.92	2.25	5.05
17	1.67	2.23	3.93	1.06	2.75	42	2.81	1.35	4.96	4.04	1.97
18	3.38	3.70	4.65	1.28	1.51	43	1.92	4.25	3.24	3.89	1.92
19	2.83	1.77	2.61	4.50	2.40	44	4.61	2.68	4.89	1.03	1.63
20	1.48	4.44	1.33	3.25	2.44	45	3.04	4.97	2.77	2.63	1.44
21	3.37	2.13	2.42	3.95	1.99	46	4.82	3.80	4.73	2.69	1.39
22	2.84	1.24	4.42	1.21	3.42	47	2.58	1.97	4.16	2.95	2.29
23	1.19	1.53	2.54	3.22	4.99	48	4.14	4.76	2.63	3.88	1.33
24	4.10	1.71	2.54	1.76	2.27	49	4.35	3.90	2.55	1.65	1.40
25	1.65	1.38	4.57	4.03	3.94	50	2.22	1.35	2.75	1.01	3.39

14.5   Consider the data sets in Table 14.6, where x_1, x_2, x_3, x_4 are the input candidates, and y is the output. Build a fuzzy model with constant consequent constituents. Select input variables from the four input candidates using the three methods based on the MSE criterion and UC, respectively, and compare their results. When the MSE criterion is employed, the whole dataset (i.e, Group A + Group B) should be used to identify the model.(Source of this exercise: M. Sugeno and T. Yasukawa, A fuzzy Logic-based Approach To Qualitative Modeling, *IEEE Transactions on Fuzzy Systems*, vol. 1, pp. 7-31, 1993.)

14.6 Use the same dataset as in the previous problem. Suppose x_1, x_2 are used as the input variables of a fuzzy model with constant consequent constituents. Given the same number of fuzzy rules, compare the MSEs of the resulting models when different types of membership functions are used.

14.7 Consider the following nonlinear time series model:

$$y(k) = (0.8 - 0.5\exp(-y^2(k-1)))$$

$$- (0.3 + 0.9\exp(-y^2(k-1)))y(k-2) + 0.1\sin(\pi y(k-1)) + e(k)$$

where $e(k)$ is a zero mean Gaussian white noise process with variance 0.1. Generate 600 simulated data points from the model. Build a TSK model using the first 400 data points and then test the model using the remaining 200 data points.

14.8 Consider the ball-and-beam system in Fig 14.12. The beam is made to rotate in a vertical plane by applying a torque at the center of rotation and the ball is free to roll along the beam. We require that the ball remain in contact with the beam.

FIGURE 14.12 The Ball and Beam System

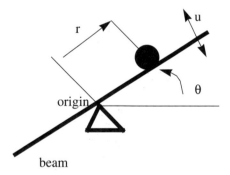

Let $\bar{x} = [x_1, x_2, x_3, x_4]^T = [r, \dot{r}, \theta, \dot{\theta}]^T$ be the state vector and $y = r$ be the output of the system. The system can be represented by the following state space model:

$$\begin{bmatrix} \dot{x}_1 \\ \dot{x}_2 \\ \dot{x}_3 \\ \dot{x}_4 \end{bmatrix} = \begin{bmatrix} x_2 \\ B(x_1 x_4^2 - G\sin x_3) \\ x_4 \\ 0 \end{bmatrix} + \begin{bmatrix} 0 \\ 0 \\ 0 \\ 1 \end{bmatrix} u \qquad \text{(EQ 14.65)}$$

$$y = x_1$$

where the control u is the acceleration of θ, and B and G are known constants. The goal of the controller is to keep the beam balanced (i.e., the angle $\theta = 0$) as well as to return the ball to its origin (i.e., the position $r = 0$). It can be shown that the control goal can be achieved if the following control law is used:

$$u(\bar{x}) = \frac{v(\bar{x}) - b(\bar{x})}{a(\bar{x})} \qquad\qquad \textbf{(EQ 14.66)}$$

where $v(\bar{x}) = 8\phi_4(\bar{x}) + 24\phi_3(\bar{x}) + 32\phi_2(\bar{x}) + 16\phi_1(\bar{x})$, $a(\bar{x}) = -BG\cos x_3$ and $b(\bar{x}) = BG x_4^2 \sin x_3$.

Suppose $B = 0.7143$ and $G = 9.81$. Generate 1000 pairs of input-output data points from (1) with $\bar{x}$ uniformly distributed in the region $U = [-5, 5] \propto [-2, 2] \propto [-\pi/4, \pi/4] \propto [-0.8, 0.8]$. Build a constant consequent constituent fuzzy model and a TSK model, respectively, using these data points to approximate the control law. Then use the two fuzzy models to balance the ball-and-beam system. Compare the performance of the two fuzzy models with that of the original control law for four different initial values: $\bar{x}(0) = [2.4, -0.1, 0.6, 0.1]^T$, $[1.6, 0.05, -0.3, -0.05]^T$, $[-1.6, -0.05, 0.3, 0.05]^T$, and $[-2.4, 0.1, -0.6, -0.1]^T$. (Hint: use SIMULINK and the Fuzzy Logic Toolbox for MATLAB.)

15 ADVANCED TOPICS OF FUZZY MODEL IDENTIFICATION

Recognition of the connections between fuzzy models and other models is important, because it allows the learning algorithms and convergence proofs (of other models) to be employed in fuzzy models. It also enables the information stored in those other models to be expressed as a set of fuzzy if-then rules which provide model interpretability.

15.1 Radial Basis Function Networks

Unlike backpropagation networks which originate from the field of biological science, *Radial Basis Function* (RBF) networks are rooted primarily in the theory of multivariable functional interpolation in high-dimensional space [480]. They were first used within the neural network community in 1988 [86]. Fig. 15.1 shows the basic structure of the RBF network (Without loss of generality the output is assumed to be a single one.) The input nodes pass the input values to the connecting arcs, and the first layer connections are not weighted. Thus each hidden node receives each input value unaltered. The hidden nodes are the radial basis function units. The second layer of connections are weighted, and the output node is a simple summation. The RBF network does not extend to more layers in a natural manner as backpropagation networks because of the different functions and parameters of the layers, but improved training and performance are directly related to this architecture.

There are a number of choices for the radial basis function R_i, but the most commonly used one is the Gaussian function, as defined by

$$R_i(\dot{x}) = \exp(\|\dot{x} - \overrightarrow{m_i}\|_2^2 / \sigma_i^2)$$ **(EQ 15.1)**

where $\dot{x}$ is an s-dimensional input vector, $\bar{m}_i$ is the center that has the same dimension as $\dot{x}$, and σ_i is the width. We have seen the same function in Section 14.5.1 (Equation 14.20) where it is used to express a fuzzy membership function.

FIGURE 15.1   *An RBF Network Architecture*

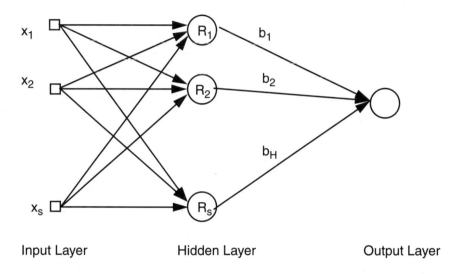

Input Layer   Hidden Layer   Output Layer

There are H ($H \leq N$), the number of training samples nodes in the hidden layer of the network. The output of the network is given by

$$y = \frac{\displaystyle\sum_{i=1}^{H} R_i(\dot{x})b_i}{\displaystyle\sum_{i=1}^{H} R_i(\dot{x})} \qquad \textbf{(EQ 15.2)}$$

Let $H = M, b_i = c_i$, and recall the factorizability property of the Gaussian function given in Section 14.4.2 (Equation 14.1). We see that Equation 15.2 is equivalent to the formula of zero-order TSK model if the rule firing strength w_i is computed by multiplication rather than min (i.e., Equations 14.5 and 14.7). Hence, an RBF network is equivalent to a zero-order TSK model that uses the product operator to calculate the matching degrees of a rule's conjunctive condition.

15.2 Modular Networks

The modular network is a neural network architecture that learns to partition a task into two or more functionally independent tasks and allocates different networks to learn each task. An appropriate task decomposition is discovered by forcing the networks comprising the architecture to *compete* to learn the training patterns. As a result of the competition, different networks learn different training patterns and, thus, learn to compute different functions. This architecture was first presented by R. Jacobs et al. and offered several advantages over a single neural network in terms of learning speed, representation capability, and the ability to deal with hardware limitations [274].

Fig 15.2 shows the basic architecture of the modular network, which is similar to the architecture of mode fusion hierarchical fuzzy control systems described in Section 9.5. The structure consists of two types of networks: *K expert networks* and a *gating network*. The expert networks compete to learn the training patterns and the gating network mediates the competition. The gating network is restricted to having as many output units as there are expert networks, and the activation of these output units must be nonnegative and sum to 1.

The input vector $\tilde{x} = [x_1, x_2, ..., x_s]^T$ is applied to the expert networks and the gating network simultaneously. Let y_i denote the output of the ith expert network and g_i denote the activation of the ith output unit of the gating network. Then the output of the entire architecture, denoted by y, is

$$y = \sum_{i=1}^{K} g_i y_i \tag{EQ 15.3}$$

with the constraints

$$0 \leq g_i \leq 1 \tag{EQ 15.4}$$

and

$$\sum_{i=1}^{K} g_i = 1 \tag{EQ 15.5}$$

FIGURE 15.2   A Modular Network Architecture (Reprinted from *Fuzzy Sets and Systems*, Vol 79, R. Langari and L. Wang, "Fuzzy Models, Modular Networks, and Hybrid Learning," pp. 141-150, © 1996, with permission from Elsevier Science.)

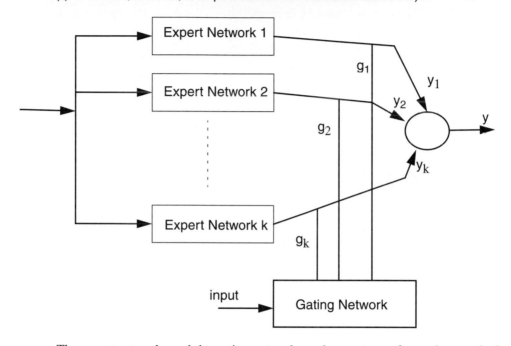

The expert networks and the gating network can be any type of neural network. Suppose now that each expert network consists of a two-layer neural network, as depicted in Fig 15.3. Note that the output unit in this architecture is assumed to be a simple summation. Thus, the output of the network is given by

$$y_i = b_{i0} + b_{i1}x_1 + \ldots + b_{is}x_s \qquad \textbf{(EQ 15.6)}$$

FIGURE 15.3   Two Layer Neural Network Constituting the *i* th Expert Network

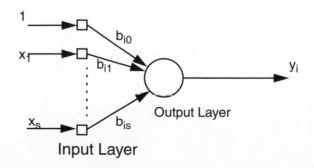

FIGURE 15.4   Three Layer Neural Network Constituting the Gating Network

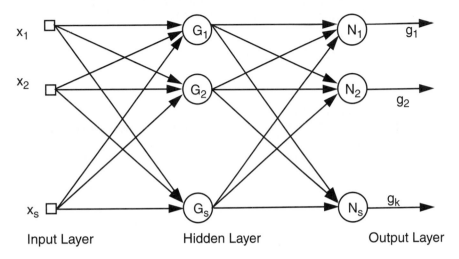

Input Layer   Hidden Layer   Output Layer

As for the gating network, we use a three-layer neural network, as shown in Fig 15.4. The connections from the input layer to the hidden layer and from the hidden layer to the output layer are not weighted. Each unit in the hidden layer is a Gaussian function (denoted as G_i). This function is assumed to have the same form as $R_i(\overset{\scriptscriptstyle\vee}{x})$ defined in Equation 15.1, that is $G_i(\overset{\scriptscriptstyle\vee}{x}) = R_i(\overset{\scriptscriptstyle\vee}{x})$. Each unit in the output layer performs a *normalization transformation* (which is called *softmax transformation* in Jacobs et al. [275]) according to

$$N_i(\overset{\scriptscriptstyle\vee}{x}) = \frac{G_i(\overset{\scriptscriptstyle\vee}{x})}{\displaystyle\sum_{l=1}^{K} G_l(\overset{\scriptscriptstyle\vee}{x})}$$ **(EQ 15.7)**

The final output of each unit is simply given by

$$g_i(\overset{\scriptscriptstyle\vee}{x}) = N_i(\overset{\scriptscriptstyle\vee}{x})$$ **(EQ 15.8)**

Clearly, g_i satisfies the constraints imposed by Equation 15.4 and Equation 15.5. Substituting Equations 15.8 and 15.6 into Equation 15.3, we have

$$y(\overset{\scriptscriptstyle\vee}{x}) = \frac{\displaystyle\sum_{i=1}^{K} G_i(\overset{\scriptscriptstyle\vee}{x})_i (b_{i0} + b_{i1}x_1 + \dots + b_{is}x_s)}{\displaystyle\sum_{i=1}^{K} G_i(\overset{\scriptscriptstyle\vee}{x})}$$ **(EQ 15.9)**

Recall that the output of the TSK model given in Equation 5.41 has a similar form to this, which is reformulated as follows

$$y(\vec{x}) = \frac{\displaystyle\sum_{i=1}^{M} w_i(\vec{x}) f_i(x_1, x_2, \ldots, x_s)}{\displaystyle\sum_{i=1}^{M} w_i(\vec{x})}$$

(EQ 15.10)

Let $f_i(x_1, x_2, \ldots, x_s) = b_{i0} + b_{i1}x_1 + \ldots + b_{is}x_s$"" $w_i(\vec{x}) = G_i(\vec{x})$, and $M = K$. Then we see that Equation 15.10 is identical to Equation 15.9.

This equivalence between modular networks and TSK-type models was recognized in Langari and Wang and may provide an alternative means for the integration of neural networks and fuzzy models [355]. Also, it provides some valuable insights into the understanding of fuzzy models. For instance, we may justify the use of the fuzzy modeling approach of neurobiological grounds. Furthermore, many concepts and tools used in modular networks can be used for the analysis and construction of fuzzy models. For example, a key concept in modular network theory is "competition": the expert networks compete to learn the training patterns and to generate the final output. The competition is achieved by minimizing a new objective function, as defined by

$$J_1(k) = \sum_{i=1}^{K} g_i(k)[y(k) - y_i(k)]^2$$

(EQ 15.11)

where $y(k)$ is the whole output of the network at time k, and $y_i(k)$ is the output of the ith expert network. This is an objective function that encourages *localization*, because it tends to devote a single expert network to each training case. In other words, each expert network in this objective function is encouraged to produce the whole of the output rather than a part. In contrast to the competitive objective function, a *cooperative* objective function may look like this

$$J_2(k) = \left[y(k) - \sum_{i=1}^{K} g_i(k) y_i(k) \right]^2$$

(EQ 15.12)

This function compares the whole output of the network with a blend of the outputs of the expert networks. In order to minimize the error, each expert network must make its output cancel the residual error that is left by the combined effects of all other experts networks. Here, instead of competing, each expert network has cooperated with others to generate the whole output. This function is *global* in the sense that it tends to lead to solutions in which many expert networks are used for each case.

The concept of cooperation is not new to fuzzy modeling because we have used the same objective function form as *J2* in the previous discussions (see, e.g., Equation 14.41). The concept of competition, however, is new, and can be used to highlight the local behavior (interpretability) of fuzzy models. In practice, it is often expected that a fuzzy model should have not only good local behavior but also good global behavior. In this situation it is advisable to use the objective functions *J*1 and *J*2 simultaneously. Thus we have the *combined* objective function

$$J(k) = \alpha(k)J_1(k) + [1 - \alpha(k)]J_2(k) \qquad \textbf{(EQ 15.13)}$$

where $0 \leq \alpha(k) \leq 1$ is a weight parameter. J. Yen and Gillespie proposed a similar approach in which *J*2 is replaced by a slightly different objective function called *local fitness function* [666]. For a modular network, this amounts to adding a *share network* to its standard architecture [307, 273].

15.3   Nonparametric Regression Estimators

15.3.1   Basic Idea

Nonparametric regression has been one of the most active research fields in statistics. For a clear and readable survey of this technique, we refer the reader to Gasser, Engel and Seifert [203]. The basic idea of nonparametric regression may be illustrated by considering the following univariate regression equation

$$y = f(x) + e \qquad \textbf{(EQ 15.14)}$$

where *e* is the observation error. Given *N* input-output observations $\{x(k), y(k)\}$, $k = 1$, 2, ..., *N*, we want to determine the unknown regression function *f*. If the form of *f* is known *a priori*, e.g., a linear function, the regression Equation 15.14 is called *parametric*. If, on the other hand, the form of *f* is unknown and has to been determined by the data themselves, then the equation is called *nonparametric*. The term "nonparametric" here thus refers to the flexible form of the regression function *f*.

    A variety of nonparametric estimators provide powerful data-analytic tools. They have been proposed in the statistical community both as stand-alone techniques and as supplements to parametric analyses. A more detailed introduction to nonparametric regression techniques can be found in the textbooks by W. Hardle and T. Hastie and R. Tibshirani [238, 240]. There is also an increased interest in the implementation of nonparametric regression in the context of neural networks [27, 76, 123, 124, 264, 548, 717].

    Below we will introduce three nonparametric estimators, known as *kernel estimator*, *local polynomial estimator*, and *B-spline estimator*. In particular, we will show that these estimators have a close connection with fuzzy models.

15.3.2   Kernel Estimator

The kernel estimator of *f*(*x*) is defined by

$$\hat{f}(x) = \frac{\sum\limits_{i=1}^{N} K\left[\dfrac{x - x(i)}{\lambda}\right] y(i)}{\sum\limits_{i=1}^{N} K\left[\dfrac{x - x(i)}{\lambda}\right]}$$ (EQ 15.15)

where $K(x)$, known as the *kernel function*, is an even function decreasing in $|x|$. The parameter λ is called the *bandwidth*, which controls the shape of the kernel function. We can visualize the action of the kernel estimator as sliding a weight function along the x-axis in short steps, each time computing the weights mean of y [240].

The kernel function K here is used for a single input variable. For a multidimensional input $\check{x} = [x_1, x_2, ..., x_s]^T$ we can use a multidimensional product kernel function [238]

$$K(\check{x}) = \prod_{j=1}^{s} K(x_j)$$ (EQ 15.16)

where

$$K(x_j) \equiv K\left[\frac{x_j - x(i)}{\lambda_j}\right]$$ (EQ 15.17)

Subsequently, we can build a multidimensional kernel estimator

$$\hat{f}(\check{x}) = \frac{\sum\limits_{i=1}^{N} \prod\limits_{j=1}^{s} K(x_j) y(i)}{\sum\limits_{i=1}^{N} \prod\limits_{j=1}^{s} K(x_j)}$$ (EQ 15.18)

The kernel estimator in equation 15.18 (also the one-dimensional estimator in equation 15.15) has retained all training samples. Alternatively, it can be constructed by using a subset of training samples. In particular, let $\{x'(i), y'(i)\}, i = 1, 2, ..., N', N' \leq N$ be a subset drawn from $\{x(k), y(k)\}, k = 1, 2,..., N$. Then the estimator becomes

$$\hat{f}(\check{x}) = \frac{\sum\limits_{i=1}^{N'} \prod\limits_{j=1}^{s} K(x_j) y'(i)}{\sum\limits_{i=1}^{N'} \prod\limits_{j=1}^{s} K(x_j)}$$ (EQ 15.19)

where $K(x_j)$ is computed by Equation 15.17 with $x(i)$ replaced by $x'(i)$. For a given input $\check{x}(k)$, the output of the regression equation is given by

$$y(k) = \hat{f}[\tilde{x}(k)] = \frac{\sum\limits_{i=1}^{N'} \prod\limits_{j=1}^{s} K[x_j(k)] y'(i)}{\sum\limits_{i=1}^{N'} \prod\limits_{j=1}^{s} K[(x_j)]} \qquad \text{(EQ 15.20)}$$

There are a number of choices for the kernel function K. A typical candidate is the Gaussian function. Suppose that the Gaussian function defined in Equation 14.23 is used. We thus have $K(x_j) = \mu_{A_{ij}}(x_j)$. Furthermore, we let $N = M$, and $y'(i) = c_i$. Then the identity between Equation 15.20 and the zero-order TSK model using product (Equation 14.24) is immediately established.

The identical relation between kernel estimators and fuzzy models is useful, because it allows us to apply many interesting theoretical results established for Kernel estimators, for example, the statistical consistency of the estimator, the convergence rate of the approximation error, and the optimal selection of kernel function parameters, to the analysis and construction of fuzzy models. Several researchers have used the existing theoretical results about kernel estimators as tools to obtain theoretical results for RBF networks, but no work has been done for fuzzy models [649].

15.3.3 Local Polynomial Estimator

The local polynomial (LP) estimator is a procedure for constructing a regression curve (surface) through a local fitting of polynomial functions of the input variables [133, 134]. We may illustrate the basic framework of the estimator by considering the univariate regression Equation 15.14. This consists of the following steps:

(i) For a given observation of the input variable, $x(k)$, $k = 1, 2, ..., N$, identify its p nearest neighbors as $\Omega[(x(k))]$.

(ii) Compute the distance of the furthest near-neighbor from $x(k)$ by

$$\Delta[x(k)] = max_{x(i) \in \Omega[x(k)]} |x(k) - x(i)| \qquad \text{(EQ 15.21)}$$

(iii) For $= 1, 2, ..., \Lambda$, let

$$w_l[x(k)] = W\left(\frac{|x(k) - x(l)|}{\Delta[x(k)]}\right) \qquad \text{(EQ 15.22)}$$

where

$$W(u) = \begin{cases} (1 - |u|^3)^3, & for \ |u| < 1 \\ 0, & for \ |u| \geq 1 \end{cases} \qquad \text{(EQ 15.23)}$$

is a *tricube function* [134]. The shape of this function is shown in Fig 15.5.

(iv) Define a polynomial function of degree d as

$$\imath(x) = \theta_0 + \theta_1 x + \theta_2 x^2 + ... + \theta_d x^d \equiv \tilde{x}^T \tilde{\theta} \qquad \text{(EQ 15.24)}$$

where $\underset{\sim}{\hat{x}} = [1, x, ..., x^d]^T$, and $\underset{\sim}{\hat{\theta}} = [\theta_0, \theta_1, ..., \theta_d]^T$. Solve the weighted least squares problem

$$\sum_{l=1}^{N} w_l[x(k)](y(l) - g[x(l)])^2 \qquad \text{(EQ 15.25)}$$

and get the estimated value of $\underset{\sim}{\hat{\theta}}$ at $x(k)$, denoted as $\underset{\sim}{\hat{\theta}}[x(k)]$.

(v) Compute the output value of the regression at $x(k)$ by

$$y(k) = \underset{\sim}{\hat{x}}^T(k)\underset{\sim}{\hat{\theta}}(k) \qquad \text{(EQ 15.26)}$$

FIGURE 15.5   An Example of Tricube Function

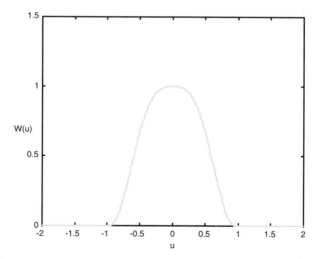

Several remarks on this estimator are in order. First, the above procedure can be extended to multiple regression by replacing the absolute operation $|.|$ in Equations 15.21 and 15.22 using a norm operation $\|.\|$ [134]. Second, the size of the neighborhood, p, is an adjustable parameter that determines how local the fitting is. The larger the p, the smoother the estimate. Third, the use of the tricube function is based on computational considerations. Since $W(u) = 0$ for $|u| > 1$, only $x(k)$ and its p nearest neighbors are required in solving the weighted squares (Equation 15.25). In fact, many other functions, such as the Gaussian function, can also give reasonable results from a local fitting point of view, but these functions cannot save computations because they become small but not zero for large u. Fourth, the degree d of the polynomial function must be selected before carrying out the weighted regression. W. Cleveland suggested that taking $d = 1$ strikes a good balance between computational ease and the need for flexibility to reproduce patterns in the data and thus be advocated in practice [133]. Therefore, for an s-dimensional input vector $\underset{\sim}{\hat{x}} = [x_1, x_2, ..., x_s]^T$, the polynomial may look like

$$g(\hat{x}) = \beta_0 + \beta_1 x_1 + \beta_2 x_2 + \ldots + \beta_s x \qquad \text{(EQ 15.27)}$$

There exist great similarities in the modeling philosophy of the LP estimator and that of the TSK model. In particular, both methods try to approximate a smooth function locally through the use of low-order polynomials that are *connected* by some weight functions. Furthermore, both methods prefer to use a linear polynomial that has the form of Equation 15.27. The weight functions employed in the two methods are not exactly the same, but they possess a common property: The functions give the highest weight at some center point (say $u = 0$), and the weights smoothly decrease as they move further away from the center.

There also exist fundamental discrepancies between the two methods. For example, the output of the model for a given input has been computed quite differently by the two methods. The LP estimator computes the output based on a single polynomial that lies in the neighborhood of the given input, while the TSK model computes the output by linearly combining the outputs of all linear polynomials. Also, the strategy for computing the parameters in the local polynomial is different in the two methods. The LP estimator computes the parameters of each polynomial *separately* by applying a locally weighted regression to the neighborhood in which the polynomial lies, while the TSK model computes the parameters of all polynomials *simultaneously* by applying an unweighted regression to the whole input-output space.

We believe that the recognition of the similarity and the discrepancy between the two methods is useful. For example, we may incorporate the strategy of locally weighted regression into the parameter estimation of the TSK model by constructing a combined objective function that has a form similar to Equation 15.13. This will highlight the local representation nature (interpretability) of the TSK model and meanwhile, maintain its global representation ability.

15.3.4   B-spline Estimator

Informally speaking, a B-spline is a polynomial function that possesses some special properties. The strict definition of B-splines requires the concept of *divided differences* and can be found in any textbook involving splines theory [69, 525].

Two things have to be specified in constructing B-splines: One is the *order*, which indicates the type of the B-splines, e.g., linear or cubic; the other is the *knots*, which divide the domain of interest into subintervals. Once the order and the knots are specified, the B-splines can be formed according to a stable and efficient recurrence relation established by Cox [140] and de Boor [70,69]. Moreover, the order (denoted as p) and the number of knots (denoted as q) also determine the number of B-splines (denoted as n), and in particular we have $n = p + q - 2$.

Fig 15.6 shows B-splines of orders $p = 1, \ldots, 4$ defined on the domain $[0, 4]$ with knots $\{0, 1, 2, 3, 4\}$.

FIGURE 15.6   B-Spines of Orders 1-4

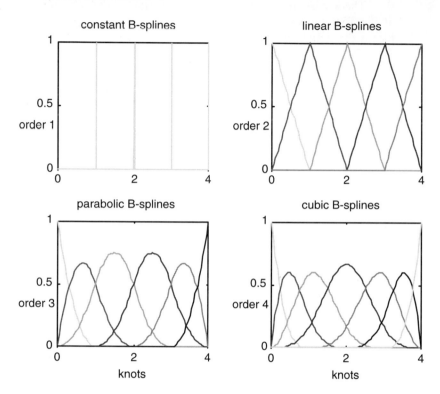

Fig 15.6 reveals an important property of B-splines: They are *locally supported*. This means that for any input lying in the domain of interest only a small number of the B-splines will have a nonzero output, and further, the output is always positive. Another important property of B-splines, which may not be visually derived from this figure, is that they form a *partition of unity*. In particular, let $B_i^p(x)$ denote the ith B-spline of the n B-splines with order p then we have

$$\sum_{i=1}^{n} B_i^p(x) = 1 \qquad \text{(EQ 15.28)}$$

This property is related to the concept of soft constrained partition discussed in Section 5.3. We will leave the comparison of these two notions as an exercise.

Having the knowledge of B-splines, we can define the B-spline estimator of the function $f(x)$ in equation 15.14 by

$$\hat{f}(x) = \sum_{i=1}^{n} B_i^p(x)\lambda_i \qquad \text{(EQ 15.29)}$$

where λ_i are real constants. Subsequently, the output of the regression can be expressed by

$$y = \hat{f}(x) = \sum_{i=1}^{n} B_i^p(x)\lambda_i \qquad \text{(EQ 15.30)}$$

If the knots in the B-splines $B_i^p(x)$ are fixed, the only unknown parameters in Equation 15.30 are the constants λ_i. Since the equation is linear in λ_i, it can easily be solved using some linear optimization method. If, on the other hand, the knots are not known and have to be included as parameters to be estimated from the data, the equation 15.30 is linear in λ_i but nonlinear in the knots. In this case some nonlinear optimization method must be used to solve the equation. Eubank proposed several computational procedures for constructing B-spline estimators with fixed knots or free knots [185].

The B-spline estimator in Equation 15.29 is proposed for a univariate regression function $f(x)$, but it can be extended to handle an s-dimensional multivariate regression function $f(\dot{x})$ by replacing $B_i^p(x)$ using a multivariate B-spline $B_i^p(\dot{x})$. Thus, we have

$$\hat{f}(\dot{x}) = \sum_{i=1}^{n} B_i^p(\dot{x})\lambda \qquad \text{(EQ 15.31)}$$

Here the s-dimensional B-splines $B_i^p(\dot{x})$ have been formed from the *tensor product* of the B-splines of the s single variables in $\dot{x}$. The concept of tensor product may be illustrated by a simple example. Suppose we have two function sets $\{\phi_1, \phi_2\}$ and $\{\psi_1, \psi_2\}$. Then the tensor product of the two sets will lead to an extended set $\{\phi_1\psi_1, \phi_1\psi_2, \phi_2\psi_1, \phi_2\psi_2\}$. Thus, if we let n_i denote the number of B-splines of the ith variable x_i, the total number of the s-dimensional B-splines in Equation 15.31 will be $n = n_1 \times n_2 \times ... \times n_s$. It can be proved that multivariate B-splines generated from the tensor product are locally supported and form a partition of unity as expressed by

$$\sum_{i=1}^{n} B_i^p(\dot{x}) = 1 \qquad \text{(EQ 15.32)}$$

Also, similar to Equation 15.30, the output for the multivariate regression is given by

$$y = \hat{f}(\dot{x}) = \sum_{i=1}^{n} B_i^p(\dot{x})\lambda \qquad \text{(EQ 15.33)}$$

Up to this point we have seen several resemblances between the B-spline estimator and the fuzzy model. For example, fuzzy membership functions are of the same local support property as B-splines. Moreover, the property of unity partition possessed by B-splines is also retained in the fuzzy model through a normalization computation (recall Equation 14.8), although most fuzzy membership functions themselves are not *automatically* endowed with it. Furthermore, if we use B-splines to express the membership functions of input variables in Equation 14.4 (i.e., the zero-order TSK model), and set $c_i = \lambda_i$ and $M = n$, then the output of the zero-order TSK model (Equation 14.9) is identical to the output the B-spline estimator (Equation 15.33). Hence, B-splines have been used as membership functions in a fuzzy model.

Finally we want to point out that the *curse of dimensionality* encountered in fuzzy models (Section 14.4.3) can also appear in the B-spline estimator. This may be viewed as another similarity between the two methods. The reason is that the number of multivariate B-splines formed by the tensor product of univariate B-splines can be very large when the dimension of input space is high. To illustrate this, suppose we have 10 inputs and the number of B-splines for each input is 2. The tensor product of the B-splines of the 10 inputs will produce $2^{10} = 1024$ multivariate B-splines with dimension 10 (We have seen the same number of fuzzy rules in Section 14.4.3 where a grid partition strategy is used to divide the input space). Such a high number of B-splines is problematic due to its high computational cost.

15.4 Other Related Modeling Techniques

There also exist several models in the field of control and signal processing which present a similar philosophy to that of fuzzy models in modeling and control of complex nonlinear systems. *Codebook prediction* [540], for example, involves fitting an *autoregressive (AR)* model to the signal locally in the state space. Under this paradigm, the signal is viewed as a collection of data patterns. These data patterns are grouped using the nearest neighbors technique as used in the local polynomial estimator. Then an AR model is fit to each of the groups. *Competitive local linear modeling* also considers several AR models each of which is associated with some data patterns [462]. In this procedure, the grouping of data patterns is not done through the nearest neighbors; instead, it is performed by a competition among the AR models as they are trained. During the training all AR model accept the same data pattern each time, but only the one that has the best generalization performance is adapted via a LMS algorithm while the rest are left intact. When the training completes, we have a group of AR models as well as a collection of data pattern associated with each model. T. Johansen and B. Foss proposed a complicated *NARMAX (nonlinear autoregressive moving average with exogenous inputs)* model through the use of several simple *ARMAX* models [286]. It is based on the decomposition of the system's operating range into a number of smaller operating regimes, and the use of linear local models (ARMAX) to describe the system within each regime. A nonlinear global model (NARMAX) is then formed by interpolating the linear local models using smooth interpolation functions, depending on the operating point. The equivalences of the method with the TSK model was established [189]. Several other models which follow the same spirit include *composite models* [541], *multi-models* [54, 284,479], and *state dependent models* [482].

15.5 Summary

In this chapter, we discussed the relationships between fuzzy models and three other modeling techniques: (1) radial basis function networks, (2) modular networks, and (3) nonparametric regression estimators. We summarize below some of the major points we have covered.

- Radial basis functions are equivalent to zero-order TSK models that use the product operator (rather than the min operator) to calculate the matching degree of conjunctive antecedent conditions.
- The modular neural network is analogous to the TSK model.
- We discussed three nonparametric regression estimators: kernel estimator, local polynomial estimator, and B-spline estimator.
- We have shown the similarities between the three nonparametric regression estimators and TSK models.
- We discussed some other modeling techniques related to fuzzy models.

Bibliographical and Historical Notes

The equivalence between RBF networks and fuzzy models with constant constituents was first independently established by Jang and Sun and Nie and Linkens and further extended in Hunt, Hass and Murray-Smith [280, 449, 262].

Due to the similarity between B-splines and TSK models, van Rijckevorsel, Harris and Brown, Wang, and others have used B-splines as antecedent membership functions of fuzzy models and neurofuzzy systems [503, 89, 239, 263, 618,672, 716].

There has been a large research effort (mainly in the statistical community) devoted to developing algorithms that can partially overcome the curse of dimensionality associated with high-dimensional data modeling problems. The *Additive model, Projection Pursuit Regression, Multivariate Adaptive Regression Splines* (MARS), and *Adaptive B-splines Model* are several typical examples of this [240,191, 306]. These algorithms are not necessarily used only for the B-spline estimator; rather, they can be applied to a wide class of nonparametric regression estimators. Further, because of the similarity between fuzzy models and regression estimators, it is expected that these algorithms can also be used to alleviate the curse of dimensionality in fuzzy modeling.

Exercises

15.1   If the constant weight b_i in the RBF network is replaced by a linear regression function of the input variables, i.e., $b_i = a_{i0} + a_{i1}x_1 + \ldots + a_{is}x_s$, the network is identical to the TSK model. Consider the following nonlinear time series modeling problem

$$y(k) = (0.8 - 0.5\exp(-y^2(k-1)))$$

$$- (0.3 + 0.9\exp(-y^2(k-1)))y(k-2) + 0.1\sin(\pi y(k-1)) + e(k)$$

where $e(k)$ is a zero mean Gaussian white noise process with variance 0.1. Generate 600 simulated data points from the model. Construct a RBF network with constant weights and a RBF network with regression weights, respectively, using the first 400 data points and then test their performance using the remaining 200 data points. What conclusion can you derive from this modeling problem?
(Hint: Compare the number of hidden units and the number of adjustable parameters in the networks.)

15.2   A key concept in modular networks theory is "competition": The expert networks compete to learn the training patterns and to generate the final output. The competition is achieved by minimizing the objective function

$$J(k) = \sum_{i=1}^{K} g_i(k)[y(k) - y_i(k)]^2$$

where $g_i(k)$ is the activation of the ith output unit of the gating network at time k, $y(k)$ is the whole output of the network, and $y_i(k)$ is the output of the ith expert network. If we view $g_i(k)$ as the normalized truth value of ith rule in a fuzzy model and $y_i(k)$ as the output of the ith rule, how will the resultant fuzzy model behave? Derive an identification algorithm based on the objective function.
(Hint: This is an objective function that encourages *localization*.)

15.3   Some types of B-splines (e.g., parabolic and cubic) are not "normal." Is this a problem when they are used to express fuzzy membership functions?

16   NEURO-FUZZY SYSTEMS

16.1 Basics of Neural Networks

Neural networks are computational models that consist of nodes that are connected by links. Each node performs a simple operation to compute its output from its input, which is transmitted through links connected to other nodes. This relatively simple computational model is named a "neural network" or "artificial neural net" because the structure is analogous to that of neural systems in human brains — nodes corresponding to neurons and links corresponding to synapses that transmit signals between neurons.

One of the major features of a Neural Network (NN) is its learning capability. While the details of learning algorithms of neural networks vary from one architecture to another, they have one thing in common — they can adjust the parameters in a neural network such that the network learns to improve its performance for a given task.

There are three types of neural network learning algorithms. The first type is *supervised learning*, which uses a set of training data to train a neural network such that the input-output mapping of the neural network becomes more and more consistent with the training data. In other words, the error between the neural net's output and the training data's target output gradually reduces. In essence, this type of learning algorithm is similar to some of the techniques for identifying fuzzy models through minimizing root mean square errors which we introduced in Chapter 14. Backpropagation is the best known supervised learning algorithm. In Section 16.3, we will discuss how to use backpropagation learning to identify or to tune parameters ina fuzzy model.

The second type of neural network learning algorithm is *reinforcement learning*. which uses positive and negative reinforcement signals about the NN's performance to adjust its parameters. The reinforcement feedback is typically much less informative than the error signal calculated in supervised learning. On the other hand, it does not require training data as supervised learning does. We will discuss the use of reinforcement learning in identifying parameters in a fuzzy model in Section 16.5.

The third type of neural network learning is unsupervised learning. These learning algorithms are similar to clustering techniques introduced in Chapter 13.

16.1.1 The Architecture of a Three-Layer Feedforward NN

While many NN architectures exist, the most widely used one is the three-layer feedforward neural network architecture, which is usually used together with backpropagation learning. The architecture, as shown in Fig 16.1, organizes its nodes into three layers: the input layer, the hidden layer, and the output layer. Each node in the input layer receives external input signal and transmits it to all nodes in the hidden layer through links.

FIGURE 16.1   The Architecture of Three-layer Feedforward NN

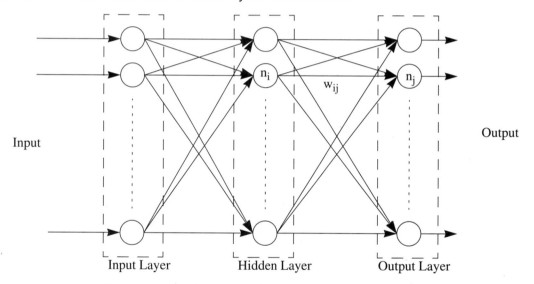

Similarly, nodes in the hidden layer receive input signals from the input layer and sends the processed signal to each node in the output layer. Obviously, nodes in the output layer generate NN's output by combining signals transmitted from the hidden layer. This NN architecture is called "feedforward" because the signals flow from the input layer to the output layer in a forward direction.[1]

The intelligence of a NN is captured by weights associated with its links. Each signal transmitted through a link is multiplied by the associated weight before it reaches the receiving node. Therefore, the weights can reinforce or inhibit signals between two nodes

---

[1] Some other NN architectures allow the signals to feedback in the reverse direction. These architectures are often used together with unsupervised learning.

(neurons). These weights are learned (i.e., automatically adjusted) from a set of training data that describes various input conditions and corresponding desired outputs. During the training phase of the NN, these weights are adjusted to minimize the difference between the NN's output and the output specified in the training data for the same input. After learning, the entire set of weights in a NN is thus said to capture a pattern exhibited in the training data. Backpropagation learning is the most widely used NN learning algorithm. For the convenience of our discussion, we will denote a weight associated with a link from node n_i to node n_j as w_{ij}, as illustrated in Fig 16.1. The signal sent from n_i is denoted x_i.

FIGURE 16.2   The Computation of a Neuron

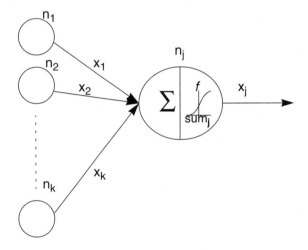

Each node in the three-layer feedforward NN architecture performs a simple two-step operation. First, it calculates a weighted sum of all input signals. For example, node n_j in Fig 16.2 calculates the following weighted sum in the first step:

$$s_j = \sum_{i=1}^{k} w_{ij} \times x_i \qquad\qquad \textbf{(EQ 16.1)}$$

In the second step, the sum calculated is fed to a predefined function f (typically a sigmoid function) to produce the output signal of the node. These two steps are illustrated in Fig 16.2. Combining these two steps, we have

$$x_j = f(s_j) = f\left(\sum_{i=1}^{k} w_{ij} \times x_i \right) \qquad\qquad \textbf{(EQ 16.2)}$$

To apply backpropagation learning, the function f needs to be continuous and differentiable. We will describe the backpropagation learning algorithm in a later section.

16.2 Neural Networks and Fuzzy Logic

Neural networks and fuzzy logic are two complimentary technologies. Neural networks can learn from data and feedback; however, understanding the knowledge or the pattern learned by the neural networks has been difficult. More specifically, it is difficult to develop an insight about the meaning associated with each neuron and each weight. Hence, neural networks are often viewed as a "black box" approach — we can understand what the box does, but not how it is done conceptually.

In contrast, fuzzy rule-based models are easy to comprehend because it uses linguistic terms (linguistic expressions in general) and the structure of if-then rules. Unlike neural networks, however, fuzzy logic does not come with a learning algorithm. The learning and identification of fuzzy models need to adopt techniques from other areas (e.g. statistics, linear system identification, etc.) as we have seen in Chapter 14. Since neural networks can learn, it is natural to marry the two technologies. This marriage has created a new term — neuro-fuzzy system. A neuro-fuzzy system can be loosely defined as a system that uses a combination of fuzzy logic and neural networks.

Neuro-fuzzy systems can be classified into three categories:

1. A fuzzy rule-based model constructed using a supervised NN learning technique.

2. A fuzzy rule-based model constructed using reinforcement-based learning.

3. A fuzzy rule-based model using NN to construct its fuzzy partition of the input space. We will introduce these neuro-fuzzy systems in the rest of this chapter.

16.3 Supervised Neural Network Learning of Fuzzy Models

16.3.1 Neuro-fuzzy Architectures

A neuro-fuzzy system describes a fuzzy rule-based model using an NN-like structure (i.e., involving nodes and links). A neuro-fuzzy system differs from an NN in four major ways. First, the nodes and links in a neuro-fuzzy system usually are comprehensible because they each correspond to a specific component in a fuzzy system. For example, the first layer of a neuro-fuzzy architecture shown in Fig 16.3 describes the antecedent membership functions of input variables. Second, a node in a neuro-fuzzy system is usually not fully connected to nodes in an adjacent layer. In fact, the connections between nodes in a neuro-fuzzy system reflect the rule structure of the system. For instance, the second-layer nodes in Fig 16.3 are connected to only two nodes from the first layer, each one of which describes a condition about an input variable. The nodes in the second layer thus perform the "AND" (conjunction) operator in fuzzy rules. Hence, the number of links connected to a node in the second layer equals the number of input variables. Third, the nodes in different layers of a neuro-fuzzy system typically preform different operations. We will elaborate on this point below. Finally, a neuro-fuzzy system typically has more layers than neural networks.

Typically, a neuro-fuzzy architecture has five to six layers of node. The functionalities associated with different layers usually includes the following:

1. Compute the matching degree to a fuzzy condition involving one variable
2. Compute the matching degree of a conjunctive fuzzy condition involving multiple variables.
3. Compute the normalized matching degree.
4. Compute the conclusion inferred by a fuzzy rule.
5. Combine the conclusion of all fuzzy rules in a model.

It is obvious that these tasks together constitute fuzzy rule-based inference. Fig 16.3 depicts a neuro-fuzzy architecture called ANFIPS whose five layers correspond exactly to the five functionalities listed above. The reason the normalization step (layer 3) is separated from the combination step (layer 5) in neuro-fuzzy architectures is due to the constraint that multilayer feedback NN architecture nodes in a layer can only receive input signals from the closest left adjacent layer. If we merge the tasks performed by layer 3 and layer 5, the merged layer would need input from two layers: layer 2 and layer 4. This would violate the connectivity principle for multilayer feedforward NN and create difficulties in applying backpropagation learning to the neuro-fuzzy system.

FIGURE 16.3   (a) ANFIS Architecture for a Two-input Sugeno Fuzzy Model with Nine Rules; (b) the Input Space that are Partitioned into Nine Fuzzy Regions

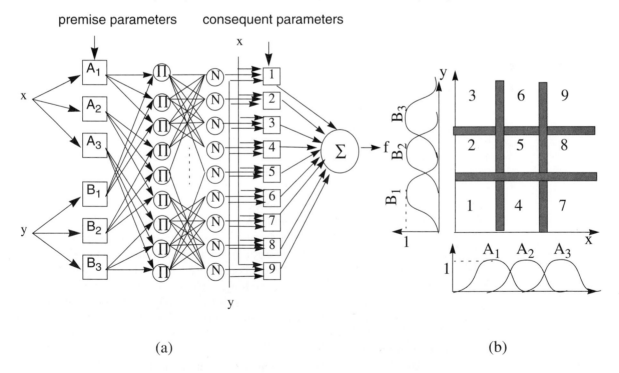

(a)   (b)

The ANFIPS architecture implements a TSK model that partitions the input space using differentiable membership functions. In fact, both the choice of TSK model and the use of differentiable antecedent membership functions are common among neuro-fuzzy architectures. A requirement for applying backpropagation learning to a feedforward NN is that the function preformed at each node needs to be differentiable. Hence, the membership functions in a neuro-fuzzy system must be differentiable. Gaussian membership functions have thus become a popular choice for neuro-fuzzy systems.

Regarding the choice of a rule-based model for neuro-fuzzy systems, we first recall that there are two types of fuzzy models based on fuzzy mapping rules (as we discussed in Chapter 5): additive models and nonadditive models. The former includes the TSK model and the SAM model, while the latter includes the Mamdani model. Most neuro-fuzzy architectures choose additive models because the max and the min operations in no-additive models are not differentiable. Those few neuro-fuzzy architectures that choose to use non-additive model (e.g., Mamdani model) replace min and max operations with differentiable functions that approximate min and max. Among the two additive models, TSK is often preferred for the simplicity of its operation.

16.3.2   Backpropagation Learning Applied to Fuzzy Modeling

In this section we will discuss how to bring the learning ability of neural networks to the construction of fuzzy models. In particular, we will apply one of the most important neural network learning algorithms, known as the *Backpropagation (BP) algorithm*, to the parameter estimation of fuzzy models.

The basic idea of backpropagation was first described by P. Werbos in his Ph.D. thesis [633], in the context of general networks with neural networks as a special case. Subsequently, it was rediscovered by D. Rumelhart, G. Hinton, and J. Williams in 1986 [511] and popularized through the publication of the seminal book entitled *Parallel Distributed Processing* [512]. A similar formulation was also derived independently by D. Parker [463] and Y. LeCun [360].

Before applying BP to a fuzzy model, let us first see how the algorithm works in a neural network. Basically, this algorithm consists of two passes through the different layers of a neural network: a forward pass and a backward pass. In the *forward pass*, an input vector is applied to the input nodes of the network, and its effect propagates through the network, layer by layer. Finally, a set of outputs is produced as the actual response of the network. During the forward pass the synaptic weights of the network are all fixed. During the *backward pass*, on the other hand, the synaptic weights are all adjusted in accordance with the *error-correction rule* (a generalization of the delta or Widrow-Hoff rule introduced in Section 6.1.3). Specifically, an error signal is the difference between the actual response of the network and a desired (target) response. This scheme of weight adjustment repeats many times until the weights no longer changes, which we refer to as *convergence* of the learning algorithm. This error signal is then propagated backward through the network, against the direction of synaptic connections — hence the name "backpropagation." The synaptic weights are adjusted so as to make the actual response of the network move closer to the desired response.

There are actually two ways to adjust the weights using backpropagation. One approach adjusts the weights based on the error signal of one input-output pair in the training data. Hence, these adjustments are made immediately after each training datum is fed into the neural network. For example, if we use a training set containing 500 input-output pairs, this mode of backpropagation algorithm adjusts the weights 500 times for each time the algorithm sweeps through the training set. Assuming that the algorithm converges after 1000 such sweeps, each weight is adjusted a total of 500,000 times. This approach is called the *pattern mode* of backpropagation learning.

The other approach, referred to as the *batch mode* of backpropagation learning, adjusts weights based on the error signal of the entire training set. Therefore, weights are adjusted once only after all the training data have been processed by the neural network. Returning to our previous example, each weight in the neural network is adjusted 1000 times by a batch mode backpropagation learning algorithm.

The batch mode of backpropagation is obtained by applying the gradient descent method to minimize the error between the neural net's output and the target outputs of the entire training set. This error is calculated by the following global error function:

$$E = \frac{1}{2}\sum_{i=1}^{N}(\hat{y}_i - y_i) \qquad \text{(EQ 16.3)}$$

where y_i denotes the output of the ith training data, $\hat{y}_i$ denotes the neural net's output for the same input, and N denotes the total number of input-output pairs in the training set. The pattern mode of backpropagation applies the gradient descent to a different error functions

$$E_i = \frac{1}{2}(\hat{y}_i - y_i)^2 \quad i = 1, 2, ..., N \qquad \text{(EQ 16.4)}$$

It has been shown that pattern BP learning approximates batch BP learning under certain condition (i.e., low learning rate).

We derive the pattern BP learning below. We first calculate the gradient of E_i in Equation 16.4 with respect to the weight w_j of a connection from a node in the middle layer to a node in the output layer:

$$\frac{dE_i}{dw_j} = (\hat{y}_i - y_i)\frac{d\hat{y}_i}{dw_j} \qquad \text{(EQ 16.5)}$$

Using Equation 16.2 to describe the process of the output layer, we have

$$\hat{y}_i = f(s) = f\left(\sum_{i=1}^{k} w_j \times x_j\right) \qquad \text{(EQ 16.6)}$$

We can then apply chain rule to calculate the derivative $\dfrac{d\hat{y}_i}{dw_j}$:

$$\frac{d\hat{y}_i}{dw_j} = \frac{ds}{dw_j} \times \frac{df}{ds} = x_j \times \frac{df}{ds} \qquad \text{(EQ 16.7)}$$

Substituting this into Equation 16.5, we get

$$\frac{dE_i}{dw_j} = (\hat{y}_i - y_i) \times x_j \times \frac{df}{ds}$$ **(EQ 16.8)**

The gradient descent method minimizes a function (E_i in our case) by adjusting each parameter (w_j in our discussion) by an amount proportional to a derivative of the function with respect to the parameter. Applying this principle to minimize E_i by adjusting the weights w_j, we get

$$w'_j - w_j = \Delta w_j = -\alpha \frac{dE_i}{dw_j}$$ **(EQ 16.9)**

where α is a parameter often referred to as *the learning rate*. Substituting Equation 16.8 into Equation 16.9, we get the formula for updating weights in pattern BP learning:

$$\Delta w_j = -\alpha(\hat{y}_i - y_i) \times x_j \times \frac{df}{ds}$$ **(EQ 16.10)**

Once we calculate the number of weight adjustments for each connection between the hidden layer and the output layer, we can view them as the error signal for the hidden layer. Using a strategy similar to the one described above, we can derive the formula for updating weights connecting to the input layer.

This algorithm becomes somewhat simple when it is applied to the fuzzy model expressed in Equation 16.12, because the latter has a relatively simple and fixed structure compared to a neural network. In the following discussions, we will apply BP learning to the parameter identification of TSK fuzzy models whose antecedent membership functions are of Gaussian type. More specifically, in the forward pass, for a given input pattern, the actual response of the model is computed directly from Equation 16.14, and the effect from the input to the output is completed through just a single propagation step. During this process, the antecedent parameters (m_{ij} and σ_{ij}) and the consequents (c_i), which amount to the weights in the neural network, are all fixed. In the backward pass, the error signal resulting from the difference between the actual output and the desired output of the model is propagated backward and the parameters m_{ij}, σ_{ij}, and c_i are adjusted using the error-correction rule. Again, the process is completed in a single propagation step. We will denote the error function at kth iteration as J(k):

$$J(k) = \frac{1}{2}(\hat{y}_i - y_i)^2$$ **(EQ 16.11)**

The error-correction rules for c_i, m_{ij}, and σ_{ij} are given by

$$c_i(k) = c_i(k-1) - \eta_1 \frac{\partial J(k)}{\partial c_i}\bigg|_{c_i = c_i(k-1)} , \quad i = 1, 2, ..., M$$ **(EQ 16.12)**

$$m_{ij}(k) = m_{ij}(k-1) - \eta_2 \frac{\partial J(k)}{\partial m_{ij}}\bigg|_{m_{ij} = m_{ij}(k-1)} \quad , \quad i = 1, 2, ..., M, \ j = 1, 2, ...m \quad \textbf{(EQ 16.13)}$$

$$\sigma_{ij}(k) = \sigma_{ij}(k-1) - \eta_3 \frac{\partial J(k)}{\partial \sigma_{ij}}\bigg|_{\sigma_{ij} = \sigma_{ij}(k-1)} \quad , \quad i = 1, 2, ..., M, \ j = 1, 2, ...m \quad \textbf{(EQ 16.14)}$$

where η_1, η_2, and η_3 are the learning-rate parameters. Note that Equation 16.12 is actually the extended expression of Equation 20.

To give a clear picture that shows how the gradients $\frac{\partial J(k)}{\partial c_i}$, $\frac{\partial J(k)}{\partial m_{ij}}$, and $\frac{\partial J(k)}{\partial \sigma_{ij}}$ are

formed, we rewrite $\hat{y}(k)$, the actual output of the model at the kth observation, in its extended form

$$\hat{y}(k) = \dot{p}(k)\dot{\theta} \equiv \sum_{i=1}^{M} v_i(k)c_i \quad \textbf{(EQ 16.15)}$$

where

$$v_i(k) \equiv \frac{\displaystyle\prod_{j=1}^{s} \mu_{A_{ij}}(x_j(k))}{\displaystyle\sum_{i=1}^{M}\prod_{j=1}^{s} \mu_{A_{ij}}(x_j(k))} = \frac{\displaystyle\prod_{j=1}^{s} \exp\left(\frac{(x_j(k)-m_{ij})^2}{\sigma_{ij}^2}\right)}{\displaystyle\sum_{i=1}^{M}\prod_{j=1}^{s} \exp\left(\frac{(x_j(k)-m_{ij})^2}{\sigma_{ij}^2}\right)} \quad \textbf{(EQ 16.16)}$$

Correspondingly, the error signal becomes

$$e(k) = y(k) - \hat{y}(k) \equiv y(k) - \sum_{i=1}^{M} v_i(k)c_i \quad \textbf{(EQ 16.17)}$$

Now differentiate $J(k)$ with respect to $\dot{\theta}$ and equate the result to zero and using the *chain rule*, we get

$$\frac{\partial J(k)}{\partial c_i} = \frac{\partial J(k)}{\partial e(k)}\frac{\partial e(k)}{\partial c_i} = -2e(k)v_i(k) \quad \textbf{(EQ 16.18)}$$

$$\frac{\partial J(k)}{\partial \sigma_{ij}} = \frac{\partial J(k)}{\partial e(k)}\frac{\partial e(k)}{\partial v_i(k)}\frac{\partial v_i(k)}{\partial \sigma_{ij}} = -2e(k)v_i(k)\left[c_i - \sum_{l=1}^{M} v_l(k)c_l\right]\left[\frac{x_j(k)-m_{ij}}{\sigma_{ij}^2}\right] \quad \textbf{(EQ 16.19)}$$

$$\frac{\partial J(k)}{\partial m_{ij}} = \frac{\partial J(k)}{\partial e(k)}\frac{\partial e(k)}{\partial v_i(k)}\frac{\partial v_i(k)}{\partial m_{ij}} = -2e(k)v_i(k)\left[c_i - \sum_{l=1}^{M} v_l(k)c_l\right]\left[\frac{(x_j(k)-m_{ij})^2}{\sigma_{ij}^3}\right] \quad \textbf{(EQ 16.20)}$$

Substituting Equations 16.18, 16.19, and 16.20 into Equations 16.12, 16.13, and 16.14, respectively, we have

$$c_i(k) = c_i(k-1) + 2\eta_1 e(k)v_i(k), \quad i = 1, 2, ..., M \qquad \textbf{(EQ 16.21)}$$

$$m_{ij}(k) = m_{ij}(k-1) + 2\eta_2 e(k)v_i(k)\left[c_i(k) - \sum_{l=1}^{M} v_l(k)c_l(k)\right]\left[\frac{x_j(k)-m_{ij}(k)}{\sigma_{ij}^2(k)}\right] \qquad \textbf{(EQ 16.22)}$$

$$i = 1, 2, ..., M, \ j = 1, 2, ..., s$$

$$\sigma_{ij}(k) = \sigma_{ij}(k-1) + 2\eta_3 e(k)v_i(k)\left[c_i(k) - \sum_{l=1}^{M} v_l(k)c_l(k)\right]\left[\frac{(x_j(k)-m_{ij}(k))^2}{\sigma_{ij}^3(k)}\right]$$

$$i = 1, 2, ..., M, \ j = 1, 2, ..., s$$

$$\textbf{(EQ 16.23)}$$

where $v_i(k)$ are computed by Equation 16.16 with m_{ij} and σ_{ij} replaced by $m_{ij}(k-1)$ and $\sigma_{ij}(k-1)$, respectively, and $e(k)$ is computed by Equation 16.17 with c_i replaced by $c_i(k-1)$.

These last three equations make up the BP algorithm for the parameter estimation of TSK fuzzy models using Gaussian antecedent membership functions. Similar equations can be constructed for other additive models. We will leave this as an exercise..

The choice of the learning-rate parameters η_1, η_2 and η_3 can have a great influence on the performance of the BP algorithm. A smaller value for η_1, η_2 and η_3 can make the changes of the parameters from one iteration to the next smoother, but this is usually attained at the cost of a slower rate of convergence. On the other hand, a larger value for η_1, η_2, and η_3 can speed up the rate of convergence, but the resulting large changes of parameters from one iteration to the next can make the algorithm unstable. Thus, great care has to be taken in choosing the learning parameters. Unfortunately, there is not a well-defined criteria for choosing these parameters. They are usually determined by trial and error.

Note that the parameter updating here has been performed after the presentation of each training example. Thus, this mode of operation is called *pattern learning*. Alternatively, the updating process may be performed after the presentation of *all* the training examples that constitute an *epoch*. This mode of operation is called *batch learning*. For a particular epoch, we define the objective function J_b as the average squared error of $e(k)$, that is,

$$J_b = \sum_{k=1}^{N} e^2(k) \qquad \textbf{(EQ 16.24)}$$

Let $c_i(n\text{-}1)$, $m_{ij}(n\text{-}1)$, and $\sigma_{ij}(n\text{-}1)$ denote the parameter estimates at the n-1th epoch. By proceeding in a similar way as for the pattern BP, we can obtain the updated estimates $c_i(n)$, $m_{ij}(n)$, and $\sigma_{ij}(n)$ at the nth epoch, as computed by

$$c_i(n+1) = c_i(n) + 2\eta_1 \sum_{k=1}^{N} e(k)v_i(k), \quad i = 1, 2, ..., M \qquad \textbf{(EQ 16.25)}$$

$$m_{ij}(n+1) = m_{ij}(n) + 2\eta_2 \sum_{k=1}^{N} e(k)v_i(k)\left[c_i(n) - \sum_{l=1}^{M} v_l(k)c_l(n)\right]\left[\frac{x_j(k) - m_{ij}(n)}{\sigma_{ij}^2(n)}\right]$$

$$i = 1, 2, ..., M, \quad j = 1, 2, ..., s$$

(EQ 16.26)

$$\sigma_{ij}(n+1) = \sigma_{ij}(n) + 2\eta_3 \sum_{k=1}^{N} e(k)v_i(k)\left[c_i(n) - \sum_{l=1}^{M} v_l(k)c_l(n)\right]\left[\frac{(x_j(k) - m_{ij}(n))^2}{\sigma_{ij}^3(n)}\right]$$

$$i = 1, 2, ..., M, \quad j = 1, 2, ..., s$$

(EQ 16.27)

where $v_i(k)$ are computed by Equation 16.16 with m_{ij} and σ_{ij} replaced by $m_{ij}(n\text{-}1)$ and $\sigma_{ij}(n\text{-}1)$, respectively, and $e(k)$ is computed by Equation 16.17 with c_i replaced by $c_i(n\text{-}1)$.

From an "on-line" operational point of view, the pattern mode of learning is preferred over the batch mode, because it requires less local storage for updating each parameter. On the other hand, the use of batch mode of learning provides a more accurate estimate of the gradients [242]. Whereas it is not clear yet whether these two modes of learning will always give rise to the same model, evidence exists that shows pattern learning approaches batch learning provided that learning-rate parameters are small [487]. Table 16.1 and 16.2 presents a summary of the pattern mode BP algorithm and batch mode BP algorithm, respectively.

TABLE 16.1 The Pattern Mode Backpropagation Algorithm Applied to TSK models with Gaussian Membership Functions

Step 1: Set $c_i(0)$, $m_{ij}(0)$, and $\sigma_{ij}(0)$ to small random numbers.

Step 2: For $k = 1, 2, ..., N$, do the following steps:

Step 3: (*Forward Pass*) Compute $v_i(k)$ using Equation 16.8, where m_{ij} and σ_{ij} are replaced by $m_{ij}(k\text{-}1)$ and $\sigma_{ij}(k\text{-}1)$, respectively; compute $\hat{y}(k)$ using Equation 16.7, where c_i is replaced by $c_i(k\text{-}1)$.

Step 4: (*Backward Pass*) Form error signal $e(k) = y(k) - \hat{y}(k)$; update c_i, m_{ij} and σ_{ij} using Equations 16.13, 16.14, and 16.15, respectively.

Step 5: If some stopping criterion is satisfied, stop; otherwise set $k = k+1$ and go to Step 2.

TABLE 16.2   The Batch Mode Backpropagation Algorithm Applied to TSK Models with Gaussian
Membership Functions

Step 1: Set $c_i(0)$, $m_{ij}(0)$, and $\sigma_{ij}(0)$ to small random numbers, and set $n = 1$.

Step 2: (Forward Pass) Compute $v_i(k)$ using Equation 16.8 for $k = 1, 2, ..., N$, where m_{ij}
and σ_{ij} are replaced by $m_{ij}(n\text{-}1)$ and $\sigma_{ij}(n\text{-}1)$, respectively; compute $\hat{y}(k)$ using Equation
16.7 for $k = 1, 2, ..., N$, where c_i is replaced by $c_i(n\text{-}1)$.

Step 3: (Backward Pass) Form error signal $e(k) = y(k) - \hat{y}(k)$ for $k = 1, 2, ..., N$; update
c_i, m_{ij} and σ_{ij} using Equations 16.17, 16.18 and 16.19, respectively.

Step 4: If some stopping criterion is satisfied, stop; otherwise set $n = n + 1$ and go to step 2.

16.4 Reinforcement-based Learning of Fuzzy Models

As we mentioned briefly in Section 16.2, one of the NN learning techniques is reinforcement learning. Reinforcement learning uses external reinforcement signals (rewards or
punishments) to adjust parameters in a system (typically a rule-based system). Reinforcement learning was motivated by learning theory in behavioral science that states people
and animals can learn to change their behavior from positive reinforcements (rewards) and
negative reinforcements (punishments). The basic principles of reinforcement learning can
be summarized as follows.

1. *External reinforcement* signals are generated based on the performance of the system
   being trained. For example, a system for balancing a pole will receive a negative reinforcement signal whenever the pole falls.

2. The external reinforcement signal is used to learn the *desirability* of recent states of the
   system being trained. A positive reinforcement will increase the desirability, while a
   negative reinforcement will decrease the desirability.

3. Because external reinforcements are generated infrequently and can be all negative or
   all positive, the learning algorithm generates a frequent *internal reinforcement* signal
   from both the external reinforcement and the change of state desirability. For example,
   changing from a less desirable state to a more desirable state without external reinforcement generates a positive internal reinforcement.

Reinforcement learning uses two neuron-like units to carry out the ideas summarized
above. It uses an Adaptive Critic Element (ACE) to generate internal reinforcement, and
an Associative Search Element (ASE) to adjust the parameters of the system being trained.
Fig 16.4 shows the basic architecture of a reinforcement learning system for training a
rule-based controller.

One of the important issues in reinforcement learning is to determine which subsets
of rules should be modified when ASE receives an international reinforcement signal at
time t. This problem is difficult to solve due to two reasons. First, the delay between the
time a rule fires and the time its effect is observed is difficult to predict, if it is predictable
at all. Second, the success or the failure of a control task is often caused by a set of rules

working together properly, not just one "hero rule" (for success) or a single "devil rule" (for failure). This problem, which occurs not only in reinforcement learning but also in other learning approaches, is often referred to as the *credit assignment problem*. To address this problem, reinforcement learning uses the recent history of a rule firing to

FIGURE 16.4 A Basic Reinforcement Learning Control System

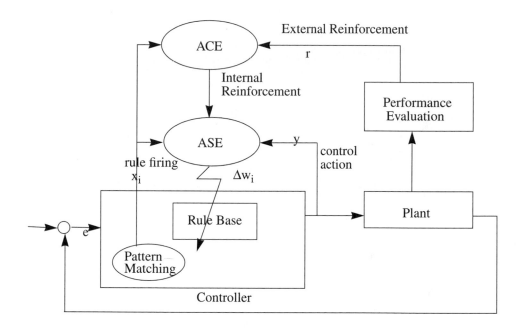

determine the degree the rule is responsible for an internal reinforcement at time t. Let us denote the firing of the ith rule at time t using $x_i(t)$. For a classical rule-based system, $x_i(t)$ is 1 if the ith rule is fired (i.e., executed) at time t, 0 otherwise. For a fuzzy rule-based system, $x_i(t)$ is the degree the ith rule is fired at time t. The degree the ith rule is responsible for the internal reinforcement at time t is determined by the rule's eligibility at t, denoted $e_i(t)$. The eligibility is computed using the following equation:

$$e_i(t+1) = \delta e_i(t+1) + (1-\delta)Y(t)x_i(t) \qquad \textbf{(EQ 16.28)}$$

where $y(t)$ denotes the control action generated by the controller at time t and δ is a constant that determines the rate the eligibility of a rule decays over time when the rule is no longer fired.

The Associative Search Element (ASE) adjusts the consequent parameter of rules using the following equation:

$$\Delta w_i(t+1) = \alpha \hat{r}(t)e_i(t) \qquad \textbf{(EQ 16.29)}$$

where α is a constant that determines the rate of change of w_i and $\Delta w_i(t+1)$ is the adjustment to at time $t+1$, i.e.,

$$w_i(t+1) = w_i(t) + \Delta w_i(t+1) \qquad \text{(EQ 16.30)}$$

Therefore, the amount of change to w_i at t is proportional to the internal reinforcement and the eligibility of ith rule at t.

The Adaptive Critic Element (ACE) needs not only to generate the internal reinforcement signal, but also to learn the desirability of regions covered by the antecedent of each rule. Let us denote the desirability of the region covered by the ith rule as υ_i. The ACE continuously updates υ_i using the following equation:

$$\upsilon_i(t+1) = \upsilon_i(t) + \beta[r(t) + \gamma\rho(t) - \rho(t-1)]\bar{x}_i(t) \qquad \text{(EQ 16.31)}$$

where β is a positive constant that determines the rate of updating, $\bar{x}_i(t)$ is a trace of the firing of the ith rule, $\rho(t)$ is the overall desirability of the plant's state at time t, and r is a learning constant between 0 and 1.

The trace $\bar{x}_i$ is similar to the eligibility trace e_i. The main difference is that e_i takes into account the control action, while $\bar{x}_i$ does not. The trace is computed as follows:

$$\bar{x}_i(t+1) = \lambda\bar{x}_i(t) + (1-\lambda)x_i(t) \qquad \text{(EQ 16.32)}$$

where λ is a degree constant similar to δ in Equation 16.28. The overall desirability of the plant's state at t is computed from the desirability of those rules fired at time t:

$$\rho(t) = \sum_{i=1}^{n} \upsilon_i(t)x_i(t) \qquad \text{(EQ 16.33)}$$

Finally, the ACE generates the internal reinforcement using the following formula:

$$\hat{r}(t) = r(t) + \gamma\rho(t) - \rho(t-1) \qquad \text{(EQ 16.34)}$$

The internal reinforcement is thus due to external reinforcement as well as self-evaluation of plant state changes. More specifically if the desirability of the plant's state increases such that $\rho(t) > \rho(t-1)/\gamma$, a positive internal reinforcement signal can be generated.

Reinforcement learning has been used to identify parameters in fuzzy controllers. Typically, it is used for identifying consequent parameters w_i in rules of the following form (i.e., zero-order TSK model):

If e is A_i and Δe is B_i THEN $y = w_i$.

Because the firing of rules in a fuzzy model is a matter of degree, the rule firing record $x_i(t)$ in Equation 16.33 becomes a number in the interval [0,1]. Consequently, the control action is computed through the normalized weighted sum scheme in the TSK model.

$$f(t) = \frac{\sum\limits_{i=1}^{n} x_i(t) w_i(t)}{\sum\limits_{i=1}^{n} x_i(t)}$$
      (EQ 16.35)

One of the important design issues in applying reinforcement learning to fuzzy logic control is the design of external reinforcement signals. For controlling inverted poles a typical reinforcement is to give a negative reinforcement signal whenever the pole falls outside a predefined angle (e.g., +/- 10°) from the target vertical position. For process control problems, however, giving a fixed negative reinforcement when a failure is detected may not be sufficient to train the controller. To be more effective, reinforcement learning often needs to use more informative feedback such as the accumulated error measure or the deviations from control objectives (e.g., max overshoot, reaching time, etc.)

16.5 Using Neural Networks to Partition the Input Space

As we discussed in Chapter 14, an important element of fuzzy rule-based modeling is the fuzzy partition of the input space. We have also discussed three major approaches to fuzzy partition: grid-based partition, tree-based partition, and scattered partition. However, the fuzzy subregions we have used to form these partition are all described in the form of "x_1 is A_{1i} and x_2 is A_{2i} ... and x_k is A_{ki}" where x_1, x_2, ...x_k are linguistic variables and A_{1i}, A_{2i}, ..A_{ki} are linguistic labels of these variables. Such a fuzzy subregion forms a fuzzy hypercubes in the k-dimensional input space. Even though these fuzzy hypercubes are easy to comprehend, they do not offer the flexibility of partitioning using arbitrary nonlinear hypersurfaces. Since the latter can be achieved using NN, it offers an alternative way for constructing a fuzzy partition.

The basic idea of using a NN to construct a fuzzy partition is simple. We only need to construct a NN that takes input variables of a fuzzy model as input and generates the degrees the input data belong to a predetermined number of fuzzy regions. Fig 16.5 depicts a NN that partitions a two-dimensional input space into three fuzzy subregions. If the NN's output (i.e., A_1, A_2, A_3), were either 0 or 1, the NN preforms a hard (nonfuzzy) partition. To preform a fuzzy partition, the NN's output needs to be a real number in the interval [0, 1]. Fig 16.5 (b) depicts the fuzzy partition generated by the NN. As shown in the figure, the boundary of each fuzzy region is a fuzzy hypersurface. Due to the flexibility of these fuzzy hypersurfaces, an NN can reduce the number of rules in a fuzzy model. We will leave this as an exercise.

FIGURE 16.5   (a) An NN that Partitions a Two-dimensional Input Space (b) The Corresponding
Fuzzy Partition

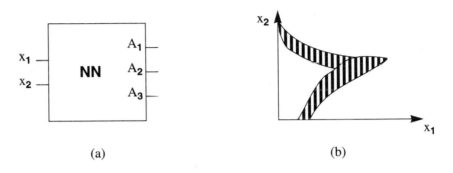

In order to train the NN to partition the input space, we first need to have training
data. The training data can be obtained in two ways: (1) by asking an expert of the problem domain to assign membership degrees to a sample set of input data, (2) by clustering a
sample set of training data using a clustering algorithm (e.g., fuzzy c-means). The former
approach is feasible only if such an expert is available. Notice that the training data used in
the latter approach contain desired input-output pairs for the fuzzy model being constructed. However, this training data cannot be used directly to train the NN for fuzzy partition because the target outputs of a fuzzy partition are the desired membership degrees in
the partition's fuzzy subregions, which are different from the fuzzy model's outputs.
Hence, the data need to be first partitioned using a clustering algorithm. The classified
dataset can then be used to train the NN for partitioning the input space.

16.6 Neuro-fuzzy Modeling Examples

In this section we provide two examples to show how the BP algorithm is used to estimate
the parameters of a fuzzy model. For simplicity, we assume that the fuzzy model has a
constant consequent constituents. That is,

$$R_i\text{: } If \quad x_1 \text{ is } A_{i1} \text{ and } x_2 \text{ is } A_{i2} \text{ and } ... \text{ and } x_m \text{ is } A_{im}$$
$$Then \text{ y is } c_i , i = 1, 2, ..., M.$$

(EQ 16.36)

where c_i are the constant constituents. The antecedent membership functions are assumed to be Gaussian functions, as defined by

$$A_{ij} = \exp\left(-\frac{(x_j - m_{ij})^2}{2\sigma_{ij}^2}\right)$$

(EQ 16.37)

In order to determine the structure (mainly the number of fuzzy rules constituting the underlying model) and the parameters of the model in an integrated manner, the BP algorithm is combined with the SVD-QR with column pivoting algorithm introduced in Chapter 14. The main steps of the process are shown in Fig 16.6. First, a candidate fuzzy model is considered, which may be exhaustive and oversized (i.e., have a large rule base) but not undersized. The parameters of the model are trained using the BP algorithm. Second, the number of important fuzzy rules in the rule base is specified using SVD. Third, the position of these fuzzy rules in the rule base is identified using a subset selection algorithm, known as SVD-QR with column pivoting algorithm. Finally, the important fuzzy rules are retained, while the less important or redundant fuzzy rules are removed from the rule base. A compact fuzzy model is then constructed using the retained fuzzy rules whose parameters are retrained using BP.

FIGURE 16.6   Fuzzy Model Identification Using a BP-SVD Combination

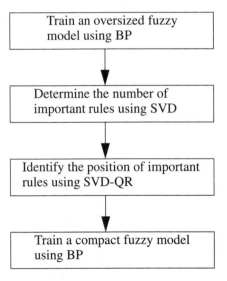

EXAMPLE 18   Consider the nonlinear function [126]

$$y = \frac{\sin(x)}{x}$$
(EQ 16.38)

We want to approximate this function using the fuzzy model described in Equation 16.36. For this purpose, we generated 100 training data points using equally spaced x values in the interval [-10, 10].

We first built a fuzzy model with 30 rules whose antecedent and consequent parameters were trained using the BP algorithm with the initial parameter values $c_i(0) \in [-1, 1]$, $m_{ij}(0) \in [-10, 10]$, and $\sigma_{ij}(0) \in [1, 2]$. These fuzzy rules were labeled as 1, 2, ..., 30 to indicate their position in the rule base. For each of the 100 input data points $x(k)$, $k = 1, 2$, ..., 100 to the model, we computed the normalized firing strengths of the 30 rules using Equation 14.8 (see Chapter 14). A 100-by-30 firing strength matrix P was then formed.

Applying SVD to P, the resulting singular values are shown in Fig 16.7. Based on the distribution of the singular values, we retained 12 rules that were used to construct a reduced fuzzy model. The position of these retained rules in the rule base was identified using the SVD-QR with column pivoting algorithm as 3, 5, 30, 16, 1, 13, 8, 24, 19, 29, 23, 14. The order of the position also indicated the importance of the associated fuzzy rules in the rule base.

The antecedent and consequent parameters of the reduced 12-rule fuzzy model were re-trained using the BP algorithm with the initial parameter values determined in the 30-rules fuzzy model.

Fig 16.8 shows the log-*MSE* (*Mean Squared Error*) curves of the 30-rule fuzzy model and the 12-rule fuzzy model over 1000 epochs of BP training. As a comparison, Fig 16.8 also shows a 12-rule fuzzy model trained using the BP algorithm with random initial parameter values $c_i(0) \in [-1, 1]$, $m_{ij}(0) \in [-10, 10]$, and $\sigma_{ij}(0) \in [1, 2]$. Table 16.3 gives the MSEs of the three fuzzy models after 1000 epochs. From Fig 16.7 and Table 16.3 we know that the reduced 12-rules model with predetermined initial parameter values has achieved its feasible parameter sets much faster than the one with random initial parameter values. This is because the BP algorithm is a nonlinear optimization algorithm and its convergence depends to a great extent on the choice of initial parameter values. The original 30-rule fuzzy model gives a small MSE compared with the two reduced 12-rule fuzzy models. This is not surprising because the former contains a larger number of fuzzy rules. However, good fit on training data points does not always assure good performance on untrained data points. A fuzzy model with a large number of rules may have a high risk of overfitting: capable of fitting training data points completely, but incapable of generalizing to untrained data points satisfactorily. We will see this point in the following example. Fig 16.9 shows the real output of the nonlinear function and the outputs of 30-rule fuzzy model and 12-rules fuzzy model with predetermined initial parameter values. Fig 16.10 shows the membership functions of the input variable x in the 30-rule fuzzy model and in the 12-rule fuzzy model with predetermined initial parameter values.

FIGURE 16.7   Distribution of Singular Values of the 100-by-30 Firing Strength Matrix

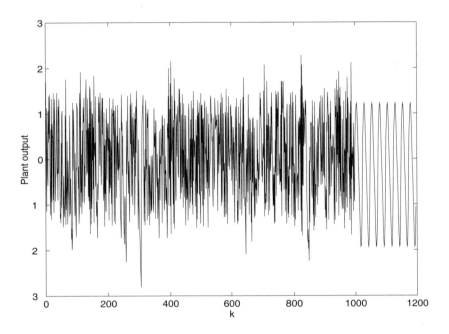

FIGURE 16.8   Log-MSE Curves for Three Fuzzy Models Trained Using the BP Algorithm

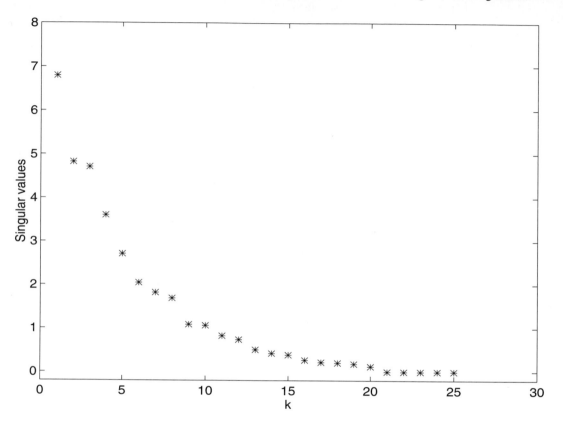

TABLE 16.3   MSEs of three Fuzzy Models After 1000 Epochs of BP Training

Fuzzy Model	MSE
Original model (30 rules)	2.7902e-6
Reduced model (12 rules, with predetermined initial values)	2.8342e-6
Reduced model (12 rules, with random initial values)	6.8712e-6

FIGURE 16.9   Comparison of Outputs of Real System and Identified Fuzzy Models

Original fuzzy model with 30 rules   Reduced fuzzy model with 12 rules

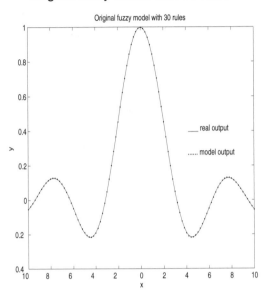

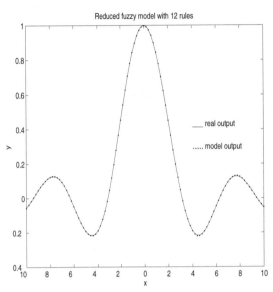

FIGURE 16.10   Membership Functions of Input *x* in the Original and the Reduced Models

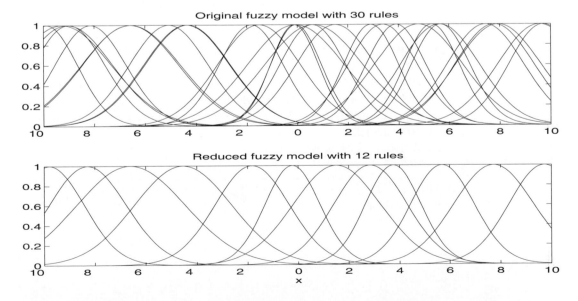

EXAMPLE 19   Consider the second-order nonlinear plant

$$y(k) = f(y(k-1), y(k-2)) + u(k)$$ **(EQ 16.39)**

where

$$f(y(k-1), y(k-2)) = \frac{y(k-1)y(k-2)[y(k-1) - 0.5]}{1 + y^2(k-1) + y^2(k-2)}$$ **(EQ 16.40)**

Our goal was to approximate the nonlinear component f using the fuzzy model shown by Equation 16.36. For this purpose, 700 simulated data points were generated from the plant model in Equation 16.39. The first 500 data points were obtained by assuming a random input signal $u(k)$ uniformly distributed in the interval [-2, 2], while the last 200 data were obtained by using a sinusoid input signal $u(k) = \sin(2\pi k/25)$.

We used the first 500 data points to construct the model with $y(k-1)$ and $y(k-2)$ as the input variables. The performance of the resulting model was tested using the remaining 200 data points. We first constructed a 30-rule fuzzy model whose antecedent and consequent parameters were trained using the BP algorithm with the initial parameter values $c_i(0) \in [-2, 2]$, $m_{ij}(0) \in [-4, 4]$, and $\sigma_{ij}(0) \in [1, 2]$. These 30 rules were labeled by the numbers 1, 2, ..., 30 which indicate the position of the rules in the rule base as well as the associated combinations of membership functions. For example, the number "1" indicates the first rule in the rule base which is associated with the membership function combination $\{A_{11}(x_1), A_{12}(x_2)\}$ (where $x_1 \equiv y(k-1)$, $x_2 \equiv y(k-2)$), while the number "30" indicates the thertieth rule in the rule base which is associated with the membership function combination $\{A_{30,1}(x_1), A_{30,2}(x_2)\}$.

For each of the 500 input data points $\{y(k-1), y(k-2)\}$, $k = 1, 2, ..., 500$ in the model, we computed the normalized firing strengths of the 30 fuzzy rules using Equation 14.8 (see Chapter 14) to form a 500-by-30 firing strength matrix P. Applying SVD to P, we obtained the singular values as shown in Fig 16.11. Based on the distribution of the singular values, we determined to retain 15 fuzzy rules to construct a reduced fuzzy model. The position of the 15 rules in the rule base was identified using the SVD-QR with column pivoting algorithm as 10, 23, 1, 6, 15, 14, 26, 29, 16, 11, 8, 20, 30, 2, 18. The order of the position also indicated the importance of the associated fuzzy rules in the rule base.

We retrained the parameters of the reduced model using the BP algorithm. The initial values of the parameters were taken as the corresponding parameter values in the original 30-rule fuzzy model.

FIGURE 16.11   Distribution of Singular Values of the 500-by-30 Firing Strength Matrix

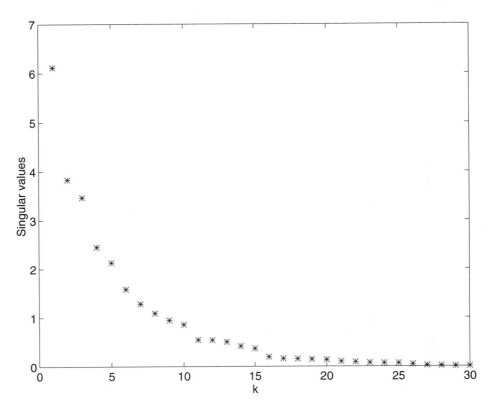

Fig 16.12 shows the log-MSE curves of the 30-rule fuzzy model and the 15-rule fuzzy model over 1000 epochs of BP training. As a comparison, Fig 16.12 also shows a 15-rule fuzzy model trained using the BP algorithm with random initial parameter values $c_i(0) \in [-2, 2]$, $m_{ij}(0) \in [-4, 4]$, and $\sigma_{ij}(0) \in [1, 2]$. It can be seen that the 15-rule model with predetermined initial parameter values shows a much faster convergence than the 15-rule model with random initial parameter values. Table 16.4 shows the MSE's of the three fuzzy models in both the training stage and the testing stage. As expected, the 30-rules fuzzy model gives the smallest MSE among the three fuzzy models in the training stage. However, the 15-rule fuzzy model with predetermined initial parameter values shows the best performance in the testing stage. This indicates that a well-trained simple fuzzy model has a better generalizing ability than a complex fuzzy model. Fig 16.13 shows the outputs of the real plant, the 30-rule fuzzy model, and the 15-rule fuzzy model with predetermined initial parameter values for the sinusoid input signal $u(k) = \sin(2\pi k/25)$ in the testing stage. Fig 16.14 shows the centers m_{ij} of the membership functions in the 30-rule fuzzy model and in the 15-rule fuzzy model with predetermined initial parameter values.

FIGURE 16.12   Log-MSE Curves for three Fuzzy Models

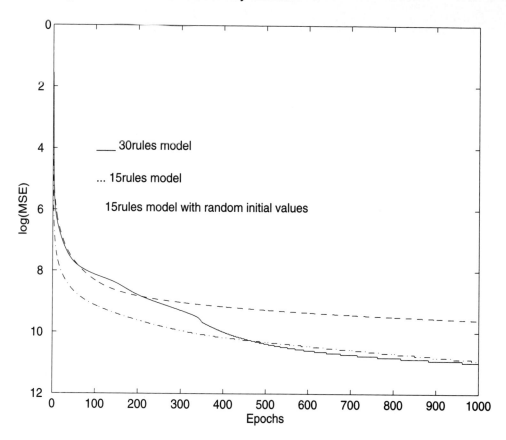

TABLE 16.4   MSE's of three Fuzzy Models for Training and Testing Data

Model	MSE (Training)	MSE (Testing)
Original model (30 rules)	1.7448e-5	2.3648e-5
Reduced model (15 rules, with predetermined initial values)	1.9375e-5	1.8914e-5
Reduced model (15 rules, with random initial values)	4.9907e-5	5.4542e-5

FIGURE 16.13   Outputs of the Real Plant and the Identified Fuzzy Models for $u(k) = sin(2\pi k/25)$

Original fuzzy model with 30 rules   Reduced fuzzy model with 15 rules

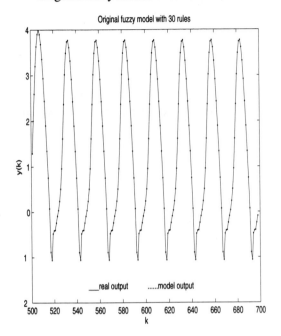
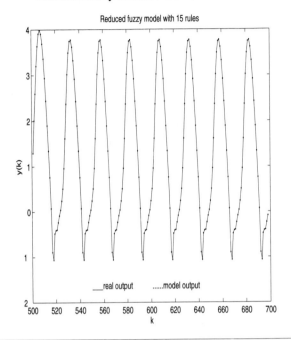

FIGURE 16.14   The Centers m_{ij} of the Gaussian Membership Functions in the 30-rule and 15-rule Fuzzy Models

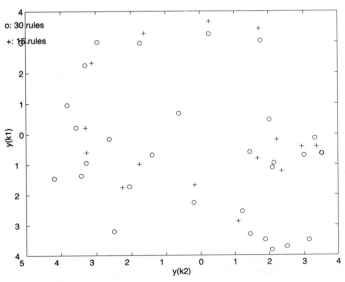

16.7 Summary

We have given an overview of neuro-fuzzy systems in this chapter. Some of the main topics covered are listed below.

- There are three types of neural network learning algorithms: supervised learning, reinforcement learning, and unsupervised learning.

- We described the basics of the three-layer feedforward neural network architecture.

- We compared the pros and cons of neural networks and fuzzy rule-based models.

- We described the ANFIPS neuro-fuzzy architecture.

- We discussed the application of the Backpropagation (BP) learning algorithm, a supervised neural network learning algorithm, to the parameter identification of fuzzy models. Both modes of the BP algorithm (i.e., the pattern mode and the batch mode) were introduced.

- Reinforcement learning uses external reinforcements to learn desirability of system states and to generate internal reinforcements, which are more frequent than the external ones.

- Reinforcement learning uses two neuron-like units: an Adaptive Critic Element (ACE) for generating internal reinforcement signals and learning the state's desirability, and an Associative Search Element (ASE) for adjusting parameters in rules.

- Neural networks can be used to learn a fuzzy partition of the input space.

- We used a function approximation problem and a nonlinear plant modeling problem to illustrate the application of the BP algorithm in neuro-fuzzy systems.

Bibliographical and Historical Notes

The BP algorithm can essentially be viewed as a gradient descent algorithm. Because of this reason, the BP algorithm sometimes is simply called gradient descent algorithm in the fuzzy community [162, 387, 451, 464, 580].

The BP algorithm provides a general and powerful tool for the parameter estimation of fuzzy models. However, the basic version of the algorithm is slow for many practical applications. To speed up the convergence and make the algorithm more practical, some variations are necessary. Indeed, considerable research has been carried out in the neural network community to accelerate the convergence of the BP algorithm, which falls roughly into two categories. The first category involves the development of heuristic techniques, which arise out of a study of the distinctive performance of the basic BP algorithm. These heuristic techniques include such ideas as varying the learning rate, using momentum, and rescaling variables [272, 511, 502, 615]. The second category has focused on the use of standard numerical optimization techniques in the BP procedure. For example, two such techniques, known as the *conjugate gradient algorithm* and the *Levenberg-Marquardt algorithm*, have been applied to the training of neural networks and shown a faster convergence than the basic BP algorithm [113, 232]. Other standard optimization techniques that have promise for neural network training are also discussed in [26, 31, 531,

542]. The motivation behind this category of research is based on the following observation: Training neural networks to minimize squared error is simply a numerical optimization problem. Because numerical optimization is a much more mature area than neural networks, it is thus reasonable to look for fast training algorithms in the large number of existing numerical optimization techniques.

Although all these studies are dedicated to the training of neural networks, they will benefit the training of fuzzy models as well. As a matter of fact, many variations of the basic BP algorithm for training fuzzy models presented in Table 7. 4 can be done based on the results obtained under these two categories of research, which in turn will improve the performance of the algorithm.

H. Takagi and I. Hayashi made pioneer contributions in developing neuro-fuzzy technology in the late 80s [574, 575, 573]. They have also worked closely with Japanese researchers in the industry to develop consumer products using neuro-fuzzy systems [451, 577]. A survey of these products is reported in [572]. P. Werbos discussed several connectsions between neural networks and fuzzy logic in [630].

R. Jang developed the ANFIS neuro-fuzzy system in the early 90s as the result of his doctoral thesis research [279, 279]. After graduation, he joined MATHWORKS to incorporate ANFIPS into MATLAB's Fuzzy Logic Toolbox. C. T. Sun extended ANFIPS to search for a more flexible fuzzy partition of the model's input space [564]. J. Jang, C. Sun and E. Mizutani later wrote one of the first texts to focus on neuro-fuzzy systems and soft computing [282].

Reinforcement learning was developed by Andrew G. Barto, Richard S. Sutton, and Charles W. Anderson in 1983 [29]. C. C. Lee was the first to use reinforcement learning to identify parameters in fuzzy logic controllers [361]. H. Berenji later extended Lee's approach to a more complex architecture called GARIC (Generalized Approximate Reasoning-based Intelligent Control) [38]. Chin-Teng Lin combined reinforcement learning and supervised learning for neuro-fuzzy controllers [383]. Y. Y. Chen has discussed the design of external reinforcement signals for fuzzy control objectives such as target rise time, settling time, and maximum overshoot [120].

Exercises

16.1   Discuss the pros and cons of neural networks, fuzzy rule-based models, and neuro-fuzzy systems.

16.2   Discuss the similatiry and the difference between B-spline's partition of unity Equation 15.28 and the constrained soft partition discussed in Section 5.3.1. Consider the guidelines of membership function design introduced in Chapter 2.

16.3   Find in the literature a neuro-fuzzy architecture using supervised learning that is different from ANFIPS. Discuss the similarities and differences between the architecture and ANFIPS.

16.4   We derived the Backpropagation (BP) algorithm for parameter estimation of the fuzzy model with constant consequent constituents. Following a similar strategy, derive the BP algorithm for the general TSK model.

16.5   Consider the dataset in Table 14.6 construct a neuro-fuzzy model using ANFIPS and the BP algorithm. You can use 60 percent of the dataset for training the remaining 40 percent should be reserved for testing. Compare the result with that of Exercise 14.5.

16.6   Consider the following nonlinear time series model:

$$y(k) = (0.8 - 0.5\exp(-y^2(k-1)))$$

$$- (0.3 + 0.9\exp(-y^2(k-1)))y(k-2) + 0.1\sin(\pi y(k-1)) + e(k)$$

where $e(k)$ is a zero mean Gaussian white noise process with variance 0.1. Generate 600 simulated data points from the model. Build a neuro-fuzzy model using the first 400 data points and then test the model using the remaining 200 data points. Compare the accuracy of the model and the computational cost for constructing it with those of Exercise 14.7.

16.7   For the ball-and-beam system described in Exercise 14.8, design and implement a reinforcement learning-based neuro-fuzzy controller. Consider two different techniques to generate external reinforcement signals: (1) A constant negative reinforcement is given only when the ball falls off the beam: (2) When the ball falls off the beam, the environment generates a negative reinforcement that is proportional to the accumulated distance between the ball's location and the center of the beam. Which one do you think is better? Compare the two feedback generation strategies empirically by applying them to the reinforcement-based neuro-fuzzy controller.

16.8   Construct a grid-based fuzzy partition that approximates the fuzzy partition in Fig 16.5. Use fuzzy subregions described in the form of "x_1 is B_i and x_2 is C_j." How many rules are needed? Reduce and increase the number of membership functions for each variable and observe the impact of these changes to the resulting fuzzy partition.

16.9   Suppose we use fuzzy c-means to partition a set of training data into five fuzzy clusters. Then we use the clustered data to train an NN for partitioning the input space. What is the relationship between the membership functions constructed by FCM and those learned by the NN? Suppose we use (hard) c-means instead of FCM. What is the relationship between the partition of c-means and that of the NN?

17   GENETIC ALGORITHMS AND FUZZY LOGIC

The identification of parameters in a fuzzy model can be viewed as an optimization problem, finding parameter values that optimize the model based on given evaluation criteria. Therefore, search and optimization techniques can be applied to parameter identification as well. This chapter introduces a family of such optimization techniques — genetic algorithms.

17.1 Basics of Genetic Algorithms

Genetic Algorithms (GA) are global search and optimization techniques modeled from natural genetics, exploring search space by incorporating a set of candidate solutions in parallel [256]. A Genetic Algorithm (GA) maintains a *population* of candidate solutions where each candidate solution is usually coded as a binary string called a *chromosome*. A chromosome, also referred to as a *genotype*, encodes a parameter set (i.e., a candidate solution) for a set of variables being optimized. Each encoded parameter in a chromosome is called a *gene*. A decoded parameter set is called a *phenotype*. A set of chromosomes forms a population, which is evaluated and ranked by a *fitness evaluation* function. The fitness evaluation function plays a critical role in GA because it provides information about how good each candidate solution is. This information guides the search of a genetic algorithm. More accurately, the fitness evaluation results determine the likelihood that a candidate solution is selected to produce candidate solutions in the next generation. The initial population is usually generated at random.

The evolution from one generation to the next one involves mainly three steps: (1) fitness evaluation, (2) selection, and (3) reproduction, as shown in Fig 17.1. First, the current population is evaluated using the fitness evaluation function and then ranked based on their fitness values. Second, GA stochastically select "parents" from the current popula-

469

tion with a bias that better chromosomes are more likely to be selected. This is accomplished using a selection probability that is determined by the fitness value or the ranking of a chromosome, as shown in Fig. 17.2. Third, the GA reproduces "children" from selected "parents" using two *genetic operations*: crossover and mutation. This cycle of evaluation, selection, and reproduction terminates when an acceptable solution is found, when a convergence criterion is met, or when a predetermined limit on the number of iterations is reached.

The crossover operation produces offspring by exchanging information between two parent chromosomes. An example of crossover operation for binary GA is shown in Fig 17.3. The mutation operation produces an offspring from a parent through a random modification of the parent. The chances that these two operations apply to a chromosome is controlled by two probabilities: the crossover probability and the mutation probability. Typically, the mutation operation has a low probability of reducing its potential interference with a legitimately progressing search.

The genetic algorithm differs from the conventional optimization techniques in that it is inherently parallel. All individuals in a population evolve simultaneously without central coordination. They operate on a set of solutions rather than on one solution, hence multiple frontiers are searched simultaneously. The only feedback used by the genetic algorithm is the fitness evaluation. The GA has been shown to be an effective search technique on a wide range of difficult optimization problems [151, 216, 256, 413].

FIGURE 17.1   Architecture of the Genetic Algorithm

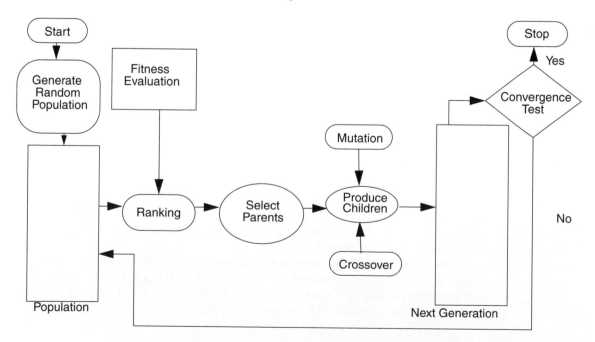

FIGURE 17.2   Selection of Parents

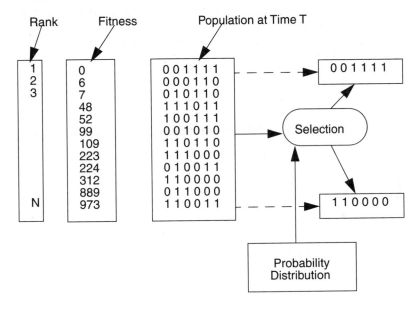

FIGURE 17.3   Binary Crossover
(From J. Yen,et al., "A Hybrid Approach to Modeling Metabolic Systems Using Algorithms and Simplex Method," *IEEE Trans. on Systems, Man, and Cybernetics,* Vol. 28, Part B, No. 2, pp. 173-191, © 1998 IEEE)

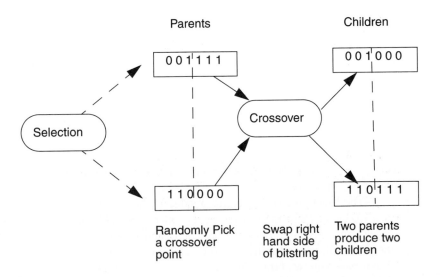

17.1.1 Binary GA Implementation

In a binary implementation of GA, a chromosome is represented by a binary string. The binary crossover operation consists of two steps. First, a randomly selected bit position is used to cut each parent chromosome (i.e., bitstring) into two substrings. The parents then exchange their right substrings to produce two new strings (i.e., their children) with the same length. The mutation operation replaces a randomly chosen bit of a chromosome with its complement (i.e., changing a "1" with a "0" or a "0" with a "1").

17.1.2 Real-coded GA

A real-coded GA is a genetic algorithm representation that uses a floating point [283]. A chromosome in real-coded GA becomes a vector of floating-point numbers. With some modifications of genetic operators, real-coded GAs have resulted in better performance than binary-coded GA for certain problems [283]. We briefly describe below some modified genetic operators recommended for real-coded GA.

The crossover operator of a real-coded GA is analogous to that of binary-coded GA except that its crossover points fall between genes (i.e., encoded parameters). Two mutation operators have been proposed for real-coded GA. A *random mutation* changes a gene with a random number in the feature's domain. A *dynamic mutation* stochastically changes a gene within an interval that narrows over time.

17.1.2.1 Dynamic Mutation of Real GA

C. Z. Janikow and Z. Michalewicz introduced a dynamic mutation operator for real-coded GA [283]. The basic idea is to gradually reduce the range of the mutation operator over time such that the GA gradually changes from "explorative" search to focused search. The dynamic mutation operator is defined as

$$v'_k = \begin{cases} v_k + \Delta(t, UB - v_k) & if \quad a \quad Random \quad digit \quad is \quad 0 \\ v_k + \Delta(t, v_k - LB) & if \quad a \quad Random \quad digit \quad is \quad 1 \end{cases} \qquad \textbf{(EQ 17.1)}$$

$$\Delta(t, y) = y \cdot \left(1 - r^{\left(1 - \frac{1}{t}\right)^b} \right) \qquad \textbf{(EQ 17.2)}$$

where r is a uniform random number from [0, 1], T is the maximal generation number, a and b are system parameters determining the degree of dependency on an iteration number (a typical choice is $b = 5$).

The function $\Delta(t, y)$ returns a value in the range of [0, y] such that the probability $\Delta(t, y)$ being close to 0 increases as t increases. This property causes this operator to space uniformly initially (when t is small), and very locally at later stages, thus increasing the probability of generation children closer to their parent. The design of dynamic mutation is based on two rationales. First, it is desirable to take large leaps in the early phase of GA search so that GA can explore as wide an area as possible. Second, it is also desirable to take smaller jumps in the later phase of GA search so that GA can direct its search toward a local or global minimum more effectively. Both are accomplished through the role that "time" t plays in $\Delta(t, y)$.

17.2 Design Issues in GA

17.2.1 Choice of Encoding

For binary GA, the choice of an encoding scheme is one of the important design decisions. Because binary GA encodes each parameter value using a binary string (i.e., a fixed length of binary numbers), we need to find a scaling function that maps the real parameter value into an integer in the interval $[0, 2^n - 1]$ where n is the length of the binary string. Such a scaling function can be easily constructed from (1) the range of the parameter values we wish GA to search, and (2) the length of the binary string. The role of the scaling function and its inverse in encoding and decoding a parameter set is shown in Fig. 17.4.

In designing GA, we usually first decide the range of each parameter value based on background knowledge of the problem being optimized. Based on the range and the desired precision of the optimal value for each parameter, we can then calculate the length of the binary string required.

EXAMPLE 20   Suppose we want to design a binary GA to optimize parameters in a fuzzy model. The range of these parameters is [-10, 100], and the desired precision of the optimal parameter is one digit after the decimal point. Find the minimal length of the binary string.

The range of the parameter is $100 - (-10) = 110$. The shortest binary string that can represent all numbers in the interval up to one digit after the decimal point is
$$\log_2(10 \times 110) = \log_2 110 = 11$$
Also notice that if we reduce the range of the parameters to 102 (e.g., [-9, 93], only 10 bits are needed. This suggests that sometimes we may wish to make minor adjustments to the parameter ranges in order to minimize the utilization of their encoded binary strings.

FIGURE 17.4   Encoding and Decoding of Search Variables

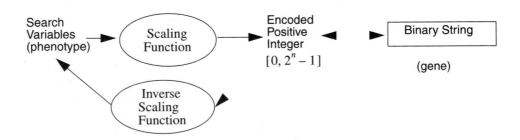

17.2.2 Selection Probability

As we mentioned in Section 17.1, the probability that a chromosome is selected to be a parent for reproduction can be based on (1) its fitness value or (2) its ranking. Most GA experiments reported in the literature recommend ranking-based selection probability, because using fitness-based selection probability often results in premature convergence. This is because the fitness of chromosomes in generation often varies widely, especially during the early stage of the search. Hence, a fitness-based selection probability may biase the selection process to such an extent that medium-rankcd chromosomes are rarely selected if their fitness values are much lower than the fitness of those ranked on the top.

17.2.3 Mutation and Crossover Probability

The design of mutation and crossover probability differs significantly for binary GA and real-coded GA. This is because their roles are different in the two GA schemes. A binary GA mainly relies on the crossover operation to produce new points in the search space. The mutation operation plays a secondary role in binary GA by introducing "stochastic randomness" into the reproduction process. In contrast, a real-coded GA primarily uses its mutation operator to generate new points in the search space. The crossover operation plays a secondary role of "exchanging parameters". Because of these differences, the crossover probability of a binary GA is usually much higher than the mutation probability. In contrast, the mutation probability of a real-coded GA is usually much larger than the crossover probability.

17.2.4 Fitness Evaluation Function

A good design of fitness evaluation function is probably the most important factor in a successful application of GA. However, its importance can be easily overlooked. When a GA does preform well in solving an optimization problem, a natural temptation is to change a GA design parameter (e.g., the population size, the crossover probability, or the mutation probability). Even though changing these parameters does affect the performance of GA, tuning GA parameters will be in vain if the fitness evaluation function does not correctly reflect the desired optimization criteria.

There are at least two types of fitness evaluation functions: fitness-based and performance-based. A fitness-based function calculates how well a model using the parameter set in a chromosome fits a set of input-output training data. In other words, the function computes the error (typically root mean square errors as used in Chapter 14) between the target's output and the model's output. A performance-based evaluation function determines how well a system using the parameter set in a chromosome achieves a set of performance objectives.

These two types of fitness evaluation functions are analogous to two types of neural network lcarning that we discussed in Chapter 16. The goal of fitness-based GA evaluation is similar to that of supervised learning (e.g., the backpropagation algorithm). The spirit of performance-based GA evaluation is similar to that of reinforcement learning. Therefore, GA can be used as an alternative to both of these two neural network learning techniques.

Not only can GA be used to solve the same problem that NN solves, it can also be used to solve problems that are difficult for NN learning. A typical example is the optimization of parameters in a model that already has a mathematical structure. Applying NN to this type of problem is difficult because it is often difficult to map the given mathematical structure to the structure of an NN. We will describe a GA application to a problem of this type in Section 17.6.

Obviously, GA and NN offer different trade-offs. Forcing a model to be described using an NN architecture and only attempting to find local minimum backpropagation learning (and supervised learning in general) are typically computationally more efficient than GA.

Finally, the two types of GA fitness evaluation functions can be combined. That is, they are not mutually exclusive. The fitness evaluation function used in the metabolic modeling problem in Section 17.6 is actually a combination of two components. The first component calculates the fitness between a candidate model's simulation result and the desired output. The second component calculates the degree the model converges, since convergence is an important performance criterion of a good metabolic model. SImilar needs to combine these two types of fitness evaluation functions can be found in other applications as well.

17.3 Improving the Convergence Rate

One of the main limitations of GA is its high computational cost due to its slow convergence rate. A common approach to improving the convergence rate of GA is to combine GA and a local search technique into a hybrid GA architecture.

While genetic Algorithms (GAs) have shown to be effective for solving a wide range of optimization problems [151], its convergence speed is typically much slower than local optimization techniques. It can only recombine good guesses, hoping that one recombination will have a better fitness than both of its parents.[1] Because of this limitation, many researchers have combined GAs with other optimization techniques to develop *hybrid genetic algorithms* [498, 551, 506, 3, 159, 433, 269, 407, 499, 497]. The purpose of such hybrid systems is to speed up the rate of convergence while retaining the ability to avoid being easily entrapped at a local optimum. Although local optimization in a hybrid often results in a faster convergence, it has been shown that too much local optimization can interfere with the search for a global optimum by drawing the genetic algorithm's attention to local optima too quickly, leading to premature convergence [405]. Thus, while local optimization might improve the speed of the analysis, it may also reduce the quality of the final solution. Thus, designing a hybrid approach for an application involves a careful analysis of these trade-offs.

To put our discussion in the bigger context, we briefly review the types of hybrid GA architectures before we discuss a specific hybrid approach that combines the GA with the simplex method.

---

[1] The underlying foundation of this search strategy is Holland's schema theory [256].

17.3.1 Local versus Global Search Methods

Optimization techniques can be classified into two categories: (1) local search methods, and (2) global search methods. A local search method uses local information about the current set of data (state) to determine a promising direction for moving some of the data set, which is used to form the next set of data. The advantage of local search techniques is that they are simple and computationally efficient. But they are easily entrapped in a local optimum. Examples of common gradient methods include steepest descent [147], Newton strategies [528], Powell's version of conjugate directions [481], and R. Hook and T. Jeeves' pattern search [260, 528]. In contrast, global search methods explore the global search space without using local information about promising search direction. Consequently, they are less likely to be trapped in local optima, but their computational cost is higher. In J. Renders' recent work [499], the distinction between trade-off in the local search methods and global search methods is referred to as the "exploitation-exploration." While the global search methods often focus on "exploration," the local search methods focus on "exploitation."

17.3.2 Types of Hybrid Architecture

Like most other global search methods, Genetic Algorithms (GAs) are not easily entrapped in local minima. On the other hand, they typically converge slowly. Many researchers have reported in the literature that combining a GA and a local search technique into a hybrid approach often produces certain benefits [405, 551, 506, 407]. This is because a hybrid approach can combine the merits of the GA with those of a local search technique. Because of the GA, a hybrid approach is less likely to be trapped in a local optimum than a local search technique. Due to its local search, a hybrid approach often converges faster than the GA does. Generally speaking, a hybrid approach can usually explore a better trade-off between computational cost and the optimality of the solution found.

Hybrid genetic algorithms can be classified into four categories: (1) pipelining hybrids, (2) asynchronous hybrids, (3) hierarchical hybrids, and (4) additional operators.[2] We briefly review each category below.

[2.] These four categories are not mutually exclusive, because a specific hybrid approach can belong to multiple categories. For instance, a pipelining hybrid could introduce additional operators in the GA phase.

17.3.3   Pipelining Hybrids

Probably the simplest and the most commonly used hybrids are the pipelining hybrids, in which the genetic algorithm and some other optimization techniques are applied sequentially — one generates data (i.e., points in the search space) used by the other. Typically, pipelining hybrids use the first search algorithm to prune or bias the initial search space such that the second algorithm will either converge more quickly or more accurately. There are three basic types of pipelining hybrid GA, as shown in Fig 17.5: (1) The GA is applied first, serving as a *preprocessor*; (2) the GA is applied last, serving as the *primary* search routine; (3) the GA interleaves with another optimization technique. This has often been referred to as a *staged* hybrid in the literature [405].

FIGURE 17.5   Pipelining Hybrid Architectures
(From J. Yen, et al., "A Hybrid Approach to Modeling Metabolic Systems Using Algorithms and Simplex Method," *IEEE Trans. on Systems, Man, and Cybernetics,* Vol. 28, Part B, No. 2, pp. 173-191, © 1998 IEEE)

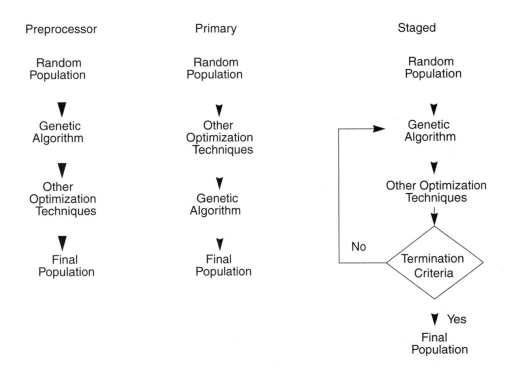

G-bit improvement is one of the pipelining hybrids that uses a simple local optimization method by searching the neighbors of the best chromosomes in each generation [216]. The original idea of G-bit improvement on binary GA is as follows:

(1) Select one or more of the best strings from the current population.

(2) Sweep bit by bit, performing successive one-bit changes to the subject string or strings, retaining the better of the last two alternatives.

(3) At the end of sweep, insert the best structure (or k-best structures) into the population and continue the normal genetic search.

For a real-coded version of G-bit improvement algorithm, the process of flipping a bit is replaced by mutating a gene (e.g., through dynamic mutation).

17.3.4 Asynchronous Hybrids

An asynchronous hybrid architecture, shown in Fig. 17.6, uses a shared population to allow a GA and other optimization processes to proceed and cooperate asynchronously. One process might work on the problem by itself for several iterations before accessing the shared population again. If its findings are better than those in the shared population, it updates the shared population. However, if the process does not make any significant improvement after some time, it returns to the shared population to see if any other processes have posted any progress.

FIGURE 17.6   Asynchronous Hybrids

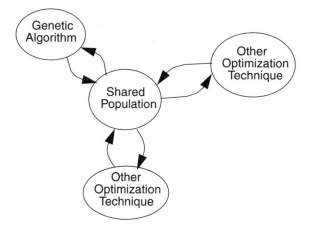

Two approaches in building asynchronous hybrids are (1) to combine a process that converges slowly with another that converges swiftly, or (2) to combine multiple processes that are each suitable for performing search in a subregion of the entire search space. The A-Teams (Asynchronous Teams) methodology describes the first kind of combination by mating the GA with Newton's Method [551].

17.3.5 Hierarchical Hybrids

A hierarchical hybrid GA uses a GA and another optimization technique at two different levels of an optimization problem. An example of hierarchical hybrid is the hybrid of the genetic algorithm and Multivariate Adaptive Regression Splines (MARS) to create the G/ SPLINES algorithm [506]. In this hierarchical hybrid, the GA searches for the best structure of a spline model at a high level, whereas the parameters of the spline model are computed using regression.

17.3.6 Additional Operators

A genetic algorithm can sometimes be improved by introducing additional reproduction operators that perform one-step (or multistep) local search. Almost all local optimization techniques can be incorporated into GA this way, since they all can be viewed as an operator that generates a "child" from one or multiple parents. Among them, the simplex method is particularly suited for this type of hybrid, since the entire population is already ranked by GA. The computation overhead introduced by a new simplex operator is thus very low.

17.4 A Simplex-GA Hybrid Approach

We developed a hybrid GA method by introducing the simplex method as an additional local search operator in the genetic algorithm [675]. The hybrid of the simplex method and the genetic algorithm applies the simplex method to the top S chromosomes in the GA population to produce S-N children. The top N chromosomes are copied to the next generation. The remaining P-S chromosomes are generated using the GA's reproduction scheme (i.e., selection, crossover, and mutation) where P is the population size in the GA. Fig depicts the reproduction stage of the hybrid approach.

The simplex method is a local search technique that uses the evaluation of the current data set to determine the promising search direction. In this section, we first review the basic simplex method. We then describe two modifications to the basic simplex method.

Before we elaborate on the simplex method, we should point out that another kind of local search method is the gradient-based method, which uses the gradient of the function being optimized as the promising search direction [337].

17.4.1 Basic Simplex Method

The basic simplex method was first introduced by W. Spendley et al. [549]. A simplex is defined by a number of points equal to one more than the number of dimensions of the search space. For an optimization problem involving N variables, the simplex method searches for an optimal solution by evaluating a set of $N + 1$ points (i.e., points forming a simplex), denoted as X_1, X_2, ..., X_{N+1}. The method continually forms new simplies by replacing the worst point in the simplex, denoted as X_w, with a new point X_r generated by reflecting X_w over the centroid X of the remaining points:

$$X_r = X + (\ \overline{X} - X_w)$$

where

$$\overline{X} = \frac{X_1 + X_2 + ... + X_{w-1} + X_{w+1} + ... + X_{N+1}}{N}$$

The new simplex is then defined by $X_1, X_2, ..., X_{w-1}, X_{w+1}, ..., X_{N+1}, X_r^3$. This cycle of evaluation and reflection iterates until the step size (i.e., $X_r - X_w$) becomes less than a predetermined value or the simplex circles around an optimum.

17.4.2　Nelder-Mead Simplex Method

J. Nelder and R. Mead developed a modification to the basic simplex method that allows the procedure to adjust its search step according to the evaluation result of the new point generated [440]. This is achieved in three ways. First, if the reflected point is very promising (i.e., better than the best point in the current simplex), a new point further along the reflection direction is generated using the equation

$$X_e = \bar{X} + \gamma (\ \bar{X} - X_w)$$
(EQ 17.3)

where γ is called the expansion coefficient ($\gamma > 1$) because the resulting simplex is expanded. Second, if the reflected point X_r is worse than the worst point in the original simplex (i.e., X_w), a new point close to the centroid on the same side of X_w is generated using the following equation

$$X_c = \bar{X} - \beta (\ \bar{X} - X_w)$$
(EQ 17.4)

where β is called the contraction coefficient ($0 < \beta < 1$) because the resulting simplex is contracted. Third, if the reflected point X_r is not worse than X_w, but is worse than the second worst point in the original simplex, a new point close to the centroid on the opposite side of X_w is generated using the contraction coefficient β:

$$X_c = \bar{X} + \beta (\ \bar{X} - X_w)$$
(EQ 17.5)

17.4.3　Probabilistic Simplex Method

In order to introduce a cost-effective exploration component into the simplex method, we have developed a stochastic variant of the simplex method, which we call the *probabilistic simplex method*. It modifies the basic simplex method to allow the distance between the centroid and the reflected point to be determined stochastically. This is achieved by combining Equations 17.3 and 17.4 into the following equation:

$$X_p = \bar{X} + \alpha (\ \bar{X} - X_w)$$
(EQ 17.6)

where α is a random variable taking its value from the interval [0, 2] based on a predetermined probability distribution. A probability distribution used in our application is a triangular probability density function that peaks at 1 and reaches zero probability at 0 and 2 respectively. If the reflected point is worse than the worst point, a probabilistic contraction operation is applied in a similar way:

$$X_p = \bar{X} - \beta (\bar{X} - X_w)$$
(EQ 17.7)

[3.] If X_r has the worst evaluation in the new simplex, replace the second worst point in the next cycle instead.

where β is a random variable taking its value from the interval $[0,1]$ with a triangular probability density function that peaks at 0.5. By introducing a stochastic component into the reflection operation and the contraction operation, the newly generated point can lie anywhere in a line segment connecting X_w and X, rather than constraining to three points in the line. This flexibility allows the probabilistic simplex to explore the search space with more freedom. It may also facilitate the fine-tuning of solutions around an optimum.

17.4.4 Partition-based versus Elite-based Hybrid Architecture

There are at least two hybrid GA architectures for introducing a new operator into GA: (1) the *partition-based hybrid GA* architecture and (2) the *elite-based hybrid GA* architecture. In a *partition-based hybrid GA architecture*, the entire population is partitioned into disjoint subgroups. A fixed number of childern are produced by each subgroup in a generation to replace a fixed number of worst chromosomes in the subgroup. Each child can be produced by a conventional GA reproduction scheme or by a new operator (i.e., a local search step). Hence, the new operator is associated with a probability that indicates the likelihood that the operator is selected in generating a child in the reproduction process. In an *elite-based hybrid architecture*, the new operator is applied to top-ranking chromosomes to generate a portion of the new population, while the remaining population is generated by a conventional GA reproduction scheme. In both architectures, a fraction of the new generation is created using the new operator, while a fraction of the remainder is produced by the conventional GA operators. In the first architecture, the new operator is applied to the entire selected population with a fixed probability. In the second architecture, the new operator is applied to top-ranking chromosomes with probability 1, and to lower ranking chromosomes with probability 0. Examples of elite-based hybrid GA include G-bit improvement on GA and our simplex-GA hybrid. The partition-based hybrid GA was introduced in J. Renders-H. Bersini's work, even though they did not give this architecture a name [498]. The choice between these two architectures can have a significant impact on the performance of a hybrid GA system[675].

We developed an alternative simplex-GA hybrid independently by applying a concurrent version of a probabilistic simplex operator to top ranking chromosomes [675]. Hence, our approach is based on the elite-based hybrid GA architecture.

17.4.5 Concurrent Simplex

A concurrent simplex is very much like the classical simplex methods with one minor difference. Instead of starting with $N + 1$ points in the simplex (where N is the number of variables to be optimized), the variant begins with $N+\Omega$ points, where $\Omega > 1$. Like a classical simplex, the best N points $X_1 ... X_N$ are selected and their centroid $\bar{X}$ is calculated. However, instead of reflecting only one point across $\bar{X}$, the concurrent simplex reflects multiple points $X_{N+1} ... X_{N+\Omega}$ across $\bar{X}$ to produce $X_{N+1}' ... X_{N+\Omega}'$. All new points are reevaluated and contraction operations are applied if needed. This process of ranking, selection, reflection, evaluation, contraction and elimination iterates like the sequential simplex method. The benefit of the concurrent simplex is that it can explore a wider search frontier. The main disadvantage is the overhead of evaluating and reflecting Ω-1 more points for every iteration. Note also that the concurrent version can incorporate any one of the three simplex methods described in previous sections.

In our simplex-GA hybrid, the concurrent simplex is applied to the top S chromosomes in the population to produce $S - N$ children. The top N chromosomes are copied to the next generation. The remaining chromosomes (i.e., $P - S$ chromosomes where P is the total population size) are generated using GA's reproduction scheme (i.e., selection, crossover, and mutation). Fig depicts the reproduction stage of this hybrid approach. The algorithm of this simplex-GA hybrid approach is summarized in Fig . The algorithm terminates when it satisfies a convergence criterion or reaches a predetermined maximal number of fitness evaluations (i.e., a maximal trial number).

FIGURE 17.7   The Architecture of a Simplex-GA Hybrid

(From J. Yen, et al., "A Hybrid Approach to Modeling Metabolic Systems Using Algorithms and Simplex Method," *IEEE Trans. on Systems, Man, and Cybernetics,* Vol. 28, Part B, No. 2, pp. 173-191, © 1998 IEEE)

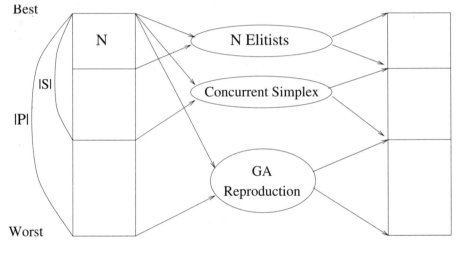

We will refer to a specific version of our architecture by the percentage of population to which the concurrent simplex operator is applied. For example, a 50 percent simplex-GA applies the concurrent simplex to the top half population. A 100 percent simplex-GA will mean that no GA reproduction is performed. Obviously, 0 percent simplex-GA is equivalent to the pure GA.

17.5 GA-based Fuzzy Model Identification

There are at least four issues involved in applying GA to fuzzy model identifications:

1. Choose the model parameters to be optimized.

2. Choose the type of GA architecture.

3. Design the encoding scheme.

4. Design the fitness evaluation function.

We discuss the first three issues in the following sections. The last issue will be addressed in Section 17.6.

17.5.1 Choose the Model Parameters to be Optimized

There are two major choices on the parameters to be optimized by GA: (1) consequent parameters, or (2) both antecedent and consequent parameters. The first approach is easier but does not attempt to search for an optimal fuzzy partition using GA. The second approach is more difficult due to a much larger search space, yet it attempts to search for an optimal fuzzy partition.

We recommend using only GA to search for consequent parameters unless

1. the dimensionality of the problem is low, and

2. the number of rules is small, and

3. finding an optimal fuzzy partition is important.

17.5.2 Choose the Type of GA Architecture

We need to first choose between binary GA and real-coded GA and then to choose between pure or hybrid GA. Since the parameters of fuzzy models are usually real numbers, you may want to give real-coded GA careful consideration.

Since hybrid GA is usually more complicated, it may not be a good choice if you do not have previous experience in using GA. However, after having success in the use of GA, you should consider improving the convergence rate using hybrid GA.

17.5.3 Design the Encoding Scheme

Fig 17.8 shows an example for encoding both antecedent parameters and consequent parameters (i.e., parameters of a consequent membership function) of a Mamdani model. Because rules in a Mamdani model usually map to a fixed set of consequent membership functions, the rule mappings (i.e., the mapping from each rule to its consequent membership function) is encoded separately from the consequent parameters. For a TSK model, because the consequent parameter for each rule is identified independently, there is no need to encode the rule mapping separately.

FIGURE 17.8   Exampleof Encoding for Mamdani Model

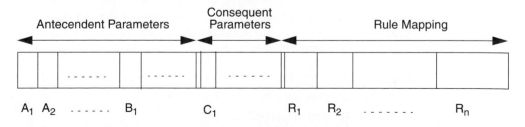

17.6   Case Study: An Application to Metabolic Modeling

Modeling or identification of a system involves first proposing a plausible model structure that captures the main behaviors of the system. The unknown model parameters are then optimized so that the model output fits the system behavior. Mathematical models have been used in many areas including biochemical engineering because a mathematical equation explicitly captures a behavior with an easily understandable form. There are cases, however, where only a partial or incomplete knowledge is available so that mathematical models are not enough to explain the behavior of a system. In that case much information exists in a qualitative or semiquantitative form. In this section, we present a strategy to incorporate such information into the mathematical equation. This strategy uses fuzzy logic-based parameters to modify mathematical models that account for partial or incomplete characteristics. The parameters introduced by the fuzzy parameters are then optimized by use of a hybrid between simplex and genetic algorithm. The resulting model, which we refer to as a *fuzzy logic-augmented model*, provides a flexible form that can simulate various system behaviors. Such fuzzy logic-augmented models are suitable for modeling without complete mechanisms.

This novel modeling strategy is applied to the enzyme kinetic modeling problems. Three enzymes in *Escherichia coli* central metabolism are used as examples: phosphoenolpyruvate (PEP) carboxylase (PPC), PEP carboxykinase (PCK), and pyruvate kinase I (PYKI). Although the kinetic mechanisms of these enzymes have been studied, only partial mechanisms are known. In particular, kinetic descriptions of the inhibitor or activator have not been integrated into the rate equations. Although detailed mechanisms can be derived by more kinetic studies at the molecular level, they are unnecessary and too expansive for the purpose of pathway modeling. Thus the fuzzy logic-augmented model developed here serves as an efficient way to construct simulators at the enzyme kinetic level for the purpose of pathway modeling. The parameters introduced by the fuzzy factors are optimized by a hybrid of simplex and genetic algorithm developed in our group [675].

17.6.1   Augmenting Algebraic Models with Fuzzy Models

The whole purpose of the modeling problem is to discover a model that explains or describes the behavior of a system. The system's behavior is represented by experimental data. Qualitative description can be obtained from the observation of experimental data. A domain expert may propose a mathematical model by understanding the underlying system mechanism or by analyzing the experimental data. Usually only the structure of the model is determined, leaving the parameters in the model to be identified later. As shown in Fig 17.9, our modeling strategy consists of cycles of four main steps: (1) choose parameters to be modeled by fuzzy logic; (2) design the structure of the fuzzy model; (3) optimize model parameters using simplex-GA hybrid; and (4) compare the model prediction with the experimental data.

FIGURE 17.9   A General Architecture for Integrating Fuzzy Models and Algebraic Models

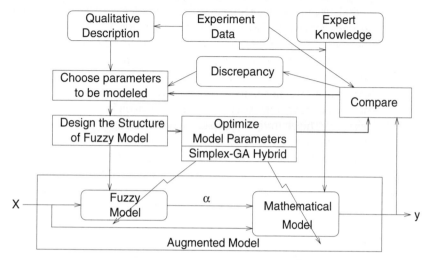

First we pick up one or more parameters to be modeled by fuzzy logic from the proposed mathematical model. Selecting a specific parameter is done by analyzing the impact of the parameter on the model output. For instance, we can use fuzzy logic to augment a mathematical model to convert a constant to a context-dependent parameter. We can also use fuzzy logic to characterize the changing qualitative behavior of the model. The second step is the design of the fuzzy model's structure. This step involves the determination of the fuzzy model inputs, the selection of a fuzzy model (e.g., the TSK model or the Mamdani model), the design of the fuzzy partition and the membership functions for each partition, and the design of the fuzzy local models. The resulting fuzzy model together with the mathematical model forms a *fuzzy logic-augmented model*, where the fuzzy model output (denoted as α) determines certain parameters in the mathematical model. This allows us to use a fuzzy model to capture qualitative description of experimental data while the main system behavior is explained by the mathematical model.

Typically, both the mathematical model and the fuzzy model contains parameters to be optimized. In this case study, the simplex-GA hybrid method, introduced in Section 17.4, is used to optimize the parameters in the mathematical model and the fuzzy model. The parameters to be optimized form a chromosome in the simplex-GA hybrid, where each gene is a parameter. The fitness of a chromosome is calculated based on the error between the output of the model it represents and the experimental data. Once the simplex-GA hybrid identifies a fuzzy logic-augmented model, we compare the model output with the experiment data. If the model output fits the experimental data well, we obtain a satisfactory model that describes the data. If there is a discrepancy between the model output and the experimental data, however, we analyze the discrepancy to choose another model parameter to be modeled by fuzzy logic, or to modify the structure of existing fuzzy models. This process iterates until a satisfactory model is discovered.

17.6.2 Enzyme Kinetic Modeling Problem

Very often, chemical reactions happen as a series of steps instead of as a single basic action. Therefore, a chemical research problem has been to capture or describe the series of steps called a *pathway* of a chemical reaction. Fig 17.10 shows the pathway of a glucose metabolic model. Over the past few decades, extensive studies have unveiled numerous functions crucial to living cells, such as metabolic pathways, enzyme actions, gene regulation, and global physiological controls. Because of the rapid progress in molecular biology, qualitative mechanisms at the molecular level are reasonably well established. These molecular mechanisms are combined to explain system behavior, most often in an intuitive manner. This intuitive approach has been successful to some extent but has rapidly become unsatisfactory as demand increases for a detailed explanation of the system behavior. Therefore, mathematical modeling of biological systems is increasingly important for the total understanding of complex behavior at the systems level. Numerous attempts have been reported to simulate or predict system behavior based on individual component models. For example, enzyme kinetic equations have been derived and assembled to model metabolic pathways [2,245, 377]; components of DNA replication and gene expression have been modeled to simulate the replication of plasmids [566]; and key aspects of cellular functions have been represented mathematically to describe overall cellular behavior [524].

FIGURE 17.10   Pathway of Glucose Metabolic Model (From J. Yen, J. Liao, B. Lee, and D. Randolph [675]. Reprinted by permission of IEEE.)

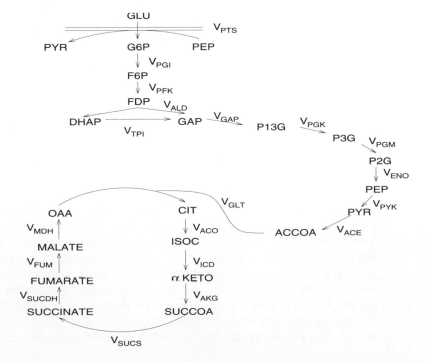

In general, the modeling of biochemical systems involves a two-level approach: a component level involving description of each molecular operation, and a system level involving interactions among all components. For metabolic systems, the components are the enzymes, which interact with each other according to the stoichiometry and enzyme kinetics. Despite tremendous progress in understanding enzyme actions in the past few decades, most enzyme kinetic studies do not aim at developing kinetic rate expressions for the purpose of pathway modeling. Therefore, kinetic equations for enzymes are often incomplete, and qualitative or semiquantitative descriptions of enzyme kinetics are common. In addition, even if the kinetic rate equations are determined, the parameters are often unavailable. This section addresses the first problem by using a fuzzy logic-augmented models and the second problem by using a simplex-GA hybrid. Our goal is to provide a general approach to incorporate qualitative or semiquantitative information into enzyme kinetic models using fuzzy logic. The resulting model is largely based on a known mechanism, which is modified by fuzzy logic to account for information that cannot be predicted by the existing mechanism.

17.6.3 Mathematical Modeling of Enzyme Kinetics

Enzymes are proteins that catalyze chemical reaction without themselves being consumed. Since the first discovery of an enzyme called diastase by Payen and Persoz in 1833, serious attempts have been made to describe the catalytic activity of enzymes in precise mathematical forms. Henri derived a mathematical equation to account for the effect of substrate concentration on the reaction velocity. Michaelis and Menton rediscovered Henri's equation later resulting in the Henri-Michaelis-Menton equation. Other atempts have been made based on different hypotheses on the mechanism. In the followings sections, we will see some of the kinetic models used in this case study. We omit the derivation of the equations since they need the understanding of the details underlying mechanisms. The derivations are easily found in the literature.

17.6.3.1 Henri-Michaelis-Menton Kinetics

Henri-Michaelis-Menton's equation explains the observation that the initial reaction rate is directly proportional to the substrate concentration, but increases in a nonlinear manner with an eventual limiting maximum rate. The overall reaction mechanism is

$$E + A \quad \leftrightarrow \quad EA \rightarrow E + P$$

where an enzyme E reacts with a substrate A forming an enzyme-substrate complex EA and dissociates to the enzyme and a product P. The equation assumes the rapid equilibrium principle; that is, E, A, and EA are at equilibrium, i.e., the rate at which EA dissociates to $E+A$ is much faster than the rate at which EA breaks down to form $E + P$. Henri-Michaelis-Menton's equation is as follows:

$$v = \frac{V_{max}}{[A][A] + k_m}$$

where $v = d[P]/dt$ and $[A]$ is the concentration of a substrate A. There are two parameters V_{max} and K_m. Fig 17.11 plots the substrate concentration versus the reaction rate of the equation. As shown in the figure, V_{max} is the maximum velocity and K_m is the substrate concentration reaching half maximum velocity.

FIGURE 17.11   Henri-Michaelis-Menton Kinetics

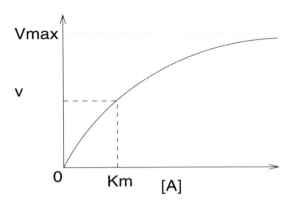

17.6.3.2   Uni Uni Kinetics (Reversible Reactions)

Uni uni kinetics involves one substrate and one product reaction as follows:

$$E + A \quad \leftrightarrow \quad EA \quad \leftrightarrow \quad EP \quad \leftrightarrow \quad E + P$$

where EP is an intermediate enzyme-product complex. As shown in the reaction mechanism, there is a reaction in a reverse direction as well as in the forward direction. We would like to know the direction of the reaction and the reaction rate. The following equation is derived using the steady-state principle:

$$v = \frac{V_{maxf}\dfrac{[A]}{k_{mA}} - V_{maxr}\dfrac{[A]}{k_{mp}}}{1 + \dfrac{[A]}{k_{mA}} + \dfrac{[A]}{k_{mp}}}$$

where there are four parameters V_{maxf}, V_{maxr}, K_{mA}, and K_{mP}.

17.6.3.3   Allosteric Enzyme Kinetics (MWC Model)

The enzymes considered so far yield normal hyperbolic velocity curves. However, in *allosteric* enzymes, the binding of one substrate results in structural or electronic changes and the velocity is no longer hyperbolic. Usually, the reaction velocity shows sigmoidal curves rather than hyperbolic curves.

At low substrate concentrations, the reaction rate is weakly responsive to increases in concentration. As substrate concentration is raised, however, a point is reached at which the reaction rate sharply rises in response to small increases in substrate concentration.

This kind of kinetic is very different from the typical hyperbolic Michaelis-Menton behavior of a normal enzyme. The following equation, called the MWC model, is used for the enzymes having n protomers. (Protomers are identical minimal units in polymeric enzymes and all allosteric enzymes are polymeric enzymes.)

$$v = V_{max} \frac{[A](1 + [A])^{n-1} + Lc[A](1 + [A])^{n-1}}{(1 + [A])^n + L(1 + c[A])^n}$$

where $[A]$ is the substrate concentration. There are three parameters: V_{max}, Lc, L and c are called conformational equilibrium constants and nonexclusive binding coefficients respectively. The MWC model is very useful and popular for discussing allosteric systems because of its simplicity in algebraic description. The equation shows different curves under various values of c and L as indicated in Fig 17.12. As L increases and c decreases, the sigmoicity of the curves increases.

FIGURE 17.12   Effect of Varying L and c on Plot of v versus $[A]$

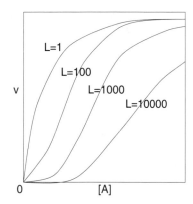 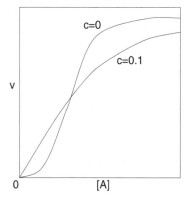

In the following sections, we describe three modeling examples for three different enzyme kinetics. In the first two examples, the fuzzy logic model is used to convert a constant to a context-dependent parameter. The last example uses fuzzy logic to change qualitative behavior.

17.6.4 Modeling of PPC Enzyme Kinetics

PPC catalyzes the following reaction:

$$CO_2 + PEP \rightarrow OAA + P_i$$

From the experimental data (shown in Fig 17.16), the following observations are made [271]: (1) the reaction rate shows a hyperbolic shape as PEP concentration increases; (2) without any activator, the reaction proceeds at a very low rate; (3) acetyl-CoA (ACoA) is a very powerful activator; (4) Fru-1,6-P_2 (FDP) exhibits no activation alone; and (5) FDP produces a strong synergistic activation with ACoA. We first describe an approach to implicitly capture these observations using an algebraic model. We then describe an alternative that explicitly captures these observations by augmenting kinetic equations with a fuzzy logic-model.

17.6.4.1 Algebraic Model

No matter what activator is used, the reaction exhibits hyperbolic behavior. However, the saturation point V_m depends on the activators. Therefore, the reaction is modeled with the following mechanistic equation:

$$V_{PPC} = V_m \frac{[PEP]}{k_m + [PEP]}$$

V_m is modeled from the two inputs ACoA and FDP since these activators change the maximum saturation rate

$$V_m = \frac{k_0 + k_1[AC_0A] + k_2[AC_0A][FDP]}{k_3 + k_4[AC_0A] + k_5[FDP]}$$

We identify seven parameters to be optimized in this model (i.e., K_m, K_0, K_1, K_2, K_3, K_4, and K_5). The resulting model using the hybrid GA is presented later.

17.6.4.2 Fuzzy Logic Augmented Model

The architecture of the fuzzy logic augmented model, as shown in Fig 17.13, consists of two components: (1) a fuzzy logic-based model that maps ACoA and FDP concentrations to the appropriate scaling factor α_{SPPC}, and (2) an algebraic equation that maps the PEP concentration and the scaling factor α_{SPPC} to the reaction rate V_{SPPC}. The situation-dependent factor α_{SPPC}, which modifies V_{max}, captures the various activation effects of ACoA and FDP:

$$V_{PPC} = \alpha_{PPC} V_{max} \frac{[PEP]}{k_m + [PEP]} \qquad \textbf{(EQ 17.8)}$$

FIGURE 17.13   The Integration of Fuzzy Model and Mathematical Modeling for V_{PPC}

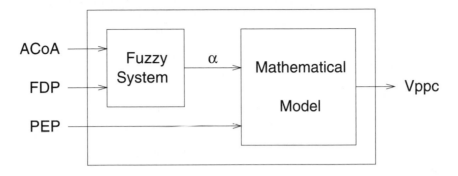

The fuzzy factor α_{SPPC} is modeled by the following four fuzzy rules:

$R1_{SPPC}$: If [ACoA] is LOW and [FDP] is LOW then $\alpha_{SPPC} = c_1$.
$R2_{SPPC}$: If [ACoA] is LOW and [FDP] is HIGH then $\alpha_{SPPC} = c_2$.
$R3_{SPPC}$: If [ACoA] is HIGH and [FDP] is LOW then $\alpha_{SPPC} = c_3$.
$R4_{SPPC}$: If [ACoA] is HIGH and [FDP] is HIGH then $\alpha_{SPPC} = c_4$.

where c_1, c_2, c_3, and c_4 are constants to be optimized. The membership functions of the fuzzy sets (i.e., LOW and HIGH), which are shown in Fig 17.14, were constructed based on experimental data (in Fig 17.16). Notice that the rules above form a special case of the TSK model. There are six model parameters to be identified: V_{max}, K_m, c_1, c_2, c_3, and c_4. We will describe our approach to identifying these parameters using a hybrid GA later.

FIGURE 17.14   The Membership Functions of the Fuzzy Sets for Modeling V_{PPC}

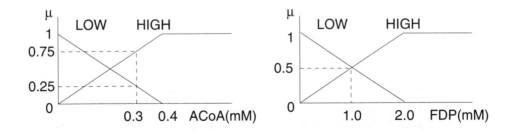

We explain below how the fuzzy model above computes the scaling parameter $\alpha\backslash SPPC$. Suppose the concentrations of ACoA, FDP, and PEP were 0.3, 1.0, and 5.0, respectively. First, the membership degrees for the fuzzy sets in rules are calculated from the membership functions shown in Fig 17.14. We get $\mu_{LOW\ ACoA}(0.3) = 0.25$, $\mu_{HIGH\ ACoA}(0.3) = 0.75$, $\mu_{LOW\ FDP}(1.0) = 0.5$, $\mu_{HIGH\ FDP}(1.0) = 0.5$. Second, the matching degree of fuzzy rules are calculated. We get $w_1 = min(0.25, 0.5) = 0.25$, $w_2 = min(0.25, 0.5) = 0.25$, $w_3 =$

$min(0.75, 0.5) = 0.5$, and $w_4 = min(0.75, 0.5) = 0.5$. Using interpolative reasoning, we obtain the value of α_{SPPC}:

$$\alpha_{PPC} = \frac{0.25 \times c_1 + 0.25 \times c_2 + 0.5 \times c_3 + 0.5 \times c_4}{0.25 + 0.25 + 0.5 + 0.5} \qquad \text{(EQ 17.9)}$$

Finally, the calculated α_{SPPC} value is used in the mathematical model as a scaling factor to get the final output V_{SPPC} using Equation 17.8:

$$V_{PPC} = \alpha_{PPC} \times V_{max} \times \frac{5.0}{k_m + 5.0}$$

FIGURE 17.15   The Chromosome in the GA for V_{SPPC} Parameter Estimation

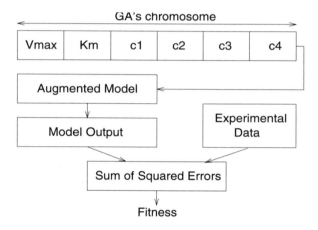

17.6.4.3 Using a Hybrid GA for the Parameter Estimation

The model parameters to be identified are directly encoded as genes in the GA's chromosome. For example, in the proposed fuzzy logic augmented model (Equation 17.9 and the fuzzy model), there are in total six parameters to be identified (i.e., V_m and K_m in the algebraic model, and c_1, c_2, c_3, c_4 in the fuzzy model). We use a hybrid GA to identify these parameters. The parameters in the algebraic model and the ones in the fuzzy model are combined to form a chromosome in the GA as shown in Fig 17.15. Since each chromosome is associated with a model, given a chromosome, the fitness of a chromosome is the error between the model's outputs and the experimental data. The error is calculated using the sum of square errors formula below:

$$Fitness(\vec{p}) = \sum_{i=1}^{M} (y_{(\vec{p}, \vec{x}_i)} - y'_i)^2 \qquad \text{(EQ 17.10)}$$

where $\vec{x}_i$ and y_i* denote the ith input vector (i.e., PEP, ACoA, and FDP concentrations) and the corresponding reaction rate V_{SPPC} observed in the experiment; $\vec{p}$ denotes the set of parameters being optimized. $y_{(\vec{p}, \vec{x}_i)}$ is the reaction rate V_{SPPC} predicted by the model with parameters $\vec{p}$ for inputs $\vec{x}$, and M is the total number of experimental data regarding the reaction rate of V_{SPPC}. Obviously, the goal of the hybrid GA is to find a parameter set that minimizes error.

17.6.4.4 Result for V_{PPC}

The hybrid GA identified the following two parameter sets for the two models: (1) $K_m = 0.499$, $K_0 = 147.64$, $K_1 = 970.29$, $K_2 = 955.55$, $K_3 = 657.02$, $K_4 = 325.01$, and $K_5 = 184.12$ for the algebraic model; and (2) $V_{max} = 0.987$, $K_m = 0.259$, $c_1 = 0.037$, $c_2 = 0.095$, $c_3 = 0.594$, and $c_4 = 0.973$ for the model augmented by fuzzy logic. The performance of the augmented model with the identified parameters is shown in Fig 17.16. To evaluate the effectiveness of the hybrid GA, we applied the pure GA to the problem of optimizing parameters for modeling V_{PPC} and compared its performance to that of the hybrid GA. The comparison is shown in Fig 17.17. The figure indicates that the simplex-GA hybrid converges about twice as fast as the GA without sacrificing the optimality of the solution found.

FIGURE 17.16   Data and Prediction of V_{PPC} Using the Augmented Model

(From "A Hybrid Genetic Algorithm and Fuzzy Logic for Metabolic Modeling." J. Yen, B. Lee, and J. C. Liao. *Proc. of the 13th National Conf. on Artificial Intelligence (AAAI'96)*, pp. 743-749. © 1996 American Association for Artificial Intelligence.)

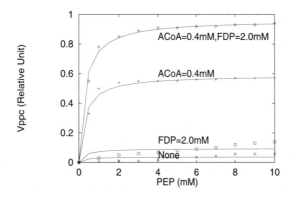

FIGURE 17.17   Performance of the Augmented Model on Modeling V_{ppc}
(From "A Hybrid Genetic Algorithm and Fuzzy Logic for Metabolic Modeling." J. Yen, B. Lee, and J. C. Liao. *Proc. of the 13th National Conf. on Artificial Intelligence (AAAI'96)*, pp. 743-749. © 1996 American Association for Artificial Intelligence.)

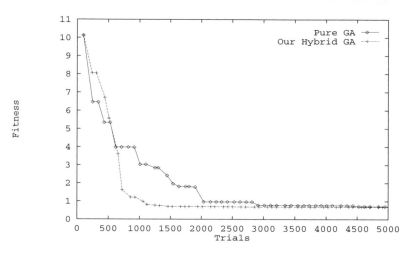

17.6.5 Modeling of PYKI Enzyme Kinetics

Allosteric enzymes show nonhyperbolic kinetic behaviors under different concentrations of metabolites. Fuzzy logic can also be used to capture activation and inhibition effects that change the qualitative behavior of enzyme kinetics. We describe an application of this technique to the kinetic modeling of the pyruvate kinase I (PYKI) reaction. The PYKI enzyme mechanism is the following:

$$PEP + ADP \rightarrow PYR + ATP$$

The reaction rate is activated by FDP with qualitative kinetic change from a sigmoidal curve to a hyperbolic one, as shown in Fig 17.19 [628]. The figure also indicates the inhibition effect on V_{max} by ACoA and ATP concentration.

17.6.5.1 Algebraic Model

We use the MWC model for PYKI enzyme modeling as follows:

$$V_{PYKI} = \frac{[PEP](1 + [PEP])^3 + Lc[PEP](1 + c[PEP])^3}{(1 + [PEP])^4 + L(1 + c[PEP])^4}$$

Since the kinetic shows a different qualitative curve with FDP concentration, we can think of c and L as functions of FDP. However, it is difficult to identify the appropriate structure of these functions for FDP activation. Moreover, the need to represent ACoA and ATP inhibition in addition to FDP activation is likely to lead to a structure that is too complex to offer many insights.

17.6.5.2 Fuzzy Logic Augmented Model

In our approach to modeling this complex reaction, we use the following fuzzy logic augmented mechanistic equation whose parameters F, L, and c are determined by fuzzy if-then rules.

$$V_{PYKI} = \alpha_{PYKI} \frac{[PEP](1 + [PEP])^3 + Lc[PEP](1 + c[PEP])^3}{(1 + [PEP])^4 + L(1 + c[PEP])^4}$$

FIGURE 17.18   The Membership Function of the Fuzzy Sets for V_{pyk} modeling
(From "A Hybrid Genetic Algorithm and Fuzzy Logic for Metabolic Modeling." J. Yen, B. Lee, and J. C. Liao. *Proc. of the 13th National Conf. on Artificial Intelligence (AAAI'96)*, pp. 743-749. © 1996 American Association for Artificial Intelligence.)

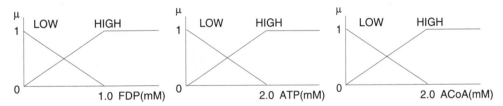

Because both the FDP activation and the inhibition of ATP and ACoA affect the saturation activation rate, we model the parameter F as a mapping from three inputs: FDP, ATP, and ACoA concentrations.

$R_1^{\alpha_{PYKI}}$: If [FDP] is LOW and [ATP] is LOW and [ACoA] is LOW then $\alpha_{PYKI} = a_1$
$R_2^{\alpha_{PYKI}}$: If [FDP] is LOW and [ATP] is LOW and [ACoA] is HIGH then $\alpha_{PYKI} = a_2$
$R_3^{\alpha_{PYKI}}$: If [FDP] is LOW and [ATP] is HIGH and [ACoA] is LOW then $\alpha_{PYKI} = a_3$
$R_4^{\alpha_{PYKI}}$: If [FDP] is LOW and [ATP] is HIGH and [ACoA] is HIGH then $\alpha_{PYKI} = a_4$
$R_5^{\alpha_{PYKI}}$: If [FDP] is HIGH and [ATP] is LOW and [ACoA] is LOW then $\alpha_{PYKI} = a_5$
$R_6^{\alpha_{PYKI}}$: If [FDP] is HIGH and [ATP] is LOW and [ACoA] is HIGH then $\alpha_{PYKI} = a_6$
$R_7^{\alpha_{PYKI}}$: If [FDP] is HIGH and [ATP] is HIGH and [ACoA] is LOW then $\alpha_{PYKI} = a_7$
$R_8^{\alpha_{PYKI}}$: If [FDP] is HIGH and [ATP] is HIGH and [ACoA] is HIGH then $\alpha_{PYKI} = a_8$

where a_1, a_2, a_3, a_4, a_5, a_6, a_7, and a_8 are constants to be optimized. However, changing only α_{PYKI} will not be sufficient because α_{PYKI} doesn't affect the qualitative behavior. The qualitative change in kinetic behavior by FDP activation is achieved by using fuzzy if-then rules for c and L.

R_1^c: If [FDP] is LOW then $c = c_1$
R_2^c: If [FDP] is HIGH then $c = c_2$
R_3^c: If [FDP] is LOW then $L = L_1$
R_4^c: If [FDP] is HIGH then $L = L_2$

where c_1, c_2, L_1, and L_2 are constants to be optimized. The membership functions for FDP, ATP, and ACoA fuzzy sets in the fuzzy model are shown in Fig 17.18, which were constructed from the experimental data. The mechanism for calculating the output V_{PYKI} is similar to that of V_{SPPC}. α_{PYKI}, c, and L are calculated first by their fuzzy models based on the fuzzy inference mechanism. The calculated values are then used in the mathematical model to get the final output V_{PYKI}.

17.6.5.3   Using a Hybrid GA for the Parameter Estimation

The model parameters to be identified are directly encoded as genes in the GA's chromosome as described in Section 17.6.4. We use the same fitness function as shown in Equation 17.10.

17.6.5.4   Result for V_{PYKI}

The hybrid GA identified the following parameter set: $a_1 = 50.4113$, $a_2 = 35.7698$, $a_3 = 47.3199$, $a_4 = 10.8140$, $a_5 = 52.3469$, $a_6 = 39.2297$, $a_7 = 50.7368$, $a_8 = 16.5200$, $L_1 = 1610.2202$, $L_2 = 6984954.4926$, $c_1 = 0.0558$, and $c_2 = 13.5644$ for the model augmented by fuzzy logic. The performance of the model with the identified parameters is shown in Fig 17.19. Our result shows that the fuzzy logic-augmented model fits the experimental data well.

17.6.6   Discussion

The fuzzy logic-based approach introduced a number of new parameters that have to be determined by data fitting. This task was made possible by using a hybrid of GA and simplex optimization. GA allows parallel search of a large parameter space without being trapped in a local minimum. The simplex method, on the other hand, greatly accelerates the convergence to a local minimum. By a proper mix of the two approaches, one can quickly converge into a local minimum and simultaneously search for other local valleys.

This case study also demonstrated a novel integration of four techniques for modeling complex systems: fuzzy logic-based modeling, mathematical modeling, the genetic algorithm, and the simplex method. The former two techniques were combined to achieve a more flexible model structure that can incorporate qualitative information explicitly into enzyme kinetics. The later two techniques were combined to obtain an efficient model parameter estimator that can avoid being easily trapped in local optima. Together, these four complementary techniques offer a promising approach for modeling complex systems when knowledge about them is incomplete.

FIGURE 17.19   Data and Model Prediction in the PYK Reaction

(From "A Hybrid Genetic Algorithm and Fuzzy Logic for Metabolic Modeling." J. Yen, B. Lee, and J. C. Liao. *Proc. of the 13th National Conf. on Artificial Intelligence (AAAI'96)*, pp. 743-749. © 1996 American Association for Artificial Intelligence.)

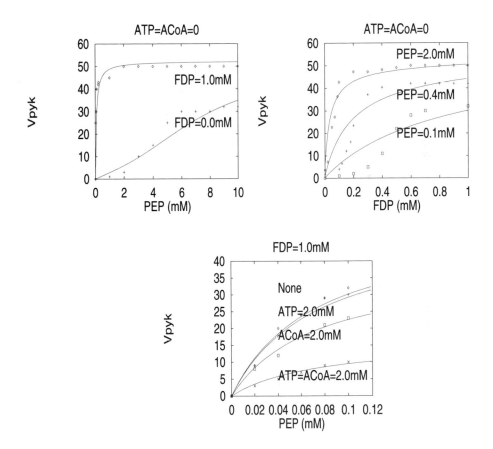

17.7 Industrial Applications

P. Bonissone, P. Khedkar, and Y. Chen at GE Corporate Research and Development developed a GA-tuned fuzzy controllers for freight train cruise control [67]. The objective of the tuning is for the controller to follow a prescribed velocity profile for a rail-based transportation system. This requires following a desired trajectory of speed, rather than maintaining speed at a fixed point in automobile cruise control. Other design objectives of the controller include providing a smooth ride and staying within the prescribed speed limits. A genetic algorithm is used to tune the fuzzy controller's performance by adjusting its parameters (the scaling factors and the membership functions) in a sequential order of significance. This approach resulted in a controller that is superior to the manually designed one, and with only modest computational effort. This makes it possible to customize automated tuning to a variety of different configurations of the route, the terrain, the power configuration, and the cargo.

J. Kingdon reviewed the application of neural networks, genetic algorithms, and fuzzy logic in fraud detection [318].

17.8 Summary

We summarize the main concepts introduced in this chapter below.

- The basics of Genetic Algorithms (GA) involve three major steps: fitness evaluation, selection, and reproduction.
- Two major kinds of reproduction operations: crossover and mutation.
- Two types of GA: binary GA and real-coded GA. They differ not only in the representation of chromosomes but also in the reproduction operators.
- Major issues in designing GA include (1) the choice of encoding scheme, (2) the choice of selection probability, (3) the design of fitness function, and (4) the choice of mutation and crossover probability.
- The slow convergence rate of GA can be improved by combining GA with other local search techniques into a hybrid genetic algorithm.
- There are four types of hybrid GA architectures for combining GA with a local search technique: pipelining hybrids, asynchronous hybrids, hierarchical hybrids, and additional operators.
- The simplex function optimization technique can be effectively combined with GA to form a hybrid optimization approach.
- We discussed several guidelines regarding the application of GA to fuzzy model identification.
- We illustrated GA-based fuzzy model identification using an application in metabolic modeling.

Bibliographical and Historical Notes

C. L. Karr applied GA to the parameter optimization of fuzzy logic controllers in the late 80s [303, 300]. His work was soon followed by P. Thrift, I. Hayashi, M. Lee, H. Takagi, R. Kruse, T. Fukuda, T. Shibata, and others [589, 452, 365, 302, 320, 195, 258]. In addition to assisting the design of fuzzy logic controllers, GA has also been applied by H. Ishibuchi et al. to the optimization of fuzzy rule-based models for classification problems [268].

Classifier systems are evolutionary computation schemes that modify rules (i.e., classifiers) based on external payoffs and a learning technique called the bucket brigade algorithm [68]. M. Valenzuela-Rendon proposed a fuzzy classifier system by introducing fuzzy rules and fuzzy inference into classifier systems [609].

M. Lee and H. Takagi explored the other direction for combining GA and fuzzy logic — using fuzzy rules to dynamically adjust GA parameters (e.g., population size, mutation probability) by analyzing the performance of GA [366]. This approach can potentially simplify the task for developing GA applications by capturing the knowledge of experienced GA designers using fuzzy rules.

In addition to optimizing fuzzy rule-based models, GA has also been applied to other problems in fuzzy logic. H. Ishibuchi et al. applied GA to solve fuzzy flowshop scheduling problems involving fuzzy due dates [269]. J. Bezdek and R. Hathaway proposed a fuzzy clustering approach by using GA to optimize the objective function of the fuzzy c-means algorithm [48]. D. Kraft, F. Petry et al. applied GA to query optimization in fuzzy information retrieval systems [339].

The case study that used the hybrid GA to optimize fuzzy rules for metabolic modeling was reported in [674]. Further details about the simplex-GA hybrid and its comparison with several other optimization techniques (including adaptive simulated annealing, G-bit improvement hybrid, and an alternative simplex-GA hybrid) can be found in [675].

Exercises

17.1   Suppose we want to use binary GA to optimize a parameter in the range [0, 200] with a precision up to one digit after the decimal point. How many bits would you use to encode the parameter?

17.2   Discuss the relationship between dynamic mutation operation in real-coded GA and simulated annealing? How are they similar? How do they differ?

17.3   Consider an implementation of the simplex-GA hybrid discussed in Section 17.4. Suppose we use a population of 50 for optimizing consequent parameters in a TSK model with two inputs and six rules. What is the minimal percentage of the simplex reproduction (i.e., what is the lowest $|S|/|P|$ ratio in Fig for this problem)?

17.4   Suppose you wish to use GA to partition a given dataset into c fuzzy clusters. The objective function of FCM has been chosen as the fitness function. Consider the following three encoding schemes being proposed:
(1) Each chromosome encodes c cluster centers.
(2) Each chromosome encodes $n \times c$ membership values.
(3) Each chromosome encodes c cluster centers and $n \times c$ membership values.
Which scheme will you choose? Describe the rationale of your choice.

17.5   Consider a fuzzy modeling problem for a two-inputs (x and y) one-output (z) system. Suppose the input variables x and y are each decomposed using five triangular membership functions (denoted A_i and B_j respectively). Design a scheme for encoding the antecedent membership parameters such that the following sum-to-one constraint is satisfied:

$$\sum_{i=1}^{5} \mu_{A_i}(x) = 1 \qquad\qquad\qquad \text{(EQ 17.11)}$$

$$\sum_{j=1}^{5} \mu_{B_j}(x) = 1 \qquad\qquad\qquad \text{(EQ 17.12)}$$

17.6   Given an optimization problem, does GA guarantee to find a global solution? Does GA guarantee to find a solution at least as good as a local search technique?

17.7   Compare the dynamic mutation operator of real-coded GA and simulated annealing. Discuss their similarities and differences?

17.8   Under what situations would you prefer GA to NN for the identification of fuzzy models? Under what situations would you prefer NN to GA?

17.9   For the simplex-GA hybrid architecture, what problems could one encounter if the percentage of the simplex operation is too high? What if the percentage is too low?

REFERENCES

[1]   L. Acar and U. Ozgüner. Design of knowledge-rich hierarchical controllers for large functional systems. *IEEE Trans. on Systems, Man and Cybernetics,* Vol. 20, pp. 791-803, 1990.

[2]   M. J. Achs and D. Garfinkel. Computer simulation of rat heart metabolism after adding glucose to the perfustate. *American Journal Physiology,* Vol. 232, pp. 175-184, 1977.

[3]   D. H. Ackley. Stochastic iterated. gentic hillclimbing. Ph.D. thesis, Carnegie Mellon University, 1987.

[4]   K. Adlassing. Fuzzy set theory in medical diagnosis. *IEEE Trans. Systems, Man and Cybernetics,* Vol. SMC-16, No. 2, 1986.

[5]   K. Adlassnig and G. Kolarz. Representation and semiautomatic acquisition of medical knowledge in CADIAG-1 and CADIAG-2. *Comp. Bio. Med. Research,* Vol. 19, pp. 63-79, 1986.

[6]   K. Adlassnig, G. Kolarz, W. Scheithauer, H. Effenberger, and G. Grabner. CADIAG: Approaches to computer-assisted medical diagnosis. *Computer Bio. Med.,* Vol. 15, No. 5, pp. 315-335, 1985.

[7]   K. Adlassing, G. Kolarz, and W. Scheithauer. Present state of the medical expert system CADIAG-2. *Methods of Inf. in Medicine,* 1985.

[8]   H. Akaike. A new look at statistical model identification. *IEEE Transactions on Automatic Control,* Vol. 19, pp. 716-723, 1974.

[9]   J. S. Albus. Outline of a theory of intelligence. *IEEE Transactions on Systems, Man, and Cybernetics,* Vol. 21, No. 3, 1991.

[10]   R. W. Albright Jr. and E. K. Fram. Microcomputer-based techniques for 3-D reconstruction and volume Measurement of computed tomographic images, In Part I: Phantom studies. *Investigative Radiology,* Vol. 23, No. 12, pp. 881-885, 1988.

[11]   C. von Altrock. Fuzzy logic applications in Europe. In *Industrial Applications of Fuzzy logic and Intelligent Systems.* (ed. J. Yen, R. Langari, L. A. Zadeh), pp. 277-310 IEEE Press, 1995.

[12]   J. Van Amerongen, H. R. van Nauta Lemke, and J. C. T. van der Veen. An autopilot for ships designed with fuzzy sets. *Proceedings of 5th IFAC Int. Conf. on Digital Computer Applications to Process Control,* 1977.

[13]   T. Anderson and M. Donath. Synthesis of reflexive behavior for a mobile robot based upon a stimulus-response paradigm. *Proc. of the SPIE,* Vol. 1007, 1988.

[14]   D. Angluin and C. Smith. Inductive inference: theory and methods. *ACM Computing Surveys,* Vol. 15, pp. 237-269, 1984.

[15]   J. Aragones and P. P. Bonissone. PRIMO: A tool for reasoning with incomplete and uncertain information. *Proc. 3rd. Int. Conf. Information Processing and Management of Uncertainity in Knowledge-Based Systems,* pp. 891-898, 1990.

[16]   L. K. Arata, A. P. Dhawan, J. P. Broderick, M. F. Gaskil-Shipley, A. V. Levy and N. N. D. Volkow. Three-dimensional anatomical model-based segmentation of MR brain images through principle axes registration. *IEEE Transactions on Biomedical Engineering,* Vol. 42, No. 11, pp. 1069-1078, November 1995.

[17]   B. Babuska, M. Setnes, U. Kaymak, and H. R. van Nauta Lemke. Rule base simplification with similarity measures. *Proceedings of the Fifth IEEE International Conference on Fuzzy Systems,* New Orleans, pp. 1642-1647, 1996.

[18]   E. Backer and A. K. Jain. A clustering performance measure based on fuzzy set decomposition. *IEEE Trans. Pattern Anal. Machine Intel,* Vol. PAMI-3, No. 1, pp. 66-75, 1981.

[19]   G. Balas, J. Doyle, K. Glover, A. Packard, and R. Smith. μ *-Analysis and Synthesis ToolBox.* Mathworks, Inc. 1991.

[20]   J. F. Baldwin and T. P. Martin. Fuzzy classes in object oriented logic programming. *Proc. IEEE Fuzzy Systems,* pp.1358-1364, 1996.

[21]   J. F. Baldwin, T. P. Martin, and B. W. Philsworth. *FRIL — Fuzzy and Evidential Reasoning in Artificial Intelligence.* John Wiley & Sons, NewYork, 1995.

[22]   J. F. Baldwin. A fuzzy relational inference language for expert systems. *Proc. of 13th. IEEE International Symposuim on Multivalued Logic,* pp. 416-423, 1983.

[23]   J. F. Baldwin and B. W. Pilsworth. Axiomatic approach to implication for approximate reasoning with fuzzy logic. *Fuzzy Sets and Systems,* Vol. 3, pp. 193-219, 1980.

[24]   W. Bandler and L. Kohout. Fuzzy power sets and fuzzy implication operators. *Fuzzy Sets and Systems,* Vol. 4, pp. 183-190, 1980.

[25]   V. E. Barker and D. E. O'Connor. Expert systems for configuration at Digital: XCON and beyond. *Communications of the ACM,* Vol. 32, No. 3, 1989.

[26] E. Barnard. Optimization for training neural nets. *IEEE Transactions on Neural Networks*, vol. 3, pp. 232-240, 1992.

[27] A. R. Barron and R. L. Barron, Statistical learning networks: a unifying view. *Proceedings of the Twentieth Symposium on Interface* (ed. Ed Wegman). Washington, DC, pp. 192-203, 1988.

[28] A. R. Barron. Predicted squared error: a criterion for automatic model selection. *Self-Organizing Methods in Modeling GMDH Type Algorithms* (ed. S. J. Farlow). Marcel Dekker, pp. 87-103, 1984.

[29] A. G. Barto, R. S. Sutton, and C. W. Anderson. Neuron-like adaptive elements that can solve difficult learning control problems. *IEEE Transactions on Systems, Man, and Cybernetics*, Vol. 13, pp. 834-846, 1983.

[30] J. Bates. The role of emotion in believable agents. *Communications of ACM,* Vol. 37, No. 7, pp. 122-125, 1992.

[31] R. Battiti. First- and second-order methods for learning: Between steepest descent and Newton's method. *Neural Computation*, Vol. 4, pp. 141-166, 1992.

[32] P. Belanger. *Control Engineering*. Saunders College Publishing, 1995.

[33] R. E. Bellman and L. A. Zadeh. Decision making in a fuzzy environment. *Management Science*, Vol. 17, pp. 141-164, 1970.

[34] D. Benachenhou. Smart trading with FRET. In *Trading on the Edge* (ed. G. J. Deboeck). pp. 207-214 John Wiley & Sons, 1994.

[35] H. R. Berenji. Machine learning in fuzzy control. In *Proceedings of the International Conference on Fuzzy Logic & Neural Networks*, pp. 231–234, 1990.

[36] H. R. Berenji and P. S. Khedkar. Clustering in product space for fuzzy inference. *IEEE Transactions on Fuzzy Systems,* 1996.

[37] H. R. Berenji, B. Chandrasekaran, D. Dubois, H. Prade, P. Smets, G. J Klir, E. Ruspini, L. A. Zadeh et al. Response to Elkan. *IEEE Expert,* Vol. 9, No. 4, pp. 9-49, August 1994.

[38] H.R. Berenji and P. Khedkar. Learning and tuning fuzzy logic controllers through reinforcements. *IEEE Transactions on Neural Networks*, 1992 .

[39] J. C. Bezdek. Fuzziness vs. Probability Again !? *IEEE Transactions on Fuzzy Systems,* Vol. 2, No. 1, pp. 1-3, February 1994.

[40] J. C. Bezdek. A convergence theorem for the Fuzzy ISODATA clustering algorithms. *IEEE Trans. Pattern Anal. Machine Intel.,* Vol. PAMI-2, No. 1, pp. 1-8, 1990.

[41] J. C. Bezdek. A physical interpretation of fuzzy ISODATA. *IEEE Trans. on System, Man, and Cybernetics,* Vol. 6, pp. 387-390, 1976.

[42] J. C. Bezdek. Cluster validity with fuzzy sets. *J. Cyber.*, Vol. 3, No. 3, pp. 58-72, 1974.

[43] J. C. Bezdek. Fuzzy mathematics in pattern classification Ph.D. thesis, Center for Applied Mathematics, Cornell University, Ithaca, 1973.

[44] J. C. Bezdek and S. Pal(eds.). *Fuzzy Models for Pattern Recognition.* IEEE Press, 1991.

[45]   J. C. Bezdek, C. Coray, R. Gunderson, and J. Watson. Detection and characterization of cluster substructure: II. Fuzzy c-varieties and convex combination thereof. *SIAM J. Appl. Math.,* Vol. 40, No. 2, pp. 358-372, 1981.

[46]   J. C. Bezdek, C. Coray, R. Gunderson, and J. Watson. Detection and characterization of cluster substructure: I. Linear structure: Fuzzy c-lines. *SIAM J. Appl. Math.,* Vol. 40, No. 2, pp. 339-357, 1981.

[47]   J. C. Bezdek and J. C. Dunn. Optimal fuzzy partitions: Aheuristic for estimating the parameters in a mixture of normal distributions, *IEEE Trans. Comp.,* C - 24, pp. 835-838, 1975.

[48]   J. C. Bezdek and R. J. Hathaway, Optimization of fuzzy clustering criteria using genetic algorithms. In *Proceedings of the 1st IEEE Conf. on Evolutionary Computation*, Orlando, FL, pp. 589- 594, June 1994.

[49]   J. C. Bezdek and R. J. Hathway. Convergence Theory for fuzzy c-means: Counterexamples and repairs. *IEEE Trans. Syst., Man, Cybern.,* Vol. SMC-17, No. 5, pp. 873-877, 1987.

[50]   J. C. Bezdek, L. Hall, L. P. Clarck. et. al. Segmentating medical images with fuzzy models: An update. *Fuzzy Information Engineering* (ed. D. Dubois, H. Prade and R. R. Yager). pp. 69-92 , 1997.

[51]   J. C. Bezdek, L. Hall, and L. P. Clark. Review of MR image segmentation techniques teniques using pattern recognition. *Medical Physics*, Vol. 20, pp 1033-1048, 1993.

[52]   J. C. Bezdek and J. D. Harris. Convex decompositions of fuzzy partitions. *Journal of Mathematical ANalysis and Applications,* Vol. 67, pp. 490-512, 1979.

[53]   S. A. Billings and W. S. F. Voon. Correlation based model validity tests for non-linear models. *International Journal of Control,* Vol. 44, No. 1, pp. 235-244, 1986.

[54]   Z. Binder, H. Fontainer, M. F. Magalhaes, and D. Baudois, About a multimodel control methodology, algorithms, multiprocessors implementation and application. *Proceedings of the Eighth IFAC Triennial World Congress*, pp. 981-986, 1981.

[55]   I. Bolch and H. Maitre. Fuzzy mathematical morphologies a comparative study. *Pattern Recognition,* Vol. 28, No. 9, pp. 1341-1387, 1995.

[56]   I. Bloch and H. Maire. Fuzzy mathematical morphology. *Annals of Mathematics and Artificial Intelligence,* Vol. 10, pp. 55-84, 1994.

[57]   D.I. Blockley. *The Nature of Structural Design and Safety*. Ellis Horwood Ltd., Chichester, 1980.

[58]   D.I. Blockley. The role of fuzzy sets in civil engineering. *Fuzzy Sets and Systems*, Vol. 2, 1979.

[59]   D.I. Blockley. Analysis of structural failures. *Proceeding of the Instituion of Civil Engineers*, Vol. 62, no. 1, 1977.

[60]   D.I. Blockley. Predicting the likelihood of structural accidents. *Proceeding of the Instituion of Civil Engineers*, London, Part 2, Vol. 59, 1975.

[61]   R. C. Bolles and M. C. Fanselow. A preceptuial defensive recuperative model of fear and pain. *Behavioral and Brain Sciences,* Vol. 3, pp. 291-301, 1980.

[62]   D. C. Bonar, K. A. Schaper, J. R. Anderson, D. A. Rottenberg, and S. C. Strother. Graphical analysis of MR feature space for measurement of CSF, grey-matter, and white-matter volumes. *Journal of Computer Assisted Tomography*, Vol. 17, No. 3, pp 461-470, 1993.

[63]   P. Bonissone. Summarization and Propagating Uncertain Information with Triangular Norms. *International Journal of Approximate Reasoning,* Vol. 1, No. 1, pp. 71-101, 1987.

[64]   P. Bonissone and K. H. Chiang. Fuzzy logic hierarchical controller for a recuperative engine: From mode selection to mode melding. In *Industrial Applications of Fuzzy Control and Intelligent Systems.* (ed. J. Yen, R. Langari, and L. A. Zadeh). IEEE Press, 1995.

[65]   P. Bonissone, S. Dutta and N. Wood. Merging strtegic and tactical planning in synamic uncetrain environments. *IEEE Trans. on Systems, Man and Cybernetics,* Vol. 24, pp. 841-862, 1994.

[66]   P. Bonissone, S. Gans, and K. S. Decker. RUM: A layered architecture for reasoning with uncertainty. *Proc. 10th. Int. Joint Conf. Artificial Intelligence,* pp. 891-898, 1987.

[67]   P. Bonissone, P. S. Khedkar, Y. Chen. Genetic algorithms for automated tuning of fuzzy controllers: A transportation application. *Fifth IEEE Int. Conf. on Fuzzy Systems,* Vol. 1, pp. 674-680, Sept. 1996.

[68]   L. B. Booker, D. E. Goldberg, and J. H. Holland, Classifier systems and genetic algorithms. *Artificial Intelligence*, Vol. 40, pp. 235-282, 1989.

[69]   C. de Boor. *A Practical Guide to Splines.* Springer-Verlag, 1978.

[70]   C. de Boor. On calculating with B-splines. *Journal of Approximation Theory*, Vol. 6, pp. 50-62, 1972.

[71]   G. Bordonga, D. Lucarella, and G. Pasi. A fuzzy object oriented data model. *Proc. of the 3rd. IEEE Int. Conf. on Fuzzy Systems,* pp. 313-318, 1994.

[72]   G. Bortolan and R. Degani. A review of some methods for ranking fuzzy subsets. *Fuzzy Sets and Systems,* Vol. 15, pp. 1-19, 1985.

[73]   P. Bosc. Subqueries in SQLf, a fuzzy database query language. *IEEE Systems, Man and Cybernetics Intelligent Systems for 21st Century,* pp. 3636-3641, 1995.

[74]   P. Bosc, M. Galibourg, and G. Hamon. Fuzzy querying with SQL: Extensions and implementation aspects. *Fuzzy Sets and Systems*, Vol. 28, pp. 333-349, 1988.

[75]   P. Bosc and O. Pivert. SQLf: A relational database language for fuzzy querying. *IEEE Trans. on Fuzzy Systems,* Vol. 3, No. 1, 1995.

[76]   L. Bottou and V. Vapnik. Local learning algorithms. *Neural Computation*, Vol. 4, pp. 888-900, 1992.

[77]   B. Bouchon. Fuzzy inference and conditional possibility distributions. *Fuzzy Sets and Systems*, Vol. 23, pp. 33-41, 1987.

[78]   G. E. P. Box and G. M. Jenkins, *Time Series Analysis: Forecasting and Control*, 2nd Ed. Holden-Day, 1976.

[79]   M. Braae and D. A. Rutherford. Theoretical and linguistic aspects of the fuzzy logic controller. *Automatica*, 15, No. 5, 1979.

[80]   M. Braae and D.A. Rutherford. Selection of parameters from a fuzzy logic controller. *Fuzzy Sets and Systems*, Vol. 2, pp. 185-199, 1979.

[81]   R. J. Brachman. What IS-A is and isn't: An analysis of taxonomic links in semantic networks. *IEEE Computer,* Vol. 16, pp. 30-37, 1983.

[82]   R. J. Brachman and J. G. Schmolze. An overview of the KL-ONE knowledge representation system. *Cognitive Science,* Vol. 9, No. 2, pp. 171-216, Aug. 1985.

[83]   J. M. Bradshow. *Software Agents.* AAAI Press, 1997.

[84]   M. M. E. Brandt, J. M. Fletcher, and T. P. Bohan. Estimation of CSF, white, and grey matter volumes from MRIs of hydrocephalic and HIV-positive subjects. *Proceedings of the SimTec and WNN 92*, pp. 643-650, 1992.

[85]   R. Brooks. A robust layered control system for a mobile robot. *Readings in Uncertain Reasoning. (*eds. *G. Shafer, J. Preal).* Morgan Kaufmann, pp. 204-213, 1990.

[86]   D. S. Broomhead and D. Lowe, Multivariable functional interpolation and adaptive networks. *Complex Systems*, Vol. 2, pp. 321-355, 1988.

[87]   C. B. Brown. A fuzzy safety measure, *Journal of the Engineering Mechanics Division.* ASCE, 106, EM4, 1980.

[88]   C. B. Brown. Entropy constructed probabilities. *Journal of the Engineering Mechanics Division.* ASCE, 106 EM4, 1980.

[89]   M. Brown and C. Harris. *Neurofuzzy Adaptive Modeling and Control.* Prentice Hall, 1994

[90]   C.B. Brown and R. S. Leonard. Subjective uncertainity analysis. presented at the ASCE National Structural Engineering Meeting, Baltimore, MD. Preprint No. 1388, pp. 19-23, April 1971.

[91]   J. Brown, R. Lia and D. Bahler. Fuzzy semantics and fuzzy constraint networks. *Proc. of the National Conf. on Artificial Intelligence,* pp. 1009-1016, 1992.

[92]   J. Browne, J. Harhen, and J. Shivnan. *Production Management Systems: A CIM Perspective.* Addison Wesley, 1988.

[93]   B. G. Buchanan and E. H. Shortliffe. *Rule-based Expert Systems — The MYCIN Experiment of the Stanford Heuristic Programming Project.* Addison-Wesley: MA, 1984.

[94]   J. J. Buckley. Sugeno type controllers are universal controllers. *Fuzzy Sets and Systems*, Vol. 53, pp. 299-303, 1993.

[95]   J. J. Buckley. The fuzzy mathematics of finance. *Fuzzy Sets and Systems*, Vol. 21, pp. 257-273, 1987.

[96]   B.P. Buckles and F.E. Petry. A fuzzy representation of data for relational databases. *Fuzzy Sets and Systems*, Vol. 7, pp. 213-226, 1982.

[97]   J. J. Buckley and H. Ying. Fuzzy controller theory: Limit theorems for linear fuzzy control rules. *Automatica,* Vol. 25, No. 3, 1989.

[98] B. P. Buckles, F. Petry, and J. Pillai. Network data models for representation of uncertainity. *Int. Journal of Fuzzy Sets and Systems,* Vol. 38, pp. 171-190, 1990.

[99] B. Buckles, H. Petry and H. J. Sachar. Functional dependency properties of fuzzy relational databases. *First International Fuzzy Systems Association Congress Abstracts,* 1985.

[100] B. Buckles, F. Petry, and H. J. Schar. Design of similarity-based relational databases. In *Fuzzy Logics in Knowledge Engineering* (eds. C. Negoita, H. Prade), pp. 1-17. North-Holland, 1986.

[101] J. J. Buckley, W. Siler, and D. M. Tucker. A fuzzy expert system. *Fuzzy Sets and Systems,* Vol. 20, pp. 1-16, 1986.

[102] R. E. Carnap. *Logical Foundations of Probability.* University of Chicago Press, 1950.

[103] D. Cayrac, D. Dubois, M. Haziza, and H. Prade. Possibility theory in fault mode effect analysis — A satellite fault diagnosis application. *Proc. of the 3rd. IEEE Int. COnf. on FUzzy Systems,* pp. 1176-1181, 1994.

[104] S. Chanas. Fuzzy sets in few classical operational research problems. In MM. Gupta and E. Sanchez eds., *Approximate Reasoning in Decision Analaysis*, North-Holland, pp. 351-363, 1982.

[105] S. Chand and S. Chiu. Robustness analysis of fuzzy control systems with application to aircraft roll control. *Proc. of the AIAA Guidance, Navigation, and Control Conf.,* 1991.

[106] C. K. Chang. On the execution of fuzzy programs using finite state machines. *IEEE Trans. Comp., C-12,* pp. 214-253, 1972.

[107] S. K. Chang and J. S. Ke. Database skeleton and its application to fuzzy query translation. *IEEE Trans. Software Eng. SE,* Vol. 4, pp. 31-43, 1978.

[108] C. L. Chang. Fuzzy topological spaces. *J. Math. Anal. & Appl.,* Vol. 24, pp. 182-190, 1968.

[109] C. L. Chang and R. C. T. Lee. *Symbolic Logic and Mechanical Theorem Proving.* Academic Press, 1972.

[110] K. S. Chang and J. S. Ke. Translation of fuzzy queries for relational database systems. *IEEE Trans. Patt. Anal. Mach. Intell,* pp. 281-294, 1979.

[111] S. L. Chang. Fuzzy mathematics, man and his environment. *IEEE Trans on Systems, Man, and Cybernetics,* pp. 93-93, 1972.

[112] S. S. L. Chang and L. A. Zadeh. Fuzzy mapping and control. *IEEE Trans. on Systems, Man, and Cybernetics,* Vol. 2, No. 1, pp. 30-35, 1972.

[113] C. Charalambous. Conjugate gradient algorithm for efficient training of artificial neural networks. *IEEE Proceedings*, Vol. 139, pp. 301-310, 1992.

[114] P. Cheeseman. An inquiry into computer understanding. *Comp. Intell.,* Vol. 4, pp. 57-142, 1988.

[115] P. Cheeseman. In defense of probability. *Proc. 9th. Int. Joint Conf. Art. Intell.,* Los Angeles, pp. 1002-1009, 1985.

[116] B. Chen, Y. Zhang, J. Yen, and W. Zhoa. Fuzzy adaptive connection admission control for real-time applications in ATM-based heterogeneous networks. *J. of Intelligent & Fuzzy Systems,* 1997.

[117]   C. T. Chen, Y. J. Chen, and C. C. Teng. Simplification of fuzzy-neural systems using similarity analysis, *IEEE Transactions on Systems, Man, and Cybernetics*, Vol. 26, pp. 344-354, 1996.

[118]   P. Chen. An algebra for a directional binary entity-relationship model. *IEEE Proc. of COMPDEC*, pp. 37-40, 1984.

[119]   P. Chen. The entity-relationship model: Toward a unified view of data. *ACM Trans. on Database Systems*, Vol. 1, pp. 9-36, 1976.

[120]   Y. Chen, K. Lin, and S. Hsu. A self-learning fuzzy controller. *IEEE International Conference on Fuzzy Systems*, 1992.

[121]   Y. T. Chen and Y. C. Shin. A surface grinding advisory system based on fuzzy logic. In *Computer Control of Manufacturing Processes, DSC-Vol 28*, ASME, 1991.

[122]   Y. Y. Chen. The global analysis of fuzzy dynamic systems. Ph.D. thesis, University of California, Berkeley, 1989.

[123]   V. Cherkassky and F. Mulier, Self-organizing networks for nonparametric regression. In *From Statistics to Neural Networks: Theory and Pattern Recognition Applications*. (ed. V. Cherkassky, J. H. Friedman, and H. Wechsler). Springer-Verlag, pp. 189-212, 1994.

[124]   V. Cherkassky, D. Gehring, and F. Mulier. Comparison of adaptive methods for function estimation from samples, *IEEE Transactions on Neural Networks*, Vol. 7, pp. 969-984, 1996.

[125]   C. Chinrungrueng and C. H. Sequin. Optimal adaptive K-means algorithm with dynamic adjustment of learning rate, *IEEE Transactions on Neural Networks*, Vol. 6, pp. 157-169, 1995.

[126]   S. L. Chiu. Fuzzy model identification based on cluster estimation. *Journal of Intelligent and Fuzzy Systems*, Vol. 2, pp. 267-278, 1994.

[127]   S. L. Chiu. Selecting input variables for fuzzy models. *Journal of Intelligent and Fuzzy Systems*, Vol. 4, No. 4, pp. 243-256, 1996.

[128]   S. L. Chui and S. Chand. Fuzzy controller design and stability analysis for an aircraft model. in *Proc. of the American Control Conf.*, Boston, MA, pp. 821-826, June 1991.

[129]   S. L. Chui and M. Togai. A fuzzy logic programming environment for real-time control. *Int. Journal of Approximate Reasoning*, Vol. 2, pp. 163-175, 1988.

[130]   H. S. Choi, D. R. Haynor, and Y. Kim. Partial volume tissue classification of multichannel magnetic resonance images — A mixed model. In *IEEE Transactions on Medical Imaging*, Vol. 10, pp 395-407, 1991.

[131]   W. J. Clancey. Heuristic Classification. *Artificial Intelligence*, Vol. 27, pp. 289-350, 1985.

[132]   M. Clark et al. MRI segmentation using fuzzy clustering techniques: integrating knowledge. In *IEEE Engineering in Medicine and Boilogy Magazine*, Vol. 13, No. 5, pp. 730-742, 1994.

[133]   W. S. Cleveland. Robust locally weighted regression and smoothing scatterplots. *Journal of the American Statistical Association*, Vol. 74, pp. 829-836, 1979.

[134]   W. S. Cleveland and S. J. Devlin. Locally weighted regression: An approach to regression analysis by local fitting, *Journal of the American Statistical Association*, Vol. 83, pp. 596-610, 1988.

[135]   H. E. Cline, W. E. Lorensen, R. Kikinis, and F. Jolsez. Three-dimentional segmentation of MR images of the head using probability and connectivity. *Journal of Computing Assistant Tomography*, Vol. 14, pp. 1037-1045, 1990.

[136]   E. F. Codd. Further normalization of the database relational model. In *Database Systems, Courant Computer Science Symposis 6*. (ed. R. Rustin). Prentice Hall, pp. 65-98, 1971.

[137]   E.F. Codd. A relational model of data for large shared data banks. *Comm. ACM,* Vol. 13, No. 6, pp. 377-387, June 1970.

[138]   V. Cross and T. Sudkamp. Patterns of rule-based inference. *International Journal of Approximate Reasoning*, Vol. 11, No. 3, pp. 235-255, October 1994.

[139]   V. Cross and T. Sudkmap. Representation and support generation in fuzzy relational database. *Proc. IEEE Aerospace and Electronics Conf.,* pp. 1136-1143, 1991.

[140]   M.G.Cox. The numerical evaluation of B-splines. *J. Inst. Maths. Applics*, Vol. 10, 1972, pp. 134-139.

[141]   A. Cumani. On a possibilistic approach to the analysis of fuzzy feedback systems. *IEEE Transactions on Systems, Man, and Cybernetics*, Vol. SMC-12, No. 3, May/June 1982.

[142]   E. Czogala and W. Pedrycz. On the concept of fuzzy probabilistic controllers. *Fuzzy Sets and Systems*, Vol. 10, pp. 109-121, 1983.

[143]   E. Czogala and W. Pedrycz. Control problems in fuzzy systems. *Fuzzy Sets and Systems,* Vol. 7, No. 3, 1982.

[144]   E. Czogala and W. Pedrycz. Fuzzy rule generation for fuzzy control. *Cybernetics and Systems,* Vol. 13, No. 3, 1982.

[145]   S. Daley and K. F. Gill. A study of fuzzy logic controller robustness using the parameter plane. *Computers in Industry*, Vol. 7, pp 511-522, 1986.

[146]   A. R. Damasio. *Descartes' Error: Emotion, Reason, and the Human Brain.* G. P. Putnam, 1994.

[147]   R. W. Daniels. *An Introduction to Numerical Methods and Optimization Techniques.* North-Holland, 1978.

[148]   W. Daugherity, B. Rathakrishnan, and J. Yen. Performance evaluation of a self-tuning fuzzy controller. *Proceedings of FUZZ-IEEE Conference,* 1992.

[149]   R. N. Dave and K. Bhaswan. Adaptive fuzzy C-shells clustering and detection of ellipses. *IEEE Trans. Neural Networks,* Vol. 3, No. 5, pp. 643-662, 1992.

[150]   T. Dean, R. Firby, and D. Miller. Hierarchical planning involving deadlines, travel time and resources. *Computational Intelligence,* Vol. 4, pp. 381-398, 1988.

[151]   K. A. Dejong. Analysis of the behaviour of a class of genetic adaptive systems. Ph.D. thesis, Department of Computer and Communication Sciences, University of Michigan, 1975.

[152]   M. Delgado and S. Moral. On the concept of possibility-probabilty consistency. *Fuzzy Sets and Systems*, Vol. 21, pp. 311-318,1987.

[153]   R. Descartes. *Passions of the Soul.* Hackett Pub. Co., 1989.

[154]   R. DeCaluwe, R. Vandenberghe, N. Van Gyseghem, and A. Van Schooten. Integrating fuzzyi-ness in database models. In *Fuzziness in Database Management Systems* (ed. P. Bosc and J. Kacprzyk), pp. 71-113, 1995.

[155]   Deveughele and B. Dubuisson. Using possibility theory in preception: An application. *Proc. of the 2nd IEEE Int. Conf. on Fuzzy Systems*, March 1993.

[156]   P. Dewilde and E. Deprettere. Singular value decomposition: An introduction. In E. Deprettere, ed., *SVD and Signal Processing: Algorithms, Applications and Architectures*, Elsevier Science Publishers, pp. 4-42, 1988.

[157]   A. P. Dhawan and S. Juvvadi. Knowledge-based analysis and understanding of medical images. *Computer Methods and Programs in Biomedicine,* Vol. 33, pp. 221-239, 1990.

[158]   L. Dorst and K. Travato. Optimal path planning by cost wave propagation in metric configura-tion space. *SPIE,* Vol. 1007, pp. 186-197, 1988.

[159]   G. Dozier, J. Bowen, and D. Bahler. Solving small and large scale constraint satisfaction prob-lems using a heuristic-based microgenetic algorithm. *Proc. of the First IEEE Conf. on Evolution-ary Computation,* Orlando, FL, June 1994.

[160]   N. R. Draper and H. Smith, *Applied Regression Analysis*, 2nd Ed., John Wiley & Sons, 1981.

[161]   M. E. Dreier. A fast, noniterative method to generate fuzzy inference rules from observed data, *Journal of Intelligent and Fuzzy Systems*, Vol. 3, pp. 181-185, 1995.

[162]   D. Driankov, H. Hellendorn, and M. Reinfrank. An *Introduction to Fuzzy Control*, Springer-Ver-lag, 1993.

[163]   D. Dubois. Quelques Outils Methodologiques Pour la Conception de Reseau de Transport. Doc-toral Dissertation, Centre d'Etudes et de Recherches de l'Ecole Nationale Superieure de l'Aero-nautictue et de l'Espace Toulouse, France, October 1977.

[164]   D. Dubois and H. Prade. Adaptive combination rules for possibility distributions. *EUFIT,* Aachen Germany, September 1994.

[165]   D. Dubois and H. Prade. Similarity-based approximate reasoning. *Computational Intelligence Imitating Life,* IEEE Press, pp. 69-80, 1994.

[166]   D. Dubois, H. Prade, and J. P. Rossazza. Vagueness, typicality and uncertainty in class hierar-chies. *Int. Journal of Intelligent Systems,* Vol. 6, pp. 167-183, 1991.

[167]   D. Dubois and H. Parade. Epistemic Entrenchment and Possibilistic Logic. *Artificial Intelli-gence*, Vol. 50, pp. 223-239, 1991.

[168]   D. Dubois and H. Prade. Resolution principles in possibilistic logic. *Int. J. Approx. Reasoning*, Vol. 4, No. 1, pp. 1-21, 1990.

[169]   D. Dubois and H. Prade. Gradual inference rules in approximate reasoning. *Tech. Report IRIT/ 90-6/R,* IRIT, Univ. P. Sabatier, Toulouse, France, 1990.

[170]   D. Dubois and H. Prade. Fuzzy sets in approximate reasoning — Part I: Inference with possibil-ity distributions. *Fuzzy Sets and Systems*, Vol. 25, 1990.

[171]   D. Dubois and H. Prade. Fuzzy rules in knowledge-based systems. *An introduction to fuzzy logic applications in intelligent systems* (ed. R. R. Yarger and L. A. Zadeh), Kluver Academic Publishers, pp. 46-67, 1990.

[172]   D. Dubois and H. Prade. An introduction to possibilistic and fuzzy logics (with discussions). In P. Smets, E. H. Mamdani, D. Dubois and H. Prade (ed.). *Non-Standard Logics for Automated Reasoning* (Academic Press, New York, 1988), pp. 287-326, reprinted in: G Shafer and J. Pearl (ed.). *Readings in Uncertain Reasoning*, pp. 742-761. Morgan Kaufmann, CA, 1990.

[173]   D. Dubois and h. Prade. Processing fuzzy temporal knowledge. *IEEE Trans. on Systems, Man, and Cybernetics,* Vol. 19, No. 4, pp. 729-744, 1989.

[174]   D. Dubois and H. Prade. Fuzzy Numbers: An Overview. In J.C. Bezdek ed., *Analysis of Fuzzy Information Vol. 1: Mathematics and Logic,* CRC Press, Boca Raton, FL., pp. 3-39, 1987.

[175]   D. Dubois and H. Prade. Necessity measures and the resolution principle. *IEEE Trans. on Systems, Man, and Cybernetics,* Vol. 17, pp. 474-478, 1987.

[176]   D. Dubois and H. Prade. *Prossibility Theory: An Approach to Computerized Processing of Uncertainity.* Plenum, 1988.

[177]   D. Dubois and H. Prade. Fuzzy logic and the generalized modus ponens revisited. *Cybernetics and Systems,* Vol. 15, pp. 293-331, 1984.

[178]   D. Dubois and H. Prade. *Fuzzy Sets and Systems—Theory and Applications.* Academic Press, 1980.

[179]   D. Dubois, J. Lang and H. Prade. Automated reasoning using possibilistic logic: semantics belief revision, and variable certainity weights. *IEEE Trans. on Knowledge and Data Engineering,* Vol. 6, No. 1, pp. 64-71, 1994.

[180]   D. Dubois, J. Lang and H. Prade. Theorem-proving under uncertianity - a possiblility theory-based approach. *Proc. IJCAI-87,* Italy, pp. 984-986, 1987.

[181]   J. C. Dunn. A fuzzy relative of the ISODATA process and its use in detecting compact well-separated clusters. *J. Cybernetics*, Vol. 3, pp. 32-57, 1973.

[182]   J. Durkin. *Expert Systems Design and Development.* Macmillan, 1994.

[183]   S. Dutta and P. P. Bonissone. Integrating case based and rule based reasoning. *Int. J. of Approximate Reasoning*, Vol. 8, pp. 163-203, 1993.

[184]   C. Elkan et al. The paradoxical success of fuzzy logic. *AAAI 93*, pp. 698-703, 1993.

[185]   C. Elkan et al. The paradoxical success of fuzzy logic (followed by 15 responses from L. A. Zadeh et al.), *IEEE Expert*, Vol. 9, No. 4, pp. 9-49, Aug. 1994.

[186]   S. Farinwata. Performance Assessment of Fuzzy Control Systems via Stability and Robustness Measures. Ph.D. Dissertation, Georgia Institute of Technology, 1993.

[187]   S. Farinwata and G. Vachtsevanos. Stability of the fuzzy logic controller. *Proceedings of the IEEE Conference on Decision and Control,* 1993.

[188]   S. J. Farlow. The GMDH algorithm. In S. J. Farlow, ed., *Self-Organizing Methods in Modeling GMDH Type Algorithms.* Marcel Dekker, pp. 1-24, 1984.

[189]   B. A. Foss and T. A. Johansen. On local and fuzzy modeling. In *Proceedings of the Third International Conference on Industrial Fuzzy Control and Intelligent Systems*, Houston, TX, pp. 80-87, December 1993

[190]   G. Franklin, D. Powell, and A. Emami-Naeini. *Feedback Control of Dynamic Systems*. 3rd Ed. Addison-Wesley, 1994.

[191]   J. H. Friedman and W. Stuetzle. Project pursuit regression. *Journal of the American Statistical Association*, Vol. 76, 1981, pp. 817-823.

[192]   K. S. Fu. *Learning Systems and Intelligent Robots*. Plenum, 1974.

[193]   K. S. Fu and J. K. Mui. A survey on image segmentation. *Pattern Recognition*, Vol. 13, pp. 3-16, 1981.

[194]   S. Fukami, M. Mizumoto, and K. Tanaka. Some considerations on fuzzy conditional inference. *Fuzzy Sets and Systems*, Vol. 4, pp. 243-273, 1980.

[195]   T. Fukuda and T. Shibata. Fuzzy-neuro-GA based intelligent robotics. In *Computational Intelligence Imitating Life* (ed. J. M. Zurada, R J. Marks II, C. J. Robinson). IEEE Press, pp. 352-363, 1994.

[196]   M. Funabashi et al. A fuzzy model-based control scheme and its application to a road tunnel ventilation system. *Proc. Int. Conf. Intdustrial Elec., Control, and Instruments,* pp. 1596-1601, 1991.

[197]   L. W. Fung and K. S. Fu. Characterization of a class of fuzzy optimal control problems. In M. M. Gupta, G. N. Saridis, and B. R. Gaines, eds. *Fuzzy Automata and Decision Processes*, North-Holland, New York, pp. 209-219, 1977.

[198]   B. R. Gains. Fuzzy and probability uncertainity logics. *Inform. and Control,* Vol. 38, pp. 154-169, 1978.

[199]   B. R. Gaines. System identification, approximation and complexity, *Int. J. General Syst.*, 3, 1976.

[200]   B. R. Gaines. Foundations of fuzzy reasoning. *Int. J. Man-Machine Studies*, Vol. 8, 1976.

[201]   B. R. Gaines and L. J. Kohout. The fuzzy decade : A bibliography of fuzzy systems and closely related topics. In M. M. Gupta, G. N. Saridis and B. R. Gaines, eds. *Fuzzy Automata and Decision Processes* North-Holland, pp. 403-490, 1977.

[202]   H. Gardner. *Frames of Mind*. Basic Books, 1993.

[203]   T. Gasser, J. Engel, and B. Seifert, Nonparametric function estimation. In *Handbook of Statistics*, Vol. 9, (ed. C. R. Rao), North-Holland, pp. 423-465, 1993.

[204]   I. Gath and A. B. Geva. Unsupervised optimal fuzzy clustering. *IEEE Trans. Pattern Anal. Machine Intell.,* Vol. PAMI-11, No. 7, pp. 773-781, 1989.

[205]   A. E. Gegov and P. M. Frank. Hierarchical fuzzy control of multivariable systcms. *Fuzzy Sets and Systems*, Vol. 72, pp. 299-310, 1995.

[206]   R. George, B. P. Buckles, and F. E. Petry. Modelling class hierarchies in the fuzzy object-oriented data model. *Fuzzy Sets and Systems,* Vol. 60, pp. 259-272, 1993.

[207]   R. Giles. The concept of grade of membership. *Fuzzy Sets and Systems*, Vol. 25, pp. 297-323, 1988.

[208]   M. de Glas. Theory of fuzzy systems. *Fuzzy Sets and Systems*, Vol. 10, No. 1, 1983.

[209]   L. Godo, R. Lopez de Mantaras, and C. Sierra. Managing linguistically experssed uncertainity in MILOR: Application to medical diagnosis. *Artifical Intelligence Communication,* Vol. 1, No. 1, pp. 14-31, 1988.

[210]   L. Godo, R. Lopez de Mantaras, C. Sierra, and A. Verdaguer. MILORD: The architecture and the management of linguistically expressed uncertainity. *Int. J. of Intelligent Systems,* Vol. 4, No. 4, pp. 471-501, 1987.

[211]   J. A. Goguen. Fuzzy robot planning. In L. A. Zadeh, K. S. Fu, K. Tanaka and M. Shimura eds. *Fuzzy Sets and their Applications ot Cognitive and Decision Processes*, Academic Press, pp. 429-447, 1975.

[212]   J. A. Goguen. Fuzzy robot planning. In L. A. Zadeh, K. S. Fu, K. Tanaka and M. Shimura, eds. *Fuzzy Sets and Their Applications ot Cognitive and Decision Processes*. Academic Press, pp. 429-447, 1975.

[213]   J. A. Goguen. The logic of inexact concepts. *Synthese*, Vol. 19, pp. 325-373, 1969.

[214]   G. H. Golub, V. Klema, and G. W. Stewart. Rank degeneracy and least squares problems. *Technical Reports TR-456*, Dept. of Computer Science, University of Maryland, College Park, MD, 1976.

[215]   G. H. Golub and C. F. Van Loan. *Matrix Computations,* 2nd. Ed. John Hopkins University Press,1989.

[216]   D. E. Goldberg. *Genetic Algorithms in Search, Optimiation and Machine Learning.* Addison-Wesley, 1989.

[217]   D. Goleman. *Emotional Intelligence.* Bantam Books, 1995.

[218]   G. H. Golub and C. F. Van Loan. *Matrix Computations*, 2nd Ed. John Hopkins University Press, 1989.

[219]   G. C. Goodwin and K. S. Sin. *Adaptive Filtering Prediction and Control.* Prentice Hall, 1984.

[220]   A. A. Grasa. *Econometric Model Selection: A New Approach.* Kluwer Academic Publishers, 1989.

[221]   G. Greng, J. Martin, R. Kikinis, O. Kobler, M. Shenton, and F. A. Jolesz. Automatic Segmentation of Dual-echo MR head data. *Information Processing in Medical Imaging* (ed. A. C. F. Cholchester and D. J. Hawks), Berlin: Springer-Verlag, 1991.

[222]   A. Gunter, M. Kopisch, and H.-J. Sebastian. Integration of knowledge-based configuration with fuzzy logic and optimization. In *Industrial Applications of Fuzzy Technology in the World* (ed.. K. Hirota, Michio Sugeno), pp. 127-170. World Scientific Publishing, 1995.

[223]   M. M. Gupta. A survey of process control applications of fuzzy set theory. In *Proceedings of 18th IEEE Conference on Decision and Control*, January 1979.

[224]   M. M. Gupta. "Fuzzy-ism," the first decade. In *Fuzzy Automata and Decision Processes*, (ed. M. M. Gupta, G. N. Saridis, and B. H. Takagi). North-Holland, pp. 5-10, 1977.

[225]   M. M. Gupta. IFAC report: Round table discussion on the estimation and control in fuzzy environments. *Automatica*, Vol. 11, pp. 209-212, 1975.

[226]   M. M. Gupta and E. H. Mamdani. Second IFAC round table on fuzzy automata and decision processes. *Automatica*, Vol. 12, pp. 219-296, 1976.

[227]   M. M. Gupta, G. N. Saridis, and B. R. Gaines. *Fuzzy Automata and Decision Processes*. North Holland, 1977.

[228]   D. E. Gustafson and W. C. Kessel. Fuzzy clustering with a fuzzy covariance matrix. *Proc. IEEE CDC,* pp. 761-766, 1979.

[229]   N. Gyseghem and R. Caluwe. Fuzzy inheritance in the UFO database model. *Proc. IEEE Fuzzy Systems,* pp. 1365-1370, 1996.

[230]   N. van Gyseghem, R. de Caluwe, and R. Vandenberghe. UFO: Uncertainity and fuzziness in an object-oriented model. *Proc. of the 2nd. IEEE Int. Conf. on Fuzzy Systems,* pp. 773-778, 1993.

[231]   M. T. Hagan and H. B. Demuth. *Neural Network Design*. PWS Publishing Company, 1996.

[232]   M. T. Hagan and M. Menhaj. Training feedforward networks with the Marquardt algorithm. *IEEE Transactions on Neural Networks*, Vol. 5, pp. 989-993, 1994.

[233]   J. Hale and S. Shenoi. Catalyzing Database Inference with Fuzzy Relations. *Third IEEE Int. Symp. On Uncertainity Modeling and Analysis,* pp. 408-413, 1995.

[234]   L. Hall and A. Kandel. *Designing fuzzy expert systems*. Verlag TUV Rheinland, 1986.

[235]   L. Hall, A. M. Bensaid, Laurence P. Clarke, Robert P. Velthuizen, Martin S. Silbiger, and James C. Bezdek. A comparison of neural network and fuzzy clustering techniques in segmenting magnetic resonance images of the brain. In *IEEE Transactions on Neural Networks*, Vol. 3, No. 5, pp 672-682, September 1992.

[236]   L. Hall, C. Li, and D. Goldgof. Knowledge-based classification and tissue labeling of MR images of human brain. *IEEE Transactions on Medical Imaging,* Vol. 12, No. 4, pp. 740-750, December 1993.

[237]   H. Hallendoorn and R. Baudrexl. Fuzzy-neural traffic control and forecasting. *Proc. Int. Conf. on Fuzzy Systems*, pp. 2187-2194, 1995.

[238]   W. Hardle. Applied Nonparametric Regression. Cambridge University Press, 1990.

[239]   C. J. Harris, C. G. Moore, and M. Brown. *Intelligent Control: Aspects of Fuzzy Logic and Neural Networks*. World Scientific, 1993.

[240]   T. J. Hastie and R. J. Tibshirani. *Generalized Additive Models*. Chapman and Hall, 1990.

[241]   P. J. Hayes. Some problems and non-problems in knowledge representation. In *Readings in Knowledge Representation*. Morgan Kaufmann Publishing, 1985.

[242]   S. Haykin. *Neural Networks: A Comprehensive Foundation*. Macmillan, 1994.

[243]   S. Haykin. *Adaptive Filter Theory*, 2nd Ed., Prentice Hall, 1991.

[244]   I. Hayashi, H. Nomura and N. Wakami. Acquisition of inference rules by neural network driven fuzzy reasoning. *Japaneese Journal of Fuzzy Theory and Systems,* Vol. 2, pp. 453-469, 1990.

[245] R. Heinrich and T. A. Rapoport. A linear steady state treatment of enzymatic chains, general properties, control and effect strength. *European Journal of Biochemistry*, Vol. 42, pp. 89-95, 1974.

[246] M. Higashi and G. J. Klir. Measures of uncertainity and information based on possibility distributions. *International Journal of General Systems*, Vol. 9, pp. 43-58, 1983.

[247] G. R. Hillman, Thomas A. Kent and J. M. Agris. Measurement of brain compartment volumes from MRI data using region growing and mixed volume methods, SPIE curves and surfaces in computer vision and graphics II, pp. 372-382, Nov. 1991.

[248] G. R. Hillman, Thomas A. Kent, Alan Kaye, Donal G. Brunder, and Hemant Tagare. Measurement of brain compartment volumes in MR using voxel composition calculations. *Journal of Computer Assisted Tomography*, Vol. 15, No. 4, pp. 640-646, 1991.

[249] K. Hirota. History of industrial applications of fuzzy logic in Japan. In J. Yen, R. Langari, L. A. Zadeh, eds., *Industrial Applications of Fuzzy Logic and Intelligent Systems*. IEEE Press, pp. 43-54, 1995.

[250] K. Hirota, Y. Arai, and S. Hachisu. Fuzzy controlled robot arm playing two-dimensional ping-pong game. *Fuzzy Sets and Systems*, Vol. 3, pp. 193-219, 1980.

[251] K. Hirota and M. Sugeno. *Industrial Applications of Fuzzy Technology in the World*. World Scientific, 1995.

[252] E. Hisdal. The philosophy issues raised by fuzzy set theory. *Fuzzy Sets Syst.,* Vol. 25, pp. 349-356, 1988.

[253] E. Hisdal. Are grades of memberships probabilities? *Fuzzy Sets Syst.,* Vol. 25, pp. 325-348, 1988.

[254] Y. C. Ho, ed. *Discrete Event Dynamic Systems: Analyzing Complexity and Performance in the Modern World*. IEEE Press. 1994.

[255] R. R. Hocking. The analysis and selection of variables in linear regression. *Biometrics*, Vol. 32, pp. 1-49, 1976.

[256] J. H. Holland. *Adaptation in Natural and Artificial Systems*. University of Michigan Press, 1975.

[257] L. P. Holmblad and J. J. Östergaard. Control of cement kiln by fuzzy logic. In *Fuzzy Information and Decision Processes*. North Holland, 1979.

[258] A. Homaifar and E. McCormick. Simultaneous design of mebership functions and rule sets for fuzzy controllers using genetic algorithms. *IEEE Transactions on Fuzzy Systems,* Vol. 3, No. 2, pp. 129-139, 1995.

[259] S. Horikawa, T. Furushi and Y. Uchikawa. On fuzzy modeling using fuzzy neural networks with the backpropagation algorithm. *IEEE Trans. on Neural Network,* Vol. 3, pp. 801-806, 1992.

[260] R. Hooke and T. A. Jeeves. Direct search solution of numetical and statistical problems. *Journal of the ACM,* Vol. 8, pp. 212-229, 1961.

[261] C. S. Hsu. *Cell-to-cell Mapping : a Method of Global Analysis for Nonlinear Systems*. Springer Verlag, 1987.

[262] K. J. Hunt and R. Haas. On the functional equivalence of fuzzy inference systems and spline-based networks, *International Journal of Neural Systems*, Vol. 6, pp. 171-184, 1995.

[263] K. J. Hunt, R. Haas, and R. Murray-Smith. Extending the functional equivalence of radial basis function networks and fuzzy inference systems, *IEEE Transactions on Neural Networks*, Vol. 7, pp. 776-781, 1996.

[264] J.-N. Hwang, S.-R. Lay, M. Maechler, R. D. Martin, and J. Schimert. Regression modeling in back-propagation and projection pursuit learning. *IEEE Transactions on Neural Networks*, Vol. 5, pp. 342-353, 1994.

[265] L. Ingber and B. Rosen. Genetic algorithms and very fast simulated annealing: A comparison. *Mathematical and Computer Modeling*, Vol. 16, No. 11, pp. 87-100, 1992.

[266] S. Isaka. Fuzzy logic applications at OMRON corporation. In *Industrial Applications of Fuzzy Logic and Intelligent Systems.* (ed. J. Yen, R. Langari, and L. Zadeh). IEEE Press, 1995.

[267] H. Ishibuchi, R. Fujioka, and H. Tanaka. Neural networks that learn from fuzzy if-then rules. *IEEE Transactions on Fuzzy Systems,* Vol. 31, No. 2, 1993.

[268] H. Ishibuchi, K. Nozaki, N. Yamamoto, and H. Tanaka. Selecting fuzzy if-then rules for classification problems using genetic algorithms. *IEEE Transactions on Fuzzy Systems,* Vol. 3, pp. 260-270, 1995.

[269] H. Ishibuchi, N. Yamamoto, T. Murata, and H. Tanaka. Genetic algorithms and neighborhood search alkgorithms for fuzzy flowshop scheduling problems. *Fuzzy Sets and Systems,* Vol. 67, no. 1, pp. 81-100, 1994.

[270] M. Ishizuka and K. S. Fu. Inexact inference for rule-based damage assessment of existing structures. *Proc. 7th. Int. Joint Conf. on Artificial Intelligence,* pp. 837-842, 1981.

[271] K. Izui, M. Taguchi, M. Morikawa, and H. Katsuki. Regulation of *escherichia coli* phosphoenolpyruvate carboxylase by mutliple effectors *in vivo*, ii. Kinetic Studies with a reaction system containing physiological concentrations of ligands. *Journal of Biochemistry*, Vol. 90, pp. 1321-1331, 1981.

[272] R. A. Jacobs. Increased rates of convergence through learning rate adaptation. *Neural Networks*, Vol. 1, pp. 295-308, 1988.

[273] R. A. Jacobs and M. J. Jordan. Learning piecewise control strategies in a modular neural network architecture, *IEEE Transactions on Systems, Man, and Cybernetics*, Vol. 23, pp. 337-345, 1993.

[274] R. A. Jacobs, M. J. Jordan, and A. G. Barto. Task decomposition through competition in a modular connectionist architecture: The what and where vision tasks. *Cognitive Science*, Vol. 15, pp. 219-250, 1991.

[275] R. A. Jacobs, M. J. Jordan, S. J. Nowlan, and G. E. Hinton. Adaptive mixtures of local experts. *Neural Computation*, vol. 3, pp. 79-87, 1991.

[276] P. Jackson. *Introduction to Expert Systems,* 2nd Ed.. Addison-Wesley, 1990.

[277] M. Jamshidi. *Large-Scale Systems: Modeling and Control.* North-Holland Series in System Science and Engineering, Elsevier Science Ltd., 1983.

[278] J. -S. R. Jang. Input selection for ANFIS learning. *Proc. 5th IEEE International Conference on Fuzzy Systems*, New Orleans, LA, pp. 1493-1499, 1996.

[279]   J. -S. R. Jang. ANFIS: Adaptive-network-based fuzzy inference systems. *IEEE Transactions on Systems, Man, and Cybernetics*,Vol. 23, pp. 665-685, 1993.

[280]   J. -S. R. Jang and C.-T. Sun. Neuro-fuzzy modeling and control. *Proceedings of the IEEE*, Vol. 83, 378-406, 1995.

[281]   J.-S. R. Jang and C.-T. Sun. Functional equivalence between radial basis function networks and fuzzy inference systems. *IEEE Transactions on Neural Networks*, Vol. 4, pp. 156-159, 1993.

[282]   J.-S. R. Jang, C.-T. Sun, and E. Mizutani. *Neuro-Fuzzy and Soft Computing*. Prentice Hall, 1997.

[283]   C. Z. Janikow and Z. Michalewicz. An experimental comparison of binary and floating point representation in genetic algorithms. *Proc. of the Fourth International Conf. on Genetic Algorithms,* pp. 31-36, San Diego, CA, July 1991.

[284]   U. Jedner and H. Unbehauen. Identification of a class of nonlinear systems by parameter estimation of a linear multi-model. In *Applied modeling and Simulation of Technological Systems* (ed. P. Borne and S. G. Tzafestas). North-Holland, 1987, pp. 11-15.

[285]   T. Johansen. Fuzzy model based control: Stability, robustness, and performance issues. *IEEE Trans. on Fuzzy Systems,* Vol. 2, pp. 221-223, 1994.

[286]   T. A. Johansen and B. A. Foss. Constructing NARMAX models using ARMAX models. *International Journal of Control*, Vol. 58, pp. 1125-1153, 1993.

[287]   B. Johnston, M. S. Atkins, B. Mackiewich and M. Anderson; Segmentation of Multiple Sclerosis Lesions in Intensity Corrected Multispectral MRI, *IEEE Transactions on Medical Imaging*, Vol. 15, No. 2, pp 154-169, 1996.

[288]   C. -F. Juang and C.-T. Lin. An on-line self-constructing neural fuzzy inference network and its applications, submitted to *IEEE Transactions on Fuzzy Systems*, 1996.

[289]   J. Kacprzyk and A. Ziolkowski. Database queries with fuzzy linguistic quantifiers. *IEEE Trans. Systems, Man and Cybernetics,* Vol. 16, pp. 474-478, 1986.

[290]   M. Kamber, R. Shinghal, L. Collins, G. S. Francis, and A. C. Evans. Model-Based 3-D Segmentation of Multiple Sclerosis Lesions in Magnetic Resonance Brain Images. *IEEE Transactions on Medical Imaging*, Vol. 14, No. 3, pp 442-453, September, 1995.

[291]   A. Kandel. *Fuzzy Expert Systems*. CRC Press, 1992.

[292]   A. Kandel. On the control and evaluation of uncertain processes. *IEEE Transactions on Systems, Man, and Cybernetics*, Vol. 25, No. 6, 1980.

[293]   A. Kandel. Application of fuzzy logic to the detection of static hazards in combinational switching systems. *Int. J. Comp. Inf. Sciences*, Vol. 3, pp. 129-139, 1974.

[294]   A. Kandel. On the properties of fuzzy switching functions. *J. Cybernetics*, Vol. 4, pp. 119-126, 1974.

[295]   A. Kandel. On minimization of fuzzy functions. *IEEE Trans. Comp.*, C-22, pp. 826-832, 1973.

[296]   A. Kandel and G. Langholz. *Hybrid Architectures for Intelligent Systems*, CRC Press, 1992

[297]   P. P. Kanjilal and D. N. Banerjee. On the application of orthogonal transformation for the design and analysis of feedforward networks. *IEEE Transactions on Neural Networks*, Vol. 6, 1995, pp. 1061-1070.

[298]   T. Kapur. Segmentation of brain tissue from magnetic resonance images. MIT AI Laboratory, Technical Report No. AITR-1566, January 1995.

[299]   C. Karr. Applying genetics to fuzzy logic. *AI Expert*, Vol. 6, pp. 38-43, 1991.

[300]   C. Karr. Genetic algorithms for fuzzy controllers. *AI Expert*, Vol. 6, pp. 26-33, 1991.

[301]   C. Karr. Design of an adaptive fuzzy logic controller using a genetic algorithm. *Proc. 4th. Int. Conf. on Genetic Algorithms,* pp. 346-353, 1991.

[302]   C. L. Karr and E. J. Gentry, Fuzzy control of pH using genetic algorithms. *IEEE Tran. Fuzzy Systems*, Vol 1, No. 1, pp. 46-53, 1993.

[303]   C. L. Karr, L. Freeman, and D. Meredith. Improved fuzzy proces control of spacecraft autonomous rendezvous using a genetic algorithm, in *Proc. of the SPIE Conf. on Intelligent Control and Adaptive Systems*, Orlando,pp. 274-283, 1989.

[304]   R. L. Kashyap, C. C. Blaydon, and K. S. Fu. Stochastic approximation. In J. M. Mendel and K. S. Fu, eds., *Adaptive, Learning and Pattern Recognition Systems*. Academic Press, pp. 329-35, 1970.

[305]   A. Kaufmann and M. M. Gupta. *Introduction to Fuzzy Arithmetic: Theory and Applications*. Van Nostrand, 1985.

[306]   T. Kavli. ASMOD-an algorithm for adaptive spline modeling of observation data. *International Journal of Control*, Vol. 58, pp. 947-967, 1993.

[307]   M. Kawato, K. Furukawa, and R. Suzuki. A hierarchical neural-network model for control and learning of voluntary movement. *Biological Cybernetics*, Vol. 57, pp. 169-185, 1987.

[308]   S. Kawamoto et al. An approach to stability analysis of second order fuzzy systems. *Proc. Fuzz-IEEE '92,* pp. 1427-1434, 1992.

[309]   J. Keller, H. Qiu, and H. Tahani. Fuzzy Integral in Image Segmentation. *Proc. NAFIPS-86,* New Orleans, LA, pp. 324-338, 1986.

[310]   J. Keller, M. R. Gary, and J. A. Givens. A fuzzy K-nearest neighbor algorithm. *IEEE Trans. Systems, Man, Cybernetics,* Vol. SMC-15, No. 4, pp. 580-585, 1985.

[311]   J. Keller et al. Fuzzy rule-based models in computer vision. *Fuzzy Modeling: Paradigms and Practice.* (ed. W. Pedrycz), 1996.

[312]   D. N. Kennedy, P. A. Filipek , and V. S. Caviness. Anatomic Segmentation and Volumetric Calculations in Nuclear Magnetic Resonance Imaging. *IEEE Transactions on Medical Imaging*, Vol. 8, No. 1, pp 1-7, March 1989.

[313]   P. P. Khargonekar, I. R. Petersen, and K. Zhou. Robust stabilization of uncertainty linear systems: Qudratic stabilizability and H^∞ control theory. *IEEE Trans. Automat. Contr.,* Vol. 35, no. 3, pp. 356-361, 1990.

[314]   L. Kiar, C. Thomsen, F. Gierris, B. Mosdal, and O. Henriksen. Tissue characterization of intracranial tumors by MR imaging. *Acta Radiologica*, Vol. 32, pp. 498-504, 1991.

[315]   W.J.M. Kickert and E.H. Mamdani. Analysis of a fuzzy logic controller. *Fuzzy Sets and Systems,* Vol. 1, No. 1, 1978, pp. 29-44.

[316]   W. M. J. Kickert and H. R. van Nauta Lemke. The application of fuzzy set theory to control a warm water process. *Automatica*, Vol. 12, No. 4, 1976.

[317]   R. E. King. Fuzzy logic control of a cement kiln pre-calciner flash furnace. *Proceedings of the IEEE Conference on Applications of Adaptive and Multivariable Control*, 1982.

[318]   J. Kingdon, Intelligent Systems for Fraud Detection. In *Genetic Algorithms and Fuzzy Logic Systems: Soft Computing Perspectives* (ed. E. Sanchez, T. Shibata, and L. A. Zadeh). World Scientific, pp. 133-154, 1997.

[319]   R. Kininis, M. E. Shenton, G. Gerig, J. Martin, M. Anderson, D. Metcalf, C. R. G. Guttmann, R. W. McCarley, W. Lorensen, H. Cline, and F. A. Jolesz. Routine quantitative analysis of brain and cerebrospinal Fluid space with MR imaging, Journal of Magnetic Resonance Imaging, Vol. 2, pp. 619-629, 1992.

[320]   J. Kinzel, F. Klawonn, and R. Kruse. Modifications of genetic algorithms for designing and optimizing fuzzy controllers. *Proc. of the 1st IEEE Conf. on Evolutionary Computations,* pp. 28-33, Orlando, FL, 1994.

[321]   J. B. Kiszka, M. M. Gupta, and P. N. Nikiforuk. Energetistic stability of fuzzy dynamic systems. *IEEE Transactions on Systems, Man, and Cybernetics*, Vol. 15, 1985.

[322]   G. Klir and T. Folger. *Fuzzy Sets, Uncertainty, and Information.* Prentice-Hall, 1989.

[323]   G. J. Klir and B. Yuan (ed), Fuzzy sets, fuzzy logic, and fuzzy systems: selected papers by Lotfi A. Zadeh, World Scientific, 1996.

[324]   M. I. Kohn, N. K. Tanna, G. T. Herman, S. M. Resnick, P. D. Mozley, R. E. Gur, A. Alavi, R. B. Zimmerman, and R. C. Gur. Analysis of brain and cerebrospinal fluid volumes with MR imaging, Part I: methods, reliability and validation. *Radiology*, Vol. 178, pp 115-123,1991.

[325]   L. J. Kohout. Automata and topology. In E. H. Mamdani and B. R. Gaines eds. *Discrete Systems and Fuzzy Reasoning, EES-MMS-DSFR-76*, Queen Mary College, University of London, (workshop proceedings), 1976.

[326]   P. K. Kokotovic and H. K. Khalil, eds. *Singular Perturbations in Systems and Control.* IEEE Press, 1986.

[327]   S. Kondo, S. Abe, H. Terai, M. Kiuchi, and H. Imahashi. Fuzzy logic controlled washing machine. *Proc. IFSA '91*, pp. 97-100, 1990.

[328]   M. Kopisch and A. Gunter. Configuration of a passenger aircraft cabin based on conceptual hierarchy (Hrsg). *5th. Int. Conf. Industrial and Engineering Applications of Artificial Intelligence and Expert Systems,* pp. 421-430, 1992.

[329]   Y. Koren. *Computer Control of Manufacturing Systems.* McGraw-Hill, 1984.

[330]   B. Kosko. *Fuzzy Engineering.* Prentice Hall, 1997.

[331] B. Kosko. Fuzzy systems as universal approximators. *IEEE Transactions on Computers*, Vol. 43, pp. 1329-1333, 1994.

[332] B. Kosko. *Neural Networks and Fuzzy Systems*. Prentice Hall, 1992.

[333] B. Kosko. Fuzziness versus probability. *Int. J. General Syst.,* Vol. 17, pp. 211-240, 1990.

[334] B. Kosko. Fuzziness versus probability. *Int. J. General Syst.,* Vol. 17, pp. 211-240, 1990.

[335] B. Kosko. Foundations of fuzzy estimation theory. Ph.D. dissertation, Department of Electrical Engineering, University of California at Irvine. Order Number 8801936, University Microfilms International, 300N. Zeeb Road, Ann Arbor, MI, 48106, 1987.

[336] B. Kosko and S. Isaka, Fuzzy Logic, Scientific American, Vol 269, No. 1, pp. 76-81, July 1993.

[337] J. Kowalik and M. R. Osborne. *Methods for Unconstrained Optimization Problems.* American Elsevier, 1968.

[338] D. H. Kraft and D. A. Buell. Fuzzy Sets and Generalized Boolean Retrieval Systems. *International Journal of Man-Machine Studies*, Vol. 19, pp. 45-56, 1983.

[339] D. H. Kraft, F. E. Petry, B. P. Buckles, and T. Sadasivan. Genetic algorithms for query optimization in information retrieval: relevance feedback. In *Genetic Algorithms and Fuzzy Logic Systems: Soft Computing Perspectives* (ed. E. Sanchez, T. Shibata, and L. A. Zadeh). World Scientific, pp. 155 - 173, 1997.

[340] R. Kruse. The strong law of large numbers for fuzzy random variables. *Information Sciences*, pp. 233-241, 1982.

[341] R. Krishnapuram. Fuzzy and possibilistic shell clustering algorithms and their application to boundary detection and surface approximation - part I. *IEEE Trans. on Fuzzy Systems,* Vol. 3, No. 1, 1995.

[342] R. Krishnapuram, H. Frigui, and O. Nasraoui. Fuzzy and possibilistic shell clustering algorithms and their application to boundary detection and surface approximation — part II. *IEEE Trans. on Fuzzy Systems,* Vol. 3, No. 1, 1995.

[343] R. Krishnapuram and J. M. Keller. A possibilistic approach to clustering. *IEEE Trans. on Fuzzy Systems,* Vol. 1, No. 2, pp. 98-110, 1993.

[344] R. Krishnapuram, O. Nasraoui, and H. Frigui. The fuzzy C spherical shells algorithms: A new approach. *IEEE Trans. on Neural Networks*, Vol. 3, No. 5, pp. 663-671, 1992.

[345] R. H. Kwong and E. W. Johnston. A variable step size LMS algorithm. *IEEE Transactions on Signal Processing*, Vol. 40, pp. 1633-1642, 1992.

[346] G. Lakoff. Hedges: A study in meaning criteria and the logic of fuzzy concepts. *J. Philos. Logic*, Vol. 2, pp. 458-508, 1973.

[347] Lancaster, F. Wilfrid, and A. J. Warner. *Information Retrieval Today.* Information Resources Press, 1993.

[348] R. Langari. Fuzzy logic control within a hierarchically structured framework. *Computational Intelligence* (eds. J. Zurada, R. Marks, and C. Robinson), pp. 293-303, IEEE Press, 1994.

[349]   R. Langari. A framework for analysis and synthesis of fuzzy linguistic control systems. Ph.D. thesis, University of California, Berkeley, 1990.

[350]   R. Langari, D. Dornfeld, and Z. Wang. Intelligent Sensing and Control in Metal Cutting (ed. M. Jamshidi, R. Lumia, J. Mullins, and M. Shahinpoor.) *Robotics and Manufacturing: Recent Trends in Research, Education, and Applications.* ASME, 1992.

[351]   R. Langari and B. B. Pate. Synthesis of nonlinear control algorithms via fuzzy logic: application in robotic motion control. *Proceedings of the International Workshop on Industrial Fuzzy Control and Intelligent Systems(IFIS 92),* TX, December 1992.

[352]   R. Langari and M. Tomizuka. Fuzzy linguistic control of arc welding (ed. E. Kannatey-Assibu, Jr. et al.). *Sensors and Controls for Manufacturing — 1988.* ASME, 1988.

[353]   R. Langari and M. Tomizuka. Analysis and synthesis of fuzzy linguistic control systems (ed. R. Shoureshi). *Intelligent Control 1990: Proceedings of the 1990 Winter Annual Meeting, DSC-Vol 23.* ASME, November 1990.

[354]   R. Langari and M. Tomizuka. Stability of fuzzy linguistic control systems. *Proceedings of the IEEE Conference on Decision and Control,* December 1990.

[355]   R. Langari and L. Wang. Fuzzy models, modular networks and hybrid learning. *Fuzzy Sets and Systems,* Vol. 79, No. 2, February 1996.

[356]   L. I. Larkin. A fuzzy logic controller for aircraft flight control. In M. Sugeno ed., *Industrial Applications of Fuzzy Control.* North Holland, 1985.

[357]   J. Latombe. *Robot Motion Planning.* Kluwer Academic, 1991.

[358]   M. Laviolette and J. W. Seaman, Jr. The efficacy of fuzzy representations of uncertainity. *IEEE Transactions on Fuzzy Systems,* Vol. 2, No. 1, pp. 4-15, February 1994.

[359]   R. N. Lea, J. Villareal, Y. K. Jani, and C. Copeland. Fuzzy logic based tether control. *North American Fuzzy Information Processing Society.* 1991.

[360]   Y. LeCun. Une procedure d'apprentissage pour reseau a seuil assymetrique. *Cognitiva,* Vol. 85, pp. 599-604. 1985.

[361]   C. C. Lee. A self-learning rule-based controller employing approximate reasoning and neural net concepts. *International Journal of Intelligent-systems,* Vol. 6, pp. 71-93, 1991.

[362]   C. C. Lee. A self-learning rule-based controller employing approximate reasoning and neural net concepts. *Int. J. of Intelligent Systems,* Vol. 5, pp. 71-93, 1991.

[363]   C. C. Lee. Fuzzy logic in control systems: Fuzzy logic controller, Parts I and II. *IEEE Transactions on Systems, Man, and Cybernetics,* Vol. 20, pp. 404-435, 1990.

[364]   D. H. Lee and M. H. Kim. Database summarization using fuzzy ISA heirarchies. *IEEE Trans. In Systems, man, and cybernetics,* Vol. 27, No. 1, pp. 68-78, 1997.

[365]   M. A. Lee and H. Takagi. Integrating design stages of fuzzy systems using genetic algorithms, in *Proc. IEEE Int. Conf. on Fuzzy Systems (FUZZ-IEEE),* pp. 612-617, San Francisco, 1993.

[366]   M. A. Lee and H. Takagi. Dynamic control of genetic algorithms using fuzzy logic techniques, in *Proc. of 5th Int. Conf. on Genetic Algorithms,* Urbana-Champaign, IL, pp. 76-83, July 1993.

[367]   M. A. Lee and H. Takagi. Integrating design stages of fuzzy systems using genetic algorithms. *Proc. Second IEEE Int. Conf. Fuzzy Systems,* pp. 612-617, 1993.

[368]   R. C. Lee. Fuzzy Logic and the resolution principle. *Journal of the Association for Computing Machinery,* Vol. 19, pp. 109-119, 1972.

[369]   S. Lee, J. Yen, and R. Krammes. Application of fuzzy logic for detecting incidents at signalized highway intersections. *Proc. Int. Conf. on Fuzzy Systems*, pp. 867-872, 1996.

[370]   L. Lesmo, L. Saitta, and P. Torasso. Learning of fuzzy production Rules for medical diagnosis. In M. M. Gupta and E. Sanchez eds. *Approximate Reasoning in Decision Analysis*, Amsterdam, pp. 249-260, 1982.

[371]   I. J. Leontaritis and S. A. Billings. Model selection and validation methods for non-linear systems. *International Journal of Control,* Vol. 45, No. 1, pp. 311-341, 1987.

[372]   K. S. Leung and W. Lam. Fuzzy concepts in expert systems. *IEEE Computer*, Sept. 1988.

[373]   K. S. Leung and W. Wam. A fuzzy expert system shell using both exact and inexact reasoning. *J. of Automated Reasoning,* Vol. 5, pp. 207-233, 1989.

[374]   C. Li, D. B. Goldgof, and L. Hall. Knowledge-based classification and tissue labeling of MRI images of human brain. *IEEE Transactions on Medical Imaging*, Vol. 12, No. 4, pp. 740-750, 1993.

[375]   Z. Liang. Tissue classification and segmentation of MR images. *IEEE Engineering in Medicine and Biology*, pp. 81-85, March 1993.

[376]   Z. Liang, J. R. MacFall and D. P. Harrington. Parameter estimation and tissue Segmentation from multispectral MR images. *IEEE Transactions on Medical Imaging*, Vol. 13, No. 3, pp. 441-449, September 1994.

[377]   J. C. Liao, E. N. Lightfoot, S. O. Jolly, and G. K. Jacobson. Appliction of characteristic reaction paths: Rate-limiting capacity of phosphofructokinase in yeast fermentation. *Biotechnology and Bioengineering,* Vol. 31, pp. 855-868, 1988.

[378]   K. O. Lim and Adolf Pfefferbaum. Segmentation of MR brain images into cerebrospinal fluid spaces, white and grey matter. *Journal of Computer Assisted Tomography*, Vol. 13, No. 4, pp. 588-593, 1989.

[379]   Y. W. Lim and S. U. Lee. On the color image segmentation algorithm based on the thresholding and the fuzzy c-Means techniques. *Pattern Recognition*, Vol. 23, No. 9, 1990.

[380]   C. -T. Lin and C. S. G. Lee. *Neural Fuzzy Systems: A Neuro-Fuzzy Synergism to Intelligent Systems*. Prentice Hall, 1996.

[381]   C. -T. Lin and C. S. G. Lee. Reinforcement structure/parameter learning for neural-network-bascd fuzzy logic control systems. *IEEE Trans. on Fuzzy Systems,* Vol. 2, 1994.

[382]   C. -T. Lin and C. S. G. Lee. Neural-network-based fuzzy logic control and decision system. *IEEE Transactions on Computer*, Vol. 40, pp. 1320-1336, 1991.

[383]   C-T. Lin and Y-C. Lu. A neural fuzzy system with linguistic teaching signals. *IEEE Transactions on Fuzzy Systems*, Vol. 3, No. 2, 1995.

[384]   L. Ljung. *System Identification: Theory for the User.* Prentice Hall, 1987.

[385]   B. Y. Lindley. Scoring rules and the inevitability of probability. *Int. Stat. Rev.*, Vol. 1, pp. 3-28, 1978.

[386]   D. T. Long, M. A. King, and B. C. Penny. 2-D versus 3-D edge detection as a basis for volume quantiation in SPECT. *Information Processing of Medical Imaging*, pp. 457-471, 1991.

[387]   A. Lotfi and A. C. Tsoi. Importance of membership functions: A comparative study on different learning methods for fuzzy inference systems. *Proceedings of the Third International Conference on Fuzzy Systems*, Orlando, FL, pp. 1791-1796, 1994.

[388]   A. de Luca and S. Termini. A Definition of a Non-probabilistic Entropy in the Setting of Fuzzy Sets Theory. *Information and Control*, Vol. 20, pp. 301-312, 1972.

[389]   P. Lundin and G. Pedersen. Pituitary macroadenomas: Assesment by MRI. *Journal of Computer Assited Tomography,* Vol. 16, No. 4, pp. 519-528, 1992.

[390]   S. Lui and H. Asada. A skill based adaptive controller for deburring robots. In *Computer Control of Manufacturing Processes, DSC-Vol 28.* ASME, 1991o

[391]   R. C. Luo and T. J. Pan. An intelligent path planning system for robot navigation in an unknown environment. *SPIE Mobil Robots IV,* pp. 316-326, 1989.

[392]   J. MacQueen. Some methods for classification and analysis of multivariate observation. In L. M. LeCun and J. Neyman, eds., *Proceedings of the 5th Berkeley Symposium on Mathematical Statistics and Probability*, University of California Press, pp. 281-297, 1967.

[393]   H. Madala and A. G. Ivakhnenko. *Inductive Learning Algorithms for Complex Systems Modeling.* CRC Press, 1993.

[394]   J. M. Maciejowsky. *Multivariable Feedback Systems.* Addison-Wesley, 1989.

[395]   P. Maeder, A. Wirsen, M. Bajc, W. Schalen, H. Sjoholm, H. Skeidsvoll, S. Cronoqvist ,and D. H. Ingvar. Volumes of chromis traumatic frontal brain lesions measured by MR imaging and CBF tomography. *Acta Radiologica,* Vol. 32, pp. 271-278, 1991.

[396]   P. Maes. Artificial life meets entertainment: Lifelike autonomous agents. *Special Issue on Novel Applications of AI Communications of ACM,* 1995.

[397]   P. Magrez and P. Smets. Fuzzy modus ponens: A new model suitable for applications in knowledge-based systems. *International Journal of Intelligent Systems*, Vol. 4, pp. 181-200, 1989.

[398]   E. H. Mamdani. Applications of fuzzy set theory to control systems: A survey (ed. M. M. Gupta, G. N. Saridis and B. R. Gaines). *Fuzzy Automata and Decision Processes*, North-Holland, pp. 77-88, 1977.

[399]   E. H. Mamdani. Application of fuzzy logic to approximate reasoning using linguistic synthesis. *Proc. 6th. Int. Symp. Multiple-Valued Logic,* IEEE 76CH1111-4C, pp. 196-202, 1976.

[400]   E. H. Mamdani. Application of fuzzy algorithms for control of simple dynamic plant. *IEEE Proceedings*, Vol. 121, No. 12, 1974.

[401]   E. H. Mamdani and S. Assilian. An experiment in linguistic synthesis with a fuzzy logic controller. *International Journal of Machine Studies*, Vol. 7, No. 1, 1975.

[402]   E. H. Mamdani and N. Baaklini. Prespective method for deriving control policy in a fuzzy-logic controller. *Electronics Lett,* Vol. 11, pp. 625-626, 1975.

[403]   E. H. Mamdani, J. J. Östergaard, and E. Lembossis. On the use of fuzzy logic for implementing rule-based control of industrial processes. In *Fuzzy Sets and Decision Analysis*, 1984.

[404]   R. L. de Mantaras and L. Valverde. New results in fuzzy clustering based on the concept of indistinguishability relation. *IEEE Transactions on Pattern Analysis and Machine Intelligence*, Vol. 10, No. 5, pp. 754-757, 1988.

[405]   K. E. Mathias, L. D. Whitley, C. Stork, and T. Kusuma. Staged hybrid genetic search for seismic data imaging. *Proc. of the First IEEE Conf. on Evolutionary Computation,* pp. 356-361, Orlando, FL, June 1994.

[406]   J. McDermott. R1: A rule-based configurer of computer systems. *Artificial Intelligence,* Vol. 19, pp. 39-88, 1982.

[407]   D. B. McGrrah and R. S. Judson. Analysis of the genetic algorithm method of molecular confomation determination. *Journal of Computational Chemistry,* Vol. 14, No. 11, pp. 1385-1395, 1993.

[408]   C. T. Meadow. *The Analysis of Information Systems: A Programmer's Introduction to Information Retrieval*. John Wiley & Sons, 1967.

[409]   J. Medina, O. Pons, and M. Vila. Gefred: A generalized model to implement fuzzy relational databases. *Information Sciences,* Vol. 47, pp. 234-254, 1994.

[410]   J. M. Mendel. Fuzzy Logic Systems for engineering: A tutorial. *Proc. of IEEE*, Vol. 83, pp. 345-377, 1995.

[411]   J. Meseguer and I. Sol. Automata in semimodule categories. *Proc. of First International Symp. on Category Theory to Computation and Control,* pp. 196-202, 1975.

[412]   Y. Meyer. *Wavelets and Operators*. Cambridge University Press. 1992.

[413]   Z. Michalewicz. *Genetic Algorithms + Data Structure = Evolution Programs*. Springer Verlag, 1992.

[414]   A. J. Miller, *Subset Selection in Regression*, Chapman and Hall, 1990.

[415]   M. Minsky. *The Society of the Mind*. Simon & Schuster, 1986.

[416]   S. Mitaim and B. Kosko. What is the best shape for a fuzzy set in function approximation? *Proceedings of the Fifth International Conference on Fuzzy Systems*, New Orleans, LA, pp. 1237-1243, 1996.

[417]   S. Miyamoto. *Fuzzy Sets in Information Retrieval and Cluster Analysis*. Theory and Decision Library, Series D. Kluwer Academic Publishers, 1990.

[418]   S. Miyamoto. Two approaches for information retrieval through fuzzy associations. *IEEE Tran. on Systems, Man, and Cybernetics*, Vol. 19, No. 1, pp. 123-130, 1989.

[419]   S. Miyamoto, T. Miyake, and K. Nakayama. Generation of a pseudothesaurus for information retrieval based on cooccurrences and fuzzy set operations. *IEEE Tran. on Systems, Man, and Cybernetics*, Vol. 13, No. 1, pp. 62-70, 1983.

[420]   S. Miyamoto and K. Nakayama. Fuzzy information retrieval based on a fuzzy pseudothesaurus. *IEEE Tran. on Systems, Man, and Cybernetics*, Vol. 16, No. 2, pp. 278-282, 1986.

[421]   M. Mizumoto. Fuzzy automata and fuzzy grammars. Ph.D. thesis, Faculty of Engineering Science, Osaka University, Osaka, Japan, 1971.

[422]   M. Mizumoto and K. Tanaka. Fuzzy-fuzzy automata. *Kybernets*, Vol. 5, pp. 107-112, 1976.

[423]   M. Mizumoto and H. J. Zimmermann. Comparision of fuzzy reasoning methods. *Fuzzy Sets and Systems,* Vol. 8, pp. 253-283, 1982.

[424]   M. Mizumoto, J. Toyoda, and K. Tanaka. General formulation of formal grammars. *Trans. Inst. Electron. Comm. Eng. (Japan)*, Vol. 54-c, pp. 600-605, 1972.

[425]   M. Mizumoto, J. Toyoda, and K. Tanaka. Some considerations on fuzzy automata. *J. Comp. Sci.*, Vol. 3, pp. 409-422, 1969.

[426]   A. Molinari and G. Pasi. A fuzzy representation of HTML documents for information retrieval systems. *Proc. of the 5th IEEE Int. Conf. on Fuzzy Systems,* New Orleans, pp. 107-112, 1996.

[427]   F. Monai and T. Chehire. Possibility assumption based truth maintenance system validation in a data fusion application. *Proc. of the 8th. Conf. of Uncertainity in Artificial Intelligence,* pp. 83-91, 1992.

[428]   J. Moody and C. J. Darken. Fast learning in networks of locally-tuned processing units. *Neural Computation*, Vol. 1, pp. 281-294, 1989.

[429]   G. C. Mouzouris and J. M. Mendel. Designing fuzzy logic systems for uncertain environments using a singular-value-QR decomposition method. *Proceedings of the Fifth IEEE International Conference on Fuzzy Systems*, New Orleans, LA, pp. 295-301, 1996.

[430]   M. Mukaidono, Z. L. Shen, and L. Y. Ding. Fundamentals of fuzzy prolog. *International Journal of Approximate Reasoning*, Vol. 3, pp. 179-193, 1989.

[431]   J. Munro. Uncertainty and fuzziness in engineering decision-making. *Proceedings, First Canadian Seminar on Systems Theory for Civil Engineers*, Calgary, 1979.

[432]   K. Murakami and M. Sugeno. An expreimental study on fuzzy parking control using a model car. In *Industrial Applications of Fuzzy Control* (ed. M. Sugeno). North Holland, 1985.

[433]   T. Murata and H. Ishibuchi. Preformance evaluation of genetic algorithms for flowshop scheduling problems. *Proc. of the first IEEE Conf. on Evolutionary Computation,* Orlando, FL, June 1994.

[434]   M. T. Musavi, W. Ahmed, K. H. Chan, K. B. Faris, and D. M. Hummels. On the training of radial basis function classifiers. *Neural Networks*, Vol. 5, pp. 595-603, 1992.

[435]   H. Nakajima, T. Sogoh and M. Aroa. Fuzzy database language and library-fuzzy extension to SQL. *Proc IEEE Second Int. Conf. on Fuzzy Systems (FUZZ-IEEE),* pp. 477-482, 1993.

[436]   M. Nakata. Dependencies in fuzzy databases: Functional dependency. *Proc. IEEE Fourth Int. Conf. on Fuzzy System and Second Int. Fuzzy Eng. Symp. FUZZ-IEEE/IFES '95*, pp. 757-764, 1995.

[437]   K.S. Narendra and K. Parthasarathy. Identification and control of dynamical systems using neural networks. *IEEE Transactions on Neural Networks*, Vol. 1, No. 1, January 1993.

[438]   K. S. Narendra and K. Parthasarathy. Identification and control of dynamical systems using neural networks. *IEEE Transactions on Neural Networks*, Vol. 1, pp. 4-27, 1990.

[439]   B. Natvig. Possibility versus probability. *Fuzzy Sets Syst.,* Vol. 10, pp. 31-36, 1983.

[440]   J. A. Nelder and R. Mead. A simplex method for function minimization. *Computer Journal,* Vol. 7, pp. 308-313, 1965.

[441]   C. V. Negoita. *Expert Systems and Fuzzy Systems*. The Benjamin/Cummings, 1985.

[442]   C. V. Negoita. On the application of the fuzzy sets separation theorem for automatic classification in information retrieval systems. *Inform. Sci,* Vol. 5, pp. 279-286, 1973.

[443]   C. V. Negoita. Information retrieval systems. Ph.D. thesis, Polytechnic Institute of Bucharest. (in Rumanian), 1969.

[444]   C. V. Negoita and P. Flondor. On fuzziness in information retrieval. *Int. J. Man-Machine Studies*, Vol. 8, pp. 711-716, 1976.

[445]   C. V. Negoita and D. A. Ralescu. Fuzzy systems and artificial intelligence. *Cybernetes*, Vol. 3, pp. 173-178, 1974.

[446]   C. V. Negoita and D. A. Ralescu. Representation Theorems for Fuzzy Concepts. *Kybernetes*, Vol. 4, pp. 169-174, 1975.

[447]   A. Newell and H. A. Simon. Computer science as empirical inquiry: Symbols and search. *Communications of the ACM,* Vol. 19, No. 3, pp. 113-126, 1976.

[448]   H. T. Nguyen. On modeling of linguistic information using random sets. *Information Science*, Vol. 34, pp. 265-274. 1984.

[449]   J. Nie and D. A. Linkens, Learning control using fuzzified self-organizing radial basis function network, *IEEE Transactions on Fuzzy Systems*, Vol. 1, pp. 280-287, 1993.

[450]   A. di Nola, W. Pedrycz, S. Sessa, and P.Z. Wang. Fuzzy relation equation under a class of triangular norms: A survey and new results. *Stochastica*, Vol. 8, No. 2, pp. 99-145, 1984.

[451]   H. Nomura, I. Hayashi, and N. Wakami, A self-tuning method of fuzzy control by descent method. *4th. IFSA Congress, Engineering Volume,* pp. 155-158, 1991.

[452]   H. Nomura, I. Hayashi, and N. Wakami, A self-tuning method of fuzzy reasoning by genetic algorithm. *Proc. of the Int. Fuzzy Systems and Intelligent Control Conf.*, Louisville, KY, 1992, pp. 236-245.

[453]   K. Nukui, M. Arakawa, M. Komido and T. Taniguchi. An auto tuning method for obtaining optimal control parameters by following changes in process characteristics. ISA Paper #91-0459.

[454]   A. Okada. Automation in arc welding: State of the art problems. *Journal of the Japan Society of Mechanical Engineers(Japanese),* March 1979.

[455]   M. Okutomi and M. Mori. Decision of robot movement by means of a potential field. *Advanced Robotics*, Vol. 1, No. 2, pp. 131-141, 1986.

[456]   A. Ortony, G. Clore, and A. Collins. *The Cognitive Structure of Emotions*. Cambridge University Press, 1988.

[457] K. Otto and V. E. Antonsson. Modelling imprecision in product design. *Proc. of 3'd Fuzzy-IEEE,* 1994.

[458] S. V. Ovchinnikov. Structure of Fuzzy Binary Relations. *Fuzzy Sets and Systems,* Vol. 6, pp. 169-195, 1981.

[459] M. Ozkan, Benoit M. Dawant, and Robert J. Maciunas. Neural-network-based segmentation of multi-modal medical images: A comparative and prospective study. *IEEE Transactions on Medical Imaging*, Vol. 12, No. 3, pp. 534-544, September 1993.

[460] N. R. Pal and S. K. Pal. A review on image segmentation techniques. *Pattern Recognition*, Vol. 26, No. 9, pp. 1277-1294, 1993.

[461] N. R. Pal and J. C. Bezdek. On cluster validity for the fuzzy c-means model. *IEEE Trans. on Fuzzy Systems,* Vol. 3, No. 3, 1995.

[462] C. J. Pantaleon-Prieto, I. Santamaria-Caballero, and A. R. Figueiras-Vidal. Competitive local linear modeling. *Signal Processing*, Vol. 49, pp. 73-83, 1996.

[463] D. B. Parker. Learning-logic: Casting the cortex of the human brain in silicon. Technical Report TR-47, Center for Computational Research in Economics and Management Science, MIT, Cambridge, MA, 1985.

[464] G. Y. Park and P. Y. Seong. Towards increasing the learning speed of gradient descent method in fuzzy system. *Fuzzy Sets and Systems*, Vol. 77, pp. 299-313, 1996.

[465] D. W. Payton, J. K. Rosenblatt, and D. M. Keirsey. Plan guided reaction. *IEEE Trans. Syst. Man Cyber.,* Vol. 20, No. 6, pp. 1370-1382, 1990.

[466] J. Pearl. Baysian Decision Methods. In *Uncertain Reasoning* (ed. G. Shaffer and J. Pearl). Morgan Kaufmann, pp. 345-352., 1990.

[467] J. Pearl. *Probabilistic Reasoning in Intelligent Systems: Networks of Plausible Inference.* Morgan Kauffman, 1988.

[468] W. Pedrycz. *Fuzzy Control and Fuzzy Systems.* John Wiley and Sons Inc., 1989.

[469] E. F. Petry. *Fuzzy Database Principles and Applications.* Kluwer Academic Publishers, 1996.

[470] E. F. Petry, B. P. Buckles, A. Yazici, and R. George. Fuzzy Information Systems. *Proc. IEEE Int. Conf. of Fuzzy Systems,* pp. 1187-1200, 1992.

[471] T. S. Perry. Lotfi A. Zadeh (cover story). *IEEE Spectrum*, pp. 32-35, June 1995.

[472] R. Pfeifer. Artificial intelligence models of emotions. *Cognitive Perspectives on Emotion and Motivation,* pp. 287-320, 1988.

[473] N. Pfluger, J. Yen, and R. Langari. A defuzzification strategy for a fuzzy logic controller employing prohibitive information in command formation. *IEEE Int. Conf. on Fuzzy Systems,* pp. 717-723, 1992.

[474] R. Picard. *Affective Computing.* MIT Press, 1997.

[475]   F. Pin and Y. Watanabe. Autonomous navigation of a mobil robot using the behaviorist theory and VLSI fuzzy inferencing chips. *Industrial applications of Fuzzy Logic and Intelligent Systems* (ed. J. Yen, R. Langari and L. Zadeh). IEEE Press, pp. 175-190, 1995.

[476]   F. Pin and Y. Watanabe. Driving a car using reflexive fuzzy behaviors. *Proc. of the 2nd. IEEE Int. Conf. on Fuzzy Systems,* pp. 1425-1430, 1993.

[477]   A. Pitsillides, Y. A. Sekercioglu, and G. Ramamurthy. Effective control of traffic flow in ATM networks using fuzzy explicit rate marking (FERM). *IEEE J. on Selected Areas in Communications,* Vol. 15, No. 2, pp. 209-225, 1997.

[478]   T. Poggio and F. Girosi. Networks for approximation and learning. *Proceedings of the IEEE,* Vol. 78, pp. 1481-1497.

[479]   M. Pottmann, H. Unbehauen ,and D. E. Seborg. Application of a general multi-model approach for identification of highly nonlinear process-a case study. *International Journal of Control,* Vol. 57, pp. 97-120, 1993.

[480]   M. J. D. Powell. Radial basis functions for multivariable interpolation: Areview *IMA Conference for the Approximation of Functions and Data*, Shrivenham, UK, pp. 143-167, 1985.

[481]   M. J. D. Powell. An efficient method for finding the minimum of a function of several variables without caculating derivaties. *Computer Journal,* Vol. 7, pp. 155-162, 1964.

[482]   M. B. Priestley. *Non-linear and Non-stationary Time Series Analysis.* Academic Press, 1988.

[483]   T. J. Procyk. A fuzzy logic learning system for a single input output plant. *Working Group Rep. 3,* Queen Mary College, University of London, UK, 1976.

[484]   T. J. Procyk and E. H. Mamdani. A linguistic self organizing process controller. *Automatica,* Vol. 15, No. 1, 1979.

[485]   M. L. Puri and D. Ralescu. Fuzzy random variables. *Journal of Mathematical Analysis and Application,* Vol. 114, pp. 409-422, 1986.

[486]   H. Prade and C. Testemale. Generalizing database relational algebra for the treatment of incomplete/uncertain information and vague queries. *Information Sciences,* Vol. 34, pp. 115-143, 1984.

[487]   S. -Z. Qin, H. -T. Su, and T. J. McAvoy. Comparison of four neural net learning methods for dynamic system identification. *IEEE Transactions on Neural Networks,* Vol. 3, pp. 122-130, 1992.

[488]   D. Ragsdale, C. Butler, B. Cox, J. Yen, and U. Pooch. A fuzzy logic approach for intelligence analysis of actual and simulated military reconnaissance mission. *IEEE Int. Conf. on Systems, Man, and Cybernetics,* 1997.

[489]   K. V. S. V. N. Raju and A. K. Majumdar. Fuzzy functional dependencies and lossless join decomposition of fuzzy relational database systems. *ACM Trans. on Database Systems*, Vol. 13, No. 2, pp. 128-167, 1988.

[490]   G. V. S. Raju, J. Zhou, and R. A. Kisner. Hierarchical fuzzy control. *International Journal of Control*, Vol. 54, No. 5, pp. 1201-1216, 1991.

[491]   P. J. Ramadge and W. M. Wonham. Supervisory control of a class of discrete event processes. *SIAM Journal of Control and Optimization*, Vol. 25, No. 1, 1987.

[492] A. Ramer. Conditional Possibility Measures. *International Journal of Cybernetics and Systems,* Vol. 20, pp. 233-247, 1989.

[493] S. Rangwala and D. Dornfeld. Sensor Integration using neural networks for intelligent tool condition monitoring. *ASME Journal of Engineering for Industry,* Vol. 112, pp. 219-228, 1990.

[494] S. S. Rao et al. Multi-objective fuzzy optimization techniques for engineering design. *Computers and Structures,* Vol. 42, No. 1, pp. 37-44, 1992.

[495] K.S. Rao and D.D. Majumdar. Application of circle criteria for stability analysis of linear SISO and MIMO systems associated with fuzzy logic controller. *IEEE Transactions on Systems, Man, and Cybernetic*, Vol. 14, No. 2, pp. 345-349, 1984.

[496] S. P. Raya. Low-level segmentation of 3-D magnetic resonance brain images — a rule-based system, *IEEE Transactions on Medical Imaging*, Vol. 9, No. 3, pp. 740-750, 1990.

[497] J. Renders. *Algorithms Genetiques et Reseaux de Nuerones: Applications a la commande de processus.* Hermes, 1994.

[498] J. Renders and H. Bersini. Hybridizing genetic algorithms with hil-climbing methods for global optimization: Two possible ways. *Proc. of the First IEEE Conf. on Evolutionary Computation,* pp. 312-317, Orlando, FL, June 1994.

[499] J. Renders and S. Flasse. Hybrid methods using genetic algorithms for global optimization. *IEEE Transactions on Systems, Man and Cybernetics,* Vol. 26, no. 2, pp. 243-258, 1996.

[500] F. C. Rhee and R. Krishnapuram. Fuzzy rule generation methods for high level computer vision. *Fuzzy Sets and Systems,* Vol. 60, pp. 245-258, 1993.

[501] J. R. Rice. *The Approximation of Functions.* Addison-Wesley, 1964.

[502] A. K. Rigler, J.M. Irvine, and T.P. Vogl, Rescaling of variables in back propagation learning, *Neural Networks*, Vol. 3, pp. 561-573, 1990.

[503] J. L. A. van Rijckevorsel. Fuzzy coding and B-splines. In *Component and Correspondence Analysis* (ed. J. L. A. van Rijckevorsel, J. de Leeuw). John Wiley & Sons, 1988.

[504] J. Rissanen. Modeling by shortest data description. *Automatica*, Vol. 14, pp. 465-471, 1978.

[505] S. Roberts and L. Tarassenko. A probabilistic resource allocating network for novelty detection. *Neural Computation*, Vol. 6, pp. 270-284, 1994.

[506] D. Rogers. Splines: A hybrid of Friedman's multivariate adaptive regression splines (MARS) algorithm with Holland's genetic algorithm. *Proc. of the Fourth Int. Conf. on Genetic Algorithms,* pp. 384-391, San Diago, CA, July 1991.

[507] I. J. Roseman, P. E. Jose, and M. S. Spindel. Appraisals of emotion-eliciting events: Testing a theory of discrete emotions. *Journal of Personality and Social Psychology,* Vol. 59, No. 5, pp. 899-915, 1990.

[508] A. Rosenfeld. The fuzzy geometry of image subsets. *Pattern Recognition Letters,* Vol. 2, pp. 311-317, 1984.

[509] A. Rosenfeld. Fuzzy graphs. In *Fuzzy Sets and Their Applications to Cognitive and Decision Processes.* (ed. L. A. Zadeh et al.). Academic Press, 1975.

[510]   R. Rovatti, R. Guerrieri, and G. Baccarani. An enhanced two-level Boolean synthesis methodology for fuzzy rules minimization. *IEEE Transactions on Fuzzy Systems*, Vol. 3, pp. 288-299, 1995.

[511]   D. Rumelhart, G. Hinton, and J. Williams. Learning internal representations by error propagation. (ed. D. Rumelhart and J. McClelland). *Parallel Distributed Processing*, pp. 318–362. MIT Press, 1986.

[512]   D. E. Rumelhart and J. L. McClelland, eds. *Parallel Distributed Processing: Explorations in the Microstructure of Cognition*, Vol. 1. MIT Press, 1986.

[513]   E. Rundensteiner, L. Hawkes, and W. Bandler. On Nearness Measures in Fuzzy Relational Data Models. *Int. Jour. Approximate Reasoning,* Vol. 3, pp. 267-298, 1989.

[514]   E. H. Ruspini. Numerical Methods for Fuzzy Clustering. *Information Sciences,* Vol. 2, pp. 319-350, 1970.

[515]   E. H. Ruspini. A new approach to fuzzy clustering. *Information and Control,* Vol. 15, pp. 22-32, 1969.

[516]   E. Ruspini, A. Saffiotti, and K. Konolige. Progress in research on autonomous vehicle motion planning. *Industrial Applications of Fuzzy Logic and Intelligent Systems.* (ed. J. Yen, R. Langari and L. Zadeh). IEEE Press, pp. 157-174, 1995.

[517]   S. Russel and P. Norvig. *Artificial Intelligence: A Modern Approach.* Prentice Hall, 1995.

[518]   E. Sacerdoti. The nonlinear nature of plane. *IJCAI,* pp. 206-214, 1975.

[519]   E. Sacerdoti. Planning in a heirarchy of abstraction spaces. *Artificial Intelligence,* Vol. 5, pp. 115-135, 1974.

[520]   S. Sanatham and R. Langari. Supervisory fuzzy control of a binary distillation column. *Proceedings of IEEE International Conference on Fuzzy Systems*, Orlando, FL, June 1994.

[521]   E. Sanchez. Solutions in composite fuzzy relation equations: Application to medical diagnosis in brouwerian Logic. In M. M. Gupta, G. N. Saridis, and B. R. Gaines eds. *Fuzzy Automata and Decision Processes,* North Holland, New York, pp. 221-234, 1977.

[522]   G. N. Saridis. Analytical formulation of the principle of increasing precision with decreasing intelligence for intelligent machines. *Automatica*, 1989.

[523]   G. N. Saridis. Toward the realization of intelligent controls. *Proc. IEEE*, pp. 1115-1133, 1979.

[524]   M. L. Schuler and M. M. Domach. Mathematical models of the growth of the individual cells. In *Foundation of Biomedical Engineering* (ed. H. W. Blanch, E. T. Papoutsakis, and G. Stephanopoulos) pp. 101, American Chemical Society, Washington, DC, 1983.

[525]   L. L. Schumaker, *Spline Functions: Basic Theory.* John Wiley & Sons, 1981.

[526]   G. Schwarz. Estimating the dimension of a model. *The Annals of Statistics*, Vol. 6, pp. 461-464, 1978.

[527]   D. G. Schwartz and G. Klir, Fuzzy logic flowers in Japan, *IEEE Spectrum*, pp. 32-35, July 1992.

[528]   H. P. Schwefel. *Numerical Optimization of Computer Models.* John Wiley& Sons, 1981.

[529]   M. Seif El-Nasr and M. Skubic. A fuzzy emotional agent for decision-making in a mobil robot. *Proc. of 1998 Int. Conf. on Fuzzy Systems* (FUZZ-IEEE '98), pp. 135-140, Anchorage, Alaska, May 1998.

[530]   G. Shafer and J. Pearl, eds. *Readings in Uncertain Reasoning*. Morgan Kaufmann, 1990.

[531]   D. F. Shanno. Recent advances in numerical techniques for large-scale optimization. In W. T. Miller, R. S. Sutton and P. J. Werbos, eds., *Neural Networks for Control*. MIT Press, 1990.

[532]   L. Shastri et al. A fuzzy logic symposuim. *IEEE Expert,* Vol. 9, No. 4, pp. 2-49, 1994.

[533]   S. Shenoi and A. Melton. An extended version of the fuzzy relational database model. *Information Sciences,* Vol. 51, pp. 35-52, 1990.

[534]   S. Shenoi and A. Melton. Proximity relations in fuzzy relational databases. *Int. Jour. Fuzzy Sets and Systems,* Vol. 31, pp. 287-296, 1989.

[535]   S. Shenoi, A. Melton, and L. Fan. Functional dependencies and normal forms in the fuzzy relational database m.odel. *Information Sciences,* Vol. 60, pp. 1-28, 1992.

[536]   E. H. Shortliffe. *Computer-Based Medical Consulations: MYCIN*. Elsevier, 1976.

[537]   D. D. Siljak. *Nonlinear Systems: The Parameter analysis and Design*. John Wiley & Sons, 1969.

[538]   P. Smets and P. Magrez. Implication in fuzzy logic. *Int. J. of Approximate Reasoning,* Vol. 1, pp. 327-347, 1987.

[539]   H. Simon. Motivational and Emotional Controls of Cognition. *Psychological Review,* Vol. 74, pp. 29-39, 1967.

[540]   A. C. Singer, G. W. Wornell, and A. V. Oppenheim. Codebook prediction: Anonlinear signal modeling paradigm. *Proceedings of the IEEE International Conference on Acoustic, Speech, and Signal Processing*, San Francisco, CA, 1992, pp. 325-328.

[541]   A. Skeppstedt, L. Ljung, and M. Millnert. Construction of composite models from observed data, *International Journal of Control*, Vol. 55, pp. 141-152, 1992.

[542]   P. P. van der Smagt. Minimization methods for training feedforward neural networks. *Neural Networks*, Vol. 7, pp. 1-11, 1994.

[543]   R. E. Smith. Measure theory on fuzzy sets. Ph. D. Thesis, University of Saskatchewan, Saskatoon, Canada, 1970.

[544]   R. Soeterboek. *Predictive Control: A Unified Approach*. Prentice Hall, 1992.

[545]   I. Sols. Aprotaciones a la teoria de topos, al algebra universal y a las mathematicas fuzzy. Ph.D Thesis, Zaragoza, Spain, 1975.

[546]   M. Sonka, S. K. Tadikonda, and S. M. Collins. Knowledge-based Interpretation of MR brain images. *IEEE Transactions on Medical Imaging,* Vol. 15, No.4, pp. 470-481, September 1994.

[547]   B. G. Song, R. J. Marks II, S. Oh, P. Arabshahi, T. P. Caudell, and J. J. Choi. Adaptive membership function fusion and annihilation in fuzzy if-then rules. *Proceedings of the Second International Conference on Fuzzy Systems*, San Francisco, CA, April, 1993, pp. 961-967.

[548]   D. F. Specht. A general regression network. *IEEE Transactions on Neural Networks*, Vol. 2, pp. 568-576, 1991.

[549]   W. Spendley, G. R. Hext, and F. R. Himsworth. Sequential application of simplex designs in optimization and evolutionary operation. *Technometrics,* Vol. 4, pp. 441-461, 1962.

[550]   M. Spong and M. Vidyasagar. *Robot Dynamics and Control.* John Wiley & Sons, 1989.

[551]   P. S. de Souza and S. N. Talukdar. Genetic algorithm in asynchronous teams. *Proc. of the Fourth Int. Conf. on Genetic Algorithms,* pp. 392-397, San Diego, CA, July 1991.

[552]   T. Sudkamp and R. J. Hammell I., Interpolation, completion, and learning fuzzy rules. *IEEE Transactions on Systems, Man, and Cybernetics*, Vol. 24, pp. 332-342, 1994.

[553]   M. Sugeno. An introductory survey of fuzzy control. *Information Science*, Vol. 36, No. 1, 1985.

[554]   M. Sugeno. *Industrial Applications of Fuzzy Control.* North-Holland, 1985.

[555]   M. Sugeno. fuzzy measures and fuzzy integrals — a survey. In M. M. Gupta, G. N. Saridis, and B. R. Gains eds. *Fuzzy Automata and Decision,* pp. 89-102, 1977.

[556]   M. Sugeno. Fuzzy Measures and fuzzy integrals — a survey. In M. M. Gupta, G. N. Saridis, and B. R. Gaines, eds., *Fuzzy Automata and Decision Processes.* North-Holland, pp. 89-102, 1977.

[557]   M. Sugeno. Inverse operation of fuzzy integrals and conditional fuzzy measures. *Trans SICE*, Vol. 11, pp. 32-37, 1975.

[558]   M. Sugeno. Theory of fuzzy integrals and its applications. Ph.D. Thesis, Tokyo Institute of Technology, Tokyo, Japan, 1974.

[559]   M. Sugeno and K. T. Kang. Structure identification of fuzzy model. *Fuzzy Sets and Systems*, Vol. 28, 1988.

[560]   M. Sugeno, T. Murofushi, J. Nishino, and H. Miwa, Helicopter flight control based on fuzzy logic, in *Proc. International Fuzzy Engineering Symposium*, Yokohama, Japan, Nov 13-15, 1991.

[561]   M. Sugeno and M. Nishida. Fuzzy control of model car. *Fuzzy Sets and Systems,* Vol. 16, pp. 103-113, 1985.

[562]   M. Sugeno and T. Yasukawa. A fuzzy-logic-based approach to qualitative modeling. *IEEE Transactions on Fuzzy Systems*, Vol. 1, No. 1, 1993.

[563]   M. Sugeno et al. Intelligent control of an unmanned helicopter based on fuzzy logic. *Proc. of American Helicopter Society Forum 51,* 1995.

[564]   C. -T. Sun. Rule-base structure identification in an adaptive-network-based fuzzy inference system, *IEEE Transactions on Fuzzy Systems*, Vol. 2, pp. 64-73, 1994.

[565]   R. F. Sutton. *Modeling Human Operators in Control System Design.* John Wiley & Sons, 1989.

[566]   D. B. Straus, W. A. Walter, and C. A. Gross. Escherichia coli heat shock gene mutants are deficient in proteolysis. *Genes Development*, Vol. 2, pp. 1851-1858, 1988.

[567]   M. Stone. Cross-validity choice and assessment of statistical predictions. *Journal of the Royal Statistical Society,* Vol. 36, B, pp. 111-133, 1974.

[568] H. Tahani and J. Keller. Information fusion in computer vision using the fuzzy integral. *IEEE T. Systems, Man, and Cybernetics,* Vol. 20, No. 3, pp. 733-741, 1990.

[569] Y. Takashi. Fuzzy database query languages and their relational completeness theorem. *IEEE Trans. On Knowledge and Data Engineering,* Vol. 5, No. 1, 1993.

[570] H. Takagi. Applications of neural networks and fuzzy logic to consumer products. In J. Yen, R. Langari, L. A. Zadeh, eds., *Industrial applications of Fuzzy logic and Intelligent Systems.* IEEE Press, pp. 93-105, 1995.

[571] H. Takagi. Survey of fuzzy logic applications in image-processing equipment. In J. Yen, R. Langari, L. A. Zadeh, eds., *Industrial Applications of Fuzzy logic and Intelligent Systems.* IEEE Press, pp. 69-92, 1995.

[572] H. Takagi. Application of neural networks and fuzzy logic to consumer products. In *Industrial Applications of Fuzzy Logic and Intelligent Systems* (ed. John Yen, R. Langari and L. Zadeh). IEEE Press, 1994.

[573] H. Takagi. Fusion technology of fuzzy theory and neural networks — survey and future directions. *1st. Int. Conf. on Fuzzy Logic and Neural Networks (IIZUKA '90),* pp. 13-26, 1990.

[574] H. Takagi and I. Hayashi. NN-driven fuzzy reasoning. *Int. J. of Approximate reasoning,* Vol. 2, pp. 191-212, 1991.

[575] H. Takagi and I. Hayashi. Artificial neural network driven fuzzy reasoning. *International Journal of Approximate Reasoning*, Vol. 5, pp. 191-212, 1991.

[576] T. Takagi and M. Sugeno. Fuzzy identification of systems and its application to modeling and control. *IEEE Transactions on Systems, Man, and Cybernetics*, Vol. 15, No. 1, 1985.

[577] H. Takagi and N. Suzuki. Application of neural-networks designed on approximate reasoning architecture to the adjustment of VTR tape-running mechanisms. *IFSA'91,* 1991.

[578] K. Tanaka and M. Sugeno. Stability analysis and design of fuzzy control systems. *Fuzzy Sets and Systems,* Vol. 45, No. 2, pp. 135-156, 1992.

[579] K. Tanaka, T. Ikeda, and H. Wang, "Fuzzy Controller Design for Stabilization of a Class of Nonlinear Systems", *Proceedings of the 34th IEEE CDC*, 1995.

[580] K. Tanaka, M. Sano, and H. Watanabe. Modeling and control of carbon monoxide concentration using a neuro-fuzzy technique. *IEEE Transactions on Fuzzy Systems*, Vol. 3, pp. 271-279, 1995.

[581] H. Tanaka, T. Tsukiyama, and K. Asai. A fuzzy system model based on the logical structure. In R. R. Yager ed. *Fuzzy Set and Possibility Theory - Recent Developments.* Pergamon Press, pp. 257-274, 1982.

[582] K. Tanaka, T. Ikeda, and H. O. Wang, Robust stabilization of a class of uncertain nonlinear systems via fuzzy control: quadratic stabilizability, H^∞ control theory, and linear matrix inequalities, *IEEE Tran. Fuzzy Systems*, Vol 4, No. 1, pp. 1-13, Feb. 1996.

[583] S. Tano. Fuzzy expert system shell — LIFE FEShell. In *Applied Research in Fuzzy Technology.* (ed. Anca Ralescu). Kluwer Academic Publishers, 1994.

[584]   E. Tazaki et al. Medical diagnosis using simplified multidimensional fuzzy reasoning. *Proc. IEEE Int. Conf. on SMC,* 1988.

[585]   T. Taxt and A. Lundervold. Multispectral analysis of the brain using magnetic resonance imaging, *IEEE Transactions on Medical Imaging*, Vol. 13, No. 3, pp 470-481, September 1994.

[586]   R. Teltz and M. A. Elbastawi. Hierarchical knowledge based control in turning. In *Computer Control of Manufacturing Processes, DSC-Vol 28.* ASME, 1991.

[587]   T. Terano, K. Asai, and M. Sugeno. *Fuzzy Systems Theory and Its Applications.* Academic Press, 1992.

[588]   T. Terano, K. Asai, and M. Sugeno. *Applied Fuzzy Systems.* Academic Press, 1989.

[589]   P. Thrift. Fuzzy logic synthesis with genetic algorithms. *Proc. of 4th. Int. Conf. on Genetic Algorithms,* pp. 509-513, 1992.

[590]   M. Togai and H. Watanabe. Expert systems on chip: an engine for real-time approximate reasoning. *IEEE Expert*, Vol. 1, No. 3, pp. 55-62, Fall 1986.

[591]   R. M. Tong. An annotated bibliography of fuzzy control. In M. Sugeno, ed. *Industrial Applications of Fuzzy Control*. North Holland, 1985.

[592]   R. M. Tong. Some properties of fuzzy feedback systems. *IEEE Transactions on Systems, Man, and Cybernetics*, Vol. 10, No. 6, 1980.

[593]   R. M. Tong. A control engineering review of fuzzy systems. *Automatica*, Vol. 13, No. 6, 1977.

[594]   R. M. Tong and P. A. Bonisonne. A Linguistic Approach to Decisionmaking with Fuzzy Sets. *IEEE Transactions on Systems, Man, and Cybernetics,* Vol. 10, pp. 716-723, 1980.

[595]   R. M. Tong, M. H. Beck, and A. Latten. Fuzzy control of the activated sludge wastewater treatment process. *Automatica*, Vol. 16, No. 6, 1980.

[596]   R. M. Tong, V. Askman, and J. Cunningham. RUBRIC an artificial intelligence approach to information retrieval. *Proc. 1st. Int. Workshop on Expert Database Systems,* 1984.

[597]   R. M. Tong. Some properties of fuzzy feedback systems. *IEEE Transactions on Systems, Man, and Cybernetics*, Vol. SMC-10, No. 6, June 1980.

[598]   P. Torasso and L. Consol. Approximate reasoning and prototypical knowledge. *Int. J. of Approximate Reasoning,* Vol. 3, pp. 157-177, 1989.

[599]   J. T. Tou and R. C. Gonzalez. *Pattern Recognition Principles*, Addison-Wesley, 1974.

[600]   E. Trillas and L. Valverde. On mode and implication in approximate reasoning. In M. M. Gupta, A. Kandel, W. Bandler and J. B. Kiszka, eds., *Approximate Reasoning in Expert Systems*. North-Holland, pp.157-166, 1985.

[601]   E. Trillas and L. Valverde. On implication and indistinguishablility in the setting of fuzzy logic. In J. Kacprzyk and R. R. Yager, eds., *Management Dicision Support Systems Using Fuzzy Sets and Possibility Theory*. Verlag TUV, pp. 198-212, 1985.

[602] Y. Tsukamoto. An approach to fuzzy reasoning method. In M. M. Gupta, R. K. Ragade, and R. R. Yager, eds. *Advances in Fuzzy Set Theory and Applications,* North-Holland, Amsterdam, pp. 137-149, 1979.

[603] J. K. Udupa, S. Samarasekera, and W. A. Barrett. Boundary detection via dynamic programming. *Proceedings of SPIE: Visualization in Biomedical Computing 1992*, Vol. 1808, pp. 33-39, 1992.

[604] J. D. Ullman. *Principles of Database and Knowledge-base Systems*. Rockville, Computer Science Press, 1989.

[605] M. Umano. Retrieval from fuzzy database by fuzzy relational algebra. *Fuzzy Information Knowledge Representation and Decision Analysis* (ed. E. Sanchez). Pergamon Press, pp. 1-6, 1984.

[606] M. Umano. FREEDOM-0: A Fuzzy Database System. In M. M. Gupta and E. Sanchez, eds., *Fuzzy Information and Decision Processes,* North-Holland, pp. 339-347, 1982.

[607] M. Umano, I. Hatono and H. Tamura. Fuzzy Database Systems. *Proc. Int. Joint Conf. Of fourth IEEE Int. Conf. on Fuzzy Systems and Second Int. Fuzzy Eng. Symp. (FUZZ-IEEE/IFES '95),* pp. 35-36, 1995.

[608] G. Vachtsevanos and S. Farinwata. Fuzzy logic control: A systematic design and performance assessment methodology. In *Fuzzy Logic: Implementation and Applications*, M. J. Patyra and D. M. Mlynek, eds., John Wiley & Sons, 1996.

[609] M. Valenzuela-Rendon. The fuzzy classifier system: A classifier system for continuously varying variables. *Proc. 4th. Int. Conf. on Genetic Algorithms,* pp. 346-353, 1991.

[610] M. W. Vannier, R. L. Butterfield, D. L. Rickman, D. M. Jordan, W. A. Murphy, R. G. Levitt, and M. Gado. Multispectral analysis of magnetic resonance images. *Radiology*, Vol. 154, pp 221-224, 1985.

[611] M. W. Vannier, T. K. Pilgram, C. M. Speidal, L. R. Neumann, D. L. Rickman, and L. D. Schertz. Validation of magnetic resonance imaging (MRI) multispectral tissue classification. *Computer Medical Imaging and Graphics*, Vol. 15, pp. 217- 223, 1991.

[612] J. Velasquez. Modeling Emotions and other motivations in synthetic agents. *Proc. of the AAAI Conference,* pp. 10-15, 1997.

[613] G. L. Vernazza, S. B. Serpico, and S. G. Dellepiane. A knowledge-based system for biomedical image processing and recognition. *IEEE Transactions on Circuits Systems*, Vol. CS-34, pp 1399-1416, 1987.

[614] M. Vidyasagar. *Nonlinear Systems Analysis*. Prentice Hall, 1978.

[615] T. P. Vogl, J. K. Mangis, A. K. Zigler, W. T. Zink, and D. L. Alkon. Accelerating the convergence of the backpropagation method. *Biological Cybernetics*, Vol. 59, pp. 256-264, 1988.

[616] A. A. Voronov. *Basic Principles of Automatic Control Theory: Special Linear and Nonlinear Systems*. Mir Publishers, 1985.

[617] N. Wakami, H. Nomura, and S. Araki, Intelligent home appliances using fuzzy technology, in *Industrial Applications of Fuzzy Technology in the World*, K. Hirota and M. Sugeno (ed), p. 215-239, World Scientific, 1995.

[618]   C.-S. Wang, W.-Y. Wang, T.-T. Lee, and P.-S. Tseng. Fuzzy B-spline membership function (BMF) and its applications in fuzzy-neural control. *IEEE Transactions on Systems, Man, and Cybernetics*, Vol. 25, pp. 841-851, 1995.

[619]   L. X. Wang. *Adaptive Fuzzy Systems and Control. Design and Stability Analysis*. Prentice Hall, 1994.

[620]   L. X. Wang and J. M. Mendel. Fuzzy basis functions, universal approximation, and orthogonal least-squares learning. *IEEE Transactions on Neural Networks*, Vol. 3, No. 5, pp. 807-814, Sept. 1992.

[621]   L. X. Wang and J. M. Mendel. Backpropagation fuzzy systems as nonlinear dynamic system identifiers. *Proc. First Int. Conf. Fuzzy Systems,* pp. 1409-1418, 1992.

[622]   L. Wang and R. Langari. Identification and control of nonlinear dynamic systems using fuzzy models. *International Journal of Intelligent Control and Systems,* Vol. 1, pp. 247-260, 1996.

[623]   L. Wang and R. Langari. Sugeno models using fuzzy discretization and EM algorithms. *Fuzzy Systems and Systems*, Vol. 82, pp. 279-288, 1996.

[624]   L. Wang and R. Langari. Building Sugeno type models using fuzzy discretization and orthogonal parametrization. *IEEE Transactions on Fuzzy Systems*, Vol. 3, pp. 454-458, 1995.

[625]   L. Wang and R. Langari. Complex system modeling via fuzzy logic. *IEEE Transactions on Systems, Man, and Cybernetics*, Vol 25+, No. 10, October 1995.

[626]   L. Wang and R. Langari. Fuzzy controller design via hyperstability approach. *Third. IEEE Int. Conf. on Fuzzy Systems,* Vol. 1, 1994.

[627]   H. O. Wang, K. Tanaka, and M. F. Griffin. An approach to fuzzy controlof nonlinear systems: Stability and design issues. *IEEE Trans. on Fuzz. Syst.*, Vol. 4, No. 1, pp. 14-23, 1996.

[628]   E. B. Waygood and B. D. Sanwal. The control of pyruvate kinases of escherichia coli. *Journal of Biology Chemistry,* Vol. 249, No. 1, pp. 265-274, 1974.

[629]   W. G. Wee and K. S. Fu. A formulation of fuzzy automata and its application as a model of learning systems. *IEEE Transactions on Systems, Man, and Cybernetics,* Vol. 5, pp. 215-223, 1969.

[630]   P. J. Werbos. Neurocontrol and fuzzy logic: connections and design. *Int. J. Approx. Reason.* Vol. 6, No. 2, pp. 185-219, Feb. 1992.

[631]   P. J. Werbos. Maximizing long-term gas industry profits in two minutes in Lotus using neural network methods. *IEEE Trans. on Systems, Man, and Cybernetics*, March 1989.

[632]   P. J. Werbos. Generalization of Backpropagation, *Neural Networks*, October 1988.

[633]   P. J. Werbos. Beyond regression: New tools for prediction and analysis in the behavioral sciences, Ph. D. Thesis, Harvard University, Cambridge, MA, November 1974.

[634]   T. Whalen. Decision making under uncertainity with various assumptions about available information. *IEEE Transactions on Systems, Man, and Cybernetics,* Vol. 14, pp. 888-900, 1984.

[635]   J. E. Whitesitt. *Boolean Algerbra and Its Applications*. Addison-Wesley, 1961.

[636]   B. Widrow and M. E. Hoff. Adaptive switching circuits. *IRE WESCON Convention Record*, 1960, pp. 96-104.

[637]   B. Widrow and S. D. Stearns. *Adaptive Signal Processing*. Prentice Hall, 1985.

[638] J. Williams. Level\5 quest. *DBMS,* pp. 33-36, 1997.

[639] D. Williams, P. Bland, L. Liu, L. Farjo, I. R. Francis, and C. R. Meyer. Liver-tumour boundary detection: human observer versus computer edge detection. *Investigative Radiology*, Vol. 24, No. 10, pp 768-775, 1989.

[640] R. Wilensky. A model for planning in complex situations. In *Readings in Planning* (ed. J. Allen, J. Hendler, and A. Tate). pp. 263-274, 1990.

[641] N. Wiener. *Extrapolation, Interpolation, and Smoothing of Stationary Time Series, with Engineering Applications*. MIT press, 1949.

[642] M. P. Windham. Cluster validity for the fuzzy c-means clustering algorithm. *IEEE Trans. Pattern Anal. Machine Intell.,* Vol. PAMI-4, No. 4, pp. 357-363, 1982.

[643] F. Wong, K. Chou, and J. Yao, Civil engineering applications, in *Int. Handbook of Fuzzy Sets and Possibility Theory*, D. Dubois and H. Prade (ed), 1999.

[644] K. L. Wood and E. K. Antonsson. Computers with imprecise parameters in enigneering design: Background and theory. *J. Mechanisms, Transmission and Automation in Design,* Vol. 111, 1989.

[645] K. L. Wood, E. K. Antonsson, and J. L. Beck. Representing imprecision in engineering design: Comparing fuzzy and probability calculus. *Research in Engineering Design,* Vol. 1, pp. 87-203, 1990.

[646] [Z. Wu and R. Leahy; An Optimal Graph Theoretic Approach to Data Clustering: Theory and Its Application to Image Segmentation, IEEE Transactions on Pattern Analysis and Machine Intelligence, Vol. 15, No. 11, pp. 1101-1113, November 1993.

[647] P. Wright et al., MOSAIC: An Open Architecture Machine Tool for Precision Manufacturing, in *Proceedings of the 1993 NSF Design nd Manufacturing Systems Conference*, Charlotte, NC, 1993.

[648] X. L. Xie and G. Beni. A validity measure for fuzzy clustering. *IEEE Trans. Pattern Anal. Machine Intell.,* Vol. PAMI-13, No. 8, pp. 841-847, 1991.

[649] L. Xu, A. Krzyzak, and A. Yuille. On radial basis function nets and kernel regression: Statistical consistency, convergence rates, and receptive field size. *Neural Networks*, Vol. 7, pp. 609-628, 1994.

[650] R. R. Yager. On the specifity of a possibility distribution. *Fuzzy Sets and Systems,* Vol. 50, pp. 279-292, 1992.

[651] R. R. Yager. On ordered weigthed aggergation operators in multicriteria decisionmaking. *IEEE Trans. on Systems, Man and Cybernetics,* Vol. 18, pp. 183-190, 1988.

[652] R. R. Yager. Using approximate reasoning to represent default knowledge. *Artificial Intelligence,* Vol. 31, pp. 99-112, 1981.

[653] R. R. Yarger. A new methodology for ordinal multiobjective decisions based on fuzzy sets. *Decision Sciences,* Vol. 12, pp. 589-600, 1981.

[654] R. R. Yager and D. P. Filev. Approximate clustering via the mountain method. *IEEE Tran. Systems, Man, and Cybernetics*, Vol. 24, No. 8, pp. 1279- 1284, 1994.

[655] R. R. Yager and D. P. Filev. *Essentials of Fuzzy Modeling and Control*, John Wiley & Sons, New York, 1994.

[656]   R. R. Yager and D. P. Filev. Unified structure and parameter identification of fuzzy models. *IEEE Trans.on Systems, Man, and Cybernetics,* Vol. 23, pp. 1198-1205, 1993.

[657]   T. Yamakawa and K. Hirota, eds. *Fuzzy Sets and Systems: Special Issue on Application of Fuzzy Logic Control in Industry.* North-Holland, 1989.

[658]   S. Yasunobu and S. Miyamoto. Atomatic train opertion by fuzzy predictive control. In M. Sugeno ed. *Industrial Applications of Fuzzy Control.* North Holland, 1985.

[659]   A. Yazici, E. Gocmen, B. P. Buckles, R. Goerge and F. E. Petry. An integrity constraint for fuzzy relational database. *Proc. IEEE Second Int. Conf. on Fuzzy Systems,* pp. 496-499, 1993.

[660]   J . T. P. Yao. NAFIP-1. Panel discussion on introduction of fuzzy sets to undergraduate engineering and science curricula. *Internaional Journal of Man-Machine Studies*, Vol. 19, pp. 5-7, 1983.

[661]   S. Yasunobu and S. Miyamoto. Automatic train operation system by predictive fuzzy control, in *Industrial Applications of Fuzzy Control*, M. Sugeno ed., pp. 1-18, Amsterdam: North Holland, 1985.

[662]   S. Yasunobu, S. Miyamoto, and H. Ihara. Fuzzy Control for Automatic Train Operation System. *Proceedings of the 4th IFAC/IFIP/IFORS International Conference on Control in Transportation Systems,* Baden-Baden, Germany, pp. 33-39, 1983.

[663]   C. Yasunobu et al. A decision-support system building tool with fuzzy logic and its application to chart technical analysis. *Adaptive Intelligent Systems: proceddings of the BANAI Workshop,* Belgium, pp. 19-32, Elsevier Science Publishers, 1993.

[664]   Z. Ye and L. Gu. A fuzzy System for trading the Shanghai stock market. *Trading on the Edge* (ed. G. J. Deboeck). pp. 207-214. John Wiley & Sons, 1994.

[665]   J. Yen. Generalizing term subsumption languages to fuzzy logic. *Proc. 12th. Int. J. Conf. on Artificial Intelligence,* pp. 472-277, 1991.

[666]   J. Yen and W. Gillespie. Integrating global and local evaluations for fuzzy model identification using genetic algorithms. *Proc. Sixth World Congress of Int. Fuzzy Systems Association (IFSA 95),* San Pablo, Brazil, 1995.

[667]   J. Yen and N. Pfluger. A fuzzy logic based extension to Payton and Rosenblatt's command fusion method for mobile robot navigation. *IEEE Trans. on Systems, Man, and Cybernetics,* Vol. 25, No. 6, pp. 971-978, June 1995.

[668]   J. Yen and L. Wang. An SVD-based fuzzy model reduction strategy. *Proceedings of the Fifth IEEE International Conference on Fuzzy Systems*, New Orleans, LA, September, 1996, pp. 835-841.

[669]   J. Yen and L. Wang. Application of statistical information criteria for optimal fuzzy model construction, IEEE Tran. on Fuzzy Systems, Vol. 6, No. 3, pp. 362-372, Aug 1998.

[670]   J. Yen and L. Wang, Granule-based Models, in *Handbook of Fuzzy Computation*, E. Ruspini, P. Bonissone, and W. Pedrycz (ed.), IOP Publishing, 1998.

[671]   J. Yen and L. Wang. Simplifying fuzzy rule-based models using orthogonal transformation methods, *IEEE Tran. on Systems, Man, and Cybernetics,* Vol 29, Part B, No. 1, Feb 1999.

[672] J. Yen, R. Langari, and L. Wang. Principal components, B-splines, and fuzzy system reduction, *International Journal of Uncertainty, Fuzziness, and Knowledge-Based Systems*, Vol. 4, pp. 561-572, 1996.

[673] J. Yen, R. Langari, and L. A. Zadeh. *Industrial Applications of Fuzzy Logic and Intelligent Systems*. IEEE Press, 1995.

[674] J. Yen, B. Lee, and J. C. Liao. A hybrid genetic algorithm and fuzzy logic for metabolic modeling. *Proc. of the 13th National Conf. on Artificial Intelligence (AAAI'96)*, pp. 743-749, Portland, Aug. 1996.

[675] J. Yen, J. Liao, B. Lee, and D. Randolph. A hybrid approach to modeling metabolic systems using algorithms and simplex method. *IEEE Trans. Systems, Man, and Cybernetics,* Vol. 28, Part B, No. 2, pp. 173-191, April 1998.

[676] S. Y. Yi and M. J. Chung. Identification of fuzzy relational model and its application to control. *Fuzzy Sets and Systems*, Vol. 59, 1993, pp. 25-33.

[677] H. Ying. Sufficient conditions on general fuzzy systems as function approximators. *Automatica*, Vol. 30, pp. 521-525, 1994.

[678] H. Ying. Fuzzy control of mean arterial pressure in postsurgical patients with sodium nitroprusside infusion, *IEEE Tran. on Biomedical Engineering*, Vol 39, No. 10, pp. 1060-1070, October 1992.

[679] H. Ying. Fuzzy control theory: a nonlinear case. *Automatica*, Vol. 26, No. 3, pp. 513-520, 1990.

[680] L. A Zadeh. Fuzzy logic = computing with words. *IEEE Trans. on Fuzzy Systems*, Vol. 4, No. 2, 1996.

[681] L. A. Zadeh. Probability theory and fuzzy logic are complementary rather than competitive. *Technometrics*, Vol. 37, No. 3, pp.. 271-276, 1995.

[682] L. A. Zadeh. Fuzzy logic, neural networks, and soft computing. *Communications of the ACM*, Vol 37(3), pp. 77-84, March 1994.

[683] L. A. Zadeh. Is probability theory sufficient for dealing with uncertainty in AI: a negative view. In Kanal, L. N. and J. F. Lemmer, eds., *Uncertainity in Artificial Intelligence*. North-Holland, 1986.

[684] L. A. Zadeh. Fuzzy probabilities. *Information Processing and Management*, Vol. 20, No. 3, pp. 363-372, 1984.

[685] L. A. Zadeh. Is possibility different from probability? *Human Systems Management*, Vol. 3, No. 2, pp. 253-254, 1983.

[686] L. A. Zadeh. Possibility theory as a basis for representation of meaning. *Proc. of the Sixth. Wittgenstein Symp.*, Kirchberg, Austria, pp. 253-261, 1982.

[687] L. A. Zadeh. Fuzzy probability and their role in decision analysis. *Proc. MIT/ONR Workshop on C3*, MIT, Cambridge, MA, 1981.

[688] L. A. Zadeh. Possibility theory and soft data analysis. In L. Cobb and R. M. Thrall, eds., *Mathematical Frontiers of the Social and Policy Sciences*. Westview Press, pp. 69-129, 1981.

[689]   L. A. Zadeh. Test-score semantics for natural languages meaning representation via PRUF. In *Empirical Semantics* (ed. B. Rieger), pp. 281-349, 1981.

[690]   L. A. Zadeh. Possibility theory as a basis for information processing and knowledge representation. *Proc. of Computer Software and Application Conference*, Chicago, 1980.

[691]   L. A. Zadeh. Possibility theory as a basis for information processing and knowledge representation. *Proc. of the Intern. Symp. on Human Communication, UNESCO,* Vienna, 1980.

[692]   L. A. Zadeh. Possibility theory and its application to the represtation and manipulation of uncertain data. In (ed. Khane, R.). *Proc. of the Workshop on Image Understanding, NESC*, Washington, DC, pp. 36-60, 1979.

[693]   L. A. Zadeh. Fuzzy sets as a basis for a theory of possibility. *Fuzzy Sets Syst.,* Vol. 1, pp. 3-28, 1978.

[694]   L. A. Zadeh. Possibility theory and its application to information analysis. *Proc. Intern. Colloq. on Information Theory, C.N.R.S.* Paris, pp. 173-182, 1978.

[695]   L. A. Zadeh. Fuzzy Sets. In J. Belzer, A. Holzman, and A. Kent eds. *Encyclopedia of Computer Science and Technology.* Marcel Dekker, 1977.

[696]   L. A. Zadeh. A fuzzy-algorithmic approach to the definition of complex or imprecise concepts. *Electronics Res. Lab. Rep. ERL-M474*, Univ. of California, Berkeley, 1974. Also in *Int. J. Man-Machine Stud.*, Vol 8, pp. 249-291, 1976.

[697]   L. A. Zadeh. Calculus of fuzzy restrictions. In L.A. Zadeh, K.S. Fu, K. Tanaka, and J. Shimara, eds., *Fuzzy Sets and Their Application to Cognitive and Decision Processes.* Academic Process, pp. 1-39, 1975.

[698]   L. A. Zadeh. A relational model for approximate reasoning. *IEEE International Conf. on Cybernetics and Society*, San Francisco, Sep. 1975.

[699]   L. A. Zadeh. The concept of a linguistic variable and its application to approximate reasoning-I, II, III. *Information Sciences*, Vol. 8, pp. 199-249, pp. 301-357; Vol. 9, pp. 43-80, 1975.

[700]   L. A. Zadeh. On the analysis of large scale systems. In *Systems Approaches and Environment Problems.* Vandenhoeck and Ruprecht, pp. 233-37, 1974.

[701]   L. A. Zadeh. Outline of a new approach to the analysis of complex systems and decision processes. *IEEE Transactions on Systems, Man, and Cybernetics*, SMC-3, 1973.

[702]   L. A. Zadeh. A rationale for fuzzy control. *Journal of Dynamic Systems, Measurement, and Control*, Vol. 3, No. 4, 1972.

[703]   L. A. Zadeh. Similarity relations and fuzzy orderings. *Infrm. Sci.*, pp.177-200, 1971.

[704]   L. A. Zadeh. Toward a theory of fuzzy systems. In *Aspects of Network and System Theory* (ed. R. E. Kalman and N. Dellaris). Rineheart and Winston, pp. 469-490, 1971.

[705]   L. A. Zadeh. Toward a theory of fuzzy systems. In *Aspects of Network and System Theory.* Rinehart and Winston, pp. 469-490, 1971.

[706]   L. A. Zadeh. Probability measures and fuzzy events. *J. of Math. Analysis and Applications,* Vol. 23, No. 2, pp. 421-427, 1968.

[707]   L. A. Zadeh. Fuzzy algorithm. *Information and Control*, Vol. 12, pp. 94-102, 1968.

[708]   L. A. Zadeh. Fuzzy sets. *Information and Control*, Vol. 8, 1965.

[709]   L. A. Zadeh and C. A. Desoer. *Linear System Theory*. McGraw-Hill, 1963.

[710]   L. A. Zadeh, K. S. Fu, K. Tanaka, and J. Shimara. *Fuzzy Sets and Their Application to Cognitive and Decision Processes,* Academic Process, New York, 1975.

[711]   S. Zaidi and M. Reguia. Fuzzy Database System: A unified Approach. *Proc. IEEE Int. Conf. On Systems, Man and Cybernetics,* pp. 794-799, 1994.

[712]   M. Zemankova. FILIP: A Fuzzy Intelligent Information System with Learning Capabilities. *Information Systems,* Vol. 14, No. 6, pp. 473-486, 1989.

[713]   M. Zemenkova and A. Kandel. Implementing imprecision in information systems. *Information Science,* Vol. 37, pp. 107-141, 1985.

[714]   M. Zemenkova and A. Kandel. *Fuzzy Relational Databases—A Key to Expert Systems*. Verlag TUV, 1984.

[715]   X. J. Zeng and M. G. Singh. Approximation accuracy analysis of fuzzy systems as function approximators, *IEEE Transactions on Fuzzy Systems*, Vol. 4, pp. 44-63, 1996.

[716]   J. Zhang and A. Knoll. Constructing fuzzy controls with B-spline models. *Proceedings of the Fifth IEEE International Conference on Fuzzy Systems*, New Orleans, LA, September, pp. 416-421, 1996.

[717]   Y. Zhao and C. G. Atkeson. Implementing projection pursuit learning. *IEEE Transactions on Neural Networks*, Vol. 7, pp. 362-373, 1996.

[718]   H. J. Zimmermann. Description and ecoptimization of fuzzy systems. *Int. J. General Syst.*, Vol. 2, pp. 209-215, 1975.

[719]   H. J. Zimmermann. Applications of fuzzy set theory to mathematical programming. *Information Sciences,* Vol. 36, pp. 29-58, 1985.

[720]   H. J. Zimmermann. *Fuzzy Set Theory — and Its Applications*. 2nd Ed. Kluwer Academic Publishers, 1991.

[721]   A. Zvieli and P. Chen. Entity-Relationship Modeling and Fuzzy Databases. *Proc. of 2nd. Int. Cnf. on Data Engineering,* pp. 320-327, 1986.

Index